CAMBRIDGE TRACTS IN MATHEMATICS

General Editors

B. BOLLOBÁS, W. FULTON, F. KIRWAN,
P. SARNAK, B. SIMON, B. TOTARO

219 Operator Analysis

CAMBRIDGE TRACTS IN MATHEMATICS

General Editors
B. BOLLOBÁS, W. FULTON, F. KIRWAN,
P. SARNAK, B. SIMON, B. TOTARO

A complete list of books in the series can be found at www.cambridge.org/mathematics.

Recent titles include the following:

183. *Period Domains over Finite and p-adic Fields.* By J.-F. Dat, S. Orlik, and M. Rapoport
184. *Algebraic Theories.* By J. Adámek, J. Rosický, and E. M. Vitale
185. *Rigidity in Higher Rank Abelian Group Actions I: Introduction and Cocycle Problem.* By A. Katok and V. Niţică
186. *Dimensions, Embeddings, and Attractors.* By J. C. Robinson
187. *Convexity: An Analytic Viewpoint.* By B. Simon
188. *Modern Approaches to the Invariant Subspace Problem.* By I. Chalendar and J. R. Partington
189. *Nonlinear Perron–Frobenius Theory.* By B. Lemmens and R. Nussbaum
190. *Jordan Structures in Geometry and Analysis.* By C.-H. Chu
191. *Malliavin Calculus for Lévy Processes and Infinite-Dimensional Brownian Motion.* By H. Osswald
192. *Normal Approximations with Malliavin Calculus.* By I. Nourdin and G. Peccati
193. *Distribution Modulo One and Diophantine Approximation.* By Y. Bugeaud
194. *Mathematics of Two-Dimensional Turbulence.* By S. Kuksin and A. Shirikyan
195. *A Universal Construction for Groups Acting Freely on Real Trees.* By I. Chiswell and T. Müller
196. *The Theory of Hardy's Z-Function.* By A. Ivić
197. *Induced Representations of Locally Compact Groups.* By E. Kaniuth and K. F. Taylor
198. *Topics in Critical Point Theory.* By K. Perera and M. Schechter
199. *Combinatorics of Minuscule Representations.* By R. M. Green
200. *Singularities of the Minimal Model Program.* By J. Kollár
201. *Coherence in Three-Dimensional Category Theory.* By N. Gurski
202. *Canonical Ramsey Theory on Polish Spaces.* By V. Kanovei, M. Sabok, and J. Zapletal
203. *A Primer on the Dirichlet Space.* By O. El-Fallah, K. Kellay, J. Mashreghi, and T. Ransford
204. *Group Cohomology and Algebraic Cycles.* By B. Totaro
205. *Ridge Functions.* By A. Pinkus
206. *Probability on Real Lie Algebras.* By U. Franz and N. Privault
207. *Auxiliary Polynomials in Number Theory.* By D. Masser
208. *Representations of Elementary Abelian p-Groups and Vector Bundles.* By D. J. Benson
209. *Non-homogeneous Random Walks.* By M. Menshikov, S. Popov and A. Wade
210. *Fourier Integrals in Classical Analysis (Second Edition).* By C. D. Sogge
211. *Eigenvalues, Multiplicities and Graphs.* By C. R. Johnson and C. M. Saiago
212. *Applications of Diophantine Approximation to Integral Points and Transcendence.* By P. Corvaja and U. Zannier
213. *Variations on a Theme of Borel.* By S. Weinberger
214. *The Mathieu Groups.* By A. A. Ivanov
215. *Slenderness I: Foundations.* By R. Dimitric
216. *Justification Logic.* By S. Artemov and M. Fitting
217. *Defocusing Nonlinear Schrödinger Equations.* By B. Dodson
218. *The Random Matrix Theory of the Classical Compact Groups.* By E. S. Meckes

Operator Analysis
Hilbert Space Methods in Complex Analysis

JIM AGLER
University of California, San Diego

JOHN EDWARD M^CCARTHY
Washington University in St. Louis

NICHOLAS YOUNG
Leeds University and Newcastle University

CAMBRIDGE
UNIVERSITY PRESS

University Printing House, Cambridge CB2 8BS, United Kingdom

One Liberty Plaza, 20th Floor, New York, NY 10006, USA

477 Williamstown Road, Port Melbourne, VIC 3207, Australia

314–321, 3rd Floor, Plot 3, Splendor Forum, Jasola District Centre,
New Delhi 110025, India

79 Anson Road, #06–04/06, Singapore 079906

Cambridge University Press is part of the University of Cambridge.

It furthers the Universitys mission by disseminating knowledge in the pursuit of
education, learning, and research at the highest international levels of excellence.

www.cambridge.org
Information on this title: www.cambridge.org/9781108485449
DOI: 10.1017/9781108751292

© Jim Agler, John Edward McCarthy, and Nicholas Young 2020

This publication is in copyright. Subject to statutory exception
and to the provisions of relevant collective licensing agreements,
no reproduction of any part may take place without the written
permission of Cambridge University Press.

First published 2020

Printed in the United Kingdom by TJ International Ltd, Padstow Cornwall

A catalogue record for this publication is available from the British Library.

ISBN 978-1-108-48544-9 Hardback

Cambridge University Press has no responsibility for the persistence or accuracy of
URLs for external or third-party internet websites referred to in this publication
and does not guarantee that any content on such websites is, or will remain,
accurate or appropriate.

To
Sarah, Suzanne, and *Зинаида*

Contents

Preface		*page* xiii
Acknowledgments		xv
PART ONE COMMUTATIVE THEORY		1

1	**The Origins of Operator-Theoretic Approaches to Function Theory**	**3**
1.1	Operators	3
1.2	Functional Calculi	4
1.3	Operators on Hilbert Space	9
1.4	The Spectral Theorem	12
1.5	Hardy Space and the Unilateral Shift	13
1.6	Invariant Subspaces of the Unilateral Shift	15
1.7	Von Neumann's Theory of Spectral Sets	16
1.8	The Schur Class and Spectral Domains	19
1.9	The Sz.-Nagy Dilation Theorem	21
1.10	Andô's Dilation Theorem	22
1.11	The Sz.-Nagy–Foias Model Theory	23
1.12	The Sarason Interpolation Theorem	25
1.13	Historical Notes	30

2	**Operator Analysis on $\mathbb{D}$: Model Formulas, Lurking Isometries, and Positivity Arguments**	**33**
2.1	Overview	33
2.2	A Model Formula for $\mathscr{S}(\mathbb{D})$	33
2.3	Reproducing Kernel Hilbert Spaces	37
2.4	Lurking Isometries	39
2.5	The Network Realization Formula (Scalar Case) via Model Theory	44
2.6	Interpolation via Model Theory	52

vii

2.7	The Müntz–Szász Interpolation Theorem	56
2.8	Positivity Arguments	61
2.9	Historical Notes	70

3	**Further Development of Models on the Disc**	**71**
3.1	A Model Formula for $\mathscr{S}_{\mathcal{B}(\mathcal{H},\mathcal{K})}(\mathbb{D})$	71
3.2	Lurking Isometries Revisited	74
3.3	The Network Realization Formula	75
3.4	Tensor Products of Hilbert Spaces	78
3.5	Tensor Products of Operators	79
3.6	Realization of Rational Matrix Functions and the McMillan Degree	81
3.7	Pick Interpolation Revisited	83
3.8	The Corona Problem	85
3.9	Historical Notes	91

4	**Operator Analysis on $\mathbb{D}^2$**	**93**
4.1	The Space of Hereditary Functions on $\mathbb{D}^2$	93
4.2	The Hereditary Functional Calculus on $\mathbb{D}^2$	95
4.3	Models on $\mathbb{D}^2$	99
4.4	Models on $\mathbb{D}^2$ via the Duality Construction	104
4.5	The Network Realization Formula for $\mathbb{D}^2$	106
4.6	Nevanlinna–Pick Interpolation on $\mathbb{D}^2$	109
4.7	Toeplitz Corona for the Bidisc	112
4.8	Operator-Valued Functions on $\mathbb{D}^2$	113
4.9	Models of Operator-Valued Functions on $\mathbb{D}^2$	116
4.10	Historical Notes	129

5	**Carathéodory–Julia Theory on the Disc and the Bidisc**	**131**
5.1	The One-Variable Results	131
5.2	The Model Approach to Regularity on $\mathbb{D}$: B-points and C-points	133
5.3	A Proof of the Carathéodory–Julia Theorem on $\mathbb{D}$ via Models	137
5.4	Pick Interpolation on the Boundary	141
5.5	Regularity, B-points and C-points on the Bidisc	143
5.6	The Missing Link	146
5.7	Historical Notes	147

6	**Herglotz and Nevanlinna Representations in Several Variables**	**148**
6.1	Overview	148
6.2	The Herglotz Representation on $\mathbb{D}^2$	149
6.3	Nevanlinna Representations on $\mathbb{H}$ via Operator Theory	153

Contents ix

6.4	The Nevanlinna Representations on $\mathbb{H}^2$	156
6.5	A Classification Scheme for Nevanlinna Representations in Two Variables	160
6.6	The Type of a Function	165
6.7	Historical Notes	168

7	**Model Theory on the Symmetrized Bidisc**	**169**
7.1	Adding Symmetry to the Fundamental Theorem for $\mathbb{D}^2$	170
7.2	How to Define Models on the Symmetrized Bidisc	173
7.3	The Network Realization Formula for G	176
7.4	The Hereditary Functional Calculus on G	177
7.5	When Is G a Spectral Domain?	182
7.6	G Spectral Implies G Complete Spectral	185
7.7	The Spectral Nevanlinna–Pick Problem	185
7.8	Historical Notes	187

8	**Spectral Sets: Three Case Studies**	**189**
8.1	Von Neumann's Inequality and the Pseudo-Hyperbolic Metric on $\mathbb{D}$	189
8.2	Spectral Domains and the Carathéodory Metric	192
8.3	Background Material	194
8.4	Lempert's Theorem	195
8.5	The Carathéodory Distance on G	203
8.6	Von Neumann's Inequality on Subvarieties of the Bidisc	205
8.7	Historical Notes	211

9	**Calcular Norms**	**213**
9.1	The Taylor Spectrum and Functional Calculus	213
9.2	Calcular Norms and Algebras	215
9.3	Halmos's Conjecture and Paulsen's Theorem	223
9.4	The Douglas–Paulsen Norm	226
9.5	The B. and F. Delyon Norm and Crouzeix's Theorem	233
9.6	The Badea–Beckermann–Crouzeix Norm	235
9.7	The Polydisc Norm	243
9.8	The Oka Extension Theorem and Calcular Norms	245
9.9	Historical Notes	252

10	**Operator Monotone Functions**	**254**
10.1	Löwner's Theorems	254
10.2	An Interlude on Linear Programming	262
10.3	Locally Matrix Monotone Functions in d Variables	269
10.4	The Löwner Class in d Variables	278

x   *Contents*

10.5	Globally Monotone Rational Functions in Two Variables	279
10.6	Historical Notes	282

PART TWO   NON-COMMUTATIVE THEORY   283

11   Motivation for Non-Commutative Functions   285

11.1	Non-Commutative Polynomials	285
11.2	Sums of Squares	287
11.3	Nullstellensatz	287
11.4	Linear Matrix Inequalities	288
11.5	The Implicit Function Theorem	289
11.6	Matrix Monotone Functions	289
11.7	The Functional Calculus	290
11.8	Historical Notes	290

12   Basic Properties of Non-Commutative Functions   291

12.1	Definition of an nc-Function	291
12.2	Locally Bounded nc-Functions Are Holomorphic	295
12.3	Nc Topologies	298
12.4	Historical Notes	300

13   Montel Theorems   302

13.1	Overview	302
13.2	Wandering Isometries	303
13.3	A Graded Montel Theorem	305
13.4	An nc Montel Theorem	309
13.5	Closed Cones	310
13.6	Historical Notes	312

14   Free Holomorphic Functions   313

14.1	The Range of a Free Holomorphic Function	313
14.2	Nc-Models and Free Realizations	315
14.3	Free Pick Interpolation	319
14.4	Free Realizations Redux	324
14.5	Extending off Varieties	326
14.6	The nc Oka–Weil Theorem	327
14.7	Additional Results	328
14.8	Historical Notes	329

15   The Implicit Function Theorem   330

15.1	The Fine Inverse Function Theorem	330
15.2	The Fat Inverse Function Theorem	332
15.3	The Implicit Function Theorem	337
15.4	The Range of an nc-Function	339

15.5	Applications of the Implicit Function Theorem in Non-Commutative Algebraic Geometry	341
15.6	Additional Results	343
15.7	Historical Notes	343
16	**Non-Commutative Functional Calculus**	344
16.1	Nc Operator Functions	344
16.2	Polynomial Approximation and Power Series	349
16.3	Extending Free Functions to Operators	354
16.4	Nc Functional Calculus in Banach Algebras	356
16.5	Historical Notes	357
	Notation	358
	Bibliography	361
	Index	372

Preface

The philosophy of this book is that Hilbert space geometry binds function theory and operator theory together, not only allowing each to aid the other but creating a rich structure that can be used to discover new phenomena. There is a "three-way street" between operator theory and function theory: sometimes one uses function theory to prove operator theorems, sometimes one uses operator theory to prove function theorems, and sometimes the theories are so interwoven that one cannot even state the theorem without using the language of both operator theory and function theory.

The main thrust of the book is to discover and prove theorems about holomorphic functions and complex geometry with the aid of Hilbert space geometry and operator theory. The holomorphic functional calculus permits us to substitute commuting operators for the variables of a holomorphic function, a step that reveals hidden properties of the function—something every function theorist should want to do.

It is remarkable how little operator theory one needs in order to prove significant facts in function theory. There will be no call here for the detailed and subtle theories of particular classes of operators, but we shall make heavy use of the functional calculus for operators. The theories of operator dilations and spectral sets also play an important role. We explain carefully what is required from these theories. The reader may either take this material on trust or consult the references that we give.

Part I of the book is devoted to holomorphic functions on domains in complex Euclidean space, beginning with scalar-valued functions on the unit disc in the complex plane, where intuition is most easily developed. Here the central notion of a Hilbert space *model* of a function is introduced, as are several types of arguments that will recur throughout the book. Gradually we build up to various domains in higher dimensions and to operator-valued functions, the latter being important for engineering applications. We do not

xiii

xiv *Preface*

aim for maximum generality, which can come at the cost of a sacrifice of elegance and impact.

Part II concerns *non-commutative functions*, that is, functions of non-commuting variables. This is a topic of relatively recent study, and one that is currently undergoing rapid development. It transpires that the Hilbert space methods of Part I are well suited to this new context, and we derive many analogs of classical theorems.

Our intended audience is graduate students and mathematicians interested in complex function theory and/or operator theory. We do not require familiarity with several complex variables, but we do assume that the reader has a basic knowledge of complex analysis and functional analysis.

The moment you buy into the functional calculus, you're ready to roll!

Acknowledgments

The catalyst for this book was two series of lectures, given by the first and second authors at Newcastle University in an LMS lecture series in April 2017. The authors would like to thank the London Mathematical Society for supporting this event. In addition, the first author was partially supported by National Science Foundation Grant DMS 1665260, the second author was partially supported by National Science Foundation Grant DMS 1565243, and the third author was partially supported by UK Engineering and Physical Sciences Research Council grants EP/K50340X/1 and EP/N03242X/1, and London Mathematical Society grants 41219 and 41730. We would like to thank the National Science Foundation, the UK Engineering and Physical Sciences Research Council, and the London Mathematical Society for their support.

Parts of the book were read in draft form by several people, who caught some mistakes. We are specially grateful to Alberto Dayan, Chris Felder, Michael Hartz, Mark Mancuso, James Pascoe, and Jeet Sampat. All remaining mistakes and typographical errors are, of course, the responsibility of the authors.

PART ONE

Commutative Theory

1

The Origins of Operator-Theoretic Approaches to Function Theory

In this chapter we present some of the highlights of operator theory over the last century and briefly describe their connections with function theory. It is intended to establish notation and provide orientation for the reader, before we go on to a detailed and systematic development in Chapter 2 (which can be understood independently of this scene-setting chapter). The reader should not attempt to master all the contents of this chapter in detail before progressing–some of them are substantial. Our choice, inevitably, is partial; we mention some other major contributions in historical notes at the end of the chapter, but we do not attempt a history of the subject.

1.1 Operators

We adopt the standard Bourbaki symbols $\mathbb{R}$ and $\mathbb{C}$ for the sets of real numbers and complex numbers, respectively, and also the notations

$$\mathbb{T} = \{z \in \mathbb{C}: |z| = 1\} \text{ and } \mathbb{D} = \{z \in \mathbb{C}: |z| < 1\}.$$

For any Banach space $\mathcal{X}$, we shall denote by ball $\mathcal{X}$ the closed unit ball of $\mathcal{X}$, that is,

$$\text{ball}\,\mathcal{X} = \{u \in \mathcal{X}: \|u\| \le 1\}.$$

If $\mathcal{X}, \mathcal{Y}$ are Banach spaces, we say that a linear transformation $T: \mathcal{X} \to \mathcal{Y}$ is *bounded* if there exists $c \in [0, \infty)$ such that

$$\|Tu\| \le c\|u\| \quad \text{for all } u \in \mathcal{X}. \tag{1.1}$$

For a bounded linear transformation $T: \mathcal{X} \to \mathcal{Y}$ we define $\|T\|$, the *norm of* T, by the formula

$$\|T\| = \sup_{u \in \text{ball}\,\mathcal{X}} \|Tu\|, \tag{1.2}$$

4   *Commutative Theory*

or equivalently, by the formula

$$\|T\| = \inf\{c \colon \|Tu\| \leq c\|u\| \quad \text{for all } u \in \mathcal{X}\}.$$

When $\mathcal{X}, \mathcal{Y}$ are Banach spaces, we denote by $\mathcal{B}(\mathcal{X})$ the set of bounded linear transformations from $\mathcal{X}$ to $\mathcal{X}$, and by $\mathcal{B}(\mathcal{X}, \mathcal{Y})$ the set of bounded linear transformations from $\mathcal{X}$ to $\mathcal{Y}$. We refer to the elements of $\mathcal{B}(\mathcal{X})$ or $\mathcal{B}(\mathcal{X}, \mathcal{Y})$ as *operators*. $\mathcal{B}(\mathcal{X})$ and $\mathcal{B}(\mathcal{X}, \mathcal{Y})$ are Banach spaces under the norm defined by equation (1.2). Furthermore, $\mathcal{B}(\mathcal{X})$ (with composition as multiplication) is a Banach algebra with identity, as it satisfies the axioms[1] $\|1\| = 1$ and

$$\|T_1 T_2\| \leq \|T_1\| \, \|T_2\|$$

for all $T_1, T_2 \in \mathcal{B}(\mathcal{X})$.

Every element of $\mathcal{B}(\mathcal{X})$ has a *spectrum*. For $T \in \mathcal{B}(\mathcal{X})$ we say that T is *invertible* if there exists $S \in \mathcal{B}(\mathcal{X})$ such that $ST = TS = 1$. The *spectrum of T*, denoted by $\sigma(T)$, is defined by

$$\sigma(T) = \{z \in \mathbb{C} \colon z - T \text{ is not invertible}\}.$$

The set $\sigma(T)$ is a nonempty compact subset of $\mathbb{C}$ provided that $\mathcal{X} \neq \{0\}$.

1.2 Functional Calculi

Informally, the term *functional calculus* refers to a rule that enables a function f to act on an operator $T \in \mathcal{B}(\mathcal{X})$ to produce an operator $f(T) \in \mathcal{B}(\mathcal{X})$. There are numerous functional calculi that have been defined in various settings. However, in these notes we shall restrict ourselves principally to the case that the functional calculus is *holomorphic*, that is, the function f is assumed to be holomorphic on a neighborhood of $\sigma(T)$. To be an honest functional calculus, the rule should satisfy the following natural conditions.

Properties of a holomorphic functional calculus

Let $\mathcal{X}$ be a Banach space and let $T \in \mathcal{B}(\mathcal{X})$.

(1) If $f(z) = 1$ for all z in a neighborhood of $\sigma(T)$, then $f(T) = 1$, the identity operator on $\mathcal{X}$.
(2) If $f(z) = z$ for all z in a neighborhood of $\sigma(T)$, then $f(T) = T$.
(3) If f, g are holomorphic in a neighborhood of $\sigma(T)$ then

$$(f + g)(T) = f(T) + g(T) \quad \text{and} \quad (fg)(T) = f(T)g(T).$$

(4) If f is holomorphic in a neighborhood of $\sigma(T)$, $\lambda \in \mathbb{C}$ and $g = \lambda f$ then $g(T) = \lambda f(T)$.

---

[1] Here the first 1 denotes the identity of $\mathcal{B}(\mathcal{X})$, which is the identity operator on $\mathcal{X}$.

The Origins of Operator-Theoretic Approaches to Function Theory   5

(5) The functional calculus has an appropriate continuity property.

For any open set U in $\mathbb{C}$, the algebra of scalar-valued holomorphic functions on U will be denoted by $\mathrm{Hol}(U)$. Observe that, for any $T \in \mathcal{B}(\mathcal{X})$, if U is a neighborhood of $\sigma(T)$ and $\tau \colon \mathrm{Hol}(U) \to \mathcal{B}(\mathcal{X})$ is defined by $\tau(f) = f(T)$, then τ is a homomorphism of algebras that satisfies $\tau(1) = 1$ and $\tau(z) = T$.

The simplest example of a functional calculus is the *polynomial calculus*. Here, for $f \in \mathbb{C}[z]$, the algebra of polynomials with complex coefficients, if

$$f(z) = \sum_{k=1}^{n} a_k z^k,$$

then $f(T)$ is defined by

$$f(T) = \sum_{k=1}^{n} a_k T^k.$$

The polynomial calculus can easily be extended to the *power series calculus*. If

$$f(z) = \sum_{k=1}^{\infty} a_k (z - a)^k$$

defines an analytic function on $\{z \in \mathbb{C} \colon |z - a| < r\}$ for some $r > 0$, then under the assumption that $\sigma(T) \subseteq \{z \in \mathbb{C} \colon |z - a| < r\}$, the operator $f(T)$ is defined by

$$f(T) = \sum_{k=1}^{\infty} a_k (T - a)^k.$$

That this series converges in the operator norm of $\mathcal{B}(\mathcal{X})$ is a simple exercise on the spectral radius formula and Cauchy's formula for the radius of convergence of a power series.

Another simple modification of the polynomial calculus is the *rational calculus*. For K, a compact subset of $\mathbb{C}$, let $\mathrm{Rat}(K)$ denote the algebra of rational functions with poles off K. If $T \in \mathcal{B}(\mathcal{X})$ and $f \in \mathrm{Rat}(\sigma(T))$, then we may write

$$f(z) = p(z)q(z)^{-1},$$

where p and q are polynomials and $q(z) \neq 0$ for all $z \in \sigma(T)$. By the spectral mapping theorem, which asserts that $q(\sigma(T)) = \sigma(q(T))$, the spectrum of $q(T)$ does not contain 0, so that $q(T)$ is invertible. Therefore, $f(T)$ can be defined by the formula

$$f(T) = p(T)q(T)^{-1}. \tag{1.3}$$

A sweet spot in the theory of functional calculi is the *Riesz–Dunford functional calculus*. Treatments of this topic, complete with proofs, are in

6 Commutative Theory

[60, VII.4; 140, section 17.2], and many other introductory texts on functional analysis. We shall just state the important results.

Fix an operator $T \in \mathcal{B}(\mathcal{X})$. For any open set U in $\mathbb{C}$ such that $\sigma(T) \subseteq U$, let Γ be a finite collection of closed rectifiable curves in $U \setminus \sigma(T)$ that winds once around each point in the spectrum of $\sigma(T)$ and no times around each point in the complement of U. If $f \in \mathrm{Hol}(U)$ then, by the Cauchy integral formula,

$$f(z) = \frac{1}{2\pi i} \int_{\Gamma} \frac{f(w)}{w - z} \, dw,$$

whenever $z \in \sigma(T)$. The Riesz–Dunford functional calculus defines $f(T)$ for $f \in \mathrm{Hol}(U)$ by the substitution $z = T$ in this formula, that is to say,

$$f(T) = \frac{1}{2\pi i} \int_{\Gamma} f(w)(w - T)^{-1} \, dw. \tag{1.4}$$

Note that this formula makes sense, since the assumption $\Gamma \subseteq \mathbb{C} \setminus \sigma(T)$ implies that $w - T$ is invertible for all $w \in \Gamma$. Moreover, the defining equation (1.4) depends neither on the choice of contour nor on the choice of U, since by Cauchy's theorem, for any vectors u, v in $\mathcal{X}$ and $\mathcal{X}^*$,

$$\langle f(T)u, v \rangle = \frac{1}{2\pi i} \int_{\Gamma} f(w) \langle (w - T)^{-1} u, v \rangle \, dw$$

is independent of Γ.

Properties of the Riesz–Dunford functional calculus

(1) If $f(z) = 1$, then $f(T) = 1$ and if $f(z) = z$, then $f(T) = T$.

(2) For any neighborhood U of $\sigma(T)$ the map

$$f \mapsto f(T)$$

is a unital algebra-homomorphism from $\mathrm{Hol}(U)$ to $\mathcal{B}(\mathcal{X})$.

(3) The calculus is consistent with the polynomial and rational calculi, that is, if U is a neighborhood of $\sigma(T)$, if r is a rational function with poles off U, and $f \in \mathrm{Hol}(U)$ is defined by $f(\lambda) = r(\lambda)$ for all $\lambda \in U$, then

$$f(T) = r(T).$$

(4) The calculus is consistent with the power series calculus, that is, if U is a disc, $\sigma(T) \subseteq U$, and $g(z)$ is the power series expansion of $f(z)$ valid in U, then $f(T) = g(T)$, where $g(T)$ is defined by the power series calculus.

(5) The calculus is continuous in the following sense. If U is a neighborhood of $\sigma(T)$, if $\{f_n\}$ is a sequence in $\mathrm{Hol}(U)$, $f \in \mathrm{Hol}(U)$, and $f_n \to f$ uniformly on compact subsets of U, then $f_n(T) \to f(T)$ in the operator norm of $\mathcal{B}(\mathcal{X})$.

The Origins of Operator-Theoretic Approaches to Function Theory 7

Another way of thinking of the Riesz–Dunford functional calculus is by way of Runge's theorem, which implies that, for any open set U in $\mathbb{C}$, every function $f \in \text{Hol}(U)$ is the limit, uniformly on compact subsets of U, of a sequence of rational functions r_n with poles off U. As a consequence, property (5) can be used to define $f(T)$ as the norm limit of $r_n(T)$.

The holomorphic functional calculus is one of the most important tools in operator theory and is central to this book. We shall frequently want to apply a holomorphic function of d complex variables to a d-tuple of pairwise commuting operators; for this we need a generalization of the Riesz–Dunford functional calculus. Such a generalization does indeed exist; the elaboration of this theory is a heroic chapter in the history of operator theory. It is more intricate than the one-variable theory and requires a significant input from the theory of several complex variables. It was developed over several decades by many mathematicians and, while "multivariable spectral theory" remains an active research topic, the foundations of the functional calculus now appear to be in a definitive form.

For a holomorphic function f of d variables and a pairwise commuting d-tuple $T = (T^1, \ldots, T^d)$ of operators on a Banach space $\mathcal{X}$, to define what we mean by $f(T)$ one strategy is again to require f to be holomorphic on the spectrum of T and to invoke an integral representation formula. To do this one must overcome some technical difficulties, the first of which is to find the appropriate notion of spectrum of T. For a start, the spectrum should contain the *joint eigenvalues* of T. We say that a non-zero vector $x \in \mathcal{X}$ is a *joint eigenvector* of T and that $\lambda = (\lambda^1, \ldots, \lambda^d)$ is a corresponding joint eigenvalue of T if $T^j x = \lambda^j x$ for $j = 1, \ldots, d$. If $\dim \mathcal{X} < \infty$ (and $\mathcal{X} \neq \{0\}$) then T has at least one joint eigenvalue. For let λ^1 be an eigenvalue of T^1 and let E_1 be the corresponding eigenspace $\{x \in \mathcal{X}: T^1 x = \lambda^1 x\}$. Since the T^j commute with T^1, the space E_1 is invariant under each T^j and the operators $T^j | E_1$ commute pairwise. Let λ^2 be an eigenvalue of $T^2 | E_1$ and let E_2 be the corresponding eigenspace $\{x \in E_1: T^2 x = \lambda^2 x\}$. Continuing in this way we arrive at a point $\lambda = (\lambda^1, \ldots, \lambda^d) \in \mathbb{C}^d$ and a non-zero subspace E_d of $\mathcal{X}$ such that $T^j x = \lambda^j x$ for $j = 1, \ldots, d$ and all $x \in E_d$. This point $\lambda \in \mathbb{C}^d$ is a joint eigenvalue of T. However, as we already know from the case $d = 1$, when $\mathcal{X}$ is infinite-dimensional there need not be any joint eigenvalues of T.

In 1953 Georgii Evgen'evich Shilov [187] constructed a functional calculus for several elements of a commutative Banach algebra. If T is a commuting d-tuple of elements of $\mathcal{B}(\mathcal{X})$, one can choose any commutative Banach subalgebra $\mathcal{A}$ of $\mathcal{B}(\mathcal{X})$ containing the elements of T and then define the spectrum of T with respect to $\mathcal{A}$ by

$$\sigma_{\mathcal{A}}(T) = \mathbb{C}^d \setminus \rho_A(T), \tag{1.5}$$

8   *Commutative Theory*

where

$$\rho_{\mathcal{A}}(T) = \left\{ \lambda \in \mathbb{C}^d : \text{ there exist } R^1, \dots, R^d \in \mathcal{A} \text{ such that} \right.$$
$$\left. (T^1 - \lambda^1)R^1 + \cdots + (T^d - \lambda^d)R^d = 1 \right\}. \qquad (1.6)$$

Any joint eigenvalue of T does belong to $\sigma_{\mathcal{A}}(T)$, for any commutative subalgebra $\mathcal{A}$ of $\mathcal{B}(\mathcal{X})$ containing the T^j (otherwise, simply apply the defining equation (1.6) of $\rho_{\mathcal{A}}(T)$ to the corresponding joint eigenvector x to get a contradiction).

Shilov's theory yields a functional calculus for a commuting tuple T, a calculus that was further developed by numerous authors; see for example [212]. A drawback of this approach is that the spectrum of a tuple is defined relative to the chosen commutative subalgebra of $\mathcal{B}(\mathcal{X})$ and is typically different for different subalgebras. The simplest choice is to take the subalgebra to be the one generated in $\mathcal{B}(\mathcal{X})$ by the operators in question; we shall call the corresponding spectrum the *algebraic spectrum* of T and denote it by $\sigma_{\text{alg}}(T)$. However, this choice often results in an unnecessarily large spectrum.

This imperfection of the Banach algebra approach was remedied in the groundbreaking papers [198, 197] by Joseph Taylor. He introduced a more geometric notion of spectrum, now called the *Taylor spectrum* of T, denoted in this book by $\sigma(T)$.

The Taylor spectrum is a subset, sometimes proper, of $\sigma_{\mathcal{A}}(T)$ for any commutative algebra $\mathcal{A}$ containing T (indeed, Taylor showed that it is a subset of the spectrum with respect to any algebra $\mathcal{A}$ with T in its center [198, lemma 1.1]). This result is important, as a smaller spectrum gives a larger class of holomorphic functions for which $f(T)$ is defined. It is now accepted that the Taylor spectrum is the right notion for the definition of a functional calculus. We define it in Section 9.1. Let us remark that, for commuting normal tuples, there is a simpler description of the Taylor spectrum: see Section 8.3.1. Note also that $\sigma(T)$ contains all the joint eigenvalues of T [67, page 22].

A definitive discussion of the various notions of spectrum for d-tuples and development of the Taylor functional calculus is given in [89], while a more accessible account can be found in [67]. The present book, on the other hand is about the *use* of the functional calculus as a tool for function theory. What the reader will need is a knowledge of the *properties* of the functional calculus, which we shall describe as we need them.

For most of the book (all except Chapter 9) the domains that we study will be polynomially convex.[2] For such a domain U, the Taylor spectrum of T is contained in U if and only if the algebraic spectrum is. Furthermore, every holomorphic function f on U can be approximated uniformly on compact

---

[2] See Definition 1.43.

The Origins of Operator-Theoretic Approaches to Function Theory 9

subsets by a sequence of polynomials p_n (this result is called the Oka-Weil theorem). The Taylor functional calculus in this case can be defined concretely by $f(T) = \lim p_n(T)$.

1.3 Operators on Hilbert Space

Operator analysis in the sense of this book concerns the special case that $\mathcal{X}$ is a complex Hilbert space $\mathcal{H}$, that is, a complete inner product space over the field of complex numbers. We assume familiarity with the basic properties of Hilbert space (see [214]). In this case, $\mathcal{B}(\mathcal{X})$ has additional structure beyond that of a Banach algebra, a structure that plays a critical role in the development of operator-analytic methods.

Definition 1.7. *For Hilbert spaces $\mathcal{H}$ and $\mathcal{K}$ and for $T \in \mathcal{B}(\mathcal{H}, \mathcal{K})$, the* adjoint *of T is the operator T^* in $\mathcal{B}(\mathcal{K}, \mathcal{H})$ given by*

$$\langle T^*v, u \rangle_{\mathcal{H}} = \langle v, Tu \rangle_{\mathcal{K}} \qquad \text{for all } u \in \mathcal{H}, v \in \mathcal{K}.$$

The adjoint satisfies the *C^*-axiom*

$$\|T^*T\| = \|T\|^2. \tag{1.8}$$

In the C^*-axiom, since $T: \mathcal{H} \to \mathcal{K}$ and $T^*: \mathcal{K} \to \mathcal{H}$, the operator T^*T is well defined and maps $\mathcal{H}$ to $\mathcal{H}$. The techniques that we shall present from operator analysis rely essentially on the meaningfulness of such algebraic expressions in T and T^* (especially on the statement (1.13)). For a general Banach space, if $T: \mathcal{X} \to \mathcal{Y}$, then $T^*: \mathcal{Y}^* \to \mathcal{X}^*$, so one cannot make sense of T^*T. For this reason, in the remainder of the book, the term *operator* will always refer to a bounded linear operator acting on a complex Hilbert space or from one complex Hilbert space to another.

The techniques of operator analysis depend essentially on the notion of *positivity*. Classically, an $n \times n$ matrix $[a_{ij}]$ is said to be *positive semi-definite* if, for all $c_1, \ldots, c_n \in \mathbb{C}$,

$$\sum_{i,j=1}^{n} a_{ij} c_j \bar{c}_i \geq 0.$$

We extend this notion to operators on Hilbert space in the following way.

Definition 1.9. *An operator $T \in \mathcal{B}(\mathcal{H})$ is said to be* positive, *written $T \geq 0$, if, for all $u \in \mathcal{H}$,*

$$\langle Tu, u \rangle_{\mathcal{H}} \geq 0.$$

There are other revealing characterizations of the positivity of an operator. See [60, theorem VIII.3.6] for a proof of the following statement.

10   *Commutative Theory*

Proposition 1.10. *For $T \in \mathcal{B}(\mathcal{H})$, the following statements are equivalent.*

(i) $T \geq 0$;

(ii) $T^* = T$ *and*

$$\sigma(T) \subseteq \{x \in \mathbb{R} : x \geq 0\};$$

(iii) $T = X^*X$ *for some Hilbert space $\mathcal{K}$ and some $X \in \mathcal{B}(\mathcal{H}, \mathcal{K})$;*

(iv) $T = X^2$ *for some $X \in \mathcal{B}(\mathcal{H})$ such that $X = X^*$.*

There is a close connection between positivity and some other important classes of operators.

Definition 1.11. *Let $\mathcal{H}, \mathcal{K}$ be Hilbert spaces. An operator $T \in \mathcal{B}(\mathcal{H}, \mathcal{K})$ is*

(i) *a* contractive operator *if $\|T\| \leq 1$;*

(ii) *an* isometric operator *if $\|Tx\| = \|x\|$ for every $x \in \mathcal{H}$.*

A simple but, in this book, all-pervading principle is the following.

Proposition 1.12. (The fundamental fact of operator analysis) *If $\mathcal{H}, \mathcal{K}$ are Hilbert spaces and $T \in \mathcal{B}(\mathcal{H}, \mathcal{K})$, then*

$$T \text{ is contractive if and only if } \quad 1 - T^*T \geq 0. \tag{1.13}$$

Proof.

$$
\begin{aligned}
\|T\| \leq 1 \quad &\Leftrightarrow \quad \|x\|^2 \geq \|Tx\|^2 \quad \text{for all } x \in \mathcal{H} \\
&\Leftrightarrow \quad \langle x, x \rangle \geq \langle Tx, Tx \rangle \quad \text{for all } x \in \mathcal{H} \\
&\Leftrightarrow \quad \langle x, x \rangle - \langle T^*Tx, x \rangle \geq 0 \quad \text{for all } x \in \mathcal{H} \\
&\Leftrightarrow \quad \langle (1 - T^*T)x, x \rangle \geq 0 \quad \text{for all } x \in \mathcal{H} \\
&\Leftrightarrow \quad 1 - T^*T \geq 0.
\end{aligned}
$$

$\square$

The importance of the equivalence (1.13) in Proposition 1.12 cannot be overstated. It enables the expression of the analytic concept of size in terms of the algebraic concept of positivity. The norm of T is bounded by 1 if and only if $1 - T^*T$ can be represented as X^*X or X^2, as in Proposition 1.10, a condition that can often be resolved by algebra.

In Proposition 1.12 the extremal case of the inequality $1 - T^*T \geq 0$ is that it hold with equality, that is $1 - T^*T = 0$.

Proposition 1.14. *If $\mathcal{H}, \mathcal{K}$ are Hilbert spaces and $T \in \mathcal{B}(\mathcal{H}, \mathcal{K})$, then*

$$T \text{ is isometric if and only if } \quad 1 - T^*T = 0. \tag{1.15}$$

The Origins of Operator-Theoretic Approaches to Function Theory 11

A slight modification of the proof of Proposition 1.12 yields the fact that T is isometric if and only if $\langle (1 - T^*T)x, x \rangle = 0$ for all vectors x. Proposition 1.14 is therefore a consequence of the following statement.

Lemma 1.16. *If Z is an operator on a Hilbert space $\mathcal{H}$ then $Z = 0$ if and only if $\langle Zx, x \rangle = 0$ for all $x \in \mathcal{H}$.*

Proof. Necessity is trivial. Conversely, suppose $\langle Zx, x \rangle = 0$ for all $x \in \mathcal{H}$. Consider any $x, y \in \mathcal{H}$ and $\lambda \in \mathbb{C}$. We have

$$
\begin{aligned}
0 &= \langle Z(x + \lambda y), x + \lambda y \rangle \\
&= \langle Zx, x \rangle + \lambda \langle Zy, x \rangle + \bar{\lambda} \langle Zx, y \rangle + |\lambda|^2 \langle Zy, y \rangle \\
&= \lambda \langle Zy, x \rangle + \bar{\lambda} \langle Zx, y \rangle.
\end{aligned}
$$

By choosing in succession $\lambda = 1$ and $\lambda = i$ we deduce that $\langle Zx, y \rangle = 0$ for all $x, y \in \mathcal{H}$, and therefore $Z = 0$. $\square$

In Lemma 1.16 we make use of our standing assumption that Hilbert spaces are vector spaces over $\mathbb{C}$. The statement would be false in a real Hilbert space (consider the operation of rotation through a right angle on $\mathbb{R}^2$).

We defined isometries to be operators that preserve lengths; an interesting consequence of Proposition 1.14 is that they also preserve inner products. If $T^*T = 1$ then, for any vectors x, y,

$$
\langle Tx, Ty \rangle = \langle T^*Tx, y \rangle = \langle x, y \rangle.
$$

For Hilbert spaces $\mathcal{H}$ and $\mathcal{K}$, there is a useful notation that allows the identification of elements of the algebraic tensor product $\mathcal{H} \otimes \mathcal{K}$ with operators from $\mathcal{K}$ into $\mathcal{H}$. If $u \in \mathcal{H}$ and $v \in \mathcal{K}$ we define $u \otimes v \in \mathcal{B}(\mathcal{K}, \mathcal{H})$ by the formula

$$
(u \otimes v)(x) = \langle x, v \rangle_{\mathcal{K}} u \qquad \text{for all } x \in \mathcal{K}. \tag{1.17}
$$

In the context of a vector space V there is an algebraic notion of *projection*. If V is expressed as the direct sum of two subspaces M and N, then one may define a linear transformation P on V by the relation

$$
P(x + y) = x
$$

for all $x \in M$ and $y \in N$. This operator P is called *the projection on M along N*; it is clearly a linear transformation on V with range M and kernel N. In the context of a Hilbert space $\mathcal{H}$, the word *projection* is usually reserved for the special case of a projection on a *closed* subspace M of $\mathcal{H}$ along its orthogonal complement. We shall denote it by P_M.

12 *Commutative Theory*

1.4 The Spectral Theorem

One of the early triumphs of operator theory was the development of the modern form of the spectral theorem, together with its myriad applications to analysis and the mathematics of physics. An operator $N \in \mathcal{B}(\mathcal{H})$ is said to be *normal* if it commutes with its adjoint, that is, if

$$N^*N = NN^*.$$

Two important special cases of normality are when N is *self-adjoint*, that is, $N^* = N$, and when N is *unitary*, that is, $N^*N = 1 = NN^*$. If N is normal, then

(1) N is self-adjoint if and only if $\sigma(N) \subseteq \mathbb{R}$, and
(2) N is unitary if and only if $\sigma(N) \subseteq \mathbb{T}$.

If $\mathcal{H}$ is a Hilbert space and Ω is a σ-algebra of subsets of a set X, then we say that E is a $\mathcal{B}(\mathcal{H})$-*valued spectral measure* defined on (X, Ω) if $E: \Omega \to \mathcal{B}(\mathcal{H})$ is a mapping satisfying

(1) for all $\Delta \in \Omega$, $E(\Delta)$ is a projection on $\mathcal{H}$;
(2) $E(\emptyset) = 0$ and $E(X) = 1$;
(3) for all $\Delta_1, \Delta_2 \in \Omega$, $E(\Delta_1 \cap \Delta_2) = E(\Delta_1)E(\Delta_2)$, and
(4) if $\{\Delta_n\}$ is a sequence of pairwise disjoint sets in Ω, then

$$E\left(\bigcup_{n=1}^{\infty} \Delta_n\right) = \sum_{n=1}^{\infty} E(\Delta_n),$$

where the sum on the right-hand side converges in the strong operator topology.[3]

Theorem 1.18. (The spectral theorem for normal operators) *If $\mathcal{H}$ is a Hilbert space and N is a normal operator on $\mathcal{H}$, then there exists a $\mathcal{B}(\mathcal{H})$-valued spectral measure E defined on the σ-algebra of Borel subsets of $\sigma(N)$ such that*

$$N = \int_{\sigma(N)} z \, dE(z). \tag{1.19}$$

See [140, section 31.6, theorem 15] or [60, theorem IX.2.2] for a proof. An immediate corollary of the spectral theorem is the following result, a cornerstone of the theory of C^*-algebras. For K a compact subset of $\mathbb{C}$, let $C(K)$ denote the algebra of continuous functions on K with supremum norm and involution $f^* = \bar{f}$, the complex conjugate of f. For a normal operator N on $\mathcal{H}$, let $C^*(N)$ denote the smallest norm-closed subalgebra of $\mathcal{B}(\mathcal{H})$ containing 1 and N and satisfying $X^* \in C^*(N)$ whenever $X \in C^*(N)$.

---

[3] A sequence (T_n) of operators converges to $T \in \mathcal{B}(\mathcal{H})$ in the strong operator topology if, for every $x \in \mathcal{H}$, $\|T_n x - T x\| \to 0$ as $n \to \infty$.

The Origins of Operator-Theoretic Approaches to Function Theory 13

Theorem 1.20. (The continuous functional calculus for normal operators)
Let $\mathcal{H}$ be a Hilbert space, N be a normal operator on $\mathcal{H}$ and E be the spectral measure of N as in Theorem 1.18. For $\varphi \in C(\sigma(N))$, define $\varphi(N) \in \varphi \in C(\sigma(N))$ by the formula

$$\varphi(N) = \int_{\sigma(N)} \varphi(z) \, dE(z). \tag{1.21}$$

The map $\rho \colon C(\sigma(N)) \to \mathcal{B}(\mathcal{H})$ given by the formula $\rho(\varphi) = \varphi(N)$ is an isometric $*$-isomorphism of $C(\sigma(N))$ onto $C^*(N)$, that is, ρ is an algebra-isomorphism satisfying $\rho(\bar{\varphi}) = \rho(\varphi)^*$ and $\|\rho(\varphi)\| = \|\varphi\|$ for all $\varphi \in \mathcal{B}(\mathcal{H})$.

This theorem illustrates the fact that not all functional calculi are holomorphic functional calculi. Nevertheless, the continuous functional calculus for normal operators agrees with the Riesz–Dunford functional calculus on functions that are holomorphic on a neighborhood of the spectrum of a normal operator. In particular, if N is a normal operator, $f \in \mathrm{Rat}(\sigma(N))$, and $f(N)$ is defined by equation (1.3), then

$$f(N) = \int_{\sigma(N)} f(z) \, dE(z). \tag{1.22}$$

1.5 Hardy Space and the Unilateral Shift

The name *Hilbert space* is a tribute to a seminal paper of 1906 by David Hilbert [115]. According to Jean Dieudonné,[4] this is one of four landmark papers in the period 1900–1906 which led to the sudden crystallization of functional analysis as a branch of analysis (the others were a paper by Ivar Fredholm on integral equations, Henri Lebesgue's thesis on integration and Maurice Fréchet's thesis on metric spaces). In this paper Hilbert made use of the space now denoted by ℓ^2. It comprises all sequences

$$a - (a_0, a_1, a_2, \ldots)$$

of complex numbers, indexed by the non-negative integers, such that

$$\|a\| \overset{\text{def}}{=} \left(\sum_{n=0}^{\infty} |a_n|^2 \right)^{\frac{1}{2}}$$

is finite. It is equipped with the inner product

$$\langle a, b \rangle = \sum_{n=0}^{\infty} a_n \overline{b_n}.$$

---

[4] In [72, chapter V] he wrote "By the depth and novelty of its ideas, it is a turning point in the history of Functional Analysis, and indeed deserves to be considered as the very first paper published in that discipline."

14 Commutative Theory

One could say that ℓ^2 is the "purest" Hilbert space, in that it exhibits the geometry of Hilbert space in the presence of minimal mathematical structure. Other Hilbert spaces typically have elements that are functions or operators of some kind, which may involve delicate notions such as measurability or analyticity, and such spaces often occur in applications. However, all separable infinite-dimensional Hilbert spaces are isomorphic ([172, section 82] or [214, theorem 4.19]), and so it is natural to try to gain insight into operators on a general Hilbert space by studying them on the "bare" Hilbert space ℓ^2.

One operator on ℓ^2 that has been much studied is the *unilateral shift S*. This is the operator defined on ℓ^2 by the formula

$$S(a_0, a_1, a_2, \ldots) = (0, a_0, a_1, \ldots). \tag{1.23}$$

This simple definition conceals a treasure trove of analytical subtlety, much of it understandable via an avatar of S dwelling on a certain Hilbert space H^2 of analytic functions on $\mathbb{D}$. The space H^2, known as Hardy space, is the Hilbert space of analytic functions on $\mathbb{D}$ that have a square-summable power series about 0. To say that a power series

$$f(z) = \sum_{n=0}^{\infty} a_n z^n \tag{1.24}$$

is square summable means that

$$\sum_{n=0}^{\infty} |a_n|^2 < \infty. \tag{1.25}$$

This condition ensures that the series (1.24) converges locally uniformly on $\mathbb{D}$ and thereby defines f as an analytic function in $\mathbb{D}$. The inner product on H^2 is defined by

$$\left\langle \sum_{n=0}^{\infty} a_n z^n, \sum_{n=0}^{\infty} b_n z^n \right\rangle = \sum_{n=0}^{\infty} a_n \bar{b}_n. \tag{1.26}$$

Condition (1.25) also ensures that the series on the right-hand side of equation (1.26) converges, so that the inner product is well defined.

Thus elements of H^2 can be regarded either as power series or as analytic functions on $\mathbb{D}$. It is useful to have a criterion for membership of H^2, and an expression for the norm, in terms of function values (as distinct from Taylor coefficients). In fact, for an analytic function f on $\mathbb{D}$ to belong to H^2, it is necessary and sufficient that

$$\limsup_{r \to 1-} \int_0^{2\pi} |f(re^{i\theta})|^2 \, d\theta < \infty. \tag{1.27}$$

The Origins of Operator-Theoretic Approaches to Function Theory   15

When this inequality holds, the lim sup is in fact a limit as $r \to 1-$, and

$$\|f\|_{H^2} = \left\{ \lim_{r \to 1-} \int_0^{2\pi} |f(re^{i\theta})|^2 \frac{d\theta}{2\pi} \right\}^{\frac{1}{2}}. \qquad (1.28)$$

One consequence of equation (1.28) (together with the maximum principle) is that, for every $\lambda \in \mathbb{D}$, the map

$$f \mapsto f(\lambda)$$

is a bounded linear functional on H^2. It follows from the Riesz–Fréchet theorem that there exists a unique element s_λ of H^2 such that, for all $f \in H^2$,

$$f(\lambda) = \langle f, s_\lambda \rangle. \qquad (1.29)$$

It is easy to check by means of the formula (1.26) for the inner product that s_λ is given by

$$s_\lambda(z) = \frac{1}{1 - \bar{\lambda}z} \qquad \text{for } z \in \mathbb{D}. \qquad (1.30)$$

The function s_λ on $\mathbb{D}$ is called the *Szegő kernel*. The property described by equation (1.29) is called the *reproducing property* of the Szegő kernel.

If M_z denotes the operator defined on H^2 by the formula

$$(M_z f)(w) = w f(w) \qquad \text{for } f \in H^2 \text{ and } w \in \mathbb{D}, \qquad (1.31)$$

and $L: \ell^2 \to H^2$ is defined by

$$L(a_0, a_1, a_2, \ldots)(z) = a_0 + a_1 z + a_2 z^2 + \cdots, \qquad (1.32)$$

then it is apparent that L is a Hilbert space isomorphism and $M_z L = LS$. In other words, if we identify ℓ^2 and H^2 by means of L, then the shift operator S becomes the operation of multiplication by the idependent variable on H^2. We may therefore recruit the rich theory of analytic functions on $\mathbb{D}$ to help us understand the unilateral shift operator.

A simple account of the basic facts about H^2 is given in [214, chapter 13, pages 157–170]. Much fuller accounts are in [84, 116].

1.6 Invariant Subspaces of the Unilateral Shift

A highly original paper of Arne Beurling [48] in 1949 exploited the relation between the shift and M_z to answer a major open question in ergodic theory at the time. The problem was to describe the closed invariant subspaces of the unilateral shift operator defined in the last section. Here an *invariant subspace* of an operator $T \in \mathcal{B}(\mathcal{H})$ is a subspace $\mathcal{N}$ of $\mathcal{H}$ such that $T\mathcal{N} \subseteq \mathcal{N}$. Much effort

16   *Commutative Theory*

has been devoted to the identification of all the closed invariant subspaces of particular operators and, more generally, classes of operators. Indeed, it is still unknown whether *every* operator on a Hilbert space of dimension at least 2 has a non-trivial closed invariant subspace.

For motivation of the notion of invariant subspace, notice that, in finite dimensions, an invariant subspace of an operator T is just the span of a set of eigenvectors of T. In infinite dimensions the eigenvectors of an operator do still play a role, but since an operator on an infinite-dimensional space need have no eigenvectors at all, they cannot be so central as in finite dimensions. The notion of invariant subspace to some extent replaces that of eigenvector.

The use of analytic functions to describe invariant subspaces has been an extraordinarily fruitful theme in operator theory. Beurling constructed the general closed invariant subspace for M_z using "inner functions" on $\mathbb{D}$ and thereby described the closed invariant subspaces of S by means of the isomorphism L of equation (1.32).

We say that a function $\varphi \colon \mathbb{D} \to \mathbb{C}$ is *inner* on $\mathbb{D}$ if φ is bounded and analytic in $\mathbb{D}$ and $|\varphi(z)| = 1$ for almost every $z \in \mathbb{T}$ with respect to Lebesgue measure on $\mathbb{T}$. If φ is an inner function on $\mathbb{D}$, then one can associate with φ a subspace of H^2 by forming

$$\varphi \mathrm{H}^2 = \{\varphi f \colon f \in \mathrm{H}^2\}.$$

Theorem 1.33. (Beurling's theorem) *A subspace $\mathcal{N}$ of H^2 is a closed invariant subspace for M_z if and only if there exists an inner function φ on $\mathbb{D}$ such that $\mathcal{N} = \varphi \mathrm{H}^2$.*

As a corollary, Beurling proved that the cyclic functions for M_z are precisely the outer functions. His readable paper [48] caused many operator theorists to become interested in function theory on the disc. Reciprocally, many function theorists began to work on problems motivated by operator-theoretic considerations. There have been many extensions and refinements of Theorem 1.33, which have underlain the development of two substantial theories: the Sz.-Nagy-Foias model theory, which we outline in Section 1.11, and the Lax-Phillips scattering theory [141].

1.7 Von Neumann's Theory of Spectral Sets

In 1951 von Neumann initiated the theory of spectral sets in the paper [210]. After observing that for any normal operator $N \in \mathcal{B}(\mathcal{H})$ and $f \in \mathrm{Rat}(\sigma(N))$, by virtue of equation (1.22),

$$\|f(N)\| = \left\| \int_{\sigma(N)} f(z)\, dE(z) \right\| \leq \max_{z \in \sigma(N)} |f(z)|,$$

von Neumann introduced the following terminology.

The Origins of Operator-Theoretic Approaches to Function Theory 17

Definition 1.34. *Let $\mathcal{H}$ be a Hilbert space, let $T \in \mathcal{B}(\mathcal{H})$ and let K be a compact subset of $\mathbb{C}$. We say that K is a* spectral set *for T if $\sigma(T) \subseteq K$ and, for all $f \in \text{Rat}(K)$,*

$$\|f(T)\| \leq \max_{z \in K} |f(z)|. \tag{1.35}$$

Thus von Neumann's observation is equivalent to the assertion that $\sigma(N)$ is a spectral set for any normal operator N.

In his paper von Neumann went on to promulgate his famous inequality, whose influence on the subsequent development of operator theory has been profound.

Theorem 1.36. (Von Neumann's inequality) *If $\mathcal{H}$ is a Hilbert space, $T \in \mathcal{B}(\mathcal{H})$, and $\|T\| \leq 1$ then, for any $p \in \mathbb{C}[z]$,*

$$\|p(T)\| \leq \max_{|z| \leq 1} |p(z)|.$$

We prove the inequality in Section 1.9 by a different method.

Corollary 1.37. *If $\mathcal{H}$ is a Hilbert space, $T \in \mathcal{B}(\mathcal{H})$ and $\|T\| \leq 1$, then $\mathbb{D}^-$ is a spectral set for T.*

Proof. Fix $T \in \mathcal{B}(\mathcal{H})$ with $\|T\| \leq 1$ and let $f \in \text{Rat}(\mathbb{D}^-)$. Since f is holomorphic on a neighborhood of $\mathbb{D}^-$, it follows by a simple power series argument that there exists $r > 1$ and a sequence of polynomials $\{p_n\}$ such that p_n converges to f uniformly on $r\mathbb{D}$. But then by the continuity of the functional calculus,

$$\|f(T)\| = \|\lim_{n \to \infty} p_n(T)\| \leq \lim_{n \to \infty} \max_{|z| \leq 1} |p_n(z)| = \max_{|z| \leq 1} |f(z)|.$$

$\square$

There are numerous ways to generalize the notion of spectral sets to several variables. One idea is to juice up the inequality (1.35) to all functions f holomorphic on a neighborhood of K.

Lemma 1.38. *Let $T \in \mathcal{B}(\mathcal{H})$ and assume that $K \subseteq \mathbb{C}$ is compact. Then K is a spectral set for T if and only if $\sigma(T) \subseteq K$ and, for all f holomorphic in a neighborhood of K,*

$$\|f(T)\| \leq \max_{z \in K} |f(z)|. \tag{1.39}$$

Proof. The proof of necessity is immediate from the observation that the elements of $\text{Rat}(K)$ are holomorphic on a neighborhood of K.

To prove the converse, assume that $\sigma(T) \subseteq K$ and inequality (1.35) holds. Fix a neighborhood U of K and $f \in \text{Hol}(U)$. By Runge's theorem, there exists

18   *Commutative Theory*

a sequence of rational functions $\{f_n\}$ in $\mathrm{Rat}(U^-)$ such that f_n converges locally uniformly to f on U. By the continuity of the functional calculus,

$$\|f(T)\| = \|\lim_{n\to\infty} f_n(T)\| \le \lim_{n\to\infty} \max_{z\in K} |f_n(z)| = \max_{z\in K} |f(z)|.$$

$\square$

There are several natural candidates for the definition of a holomorphic function (possibly vector valued) of several variables; fortunately though, they all turn out to be equivalent. This issue is discussed in detail in [136, sections 0.2 and 1.2]. We shall adopt the following.

Definition 1.40. *Let $\mathcal{X}$ be a Banach space, let Ω be an open set in $\mathbb{C}^d$ for some positive integer d, and let $f \colon \Omega \to \mathcal{X}$ be a mapping. We say that f is differentiable at a point $z_0 \in \Omega$ if there exists a linear map $Df(z_0)\colon \mathbb{C}^d \to \mathcal{X}$ such that*

$$f(z) - f(z_0) = Df(z_0)(z - z_0) + o(\|z - z_0\|) \quad \text{as } z \to z_0, z \in \Omega.$$

The linear map $Df(z_0)$, when there is one, is unique and is called the derivative *or* Fréchet derivative *of f at z_0. It is also denoted by $f'(z_0)$. The mapping f is* holomorphic *on Ω if it is differentiable at every point of Ω.*

It is clear that if f is holomorphic then the function of one variable

$$\lambda \mapsto f(\lambda, z_2, z_3, \ldots, z_d)$$

is analytic in λ for any fixed values of $z_2, \ldots, z_d$, and likewise for each of the d co-ordinates. One can summarize this statement by saying that holomorphic maps are holomorphic in each variable separately. Remarkably enough, the converse statement is also true: any separately holomorphic map on an open set in $\mathbb{C}^d$ is holomorphic. This is a theorem of Hartogs, which caused a surprise when it was proved in 1906. Another criterion for a function to be holomorphic is that it have a convergent power series expansion in a neighborhood of every point.

Lemma 1.38 suggests the following definition of spectral sets in several variables. For a compact set K in $\mathbb{C}^d$, let $\mathrm{Hol}(K)$ denote the set of functions holomorphic on a neighborhood of K (the neighborhood can depend on the function).

Definition 1.41. *Let $\mathcal{H}$ be a Hilbert space, let T be a pairwise commuting d-tuple of operators on $\mathcal{H}$ and let K be a compact subset of $\mathbb{C}^d$. We say that K is a spectral set for T if $\sigma(T) \subseteq K$ and, for all $f \in \mathrm{Hol}(K)$,*

$$\|f(T)\| \le \max_{z\in K} |f(z)|. \tag{1.42}$$

The Origins of Operator-Theoretic Approaches to Function Theory 19

We give another version of the definition, valid for polynomially convex compact sets, which involves only the algebraic spectrum of T, as defined in Section 1.2.

Definition 1.43. *A compact subset K of $\mathbb{C}^d$ is said to be* polynomially convex *if, for every $\lambda \in \mathbb{C}^d \setminus K$, there exists a polynomial f in d variables such that $|f| \leq 1$ on K and $|f(\lambda)| > 1$. The* polynomial hull *of a bounded subset A of $\mathbb{C}^d$ is the intersection of all polynomially convex compact sets in $\mathbb{C}^d$ that contain A.*

An open set U in $\mathbb{C}^d$ is said to be polynomially convex *if, for every compact subset K of U, the polynomial hull of K is contained in U.*

Observe that the polynomial hull of a compact set is a compact polynomially convex set.

Definition 1.44. *Let $\mathcal{H}$ be a Hilbert space, let T be a pairwise commuting d-tuple of operators on $\mathcal{H}$ and let K be a polynomially convex compact subset of $\mathbb{C}^d$. We say that K is a* spectral set *for T if $\sigma_{\mathrm{alg}}(T) \subseteq K$ and, for all $f \in \mathrm{Hol}(K)$,*

$$\|f(T)\| \leq \max_{z \in K} |f(z)|. \qquad (1.45)$$

Just as long as $\sigma(T) \subseteq K$, Lemma 1.38 asserts that there is no difference between the conditions (1.35) and (1.39) in one variable. However, in several variables this statement is true if and only if K is rationally convex. With the exception of Chapter 9, we shall be dealing with functions on polynomially convex sets, in which case K is a spectral set for T if and only if inequality (1.39) holds for polynomials.

1.8 The Schur Class and Spectral Domains

For any open set U in $\mathbb{C}^d$, we denote by $\mathrm{H}^\infty(U)$ the Banach algebra of bounded holomorphic functions on U equipped with the norm $\|\ \|_U$ defined by

$$\|\varphi\|_U = \sup_{\lambda \in U} |\varphi(\lambda)| \qquad \text{for all } \varphi \in \mathrm{H}^\infty(U).$$

We define the *Schur class of U*, $\mathscr{S}(U)$, by

$$\mathscr{S}(U) = \mathrm{ball}\,\mathrm{H}^\infty(U).$$

Schur himself, in [184, 185], studied the class $\mathscr{S}(\mathbb{D})$ of analytic functions in $\mathbb{D}$ bounded in modulus by 1.

We shall adopt a slight change of perspective on the notion of spectral set described in the preceding section. Instead of fixing K and assuming that

20 *Commutative Theory*

$\sigma(T) \subseteq K$ and f is holomorphic on a neighborhood of K, let us fix an open set U and assume that $\sigma(T) \subseteq U$ and f is holomorphic on U. This leads to the following definition.

Definition 1.46. *If T is a d-tuple of pairwise commuting operators and U is an open set in $\mathbb{C}^d$, we say that U is a* spectral domain *for T if $\sigma(T) \subseteq U$ and, for all $\varphi \in \mathscr{S}(U)$,*

$$\|\varphi(T)\| \leq 1.$$

Thus K is a spectral set for T if and only if every neighborhood of K is a spectral domain for T. Lemma 1.38 is equivalent via a simple approximation argument to the assertion that $\mathbb{D}$ is a spectral domain for any contraction with spectrum in $\mathbb{D}$. Theorem 1.36 can be recast in the following form.

Theorem 1.47. (Schur form of von Neumann's inequality) *If $\mathcal{H}$ is a Hilbert space, $T \in \mathcal{B}(\mathcal{H})$, $\|T\| \leq 1$, and $\sigma(T) \subseteq \mathbb{D}$, then*

$$\varphi \in \mathscr{S}(\mathbb{D}) \implies \|\varphi(T)\| \leq 1.$$

We close this section with the formalization of a problem that remains one of the most fundamental in operator theory.

Problem 1.48. (The first Holy Grail of spectral set theory) Given a compact set K or a domain U in $\mathbb{C}^d$, find simple necessary and sufficient conditions on a tuple T for K or U to be a spectral set or spectral domain respectively for T.

How would success in this quest provide value to function theory? In proving his inequality, von Neumann found the first Holy Grail for $\mathbb{D}$, namely, $\|T\| \leq 1$. What does this mean in the simple case

$$T = \begin{bmatrix} 0 & 1 \\ 0 & 0 \end{bmatrix}?$$

Clearly, $\|T\| \leq 1$. Therefore, by the Schur form of von Neumann's inequality, if $\varphi \in \mathscr{S}(\mathbb{D})$, then $\|\varphi(T)\| \leq 1$. But if $\varphi \in \mathscr{S}(\mathbb{D})$ and $\varphi(0) = 0$, then

$$\varphi(T) = \begin{bmatrix} 0 & \varphi'(0) \\ 0 & 0 \end{bmatrix},$$

which obviously has norm equal to $|\varphi'(0)|$. Thus von Neumann's inequality in this simple case is the assertion that, for all $\varphi \in \mathscr{S}(\mathbb{D})$,

$$\varphi(0) = 0 \implies |\varphi'(0)| \leq 1.$$

This is the classical Schwarz lemma, which appears in every introductory course in complex analysis. We shall see many, much less trivial examples in these notes of how finding the first Holy Grail for a domain U leads to interesting function theory on U.

The Origins of Operator-Theoretic Approaches to Function Theory   21

1.9 The Sz.-Nagy Dilation Theorem

Two years after the appearance of von Neumann's inequality, in the 1953 paper [193] Béla Szőkefalvi-Nagy published his remarkable dilation theorem, which was the starting point for a major subfield of operator theory.

Definition 1.49. *If $\mathcal{H}$ is a Hilbert space and $T \in \mathcal{B}(\mathcal{H})$, a power dilation of T is an operator V on a Hilbert space $\mathcal{K} \supseteq \mathcal{H}$ such that $T^n = P_{\mathcal{H}} V^n | \mathcal{H}$ for all integers $n \geq 0$.*

Theorem 1.50. (Sz.-Nagy dilation theorem) *If $\mathcal{H}$ is a Hilbert space, $T \in \mathcal{B}(\mathcal{H})$, and $\|T\| \leq 1$, then T has a* unitary power dilation, *that is, it has a power dilation to a unitary operator on some superspace of $\mathcal{K}$.*

Sz.-Nagy's theorem gave much deeper insight into von Neumann's inequality than the original proof, which had used measure theory (which was not widely understood at the time) and a hefty dose of classical function theory.

1.9.1 Sz.-Nagy's Proof of von Neumann's Inequality

Let $\mathcal{H}$ be a Hilbert space, $T \in \mathcal{B}(\mathcal{H})$, and $\|T\| \leq 1$. By Sz.-Nagy's theorem, there exists a unitary power dilation U of T. As noted earlier, the spectral theorem implies that $\mathbb{T}$ is a spectral set for U, that is,

$$\|p(U)\| \leq \max_{\lambda \in \mathbb{T}} |p(\lambda)|$$

for all polynomials p. But by the maximum principle, if p is a polynomial, then $\max_{\lambda \in \mathbb{T}} |p(\lambda)| = \max_{\lambda \in \mathbb{D}^-} |p(\lambda)|$. Therefore,

$$\|p(U)\| \leq \max_{\lambda \in \mathbb{D}^-} |p(\lambda)|$$

for all polynomials p. By taking linear combinations of the power dilation conditions $T^n = P_{\mathcal{H}} U^n | \mathcal{H}$ for $n \geq 0$ one deduces that

$$p(T) = P_{\mathcal{H}} \, p(U) \, |\mathcal{H}$$

for all polynomials p. Therefore,

$$\|p(T)\| = \|P_{\mathcal{H}} \, p(U) \, |\mathcal{H}\| \leq \|p(U)\| \leq \max_{\lambda \in \mathbb{D}^-} |p(\lambda)|$$

for all polynomials p. $\qquad\qquad\square$

Thus dilations yield spectral sets. William Arveson in his 1969 article [36] and 1972 article [37] developed a converse to Sz.-Nagy's dilation theorem, one that has had numerous applications throughout operator theory and the theory of C^*-algebras. Arveson introduced a powerful variant of the idea of a

22 *Commutative Theory*

spectral set. In order to describe it, we observe that if A_{ij} is a bounded linear operator on $\mathcal{H}$ for $i, j = 1, \ldots, n$, then the operator matrix

$$A = \left[A_{ij} \right]_{i,j=1}^{n}$$

defines a bounded linear operator on $\mathcal{H}^n$ in a natural way, and therefore the operator norm $\|A\|$ is well defined.

We introduce the notation ∂X for the topological boundary of a subset X of $\mathbb{C}^d$.

Definition 1.51. *Let T be a d-tuple of pairwise commuting operators acting on a Hilbert space $\mathcal{H}$ and let K be a polynomially convex compact subset of $\mathbb{C}^d$ containing $\sigma_{\mathrm{alg}}(T)$. Then K is said to be a* complete spectral set *for T if, for each positive integer n and each $n \times n$ matrix of functions $f = [f_{ij}]$ analytic on a neighborhood of K,*

$$\| [f_{ij}(T)] \| \leq \max_{\lambda \in K} \| [f_{ij}(\lambda)] \|.$$

T is said to dilate to the boundary of K *if there exists a Hilbert space $\mathcal{K} \supseteq \mathcal{H}$ and a pairwise commuting d-tuple of normal operators N acting on $\mathcal{K}$ with $\sigma(N) \subseteq \partial K$ and such that*

$$f(T) = P_{\mathcal{H}} f(N) | \mathcal{H},$$

whenever f is holomorphic on a neighborhood of K.

By mimicking Sz.-Nagy's proof of von Neumann's inequality, one can easily show that, if T dilates to the boundary of K, then K is a complete spectral set for T. Arveson proved the converse of this statement [37].

Theorem 1.52. (Arveson's dilation theorem) *K is a complete spectral set for T if and only if T has a dilation to the boundary of K.*

Problem 1.53. (The second Holy Grail of spectral set theory) For which compact sets $K \subseteq \mathbb{C}^d$ is it true that K is a complete spectral set for T whenever K is a spectral set for T?

1.10 Andô's Dilation Theorem

In 1963 Tsuyoshi Andô [33] extended Sz.-Nagy's dilation theorem to pairs of commuting contractions.

Theorem 1.54. (Andô's dilation theorem) *If $\mathcal{H}$ is a Hilbert space and $T = (T^1, T^2)$ is a commuting pair of contractions acting on $\mathcal{H}$, then T has a unitary power dilation, that is, there exists a Hilbert space $\mathcal{K} \supseteq \mathcal{H}$ and a pair of commuting unitary operators $U = (U^1, U^2)$ acting on $\mathcal{K}$ such that*

The Origins of Operator-Theoretic Approaches to Function Theory 23

$$(T^1)^{n_1}(T^2)^{n_2} = P_{\mathcal{H}}\big((U^1)^{n_1}(U^2)^{n_2}\big)|\mathcal{H}$$

for all integers $n_1, n_2 \geq 0$.

For a proof, see, for example [195, theorem I.6.4] or [11, theorem 10.26]. Clearly, by the method of Sz.-Nagy's proof of von Neumann's inequality, Andô's dilation theorem has the following corollary.

Theorem 1.55. (Andô's inequality) *If $\mathcal{H}$ is a Hilbert space, $T = (T^1, T^2)$ is a commuting pair of contractions acting on $\mathcal{H}$ and $\sigma_{\mathrm{alg}}(T) \subseteq \mathbb{D}^2$, then*

$$\varphi \in \mathscr{S}(\mathbb{D}^2) \implies \|\varphi(T)\| \leq 1.$$

This beautiful theorem obtained both of the Holy Grails of spectral set theory simultaneously. One shows, as in Sz.-Nagy's proof of von Neumann's inequality, that $(\mathbb{D}^2)^-$ is a spectral set for a commuting pair T if and only if $\|T^1\|, \|T^2\| \leq 1$ (Holy Grail 1). Andô's dilation theorem also enables us to prove a result (Theorem 4.93) from which it is easy to deduce that $(\mathbb{D}^2)^-$ is a complete spectral set for T whenever $(\mathbb{D}^2)^-$ is a spectral set for T (Holy Grail 2) (see Corollary 4.108).

1.11 The Sz.-Nagy–Foias Model Theory

After the discovery of his dilation theorem, Sz.-Nagy worked with Ciprian Foias to exploit the theorem and thereby develop a model theory for contractions on Hilbert space through the use of function theory on the unit disc. In 1967 Sz.-Nagy and Foias published *Analyse harmonique des opérateurs de l'espace de Hilbert*, a self-contained treatment of dozens of their papers. This is certainly one of the most influential works ever published in operator theory. The first English-language edition was [194]; an updated version by H. Bercovici and L. Kérchy is [195].

The idea of the Sz.-Nagy–Foias model is to represent a general contraction on Hilbert space as (up to unitary equivalence) a "piece" of a concrete operator on a Hilbert space of functions on $\mathbb{T}$, to which classical *analyse harmonique* could be applied.

One of the highlights of their theory was the introduction of the *characteristic operator function of a contraction*. For T a contraction acting on a Hilbert space $\mathcal{H}$, let

$$\mathcal{D}_T = \left(\mathrm{ran}(1 - T^*T)^{\frac{1}{2}}\right)^- \quad \text{and} \quad \mathcal{D}_{T^*} = \left(\mathrm{ran}(1 - TT^*)^{\frac{1}{2}}\right)^-.$$

The characteristic operator function of T is the $\mathcal{B}(\mathcal{D}_T, \mathcal{D}_{T^*})$-valued function Θ_T defined on $\mathbb{D}$ by

$$\Theta_T(\lambda) = -T + \lambda(1 - TT^*)^{\frac{1}{2}}(1 - \lambda T^*)^{-1}(1 - T^*T)^{\frac{1}{2}}. \tag{1.56}$$

24 *Commutative Theory*

Θ_T is used to define the space of functions on $\mathbb{T}$ on which the "model" of T acts. The model itself is closely related to the shift operator.

Sz.-Nagy and Foias initiated a substantial program aimed at solving the invariant subspace problem by extending the classical factorization theory for scalar-valued Schur functions on the disc to operator-valued functions. Although this program has had some successes, the invariant subspace problem itself remains stubbornly open.

In 1974 J. William Helton [101] pointed out that the characteristic operator function of a contraction was a special case (albeit with a more refined structure) of the network realization formula widely used by electrical engineers to represent rational functions, typically transfer functions or impedance functions of linear systems or circuits. The formalism became even more important in the 1980s in the then newly developing branch of control theory now known as H^∞-control. The realization formula provides a representation of operator-valued Schur functions. For Hilbert spaces $\mathcal{H}, \mathcal{K}$, let $H^\infty_{\mathcal{B}(\mathcal{H},\mathcal{K})}(\mathbb{D})$ denote the Banach space of bounded analytic $\mathcal{B}(\mathcal{H},\mathcal{K})$-valued functions on $\mathbb{D}$ equipped with the norm

$$\|\varphi\|_{\mathbb{D}} = \sup_{z \in \mathbb{D}} \|\varphi(z)\|, \tag{1.57}$$

and then define $\mathscr{S}_{\mathcal{B}(\mathcal{H},\mathcal{K})}(\mathbb{D})$, the $\mathcal{B}(\mathcal{H},\mathcal{K})$-*valued Schur class on* $\mathbb{D}$, by

$$\mathscr{S}_{\mathcal{B}(\mathcal{H},\mathcal{K})}(\mathbb{D}) = \text{ball}\, H^\infty_{\mathcal{B}(\mathcal{H},\mathcal{K})}(\mathbb{D}).$$

Of course, when $\mathcal{K} = \mathcal{H}$, we abbreviate $H^\infty_{\mathcal{B}(\mathcal{H},\mathcal{K})}(\mathbb{D})$ to $H^\infty_{\mathcal{B}(\mathcal{H})}(\mathbb{D})$. We give a brief explanation of the idea of a network realization in Section 2.5.

Theorem 1.58. (Network realization formula) *A mapping* $\varphi \colon \mathbb{D} \to \mathcal{B}(\mathcal{H},\mathcal{K})$ *belongs to* $\mathscr{S}_{\mathcal{B}(\mathcal{H},\mathcal{K})}(\mathbb{D})$ *if and only if there exist a Hilbert space* $\mathcal{M}$ *and an isometry*

$$V = \begin{bmatrix} A & B \\ C & D \end{bmatrix} : \mathcal{H} \oplus \mathcal{M} \to \mathcal{K} \oplus \mathcal{M},$$

such that, for all $\lambda \in \mathbb{D}$,

$$\varphi(\lambda) = A + B\lambda(1 - D\lambda)^{-1}C. \tag{1.59}$$

Helton's seminal observation created new synergies between the operator-theoretic function theory and mathematical control theory communities that continue to the present.

Remarks 1.60.

(i) If $\mathcal{H} = \mathcal{K}$ and

$$A = -T, \quad B = (1 - TT^*)^{\frac{1}{2}}, \quad C = (1 - T^*T)^{\frac{1}{2}}, \quad \text{and} \quad D = T^*,$$

The Origins of Operator-Theoretic Approaches to Function Theory 25

then equation (1.59) becomes equation (1.56), and V is not just isometric, it is unitary.

(ii) In the special case when $\dim \mathcal{H} = \dim \mathcal{K} < \infty$, $\dim \mathcal{M} < \infty$, and φ is defined by equation (1.59), the condition that V be isometric implies that φ is a rational inner function, that is, $\varphi(\tau)$ is unitary for all $\tau \in \mathbb{T}$ and $\det \varphi$ is a scalar rational inner function (a finite Blaschke product). Conversely, if $\varphi \in H^{\infty}_{\mathcal{B}(\mathcal{H},\mathcal{K})}$ is a rational inner function, then there exists $\mathcal{M}$ with $\dim \mathcal{M} < \infty$ and an isometric V such that equation (1.59) holds.

(iii) In the case when $\mathcal{H} = \mathbb{C}$, we may assume that

$$V = \begin{bmatrix} A & B \\ C & D \end{bmatrix} = \begin{bmatrix} a & 1 \otimes \beta \\ \gamma \otimes 1 & D \end{bmatrix},$$

where $a \in \mathbb{C}$, $\beta \in \mathcal{M}$, $\gamma \in \mathcal{M}$, and $D \in \mathcal{B}(\mathcal{M})$. In this case the realization formula (1.59) becomes

$$\varphi(\lambda) = a + \lambda \big\langle (1 - D\lambda)^{-1}\gamma, \beta \big\rangle_{\mathcal{M}}.$$

1.12 The Sarason Interpolation Theorem

In 1967 Donald Sarason added another ingredient to the mix with the influential paper "Generalized Interpolation in H$^\infty$" [180]. In it he exploited yet another way to think about Schur-class functions from an operator-theoretic perspective.

For $\varphi \in H^{\infty}(\mathbb{D})$, we denote by M_φ the *multiplication operator*, defined on H^2 by the formula

$$M_\varphi f = \varphi f \qquad \text{for } f \in H^2. \tag{1.61}$$

The following lemma asserts that this formula does indeed define a bounded operator.

Lemma 1.62. *If $\varphi \in H^{\infty}(\mathbb{D})$ and M_φ is defined by equation (1.61), then $M_\varphi f \in H^2$ whenever $f \in H^2$. Furthermore, M_φ (as an operator acting on H^2) has norm less than or equal to $\|\varphi\|_{\mathbb{D}}$.*

Proof. By equation (1.28), if $f \in H^2$, then

$$\|M_\varphi f\|^2 = \lim_{r \to 1-} \int |\varphi(re^{it}) f(re^{it})|^2 \frac{dt}{2\pi}$$

$$\leq \|\varphi\|^2_{\mathbb{D}} \lim_{r \to 1-} \int |f(re^{it})|^2 \frac{dt}{2\pi}$$

$$= \|\varphi\|^2_{\mathbb{D}} \|f\|^2_{H^2}.$$

Therefore, $\|M_\varphi\| \leq \|\varphi\|_{\mathbb{D}}$. $\qquad \square$

26 Commutative Theory

Lemma 1.62 suggests the following natural question: given $T \in \mathcal{B}(\mathrm{H}^2)$, when is $T = M_\varphi$ for some $\varphi \in \mathrm{H}^\infty(\mathbb{D})$?

Theorem 1.63. *Let* $T \colon \mathcal{B}(\mathcal{H}) \to \mathcal{B}(\mathcal{H})$ *be a function. Then*

$$T = M_\varphi \qquad \text{for some } \varphi \in \mathrm{H}^\infty(\mathbb{D}) \tag{1.64}$$

if and only if $T \in \mathcal{B}(\mathrm{H}^2)$ *and* T *commutes with* M_z. *Furthermore, when equation* (1.64) *holds,*

$$\|T\| = \|\varphi\|_\mathbb{D}.$$

In particular, $\mathscr{S}(\mathbb{D})$ *corresponds naturally to the set of contractions on* H^2 *that commute with* M_z.

An illuminating proof of Theorem 1.63 relies on the fact that H^2 is a reproducing kernel Hilbert space and exploits a property of the reproducing kernel functions s_λ: they are eigenvectors of all adjoints of multiplication operators.

Lemma 1.65. *If* $\varphi \in \mathrm{H}^\infty(\mathbb{D})$ *and* $\lambda \in \mathbb{D}$, *then*

$$M_\varphi^* s_\lambda = \overline{\varphi(\lambda)} \, s_\lambda. \tag{1.66}$$

Proof. Fix $\varphi \in \mathrm{H}^\infty(\mathbb{D})$ and $\lambda \in \mathbb{D}$. If $f \in \mathrm{H}^2$, then

$$
\begin{aligned}
\langle f, M_\varphi^* s_\lambda \rangle &= \langle M_\varphi f, s_\lambda \rangle \\
&= \langle \varphi f, s_\lambda \rangle \\
&= \varphi(\lambda) f(\lambda) \\
&= \varphi(\lambda) \langle f, s_\lambda \rangle \\
&= \left\langle f, \overline{\varphi(\lambda)} \, s_\lambda \right\rangle.
\end{aligned}
$$

Since f is arbitrary, it follows that equation (1.66) holds. $\square$

Furthermore, the reproducing kernel functions s_λ are the *only* eigenvectors for all adjoints of multiplication operators.

Lemma 1.67. *If* $\lambda \in \mathbb{D}$, *then*

$$\ker(\bar{\lambda} - M_z^*) = \mathbb{C} s_\lambda.$$

Proof. By Lemma 1.65, $M_\lambda^* s_\lambda = \bar{\lambda} s_\lambda$. Therefore, $\ker(\bar{\lambda} - M_z^*) \supseteq \mathbb{C} s_\lambda$. To prove the reverse inclusion, we show that if $f \in \mathrm{H}^2$ and $f \perp \mathbb{C} s_\lambda$, then $f \perp \ker(\bar{\lambda} - M_z^*)$. But if $f \in \mathrm{H}^2$ and $f \perp \mathbb{C} s_\lambda$, then $f(\lambda) = 0$. Hence there exists a holomorphic function g on $\mathbb{D}$ such that $f(z) = (\lambda - z)g(z)$. By the criterion (1.27), $g \in \mathrm{H}^2$. This implies that $f \in \mathrm{ran}(\lambda - M_z)$, or equivalently, $f \perp \ker(\bar{\lambda} - M_z^*)$. $\square$

The Origins of Operator-Theoretic Approaches to Function Theory 27

Armed with these lemmas, one can easily prove Theorem 1.63.

Proof. First assume that $\varphi \in H^\infty(\mathbb{D})$ and $T = M_\varphi$. Lemma 1.62 implies that $T \in \mathcal{B}(H^2)$. Also, if $f \in H^2$ and $z \in \mathbb{D}$, then

$$((T M_z)f)(z) = \varphi(z)zf(z) = z\varphi(z)f(z) = ((M_z T)f)(z),$$

that is, T commutes with M_z.

Conversely, assume that $T \in \mathcal{B}(H^2)$ and $T M_z = M_z T$. If $\lambda \in \mathbb{D}$, then Lemma 1.65 implies that

$$M_z^*(T^* s_\lambda) = T^*(M_z^* s_\lambda) = \bar{\lambda} T^* s_\lambda.$$

Consequently, Lemma 1.67 implies that for each $\lambda \in \mathbb{D}$, there exists $\varphi(\lambda) \in \mathbb{C}$ such that

$$T^* s_\lambda = \overline{\varphi(\lambda)} s_\lambda. \tag{1.68}$$

We claim that $\varphi \in H^\infty(\mathbb{D})$. First observe that if $f \in H^2$ and $\lambda \in \mathbb{D}$,

$$(Tf)(\lambda) = \langle Tf, s_\lambda \rangle = \langle f, T^* s_\lambda \rangle = \left\langle f, \overline{\varphi(\lambda)} s_\lambda \right\rangle = \varphi(\lambda) \langle f, s_\lambda \rangle$$

$$= \varphi(\lambda) f(\lambda). \tag{1.69}$$

In particular, if we let f be the constant function 1, we see that $\varphi = T1 \in H^2$, so that φ is analytic in $\mathbb{D}$. To see that φ is bounded, note that equation (1.68) implies that if $\lambda \in \mathbb{C}$, then $|\varphi(\lambda)| \leq \|T^*\| = \|T\|$. Hence $\varphi \in H^\infty$ and

$$\|\varphi\|_\mathbb{D} \leq \|T\|. \tag{1.70}$$

There remains to show that $\|T\| = \|\varphi\|_\mathbb{D}$. This follows from the inequality (1.70) and the observation that Lemma 1.62 implies that $\|T\| \leq \|\varphi\|_\mathbb{D}$. This concludes the proof of Theorem 1.63. $\qquad\square$

In [180], Sarason proved a generalization of Theorem 1.63, Theorem 1.75. If $\mathcal{H}$ is a Hilbert space and E is a subset of $\mathcal{H}$, we denote by $\mathcal{H} \ominus E$ the orthogonal complement of E in $\mathcal{H}$:

$$\mathcal{H} \ominus E \overset{\text{def}}{=} \{x \in \mathcal{H} \colon \langle x, y \rangle = 0 \text{ for all } y \in E\}.$$

When the space $\mathcal{H}$ is understood, it is customary to write $E^\perp$ instead of $\mathcal{H} \ominus E$. Sarason's work involved the case that $\mathcal{H} = H^2$ and $E = \varphi H^2$ for some inner function φ.[5]

The following lemma gives a concrete description of $H^2 \ominus \varphi H^2$ in the case that φ is a finite Blaschke product.

---

[5] So that, by Beurling's theorem, E is the general M_z-invariant subspace of H^2.

28 *Commutative Theory*

Lemma 1.71. *If B is the rational inner function*

$$B(z) = \prod_{j=1}^{n} \frac{z - \lambda_j}{1 - \bar{\lambda}_j z}, \tag{1.72}$$

where $\lambda_1, \ldots, \lambda_n$ are distinct points in $\mathbb{D}$, then

$$\mathrm{H}^2 \ominus B\mathrm{H}^2 = \mathrm{span}\{s_{\lambda_1}, s_{\lambda_2}, \ldots, s_{\lambda_n}\}. \tag{1.73}$$

Proof. For each j, by the reproducing property of the Szegő kernel, for any $f \in \mathrm{H}^2$,

$$\langle Bf, s_{\lambda_j} \rangle_{\mathrm{H}^2} = B(\lambda_j) f(\lambda_j) = 0.$$

Thus $s_{\lambda_j} \in \mathrm{H}^2 \ominus B\mathrm{H}^2$, and so

$$\mathrm{span}\{s_{\lambda_1}, s_{\lambda_2}, \ldots, s_{\lambda_n}\} \subseteq \mathrm{H}^2 \ominus B\mathrm{H}^2. \tag{1.74}$$

To prove the reverse inclusion, consider any function $f \in \mathrm{H}^2$ orthogonal to the left-hand side of the relation (1.74). Then $\langle f, s_{\lambda_j} \rangle = 0$ for each j, which is to say that f vanishes at all the points $\lambda_1, \ldots, \lambda_n$. It follows that the function

$$g(z) \overset{\text{def}}{=} \frac{f(z)}{B(z)} = \frac{(1 - \bar{\lambda}_1 z) \ldots (1 - \bar{\lambda}_n z) f(z)}{(z - \lambda_1) \ldots (z - \lambda_n)}$$

also belongs to H^2. Thus $f = Bg \in B\mathrm{H}^2$. Hence

$$\mathrm{span}\{s_{\lambda_1}, s_{\lambda_2}, \ldots, s_{\lambda_n}\}^{\perp} \subseteq B\mathrm{H}^2,$$

and therefore equation (1.73) holds. $\qquad\square$

Theorem 1.75. (Sarason's theorem) *If φ is an inner function, $\mathcal{M} = \mathrm{H}^2 \ominus \varphi \mathrm{H}^2$, and T is a bounded operator on $\mathcal{M}$, then T commutes with $P_{\mathcal{M}} M_z | \mathcal{M}$ if and only if there exists $\psi \in \mathrm{H}^{\infty}$ such that*

$$T = P_{\mathcal{M}} M_{\psi} | \mathcal{M}.$$

Furthermore, ψ may be chosen with $\|\psi\|_{\infty} = \|T\|$.

Sarason showed that his theorem encodes, unifies, and extends classical interpolation and moment theorems on the disc. We illustrate Sarason's idea by giving a proof in his style of the classical Pick interpolation theorem. To this end we need two lemmas; the first one is proved just like Lemma 1.65.

Lemma 1.76. *If B is as in Lemma 1.71, $\mathcal{M} = \mathrm{H}^2 \ominus B\mathrm{H}^2$ and $\psi \in \mathrm{H}^{\infty}$, then*

$$\left(P_{\mathcal{M}} M_{\psi} | \mathcal{M}\right)^* s_{\lambda_i} = \overline{\psi(\lambda_i)} \, s_{\lambda_i} \qquad \text{for } i = 1, \ldots, n. \tag{1.77}$$

The Origins of Operator-Theoretic Approaches to Function Theory 29

Lemma 1.78. *If B, $\mathcal{M}$, and ψ are as in Lemma 1.76, then*

$$\| P_{\mathcal{M}} M_\psi | \mathcal{M} \| \leq 1$$

if and only if

$$\left[\frac{1 - \overline{\psi(\lambda_i)} \psi(\lambda_j)}{1 - \overline{\lambda_i} \lambda_j} \right] \geq 0. \tag{1.79}$$

Proof. Let $T = P_{\mathcal{M}} M_\psi | \mathcal{M}$, an operator on $\mathcal{M}$. By Lemma 1.71, $\mathcal{M}$ is the span of $s_{\lambda_1}, \ldots, s_{\lambda_n}$. Since $\|T^*\| = \|T\|$, it follows from Proposition 1.12 that $\|T\| \leq 1$ if and only if $1 - TT^* \geq 0$, which is to say that

$$\left\langle (1 - TT^*) \sum_i c_i s_{\lambda_i}, \sum_j c_j s_{\lambda_j} \right\rangle \geq 0 \quad \text{for all } c_1, \ldots, c_n \in \mathbb{C},$$

or equivalently,

$$\left\langle \sum_i c_i s_{\lambda_i}, \sum_j c_j s_{\lambda_j} \right\rangle - \left\langle T^* \sum_i c_i s_{\lambda_i}, T^* \sum_j c_j s_{\lambda_j} \right\rangle \geq 0 \quad \text{for all } c_1, \ldots, c_n \in \mathbb{C}. \tag{1.80}$$

By Lemma 1.76,

$$T^* \sum_i c_i s_{\lambda_i} = \sum_i c_i \overline{\psi(\lambda_i)} s_{\lambda_i}.$$

By the reproducing property (1.29) of the Szegő kernel,

$$\langle s_{\lambda_i}, s_{\lambda_j} \rangle = s_{\lambda_i}(\lambda_j) = \frac{1}{1 - \overline{\lambda_i} \lambda_j}.$$

These facts imply that inequality (1.80) is equivalent to the statement

$$\sum_{i,j} \frac{1 - \overline{\psi(\lambda_i)} \psi(\lambda_j)}{1 - \overline{\lambda_i} \lambda_j} c_i \overline{c_j} \geq 0 \quad \text{for all } c_1, \ldots, c_n \in \mathbb{C}.$$

This statement is equivalent to the condition (1.79), and so the lemma is proved. $\square$

The following theorem of Georg Pick was proved in 1916 and impacted the development of function theory throughout the twentieth century. We shall give a complete proof of the theorem in Theorem 2.54. Here, we will show how it can be deduced from Sarason's theorem, Theorem 1.75.

Theorem 1.81. (Pick interpolation theorem) *Let $\lambda_1, \ldots, \lambda_n$ be distinct points in $\mathbb{D}$ and let $z_1, \ldots, z_n \in \mathbb{C}$. There exists $\varphi \in \mathscr{S}(\mathbb{D})$ such that $\varphi(\lambda_i) = z_i$ for $i = 1, \ldots, n$ if and only if the $n \times n$ matrix P defined by*

30   *Commutative Theory*

$$P = \left[\frac{1 - \bar{z}_i z_j}{1 - \bar{\lambda}_i \lambda_j} \right]$$

is positive semi-definite.

Proof. First assume that $\varphi \in \mathscr{S}(\mathbb{D})$ and $\varphi(\lambda_i) = z_i$ for $i = 1,\ldots,n$. By Theorem 1.63, $\|M_\varphi\| \leq 1$. Hence

$$\| P_{\mathcal{M}} M_\varphi | \mathcal{M} \| \leq \| M_\varphi \| \leq 1.$$

Lemma 1.78 now implies that $P \geq 0$.

Conversely, assume that $P \geq 0$. Choose a polynomial p satisfying $p(\lambda_i) = z_i, i = 1,\ldots,n$. As $P \geq 0$, Lemma 1.78 implies that $\| P_{\mathcal{M}} M_p | \mathcal{M} \| \leq 1$. But then Sarason's theorem, Theorem 1.75, applied with $T = P_{\mathcal{M}} M_p | \mathcal{M}$ implies that there exists $\varphi \in \mathscr{S}(\mathbb{D})$ with

$$P_{\mathcal{M}} M_p | \mathcal{M} = P_{\mathcal{M}} M_\varphi | \mathcal{M}.$$

Taking adjoints in this equation and applying each side to s_{λ_i}, we come, with the aid of Lemma 1.76, to the conclusion that $\varphi(\lambda_i) = p(\lambda_i) = z_i$ for each i. $\square$

1.13 Historical Notes

Section 1.3. Hilbert space, in the sense of a complete inner product space, as it is now understood, was introduced in 1951 by J. von Neumann in [210]. Von Neumann required a Hilbert space to be separable and infinite-dimensional, but subsequent authors have not followed him in this. Hilbert's own work that inspired the name [115] made use of the space now known as ℓ^2, sometimes called *Hilbert co-ordinate space* [172, section 82].

Section 1.4, the spectral theorem. Hilbert proved the first version of the spectral theorem in 1906 [115]. In modern terminology, it was for compact self-adjoint operators on ℓ^2. Too many mathematicians to mention have contributed to the subsequent development of spectral theory, but prominent among them in the early years were von Neumann and Marshall Stone (see Dieudonné [72, chapter VII]). Independently, around 1930, von Neumann and Stone proved a spectral theorem for general self-adjoint (possibly unbounded) operators. Stone's book [191] remained the reference book on the subject for many years. See Section 6.4 for some remarks on Stone's proof.

Section 1.6, Beurling's theorem. Original as it was, Beurling's theorem did have precursors. Two results of V. I. Smirnov [188, 189] in 1929 and 1932 together imply Beurling's theorem. However, in the words of Nikolaĭ Nikol'skiĭ and Vasyunin in an interesting discussion [157, Introduction],

The Origins of Operator-Theoretic Approaches to Function Theory 31

"Unfortunately, in Smirnov's time the very notion of a bounded operator was not completely formulated." Peter Lax [139] generalized Beurling's theorem to H^2 spaces of vector-valued functions in the course of applying the ideas to the important topic of scattering theory.

Section 1.9, the Nagy dilation theorem. A precursor of this result was due to M. A. Naimark [154]. He proved that, roughly speaking, a countably additive function on Borel sets, taking values in the set of positive operators in $\mathcal{B}(\mathcal{H})$, can be lifted to a spectral measure with values in $\mathcal{B}(\mathcal{K})$ for some superspace $\mathcal{K}$ of $\mathcal{H}$. Sz.-Nagy used Naimark's result in his original proof of his (Nagy's) dilation theorem. The first chapter of Sz.-Nagy and Foias's book [194] contains several proofs of the Nagy dilation theorem.

Section 1.11, models of operators. The ideas of model operators and characteristic operator functions seem to be due in the first place to M. S. Livšic, M. S. Brodskiĭ, and Yu. L. Šmulyan. Sz.-Nagy and Foias themselves, in the foreword to the original French edition of their book, given in English translation in [194], wrote "The recent rapid progress on this field was stimulated largely by work of mathematicians in the USSR (M. G. Krein, M. S. Livšic, M. S. Brodskiĭ etc) and in the USA (N. Wiener, H. Helson, D. Lowdenslager, P. Masani etc)." See also their notes to [194, chapter VI] for references and more history. A parallel line of development was the theory of *de Branges spaces* (de Branges and Rovnyak [69]), which played a part in the proof by de Branges of the long-standing Bieberbach conjecture [68].

The network realization formula, Theorem 1.58. Since the 1960s the standard way for an engineer to represent a rational function has been by a realization formula. Engineers prefer the notation

$$\varphi(z) = D + Cz(1 - Az)^{-1}B$$

for discrete-time systems and

$$\varphi(s) = D + Cs(1 - As)^{-1}B$$

for continuous-time systems. Here A, B, C, D are finite matrices. For computational purposes the realization formula provides greater flexibility than representation by the coefficients of the numerator and denominator of a rational function. Fifty years of experience of computation is reflected in many fast, efficient, and easily-used routines in engineering packages such as Matlab.

In the 1970s, the concern of engineers was more algebraic than analytic. According to Helton [101], "Mathematically speaking, the theory developed [by engineers] is a canonical models theory for classifying operators up to similarity rather than unitary equivalence as in Nagy and Foias." The connection between the operator norm of the $ABCD$ matrix and the supremum

32   *Commutative Theory*

norm of φ (as contained in [194]) became important for engineers with the development of the theory of H^∞ control in the 1980s.

Section 1.12. Sarason's theorem has been generalized by many authors. In particular, the theories of *commutant lifting* and *contractive intertwining dilations*, due principally to Ciprian Foias and the Romanian school [90], are widely studied tools in operator theory that derived from Sarason's insights. Independently, but at about the same time as Sarason's great paper [180], Vadym Movsesovich Adamyan, Damir Zyamovich Arov, and Mark Grigorievich Krein, the leaders of the Odessa school of analysis, developed a theory based on Hankel operators, which produced many function-theoretic results by operator-theoretic means. They wrote a series of profound papers, including [3, 4], which covered much of the same territory as Sarason's paper and in some ways went further.

2

Operator Analysis on $\mathbb{D}$: Model Formulas, Lurking Isometries, and Positivity Arguments

2.1 Overview

In this chapter we introduce the concept of a *model formula* for a Schur-class function defined on $\mathbb{D}$ (see Definition 2.7). A model formula represents the general element in a class of holomorphic functions (in this case $\mathscr{S}(\mathbb{D})$) in terms of a Hilbert space structure. In consequence, model formulas permit the use of Hilbert space geometry and operator theory in the study of holomorphic functions.

We introduce two other tools of special power in the operator analysis toolbox: *lurking isometry* and *positivity* arguments. We illustrate their efficacy by deriving four theorems from the previous chapter: von Neumann's inequality, the Sz.-Nagy dilation theorem, the network realization formula, and the Pick interpolation theorem.

Unlike the classical proofs of these theorems, briefly described in Chapter 1, the approaches we present here using model formulas, lurking isometries, and positivity arguments are primarily algebraic in nature. As we shall see in subsequent chapters, they can be modified in simple ways for both the discovery and the proof of new theorems about holomorphic functions of more than one variable.

2.2 A Model Formula for $\mathscr{S}(\mathbb{D})$

Recall that the Schur class of $\mathbb{D}$, which we denote by $\mathscr{S}(\mathbb{D})$, is the set of holomorphic functions on $\mathbb{D}$ that are bounded by 1 in modulus. The model formula for Schur class functions on $\mathbb{D}$, which we shall shortly derive, is an immediate corollary of Lemma 2.2 and Theorem 2.5. The first lemma requires the notion of a *positive semi-definite kernel*. A *kernel* on a set Ω is a complex-valued function on $\Omega \times \Omega$.

34 — Commutative Theory

Definition 2.1. *If Ω is a set, and A is a kernel on Ω, then we say that A is a positive semi-definite kernel on Ω if for all $n \geq 1$ and all choices of points $\lambda_1, \lambda_2, \ldots, \lambda_n \in \Omega$, the $n \times n$ matrix $[A(\lambda_j, \lambda_i)]$ is positive semi-definite.*

The following is one of the main results in Pick's seminal paper [163], although he did not express it in the language of kernels.

Lemma 2.2. (Pick's lemma) *If $\varphi \in \mathscr{S}(\mathbb{D})$, then the function A defined on $\mathbb{D} \times \mathbb{D}$ by the formula*

$$A(\lambda, \mu) = \frac{1 - \overline{\varphi(\mu)}\varphi(\lambda)}{1 - \bar{\mu}\lambda} \tag{2.3}$$

is a positive semi-definite kernel on $\mathbb{D}$.

There are many proofs of Pick's lemma. Here is one in the spirit of operator analysis; it uses the Szegő kernel function defined in equation (1.30) and the "fundamental fact," Proposition 1.12.

Proof. Let $\varphi \in \mathscr{S}(\mathbb{D})$. By Theorem 1.63, $\|M_\varphi\| \leq 1$. Therefore, as $\|M_\varphi^*\| = \|M_\varphi\|$, the fundamental fact implies that $1 - M_\varphi M_\varphi^* \geq 0$.

We wish to show that A is positive semi-definite on $\mathbb{D}$. Accordingly, fix $n \geq 1$ and $\lambda_1, \lambda_2, \ldots, \lambda_n \in \mathbb{D}$. We need to show that the $n \times n$ matrix $[A(\lambda_j, \lambda_i)]$ is positive semi-definite. Accordingly, consider $c_1, \ldots, c_n \in \mathbb{C}$. Then

$$
\begin{aligned}
\sum_{i,j=1}^n A(\lambda_j, \lambda_i) c_j \bar{c}_i &= \sum_{i,j=1}^n \frac{1 - \overline{\varphi(\lambda_i)}\varphi(\lambda_j)}{1 - \bar{\lambda}_i \lambda_j} c_j \bar{c}_i \\
&= \sum_{i,j=1}^n \left(1 - \overline{\varphi(\lambda_i)}\varphi(\lambda_j)\right) \langle s_{\lambda_i}, s_{\lambda_j} \rangle c_j \bar{c}_i \\
&= \sum_{i,j=1}^n \langle s_{\lambda_i}, s_{\lambda_j} \rangle c_j \bar{c}_i - \sum_{i,j=1}^n \overline{\varphi(\lambda_i)}\varphi(\lambda_j) \langle s_{\lambda_i}, s_{\lambda_j} \rangle c_j \bar{c}_i \\
&= \sum_{i,j=1}^n \langle s_{\lambda_i}, s_{\lambda_j} \rangle c_j \bar{c}_i - \sum_{i,j=1}^n \langle \overline{\varphi(\lambda_i)} s_{\lambda_i}, \overline{\varphi(\lambda_j)} s_{\lambda_j} \rangle c_j \bar{c}_i.
\end{aligned}
$$

By Lemma 1.65,

$$\overline{\varphi(\lambda_i)} s_{\lambda_i} = M_\varphi^* s_{\lambda_i},$$

and so

$$\sum_{i,j=1}^{n} A(\lambda_j, \lambda_i) c_j \bar{c}_i = \sum_{i,j=1}^{n} \langle s_{\lambda_i}, s_{\lambda_j} \rangle c_j \bar{c}_i - \sum_{i,j=1}^{n} \langle M_\varphi^* s_{\lambda_i}, M_\varphi^* s_{\lambda_j} \rangle c_j \bar{c}_i$$

$$= \left\langle (1 - M_\varphi M_\varphi^*) \sum_{i=1}^{n} \bar{c}_i s_{\lambda_i}, \sum_{i=1}^{n} \bar{c}_i s_{\lambda_i} \right\rangle$$

$$\geq 0.$$

As the last inequality holds for all $c_1, \ldots, c_n$, it follows that $[A(\lambda_j, \lambda_i)]$ is positive semi-definite. $\qquad\square$

Recall that if $A = [a_{ij}]$ is an $n \times n$ matrix with entries in $\mathbb{C}$, then a corollary of the Cholesky decomposition from linear algebra is that A is positive semi-definite if and only if A can be represented as the gramian matrix of n vectors in a Hilbert space. Here the *gramian* of a sequence $u_1, \ldots, u_n$ of vectors in a Hilbert space $\mathcal{M}$ is the $n \times n$ matrix $[a_{ij}]$ where

$$a_{ij} = \langle u_j, u_i \rangle_{\mathcal{M}} \qquad \text{for } i, j = 1, \ldots, n. \tag{2.4}$$

Furthermore, we may choose $\mathcal{M}$ to satisfy $\dim \mathcal{M} = \operatorname{rank} A$.

The second step in our derivation of a model formula for $\mathscr{S}(\mathbb{D})$ is due to E. H. Moore [152]. A substantial generalization of the representation given in equation (2.4), it gives a representation of positive semi-definite kernels in terms of Hilbert space.

Theorem 2.5. (Moore's theorem) *If Ω is a set and A is a kernel on Ω, then A is positive semi-definite on Ω if and only if there exists a Hilbert space $\mathcal{M}$ and a function $u \colon \Omega \to \mathcal{M}$ satisfying*

$$A(\lambda, \mu) = \langle u(\lambda), u(\mu) \rangle_{\mathcal{M}} \qquad \text{for } \lambda, \mu \in \Omega. \tag{2.6}$$

Furthermore, if A is positive semi-definite, then we may choose $\mathcal{M}$ and u satisfying the additional property that $\{u(\lambda) \colon \lambda \in \Omega\}$ has dense linear span in $\mathcal{M}$.

Proof. First assume that A is positive semi-definite. For each fixed $\lambda \in \Omega$ define a function $A_\lambda \colon \Omega \to \mathbb{C}$ by the formula

$$A_\lambda(\mu) = A(\lambda, \mu) \qquad \text{for } \mu \in \Omega,$$

and let $\mathcal{V}$ denote the vector space of all functions F on Ω that have the form

$$F = \sum_{j=1}^{n} c_j A_{\lambda_j},$$

36   *Commutative Theory*

where n is a positive integer, $c_1, \ldots, c_n \in \mathbb{C}$, and $\lambda_1, \ldots, \lambda_n \in \Omega$. Define a sesquilinear form $\langle \cdot, \cdot \rangle_{\mathcal{V}}$ on $\mathcal{V}$ by the formula

$$\left\langle \sum_{j=1}^{n} c_j A_{\lambda_j}, \sum_{i=1}^{m} d_i A_{\mu_i} \right\rangle_{\mathcal{V}} = \sum_{j=1}^{n} \sum_{i=1}^{m} c_j \bar{d}_i A(\lambda_j, \mu_i).$$

Since A is assumed to be positive semi-definite, so also is $\langle \cdot, \cdot \rangle_{\mathcal{V}}$. Let $\mathcal{M}_0$ be the vector space $\mathcal{V}/\mathcal{I}$ where $\mathcal{I} = \{F \in \mathcal{V} : \langle F, F \rangle_{\mathcal{V}} = 0\}$; then $\langle \cdot, \cdot \rangle_{\mathcal{V}}$ gives rise to a well defined inner product $\langle \cdot, \cdot \rangle_{\mathcal{M}_0}$ on $\mathcal{M}_0$ via the formula

$$\langle F + \mathcal{I}, G + \mathcal{I} \rangle_{\mathcal{M}_0} = \langle F, G \rangle_{\mathcal{V}}.$$

Finally, let $\mathcal{M}$ denote the completion of $\mathcal{M}_0$ and let $\langle \cdot, \cdot \rangle_{\mathcal{M}}$ denote the inner product induced on $\mathcal{M}$ by $\langle \cdot, \cdot \rangle_{\mathcal{M}_0}$.

With the constructions in the previous paragraph we have

$$A(\lambda, \mu) = \langle A_\lambda, A_\mu \rangle_{\mathcal{V}} = \langle A_\lambda + \mathcal{I}, A_\mu + \mathcal{I} \rangle_{\mathcal{M}_0} = \langle A_\lambda + \mathcal{I}, A_\mu + \mathcal{I} \rangle_{\mathcal{M}}$$

for all $\lambda, \mu \in \Omega$. Therefore, equation (2.6) holds if we define $u \colon \Omega \to \mathcal{M}$ by the formula

$$u(\lambda) = A_\lambda + \mathcal{I} \quad \text{for} \quad \lambda \in \Omega.$$

Also, since by construction, $\{A_\lambda + \mathcal{I} \colon \lambda \in \Omega\}$ has dense linear span in $\mathcal{M}$, $\{u(\lambda) \colon \lambda \in \Omega\}$ has dense linear span in $\mathcal{M}$.

Conversely, assume that $u \colon \Omega \to \mathcal{M}$ and equation (2.6) holds. To see that A is positive semi-definite, fix a positive integer n, $c_1, \ldots, c_n \in \mathbb{C}$, and $\lambda_1, \ldots, \lambda_n \in \Omega$. Then

$$\sum_{i,j=1}^{n} A(\lambda_j, \lambda_i) c_j \bar{c}_i = \sum_{i,j=1}^{n} \langle u(\lambda_j), u(\lambda_i) \rangle_{\mathcal{M}} \, c_j \bar{c}_i$$

$$= \left\langle \sum_{j=1}^{n} c_j u(\lambda_j), \sum_{i=1}^{n} c_i u(\lambda_i) \right\rangle_{\mathcal{M}}$$

$$= \left\| \sum_{j=1}^{n} c_j u(\lambda_j) \right\|_{\mathcal{M}}^{2}$$

$$\geq 0.$$

$\square$

Definition 2.7. (Definition of model) *Let $\varphi \colon \mathbb{D} \to \mathbb{C}$ be a function. A* model *for φ is a pair $(\mathcal{M}, u)$ where $\mathcal{M}$ is a Hilbert space and $u \colon \mathbb{D} \to \mathcal{M}$ is a mapping such that, for all $\lambda, \mu \in \mathbb{D}$,*

$$1 - \overline{\varphi(\mu)} \varphi(\lambda) = \langle (1 - \bar{\mu}\lambda) u(\lambda), u(\mu) \rangle_{\mathcal{M}}. \tag{2.8}$$

Equation (2.8) will be called a model formula *for φ.*

Operator Analysis on $\mathbb{D}$ 37

Taken together, Lemma 2.2 and Theorem 2.5 imply the following theorem.

Theorem 2.9. (The fundamental theorem for $\mathscr{S}(\mathbb{D})$**)** *Every function in* $\mathscr{S}(\mathbb{D})$ *has a model.*

Proof. Let $\varphi \in \mathscr{S}(\mathbb{D})$ and let the kernel A on $\mathbb{D}$ be defined by equation (2.3). By Lemma 2.2, A is positive semi-definite. Hence, by Moore's theorem, there exists a Hilbert space $\mathcal{M}$ and a map $u \colon \mathbb{D} \to \mathcal{M}$ such that

$$A(\lambda, \mu) = \langle u(\lambda), u(\mu) \rangle_{\mathcal{M}} \qquad \text{for all } \lambda, \mu \in \mathbb{D}.$$

Therefore

$$\frac{1 - \overline{\varphi(\mu)}\varphi(\lambda)}{1 - \bar{\mu}\lambda} = \langle u(\lambda), u(\mu) \rangle_{\mathcal{M}},$$

which implies the model formula (2.8). $\qquad\square$

Theorem 2.9 opens the door to the use of models to prove facts about elements of $\mathscr{S}(\mathbb{D})$ as well as the reverse process—the use of established function theory for $\mathscr{S}(\mathbb{D})$ to prove facts about models. We refer to this type of operator analysis, loosely, as *model*[1] *theory* on $\mathbb{D}$.

Remarks 2.10.

(i) We say that a model $(\mathcal{M}, u)$ for a function in $\mathscr{S}(\mathbb{D})$ is *taut* if $\{u_\lambda \colon \lambda \in \mathbb{D}\}$ has dense linear span in $\mathcal{M}$. Note that if $(\mathcal{M}, u)$ is a model for φ and $\mathcal{M}_0$ denotes the closed linear span of $\operatorname{ran} u$ in $\mathcal{M}$, then $(\mathcal{M}_0, u)$ is a taut model for φ.

(ii) If $(\mathcal{M}, u)$ is a model for φ and $\mathcal{N}$ is a Hilbert space, then we may define a second model $(\mathcal{M}^\sim, u^\sim)$ for φ by setting $\mathcal{M}^\sim = \mathcal{M} \oplus \mathcal{N}$ and $u_\lambda^\sim = u_\lambda \oplus 0$. We refer to $(\mathcal{M}^\sim, u^\sim)$ as the *trivial extension* of $(\mathcal{M}, u)$ by $\mathcal{N}$.

(iii) We say that a model $(\mathcal{M}, u)$ for $\mathscr{S}(\mathbb{D})$ is *holomorphic* if u is a holomorphic $\mathcal{M}$-valued function on $\mathbb{D}$.

(iv) When Ω is a set, $\mathcal{H}$ is a Hilbert space and $u \colon \Omega \to \mathcal{H}$ is a function, we often write u_λ for $u(\lambda)$. When the elements of $\mathcal{H}$ are functions in their own right this allows one to avoid expressions like $u(\lambda)(z)$.

2.3 Reproducing Kernel Hilbert Spaces

Another important consequence of Moore's theorem (Theorem 2.5) is the fact that there is a close correspondence between Hilbert spaces of functions and positive semi-definite kernels. We encountered one instance of this

---

[1] This use of the word *model* is different from that of Sz.-Nagy and Foias, as described in Section 1.11 and also from the usage of mathematical logicians.

Commutative Theory

correspondence in Section 1.5 (the space H^2 and the Szegő kernel). In general, consider a set Ω and a Hilbert space $\mathcal{H}$ of complex-valued functions on Ω. By the last phrase we mean that[2]

(i) for each $\lambda \in \Omega$ and $f \in \mathcal{H}$, $f(\lambda)$ is a well-defined complex number,
(ii) for each $\lambda \in \Omega$, the linear functional $f \mapsto f(\lambda)$ is continuous on $\mathcal{H}$, and
(iii) if $f, g \in \mathcal{H}$ and $f(\lambda) = g(\lambda)$ for all $\lambda \in \Omega$, then $f = g$.

Property (ii) and the Riesz–Fréchet theorem together imply that, for every $\lambda \in \Omega$, there exists a unique vector $k_\lambda \in \mathcal{H}$ such that

$$f(\lambda) = \langle f, k_\lambda \rangle_{\mathcal{H}} \qquad \text{for all } f \in \mathcal{H}. \tag{2.11}$$

Define $k \colon \Omega \times \Omega \to \mathbb{C}$ by

$$k(\lambda, \mu) = \langle k_\mu, k_\lambda \rangle_{\mathcal{H}} \qquad \text{for all } \lambda, \mu \in \Omega. \tag{2.12}$$

Then k is a kernel on Ω. For any μ, by equations (2.11) and (2.12), $k_\mu = k(\cdot, \mu)$, so that $k(\cdot, \mu) \in \mathcal{H}$ and the "reproducing property" (2.11) can be written

$$f(\lambda) = \langle f, k(\cdot, \lambda) \rangle_{\mathcal{H}} \qquad \text{for all } f \in \mathcal{H} \text{ and } \lambda \in \Omega. \tag{2.13}$$

In view of this relation, k is called the *reproducing kernel* of $\mathcal{H}$ and $\mathcal{H}$ is called a *reproducing kernel Hilbert space*. Reproducing kernels are positive semi-definite; this is because for any $n \geq 1$, $c_1, \ldots, c_n \in \mathbb{C}$ and $\lambda_1, \ldots, \lambda_n$,

$$0 \leq \left\| \sum_j c_j k(\cdot, \lambda_j) \right\|_{\mathcal{H}}^2 = \left\langle \sum_j c_j k_{\lambda_j}, \sum_i c_i k_{\lambda_i} \right\rangle = \sum_{i,j} \bar{c}_i c_j k(\lambda_i, \lambda_j).$$

Moore's theorem provides a converse statement: every positive semi-definite kernel is the reproducing kernel of a unique Hilbert space of functions.

Theorem 2.14. *Let k be a positive semi-definite kernel on a set Ω. There exists a unique Hilbert space $\mathcal{H}$ of functions on Ω such that*

(i) $k(\cdot, \lambda) \in \mathcal{H}$ *for every $\lambda \in \Omega$, and*
(ii) $f(\lambda) = \langle f, k(\cdot, \lambda) \rangle_{\mathcal{H}}$ *for all $f \in \mathcal{H}$ and $\lambda \in \Omega$.*

Proof. It is easy to see that the kernel $(\lambda, \mu) \mapsto k(\mu, \lambda)$ is also positive semi-definite on Ω. Hence by Theorem 2.5, there exists a Hilbert space $\mathcal{M}$ and a map $u \colon \Omega \to \mathcal{M}$ such that

$$k(\lambda, \mu) = \langle u(\mu), u(\lambda) \rangle_{\mathcal{M}} \qquad \text{for all } \lambda, \mu \in \Omega \tag{2.15}$$

and, furthermore, $\{u(\lambda) \colon \lambda \in \Omega\}$ has dense linear span in $\mathcal{M}$.

---

[2] Observe that H^2 is *not* a Hilbert space of functions on $\mathbb{T}$, though it *is* a Hilbert space of functions on $\mathbb{D}$.

Operator Analysis on $\mathbb{D}$ 39

For each $x \in \mathcal{M}$, let $\hat{x}: \Omega \to \mathbb{C}$ be defined by

$$\hat{x}(\lambda) = \langle x, u(\lambda) \rangle_{\mathcal{M}} \qquad \text{for } \lambda \in \Omega \tag{2.16}$$

and let $\mathcal{H} = \{\hat{x}: x \in \mathcal{M}\}$. Thus $\mathcal{H}$ is a vector space of complex-valued functions on Ω and $x \mapsto \hat{x}$ is an injective linear map from $\mathcal{M}$ to $\mathcal{H}$. By virtue of equations (2.15) and (2.16), for all $\mu \in \Omega$,

$$k_\mu = \widehat{u(\mu)}, \tag{2.17}$$

and therefore $k_\mu \in \mathcal{H}$, so that property (i) holds. Endow $\mathcal{H}$ with the inner product induced by the map $x \mapsto \hat{x}$, that is,

$$\langle \hat{x}, \hat{y} \rangle_{\mathcal{H}} \overset{\text{def}}{=} \langle x, y \rangle_{\mathcal{M}} \qquad \text{for all } x, y \in \mathcal{M}.$$

The map $x \mapsto \hat{x}$ is now an isomorphism of the Hilbert spaces $\mathcal{M}$ and $\mathcal{H}$. To see that property (ii) holds, consider any $f \in \mathcal{H}$. By definition, $f = \hat{x}$ for some $x \in \mathcal{M}$. In view of equation (2.17), for any $\mu \in \Omega$,

$$\langle f, k_\mu \rangle_{\mathcal{H}} = \langle \hat{x}, \widehat{u(\mu)} \rangle_{\mathcal{H}} = \langle x, u(\mu) \rangle_{\mathcal{M}} = \hat{x}(\mu) = f(\mu).$$

That is, k is the reproducing kernel for $\mathcal{H}$.

Suppose that $\mathcal{H}'$ is a second Hilbert space of functions on Ω satisfying (i) and (ii). Both $\mathcal{H}$ and $\mathcal{H}'$ contain the closed linear span of the functions $k(\cdot, \lambda)$, $\lambda \in \Omega$. Property (ii) implies that the orthogonal complement of the span consists of the zero function, and therefore that $\mathcal{H}$ and $\mathcal{H}'$ are equal to $(\operatorname{ran} u)^{\perp}$. Thus the Hilbert function space $\mathcal{H}$ with reproducing kernel k is unique. $\qquad\square$

2.4 Lurking Isometries

A *lurking isometry argument* is a line of reasoning based on a structure that contains a hidden isometry. The apt and colorful terminology is due to Joseph Ball. For a Hilbert space $\mathcal{H}$ and $S \subseteq \mathcal{H}$, we denote by $[S]$ the closed linear span of S, and if $\{u_\lambda: \lambda \in \Omega\}$ is an indexed family of vectors in $\mathcal{H}$, we denote by $[u_\lambda: \lambda \in \Omega]$ the closed linear span in $\mathcal{H}$ of $\{u_\lambda: \lambda \in \Omega\}$.

All lurking isometry arguments have a step wherein the following lemma is applied.

Lemma 2.18. **(Lurking isometry lemma)** *Let* $\mathcal{H}$ *and* $\mathcal{K}$ *be Hilbert spaces and let* Ω *be a set. If* $u: \Omega \to \mathcal{H}$ *and* $v: \Omega \to \mathcal{K}$ *are functions, then*

$$\langle v_\lambda, v_\mu \rangle_{\mathcal{K}} = \langle u_\lambda, u_\mu \rangle_{\mathcal{H}} \qquad \text{for all } \lambda, \mu \in \Omega \tag{2.19}$$

if and only if there exists a linear isometry $V: [\operatorname{ran} u] \to \mathcal{K}$ *such that*

$$V u_\lambda = v_\lambda \qquad \text{for all } \lambda \in \Omega. \tag{2.20}$$

40 *Commutative Theory*

Proof. First assume that equation (2.19) holds. Fix a positive integer n, scalars $c_1, \ldots, c_n \in \mathbb{C}$, and points $\lambda_1, \ldots, \lambda_n \in \Omega$. Then

$$
\left\| \sum_{j=1}^{n} c_j v_{\lambda_j} \right\|_{\mathcal{K}}^2 = \left\langle \sum_{j=1}^{n} c_j v_{\lambda_j}, \sum_{i=1}^{n} c_i v_{\lambda_i} \right\rangle_{\mathcal{K}}
$$

$$
= \sum_{i,j=1}^{n} c_j \bar{c}_i \left\langle v_{\lambda_j}, v_{\lambda_i} \right\rangle_{\mathcal{K}}
$$

$$
= \sum_{i,j=1}^{n} c_j \bar{c}_i \left\langle u_{\lambda_j}, u_{\lambda_i} \right\rangle_{\mathcal{H}}
$$

$$
= \left\langle \sum_{j=1}^{n} c_j u_{\lambda_j}, \sum_{i=1}^{n} c_i u_{\lambda_i} \right\rangle_{\mathcal{H}}
$$

$$
= \left\| \sum_{j=1}^{n} c_j u_{\lambda_j} \right\|_{\mathcal{H}}^2 .
$$

We may therefore define an isometric linear map V_0 on the linear span of $\operatorname{ran} u$ into $\mathcal{K}$ by the formula,

$$
V_0 \sum_{j=1}^{n} c_j u_{\lambda_j} = \sum_{j=1}^{n} c_j v_{\lambda_j}.
$$

Since V_0 is isometric, V_0 extends by continuity to a linear isometry $V \colon [\operatorname{ran} u] \to \mathcal{K}$ satisfying equation (2.20).

Now assume that $V \colon [\operatorname{ran} u] \to \mathcal{K}$ is a linear isometry satisfying equation (2.20). By Proposition 1.14, $V^*V = 1$, and so, for all $\lambda, \mu \in \Omega$,

$$
\langle v_\lambda, v_\mu \rangle_{\mathcal{K}} = \langle V u_\lambda, V u_\mu \rangle_{\mathcal{H}} = \langle V^* V u_\lambda, u_\mu \rangle_{\mathcal{H}} = \langle u_\lambda, u_\mu \rangle_{\mathcal{H}},
$$

which is equation (2.19). $\qquad\square$

To illustrate the power of Lemma 2.18, let us use it to prove that models for $\mathscr{S}(\mathbb{D})$ are automatically holomorphic.

Proposition 2.21. (Automatic holomorphy of models) *Let $\varphi \colon \mathbb{D} \to \mathbb{C}$ be a function and suppose that φ has a model $(\mathcal{M}, u)$. Then $(\mathcal{M}, u)$ is holomorphic.*

Proof. Assume that $\varphi \colon \mathbb{D} \to \mathbb{C}$ is a function for which $(\mathcal{M}, u)$ is a model.

Reshuffle step

The formula (2.8) can be *reshuffled*[3] to

$$\varphi(\lambda)\overline{\varphi(\mu)} + \langle u_\lambda, u_\mu \rangle = 1 + \langle \lambda u_\lambda, \mu u_\mu \rangle \quad \text{for all } \lambda, \mu \in \mathbb{D},$$

which in turn may be rewritten as

$$\left\langle \begin{pmatrix} \varphi(\lambda) \\ u_\lambda \end{pmatrix}, \begin{pmatrix} \varphi(\mu) \\ u_\mu \end{pmatrix} \right\rangle_{\mathbb{C}\oplus\mathcal{M}} = \left\langle \begin{pmatrix} 1 \\ \lambda u_\lambda \end{pmatrix}, \begin{pmatrix} 1 \\ \mu u_\mu \end{pmatrix} \right\rangle_{\mathbb{C}\oplus\mathcal{M}} \quad \text{for all } \lambda, \mu \in \mathbb{D}.$$

Therefore, Lemma 2.18 implies that there exists a linear isometry

$$V \colon [1 \oplus \lambda u_\lambda \colon \lambda \in \mathbb{D}] \to \mathbb{C} \oplus \mathcal{M} \tag{2.22}$$

such that

$$V\begin{pmatrix} 1 \\ \lambda u_\lambda \end{pmatrix} = \begin{pmatrix} \varphi(\lambda) \\ u_\lambda \end{pmatrix} \quad \text{for all } \lambda \in \mathbb{D}. \tag{2.23}$$

Extension step

If we define $W \in \mathcal{B}(\mathbb{C} \oplus \mathcal{M})$ by setting $W = V$ on $[1 \oplus \lambda u_\lambda \colon \lambda \in \mathbb{D}]$ and $W = 0$ on the orthogonal complement of $[1 \oplus \lambda u_\lambda \colon \lambda \in \mathbb{D}]$ in $\mathbb{C} \oplus \mathcal{M}$, then W is a partial isometry[4] (in particular, W is a contraction) such that

$$W\begin{pmatrix} 1 \\ \lambda u_\lambda \end{pmatrix} = \begin{pmatrix} \varphi(\lambda) \\ u_\lambda \end{pmatrix} \quad \text{for all } \lambda \in \mathbb{D}.$$

If we represent W as a 2×2 block matrix,

$$W = \begin{bmatrix} a & 1 \otimes \beta \\ \gamma \otimes 1 & D \end{bmatrix} \colon \mathbb{C} \oplus \mathcal{M} \to \mathbb{C} \oplus \mathcal{M}, \tag{2.24}$$

then

$$\begin{pmatrix} \varphi(\lambda) \\ u_\lambda \end{pmatrix} = W\begin{pmatrix} 1 \\ \lambda u_\lambda \end{pmatrix} = \begin{bmatrix} a & 1 \otimes \beta \\ \gamma \otimes 1 & D \end{bmatrix}\begin{pmatrix} 1 \\ \lambda u_\lambda \end{pmatrix} \quad \text{for all } \lambda \in \mathbb{D}. \tag{2.25}$$

In particular,

$$u_\lambda = (\gamma \otimes 1)1 + D\lambda u_\lambda = \gamma + D\lambda u_\lambda \quad \text{for all } \lambda \in \mathbb{D}.$$

As W is a contraction, so also is D. Consequently,[5] we may solve this last equation for u_λ to obtain the equation

$$u_\lambda = (1 - D\lambda)^{-1}\gamma \quad \text{for all } \lambda \in \mathbb{D}, \tag{2.26}$$

which implies that u is holomorphic on $\mathbb{D}$. $\qquad\square$

---

[3] All lurking isometry arguments have a reshuffle.

[4] $W \in \mathcal{B}(\mathcal{H})$ is a *partial isometry* if there exists a closed subspace $\mathcal{N} \subseteq \mathcal{H}$ such that W is isometric on $\mathcal{N}$ and $W = 0$ on $\mathcal{N}^\perp$.

[5] If $\mathcal{X}$ is a Banach space, $T \in \mathcal{B}(\mathcal{X})$ and $\|T\| < 1$, then $1 - T$ is invertible. Since $\|D\| \leq 1$, $\|\lambda D\| < 1$ for all $\lambda \in \mathbb{D}$. Therefore, $1 - \lambda D$ is invertible for all $\lambda \in \mathbb{D}$.

42 *Commutative Theory*

For some readers the implication in the last sentence of the proof of Proposition 2.21 will be evident, but others will ask why equation (2.26) implies that u is holomorphic. There are at least two easy ways to answer this question. One way is to expand the right-hand side of equation (2.26) in a geometric series, which is a power series in λ convergent in $\mathbb{D}$. A second way is to prove the defining property of holomorphy (as in Definition 1.40); this method is better adapted to other situations which we shall encounter shortly. We shall therefore give a proof that applies with minimal changes when $\mathbb{D}$ is replaced by other domains.

For this purpose we need to know about the continuity of the inversion map on $\mathcal{B}(\mathcal{H})$. See [53, proposition I.2.6] for the following.

Lemma 2.27. *Let E be a Banach space and let G denote the set of invertible operators on E. The map $A \mapsto A^{-1}$ is continuous as a self-map of $(G, \|\cdot\|_{\mathcal{B}(E)})$.*

We also need an identity that, though simple, is significant enough to have a name.

Proposition 2.28. (The resolvent identity) *Let X and Y be operators on a Hilbert space $\mathcal{H}$. If $1 - X$ and $1 - Y$ are invertible then*

$$(1 - X)^{-1} - (1 - Y)^{-1} = (1 - X)^{-1}(X - Y)(1 - Y)^{-1}. \qquad (2.29)$$

Proof. Evidently

$$(1 - Y) - (1 - X) = X - Y.$$

Premultiply both sides by $(1 - X)^{-1}$ and postmultiply by $(1 - Y)^{-1}$ to obtain equation (2.29). $\qquad \square$

The holomorphy of u_λ is now easy.

Lemma 2.30. *The map $u \colon \mathbb{D} \to \mathcal{M}$ defined by equation (2.26) is holomorphic.*

Proof. Consider a point $\lambda_0 \in \mathbb{D}$; we must prove that u is differentiable at λ_0. In the resolvent identity, equation (2.29), let $X = D\lambda$ and $Y = D\lambda_0$ and divide through by $\lambda - \lambda_0$ to obtain, for any $\lambda \in \mathbb{D} \setminus \{\lambda_0\}$,

$$\frac{u(\lambda) - u(\lambda_0)}{\lambda - \lambda_0} = (1 - D\lambda)^{-1} D (1 - D\lambda_0)^{-1} \gamma.$$

Let $\lambda \to \lambda_0$ and use Lemma 2.27 to deduce that $u'(\lambda_0)$ exists (and equals $D(1 - D\lambda_0)^{-2}\gamma$). It follows that u is holomorphic on $\mathbb{D}$. $\qquad \square$

Remark 2.31. (The extension step) In the proof of Proposition 2.21, following a reshuffle, we encountered a lurking isometry V defined on the subspace $[1 \oplus \lambda u_\lambda \colon \lambda \in \mathbb{D}]$ of $\mathbb{C} \oplus \mathcal{M}$. We then extended V to an operator W on the

Operator Analysis on $\mathbb{D}$

whole of $\mathbb{C} \oplus \mathcal{M}$. A step of this type occurs in virtually all lurking isometry arguments; we call it *the extension step*. In this step we have some flexibility to choose a form of the extension that is convenient for the purpose at hand. For example, instead of choosing W to be 0 on the orthocomplement of the domain of V, we could choose it to make subsequent calculations easier. One instance of this is in the derivation of the network realization formula (2.38), where it is useful to choose W to be an isometry. In other instances, we might want the extension to be unitary.

The only obstruction to the extension of a partially defined isometry V to an isometry or a unitary is a dimensional one. If W is an isometry that extends V, then W maps $(\operatorname{dom} V)^{\perp}$ to $(\operatorname{ran} V)^{\perp}$ isometrically. Therefore if $m = \dim(\operatorname{dom} V)^{\perp}$ and $n = \dim(\operatorname{ran} V)^{\perp}$, then V has an isometric extension if and only if $m \leq n$, and V has a unitary extension if and only if $m = n$.

If $\dim \mathcal{M}$ is finite, then necessarily $m = n$, and there is a unitary choice for W. (We exploit this fact in the proof of Theorem 2.54). One trick that allows us to produce a unitary extension, even when $\mathcal{M}$ is infinite-dimensional, is to use a trivial extension, as in Remark 2.10(ii), where the space $\mathcal{N}$ is chosen with infinite dimension greater than or equal to $\dim \mathcal{M}$. The result will be that the (new) dimensions of $(\operatorname{dom} V)^{\perp}$ and $(\operatorname{ran} V)^{\perp}$ are both equal to $\dim \mathcal{N}$, and consequently there *is* a unitary extension. This device is employed in the proof of Theorem 2.57 and in that of Lemma 4.43. There is a price to be paid, however: the new summand, and the choice of unitary extension, introduce *noise* into the model, that is, information that is not intrinsic to the object being modeled.

As a corollary to Proposition 2.21 we obtain a converse to the fundamental theorem.

Proposition 2.32. (Converse to the fundamental theorem for $\mathscr{S}(\mathbb{D})$) *If $\varphi \colon \mathbb{D} \to \mathbb{C}$ is a function that has a model on $\mathbb{D}$, then $\varphi \in \mathscr{S}(\mathbb{D})$.*

Proof. Assume that $\varphi \colon \mathbb{D} \to \mathbb{C}$ has a model $(\mathcal{M}, u)$ on $\mathbb{D}$. We first show that φ is holomorphic on $\mathbb{D}$. If $\varphi(\mu) = 0$ for all $\mu \in \mathbb{D}$, then clearly, φ is holomorphic on $\mathbb{D}$. Otherwise, choose $\mu \in \mathbb{D}$ such that $\varphi(\mu) \neq 0$ and solve the model formula (2.8) for $\psi(\lambda)$ to obtain

$$\varphi(\lambda) = \overline{\varphi(\mu)}^{-1} \left(1 - \langle (1 - \bar{\mu}\lambda)\, u_\lambda, u_\mu \rangle_{\mathcal{M}} \right) \quad \text{for all } \lambda \in \mathbb{D}.$$

As Proposition 2.21 states that $(\mathcal{M}, u)$ is holomorphic, this equation implies that φ is holomorphic on $\mathbb{D}$.

To see that $\varphi \in \mathscr{S}(\mathbb{D})$, notice that, by the model formula (2.8) with $\mu = \lambda$,

$$1 - |\varphi(\lambda)|^2 = (1 - |\lambda|^2)\, \|u_\lambda\|_{\mathcal{M}}^2 \geq 0 \quad \text{for all } \lambda \in \mathbb{D}.$$

Hence $\|\varphi\|_{\mathbb{D}} \leq 1$. Since φ is holomorphic and $\|\varphi\|_{\mathbb{D}} \leq 1$, $\varphi \in \mathscr{S}(\mathbb{D})$. $\qquad\square$

44 *Commutative Theory*

Another consequence of the lurking isometry lemma is the uniqueness of taut models up to isomorphism.

Proposition 2.33. *Let $(\mathcal{M}, u)$ and $(\mathcal{N}, v)$ be taut models of a function $\varphi \in \mathscr{S}(\mathbb{D})$. There exists an isomorphism of Hilbert spaces[6] $V: \mathcal{M} \to \mathcal{N}$ such that $Vu_\lambda = v_\lambda$ for all $\lambda \in \mathbb{D}$.*

The fact that both $(\mathcal{M}, u)$ and $(\mathcal{N}, v)$ are taut models of φ implies that $\langle u_\lambda, u_\mu \rangle = \langle v_\lambda, v_\mu \rangle$ for all $\lambda, \mu \in \mathbb{D}$. The lurking isometry lemma now provides the required isometry V. The tautness of the two models implies that V is an isomorphism of the Hilbert spaces $\mathcal{M}$ and $\mathcal{N}$.

2.5 The Network Realization Formula (Scalar Case) via Model Theory

A *realization formula* in engineering theory provides a recipe for the construction of a system with prescribed behavior. Here a *system* is a physical or mathematical object that, at every point in time, accepts inputs and produces outputs. It might be, for example, an electrical or mechanical network. An important class of systems comprises those that can be described by linear constant-coefficient ordinary differential or difference equations. For these, one can solve the system equations by means of the Laplace or z-transform,[7] and therefore, the action of the system is described by a matrix of functions of the frequency variable. This matrix-valued function is called the *transfer function* of the system. The *realization problem* for a class $\mathcal{C}$ of systems is to describe the set of transfer functions that can arise for systems in $\mathcal{C}$. A detailed account of the engineering theory of linear systems can be found in the book [125].

Consider, for example, the class $\mathcal{C}$ of discrete-time systems Σ, which can be constructed by the connection of an arbitrary finite number of linear ideal circuit elements of five basic types: taps, summing junctions, amplifiers, differentiators, and integrators. Such a system Σ is *time-invariant*, in an obvious sense (its behavior is the same on Tuesday as on Monday). It is easy to show that Σ can be represented by difference equations[8] of the form

$$x_{n+1} = Dx_n + Cu_n,$$
$$y_n = Bx_n + Au_n \tag{2.34}$$

---

[6] This means that $V^*V = 1_{\mathcal{M}}$ and $VV^* = 1_{\mathcal{N}}$.
[7] The z-transform of the sequence $(x_n)_{n \geq 0}$ is the power series $\sum_n x_n z^n$.
[8] In the engineering literature the roles of A and D and of B and C are reversed.

Operator Analysis on $\mathbb{D}$

for $n = 0, 1, \ldots$. Here $(u_n)_{n \geq 0}$ is a sequence of inputs to the system, $(y_n)_{n \geq 0}$ is the corresponding sequence of outputs, and for each n, x_n is the state of the system at time n. The vectors u_n, x_n, y_n are assumed to belong real vector spaces $\mathcal{H}, \mathcal{M}, \mathcal{K}$, respectively, and A, B, C, and D are real matrices of suitable types. If we assume that the space $\mathcal{M}$ is parametrized in such a way that the initial state x_0 is zero, then on multiplying the equations (2.34) by z^n and summing we find that

$$\sum_n x_n z^n = (1 - zD)^{-1} zC \sum_n u_n z^n,$$

$$\sum_n y_n z^n = B \sum_n x_n z^n + A \sum_n u_n z^n,$$

whence

$$\sum_n y_n z^n = [A + B(1 - zD)^{-1} zC] \sum_n u_n z^n.$$

This relation is expressed by the statement that the transfer function of Σ is

$$G(z) = A + B(1 - zD)^{-1} zC. \tag{2.35}$$

Thus the transfer function of any system in $\mathcal{C}$ is a matrix of rational functions with real coefficients. A basic realization theorem [125, chapter 2] states that the converse is also true: for any real-rational matrix-valued function G, there exist real matrices A, B, C, and D of suitable (finite) types such that equation (2.35) holds for all complex numbers z other than the poles of G. The equation (2.35), the 4-tuple (A, B, C, D) and the block matrix $\begin{bmatrix} A & B \\ C & D \end{bmatrix}$ are all called *realizations* of the function G. The reason for the appellation *realization* is that, once A, B, C, and D are known, it is straightforward to construct a network in the class $\mathcal{C}$, which is described by the equations (2.34) and consequently has transfer function G.

Systems in $\mathcal{C}$ have *rational* transfer functions, but other discrete-time time-invariant linear systems have non-rational transfer functions. For example, systems containing delays have transfer functions that contain exponentials. Discrete-time time invariant linear systems of interest to engineers usually have transfer functions that are meromorphic in the plane.

This realization theorem is purely algebraic in nature. Analysis enters the picture when we consider stability and energy balance. Roughly speaking, a physical system is stable if it does not fail catastrophically (e.g., burn out, flood, or explode). Mathematically, there are several notions of stability [125, section 2.6]; for expository purposes, we shall simplify and say that a discrete-time system described by a transfer function G is *stable* if G is analytic in $\mathbb{D}$.

Again to simplify somewhat, a physical system is *passive* if it contains no internal power source. For such a system, since energy is conserved, energy at

46 *Commutative Theory*

the output cannot exceed energy at the input. Mathematically, a discrete-time system with transfer function G is stable and passive if G belongs to the Schur class $\mathscr{S}_{\mathcal{B}(\mathcal{H},\mathcal{K})}(\mathbb{D})$. We shall treat such systems in Theorem 3.16.

When $\mathcal{H} = \mathcal{K} = \mathbb{C}$, Theorem 1.58 becomes the following result (cf. Remark 1.60(iii)).

Theorem 2.36. (Network realization formula when $\mathcal{H} = \mathcal{K} = \mathbb{C}$) *The following conditions are equivalent for a function φ on $\mathbb{D}$.*

(i) $\varphi \in \mathscr{S}(\mathbb{D})$;

(ii) *there exist a Hilbert space $\mathcal{M}$ and a contraction*

$$V = \begin{bmatrix} a & 1 \otimes \beta \\ \gamma \otimes 1 & D \end{bmatrix} : \mathbb{C} \oplus \mathcal{M} \to \mathbb{C} \oplus \mathcal{M} \tag{2.37}$$

such that

$$\varphi(\lambda) = a + \lambda \left\langle (1 - D\lambda)^{-1} \gamma, \beta \right\rangle \qquad \text{for all } \lambda \in \mathbb{D}; \tag{2.38}$$

(iii) *there exist a Hilbert space $\mathcal{M}$ and an isometry V of the form* (2.37) *such that the formula* (2.38) *holds.*

Proof. Since isometries are contractions, (iii) trivially implies (ii).

(i)$\Rightarrow$(iii). Assume that $\varphi \in \mathscr{S}(\mathbb{D})$. By the fundamental theorem for $\mathscr{S}(\mathbb{D})$, Theorem 2.9, φ has a model $(\mathcal{M}, u)$, and by Remark 2.10(i) we may assume that $(\mathcal{M}, u)$ is taut, that is, [ran u] is dense in $\mathcal{M}$.

Claim 2.39. $[1 \oplus \lambda u_\lambda \colon \lambda \in \mathbb{D}] = \mathbb{C} \oplus \mathcal{M}$.

To prove this claim, assume that $a \oplus v \in \mathbb{C} \oplus \mathcal{M}$ and

$$1 \oplus \lambda u_\lambda \perp a \oplus v \qquad \text{for all } \lambda \in \mathbb{D}.$$

This implies that $\bar{a} + \lambda \langle u_\lambda, v \rangle = 0$ for all $\lambda \in \mathbb{D}$. The choice $\lambda = 0$ shows that $a = 0$. It then follows that $\langle u_\lambda, v \rangle = 0$ for all $\lambda \in \mathbb{D} \setminus \{0\}$. Proposition 2.21 implies that u is holomorphic in $\mathbb{D}$, and therefore, $\langle u_\lambda, v \rangle = 0$ for all $\lambda \in \mathbb{D}$. As $(\mathcal{M}, u)$ is taut, this implies that $v = 0$. To summarize, we have shown that, if $a \oplus v \in \mathbb{C} \oplus \mathcal{M}$ and $a \oplus v \perp [1 \oplus \lambda u_\lambda \colon \lambda \in \mathbb{D}]$, then $a \oplus v = 0$. This proves the claim.

Exactly as in the proof of Proposition 2.21, the model formula for φ together with the lurking isometry lemma imply that there exists an isometry

$$V \colon [1 \oplus \lambda u_\lambda \colon \lambda \in \mathbb{D}] \to \mathbb{C} \oplus \mathcal{M}$$

such that

$$V \begin{pmatrix} 1 \\ \lambda u_\lambda \end{pmatrix} = \begin{pmatrix} \varphi(\lambda) \\ u_\lambda \end{pmatrix} \qquad \text{for all } \lambda \in \mathbb{D}.$$

Operator Analysis on $\mathbb{D}$ 47

In view of Claim 2.39, V is an isometry on the whole of $\mathbb{C} \oplus \mathcal{M}$. Write V as a 2×2 block matrix

$$V = \begin{bmatrix} a & 1 \otimes \beta \\ \gamma \otimes 1 & D \end{bmatrix}$$

for some $\alpha, \beta \in \mathcal{M}$ and $D \in \mathcal{B}(\mathcal{M})$. As in the proof of Proposition 2.21 (see equations (2.24) and (2.25)) we obtain the equations

$$\varphi(\lambda) = a + \langle \lambda u_\lambda, \beta \rangle$$

and

$$u_\lambda = \gamma + D \lambda u_\lambda \qquad \text{for all } \lambda \in \mathbb{D}.$$

Using the second equation to eliminate u_λ from the first equation we obtain the realization formula (2.38).

(ii)$\Rightarrow$(i). Assume that V is a contraction given by equation (2.37) and φ is given by equation (2.38). Define $u \colon \mathbb{D} \to \mathcal{M}$ by

$$u_\lambda = (1 - D\lambda)^{-1} \gamma \qquad \text{for all } \lambda \in \mathbb{D}.$$

By Lemma 2.30, u is holomorphic, and so therefore is φ. Moreover, for $\lambda \in \mathbb{D}$,

$$\begin{aligned} V \begin{pmatrix} 1 \\ \lambda u_\lambda \end{pmatrix} &= \begin{bmatrix} a & 1 \otimes \beta \\ \gamma \otimes 1 & D \end{bmatrix} \begin{pmatrix} 1 \\ \lambda u_\lambda \end{pmatrix} \\ &= \begin{pmatrix} a + \langle \lambda u_\lambda, \beta \rangle_\mathcal{M} \\ \gamma + D \lambda u_\lambda \end{pmatrix} \\ &= \begin{pmatrix} \varphi(\lambda) \\ u_\lambda \end{pmatrix}. \end{aligned}$$

Consequently, as V is contractive, for all $\lambda \in \mathbb{D}$,

$$\left\| \begin{pmatrix} \varphi(\lambda) \\ u_\lambda \end{pmatrix} \right\| \leq \left\| \begin{pmatrix} 1 \\ \lambda u_\lambda \end{pmatrix} \right\|, \tag{2.40}$$

which is to say that

$$|\varphi(\lambda)|^2 + \|u_\lambda\|^2 \leq 1 + \|\lambda u_\lambda\|^2.$$

Since, clearly, $-\|u_\lambda\|^2 \leq -\|\lambda u_\lambda\|^2$, it follows that $|\varphi(\lambda)| \leq 1$ for all $\lambda \in \mathbb{D}$. Hence $\varphi \in \mathcal{S}(\mathbb{D})$. $\qquad \square$

Remark 2.41. (i) The proof of Theorem 2.36 shows that, if a function φ on $\mathbb{D}$ has a model $(\mathcal{M}, u)$, then φ has a realization of the form (2.38) in which $\beta, \gamma \in \mathcal{M}$ and $D \in \mathcal{B}(\mathcal{M})$.

48 *Commutative Theory*

(ii) We shall say that a realization (2.38) of a function φ is *isometric* or *contractive* if the block matrix V in equation (2.37) is isometric or contractive respectively.

We have shown that, if $G \in \mathscr{S}(\mathbb{D})$, we can "realize" G as the transfer function of a (stable passive) discrete-time time-invariant linear system (or network) given by the equations (2.34). Here both inputs and outputs are scalars. Of course when $\mathcal{M}$ (or $\mathcal{H}$ or $\mathcal{K}$) are infinite-dimensional, it is open to question to what extent the mathematical system (2.34) can be realized physically.

Engineers call $\mathcal{M}$ the *state space* of the realization formula (2.38) or the system (2.34). We shall call $\mathcal{M}$ the *model space* of the realization (2.38).

Let us work out a simple example in order to develop some intuition for contractive and isometric realizations, particularly of rational functions in the Schur class.

Example 2.42. Consider the function $\varphi(\lambda) = \frac{1}{2}\lambda$. Certainly $\varphi \in \mathscr{S}(\mathbb{D})$. We shall calculate explicitly a model for φ, an isometric realization and a contractive realization for φ. Any model $(\mathcal{M}, u)$ for φ satisfies

$$1 - \tfrac{1}{4}\bar{\mu}\lambda = (1 - \bar{\mu}\lambda)\langle u_\lambda, u_\mu \rangle_{\mathcal{M}} \qquad \text{for all } \lambda, \mu \in \mathbb{D}. \qquad (2.43)$$

We may recognize the appearance of the Szegő kernel (1.30) in the expression for $\langle u_\lambda, u_\mu \rangle_{\mathcal{M}}$; recall that

$$\langle s_{\bar{\lambda}}, s_{\bar{\mu}} \rangle_{\mathrm{H}^2(\mathbb{D})} = \frac{1}{1 - \bar{\mu}\lambda}.$$

Accordingly, let us guess that we can choose $\mathcal{M} = \mathrm{H}^2(\mathbb{D})$ and, for some choice of scalars c_1, c_2,

$$u_\lambda(z) = c_1 + c_2 s_{\bar{\lambda}}(z) \qquad \text{for } \lambda, z \in \mathbb{D}.$$

Calculation shows that the relation (2.43) does then hold if we choose c_1, c_2 suitably. We obtain the model $(\mathrm{H}^2(\mathbb{D}), u)$ of φ, where

$$u_\lambda = 1 - \tfrac{1}{2}\sqrt{3} + \tfrac{1}{2}\sqrt{3}s_{\bar{\lambda}} \qquad \text{for all } \lambda \in \mathbb{D}. \qquad (2.44)$$

To find an isometric realization of φ we follow the method of proof that (i) implies (iii) in Theorem 2.36. We must find an isometry V such that

$$V \begin{pmatrix} 1 \\ \lambda u_\lambda \end{pmatrix} = \begin{pmatrix} \varphi(\lambda) \\ u_\lambda \end{pmatrix} \qquad \text{for all } \lambda \in \mathbb{D}.$$

Operator Analysis on $\mathbb{D}$ — 49

A little further calculation, with u_λ as in equation (2.44), shows that, in the notation of Theorem 2.36,

$$V = \begin{bmatrix} 0 & 1 \otimes \frac{1}{2}\mathbf{1} \\ \mathbf{1} \otimes 1 & M_z - c_1 z \otimes \mathbf{1} \end{bmatrix},$$

where $c_1 = 1 - \frac{1}{2}\sqrt{3}$, $\mathbf{1}$ denotes the constant function equal to 1 on $\mathbb{D}$ and M_z denotes the shift operator on $\mathrm{H}^2(\mathbb{D})$ (see equation (1.31)). One may then check that $V^*V = 1$.

We could stop at this point, since isometric realizations are also contractive realizations. However, one point of this example is to illustrate the fact that contractive realizations can be markedly more economical than isometric ones, in the sense of having a much smaller model space. It is easy to construct, with bare hands, an isometric realization of $\frac{1}{2}\lambda$ having a *one-dimensional* model space. Indeed,

$$V = \begin{bmatrix} 0 & 1/(2\gamma) \\ \gamma & 0 \end{bmatrix} : \mathbb{C}^2 \to \mathbb{C}^2$$

yields a contractive realization of $\frac{1}{2}\lambda$ for any choice of γ such that $\frac{1}{2} \le |\gamma| \le 1$. Contractive realizations can contain less noise than isometric ones.

A good exercise is to find an isometric model of the zero function having model space of the smallest possible dimension.

2.5.1 Finite Blaschke Products

An important special case of the realization formula occurs when $\dim \mathcal{M} < \infty$. Recall that a *Blaschke product of degree n* is a function $B \in \mathrm{Hol}(\mathbb{D})$ that has the form

$$B(\lambda) = \tau \prod_{i=1}^{n} \frac{\lambda - \alpha_i}{1 - \alpha_i \lambda} \qquad \text{for all } \lambda \in \mathbb{D}, \tag{2.45}$$

where $\tau \in \mathbb{T}$ and $\alpha_1, \ldots, \alpha_n \in \mathbb{D}$. A *finite Blaschke product* is a Blaschke product of degree n for some positive integer n. Blaschke products are inner functions (i.e., $|B(\lambda)| = 1$ for all $\lambda \in \mathbb{T}$), in particular they are elements of $\mathscr{S}(\mathbb{D})$, and so have models. One could ask how the degree of a Blaschke product is reflected in the properties of its models. We shall probe this question in Proposition 2.51.

To this end we introduce the notation (for $\alpha \in \mathbb{D}$)

$$b_\alpha(\lambda) = \frac{\lambda - \alpha}{1 - \bar{\alpha}\lambda}, \qquad q_\alpha(\lambda) = \frac{(1 - |\alpha|^2)^{\frac{1}{2}}}{1 - \bar{\alpha}\lambda} \qquad \text{for all } \lambda \in \mathbb{D}.$$

50   *Commutative Theory*

The functions b_α, q_α are called the *Blaschke factor* and *normalized Szegő kernel*, respectively, corresponding to $\alpha \in \mathbb{D}$. Notice the identity

$$1 - \overline{b_\alpha(\mu)}b_\alpha(\lambda) = (1 - \bar{\mu}\lambda)\overline{q_\alpha(\mu)}q_\alpha(\lambda), \tag{2.46}$$

valid for all $\alpha, \mu, \lambda \in \mathbb{D}$. More generally,

Lemma 2.47. *Let $\alpha_1, \ldots, \alpha_n \in \mathbb{D}$ and let B be the finite Blaschke product (2.45). For $j = 1, \ldots, n$ let*

$$e_j \overset{\text{def}}{=} q_{\alpha_j} \prod_{1 \leq m \leq j-1} b_{\alpha_m}. \tag{2.48}$$

The functions $e_1, \ldots, e_n$ are orthonormal in H^2 and, for all $\lambda, \mu \in \mathbb{D}$,

$$\frac{1 - \overline{B(\mu)}B(\lambda)}{1 - \bar{\mu}\lambda} = \sum_{j=1}^{n} \overline{e_j(\mu)}e_j(\lambda). \tag{2.49}$$

Proof. Since $|b_\alpha(z)| = 1$ for all $\alpha \in \mathbb{D}$ and $z \in \mathbb{T}$, clearly $\langle b_\alpha f, b_\alpha g \rangle = \langle f, g \rangle$ for any $f, g \in \mathrm{H}^2$. Thus if $1 \leq j \leq i \leq n$,

$$\begin{aligned}
\langle e_i, e_j \rangle &= \langle b_{\alpha_1} \ldots b_{\alpha_{i-1}} q_{\alpha_i}, b_{\alpha_1} \ldots b_{\alpha_{j-1}} q_{\alpha_j} \rangle \\
&= \langle b_{\alpha_j} \ldots b_{\alpha_{i-1}} q_{\alpha_i}, q_{\alpha_j} \rangle \\
&= (1 - |\alpha_j|^2)^{\frac{1}{2}} b_{\alpha_j} \ldots b_{\alpha_{i-1}} q_{\alpha_i}(\alpha_j),
\end{aligned}$$

which is 1 if $i = j$ and 0 if $i > j$. Hence $e_1, \ldots, e_n$ are orthonormal in H^2.

When $n = 2$, in view of the identity (2.46),

$$\begin{aligned}
1 - \overline{B(\mu)}B(\lambda) &= 1 - \overline{b_{\alpha_1}(\mu)b_{\alpha_2}(\mu)}b_{\alpha_2}(\lambda)b_{\alpha_1}(\lambda) \\
&= 1 - \overline{b_{\alpha_1}(\mu)}b_{\alpha_1}(\lambda) + \overline{b_{\alpha_1}(\mu)}\left(1 - \overline{b_{\alpha_2}(\mu)}b_{\alpha_2}(\lambda)\right)b_{\alpha_1}(\lambda) \\
&= (1 - \bar{\mu}\lambda)\left[\overline{q_{\alpha_1}(\mu)}q_{\alpha_1}(\lambda) + \overline{b_{\alpha_1}(\mu)}\overline{q_{\alpha_2}(\mu)}q_{\alpha_2}(\lambda)b_{\alpha_1}(\lambda)\right] \\
&= (1 - \bar{\mu}\lambda)\left[\overline{e_1(\mu)}e_1(\lambda) + \overline{e_2(\mu)}e_1(\lambda)\right].
\end{aligned}$$

This calculation can be extended to prove the identity (2.49) for general n. $\quad\square$

An immediate consequence of the lemma is the following.

Proposition 2.50. *A Blaschke product of degree n has a model in which the model space has dimension n.*

Proof. For B as in Lemma 2.47, define $u\colon \mathbb{D} \to \mathbb{C}^n$ by

$$u_\lambda = (e_1(\lambda), e_2(\lambda), \ldots, e_n(\lambda)).$$

Operator Analysis on $\mathbb{D}$ 51

By equation (2.49),

$$1 - \overline{B(\mu)}B(\lambda) = (1 - \bar{\mu}\lambda)\langle u_\lambda, u_\mu\rangle_{\mathbb{C}^n} \qquad \text{for all } \lambda, \mu \in \mathbb{D}.$$

Thus $(\mathbb{C}^n, u)$ is a model of B. $\square$

We can say more about the relation between the degree of a finite Blaschke product and the dimensions of model spaces.

Proposition 2.51. *If B is a Blaschke product of degree n and $(\mathcal{M}, u)$ is a model for B then $n \leq \dim \mathcal{M}$. Moreover, if $(\mathcal{M}, u)$ is a taut model for B, then $n = \dim \mathcal{M}$.*

Proof. Let $(\mathcal{M}, u)$ be a model for a rational function φ on $\mathbb{D}$ of degree n. By Remark 2.41, φ has a realization

$$\varphi(\lambda) = a + \lambda\langle(1 - D\lambda)^{-1}\gamma, \beta\rangle \qquad \text{for all } \lambda \in \mathbb{D}, \tag{2.52}$$

where $\beta, \gamma \in \mathcal{M}$ and $D \in \mathcal{B}(\mathcal{M})$. If $\dim \mathcal{M} = \infty$ then trivially $n \leq \dim \mathcal{M}$. Suppose $\dim \mathcal{M} = m < \infty$. Since the right-hand side in equation (2.52) can be written as a rational expression with denominator $\det(1 - D\lambda)$, which has degree m, and numerator of degree at most m, it follows that $n \leq m$. In particular the conclusion holds when φ is a Blaschke product of degree n.

Let B be the finite Blaschke product (2.45). Note first that the model of B given by Lemma 2.47, say $(\mathbb{C}^n, v)$ is taut, for otherwise, replacing $\mathbb{C}^n$ by the span of $\operatorname{ran} v$, we obtain a model of B with model space of dimension less than n, contrary to the first statement of the proposition. Now consider any taut model $(\mathcal{M}, u)$ for B. By Proposition 2.33, the Hilbert spaces $\mathcal{M}$ and $\mathbb{C}^n$ are isomorphic, that is, $\dim \mathcal{M} = n$. $\square$

To conclude the discussion of finite-dimensional models, we point out that they exist *only* for Blaschke products.

Proposition 2.53. *If $\varphi \in \mathscr{S}(\mathbb{D})$ has a model with model space of finite dimension n then φ is a Blaschke product of degree at most n.*

Proof. Let $(\mathcal{M}, u)$ be a model of φ such that $\dim \mathcal{M} = n$. As in the proof of Theorem 2.36, there exists an isometry V on $\mathbb{C} \oplus \mathcal{M}$ as in equation (2.37) such that φ is given by the realization formula (2.38), $u_\lambda = (1 - D\lambda)^{-1}\gamma$ and

$$V\begin{pmatrix} 1 \\ \lambda u_\lambda \end{pmatrix} = \begin{pmatrix} \varphi(\lambda) \\ u_\lambda \end{pmatrix}$$

for all $\lambda \in \mathbb{D}$. Since V is an isometry,

$$\left\| \begin{pmatrix} 1 \\ \lambda u_\lambda \end{pmatrix} \right\| = \left\| \begin{pmatrix} \varphi(\lambda) \\ u_\lambda \end{pmatrix} \right\|,$$

52 *Commutative Theory*

and therefore,

$$|\varphi(\lambda)|^2 = 1 - (1 - |\lambda|^2)\|(1 - D\lambda)^{-1}\gamma\|^2 \qquad \text{for all } \lambda \in \mathbb{D}.$$

Let λ tend to a point of $\mathbb{T}$ in this equation. Note that $(1 - D\lambda)^{-1}\gamma$ is analytic in λ at all but a finite number of points of $\mathbb{C}$. It follows that, for all but finitely many points $\lambda \in \mathbb{T}$, we have $|\varphi(\lambda)| = 1$. Thus φ is a rational inner function of degree at most n. $\qquad\square$

2.6 Interpolation via Model Theory

In this section we will prove a refinement of the Pick interpolation theorem (Theorem 1.81), in which the function-theoretic meaning of the rank of the Pick matrix and the precise nature of the solution in the extremal case are identified. We prove a second refinement as well, in which the finite set of nodes $\lambda_1, \ldots, \lambda_n$ is replaced with a general set $\Lambda \subseteq \mathbb{D}$.

2.6.1 The Proof of Pick's Interpolation Theorem

The following is often called the *Nevanlinna–Pick problem*: given n distinct points $\lambda_1, \ldots, \lambda_n \in \mathbb{D}$ and points $z_1, \ldots, z_n \in \mathbb{C}$,

(i) determine whether there exists a function $\varphi \in \mathscr{S}(\mathbb{D})$ such that $\varphi(\lambda_i) = z_i$ for $i = 1, \ldots, n$, and

(ii) if so, give a construction of *all* such interpolating functions φ.

Theorem 1.81 is Pick's solution of part (i) of the problem, while Rolf Nevanlinna in 1929 [156] gave a recursive construction of the general interpolating function (when there is one). In Chapter 1 we showed how Pick's theorem follows from the Sarason interpolation theorem. Now we shall prove the theorem without appealing to Sarason's result, using the tools we have developed.

Theorem 2.54. *Let n be a positive integer, let points $\lambda_1, \ldots, \lambda_n \in \mathbb{D}$ and $z_1, \ldots, z_n \in \mathbb{C}$ be given. Define an $n \times n$ matrix P by*

$$P = \left[\frac{1 - \bar{z}_i z_j}{1 - \bar{\lambda}_i \lambda_j} \right].$$

If $\varphi \in \mathscr{S}(\mathbb{D})$ and $\varphi(\lambda_j) = z_j$ for $j = 1, \ldots, n$, then $P \geq 0$. Conversely, if $P \geq 0$, then there exists $\varphi \in \mathscr{S}(\mathbb{D})$ such that $\varphi(\lambda_j) = z_j$ for $j = 1, \ldots, n$. Furthermore, if $P \geq 0$ and $\mathrm{rank}\, P = r$, then φ may be chosen to be a Blaschke product of degree r.

Operator Analysis on $\mathbb{D}$ 53

Proof. First assume that $\varphi \in \mathscr{S}(\mathbb{D})$ and $\varphi(\lambda_j) = z_j$ for $j = 1, \ldots, n$. That $P \geq 0$ follows immediately from Pick's lemma (Lemma 2.2).

Conversely, let data $\lambda_1 \ldots, \lambda_n \in \mathbb{D}$ and $z_1, \ldots, z_n \in \mathbb{C}$ be given with the property that $P \geq 0$ and rank $P = r$. By the remark following equation (2.4), there exists a Hilbert space $\mathcal{M}$ and vectors $u_1, \ldots u_n \in \mathcal{M}$ such that $\dim \mathcal{M} = r$ and

$$P = \left[\langle u_j, u_i \rangle_{\mathcal{M}} \right],$$

or equivalently,

$$1 - \overline{z}_i z_j = \left\langle (1 - \overline{\lambda}_i \lambda_j) u_j, u_i \right\rangle_{\mathcal{M}} \qquad \text{for } i, j = 1, \ldots, n.$$

This last formula reshuffles to

$$\overline{z}_i z_j + \langle u_j, u_i \rangle_{\mathcal{M}} = 1 + \langle \lambda_j u_j, \lambda_i u_i \rangle_{\mathcal{M}} \qquad \text{for } i, j = 1, \ldots, n,$$

which implies that

$$\left\langle \begin{pmatrix} z_j \\ u_j \end{pmatrix}, \begin{pmatrix} z_i \\ u_i \end{pmatrix} \right\rangle_{\mathbb{C} \oplus \mathcal{M}} = \left\langle \begin{pmatrix} 1 \\ \lambda_j u_j \end{pmatrix}, \begin{pmatrix} 1 \\ \lambda_i u_i \end{pmatrix} \right\rangle_{\mathbb{C} \oplus \mathcal{M}} \qquad \text{for } i, j = 1, \ldots, n.$$

Consequently, by Lemma 2.18 and Remark 2.31 there exists a unitary

$$V = \begin{bmatrix} a & 1 \otimes \beta \\ \gamma \otimes 1 & D \end{bmatrix} : \mathbb{C} \oplus \mathcal{M} \to \mathbb{C} \oplus \mathcal{M}$$

such that

$$\begin{bmatrix} a & 1 \otimes \beta \\ \gamma \otimes 1 & D \end{bmatrix} \begin{pmatrix} 1 \\ \lambda_j u_j \end{pmatrix} = \begin{pmatrix} z_j \\ u_j \end{pmatrix} \qquad \text{for } j = 1, \ldots, n,$$

or equivalently,

$$z_j = a + \lambda_j \langle u_j, \beta \rangle \quad \text{and} \quad u_j = \gamma + \lambda_j D u_j \qquad \text{for } j - 1, \ldots, n.$$

Solving the second equation above for u_j and substituting into the first equation we deduce that

$$z_j = a + \lambda_j \left\langle (1 - D\lambda_j)^{-1} \gamma, \beta \right\rangle_{\mathcal{M}} \qquad \text{for } j = 1, \ldots, n. \tag{2.55}$$

To complete the proof of Theorem 2.54 define φ by the formula

$$\varphi(\lambda) = a + \lambda \langle (1 - D\lambda)^{-1} \gamma, \beta \rangle_{\mathcal{M}} \qquad \text{for } \lambda \in \mathbb{D}.$$

As V is an isometry and $\dim \mathcal{M} = r$, Proposition 2.51 implies that φ is a Blaschke product of degree r, and equation (2.55) implies that $\varphi(\lambda_j) = z_j$ for $j = 1, \ldots, n$. $\qquad \square$

54 *Commutative Theory*

2.6.2 Pick Interpolation on General Subsets of $\mathbb{D}$

As we shall see, the proof of Theorem 2.54 that we just presented is amenable to substantial generalization. A straightforward way to generalize the theorem is to replace the set of nodes $\{\lambda_1, \lambda_2, \ldots, \lambda_n\}$ with an arbitrary subset Λ of $\mathbb{D}$, to replace the targets $\{z_1, z_2, \ldots, z_n\}$ with a function $z\colon \Lambda \to \mathbb{C}$ and to replace the $n \times n$ matrix

$$P = \left[\frac{1 - \bar{z}_i z_j}{1 - \bar{\lambda}_i \lambda_j} \right]$$

with the kernel P on Λ defined by

$$P(\lambda, \mu) = \frac{1 - \overline{z(\mu)}z(\lambda)}{1 - \bar{\mu}\lambda} \qquad \text{for all } \lambda, \mu \in \Lambda. \tag{2.56}$$

Theorem 2.57. *Let data $\Lambda \subseteq \mathbb{D}$ and $z\colon \Lambda \to \mathbb{C}$ be given, and define a kernel P on Λ by equation (2.56). There exists $\varphi \in \mathscr{S}(\mathbb{D})$ such that $\varphi(\lambda) = z(\lambda)$ for all $\lambda \in \Lambda$ if and only if P is positive semi-definite on Λ.*

Proof. Assume that $\Lambda \subseteq \mathbb{D}$ and $z\colon \Lambda \to \mathbb{C}$ and P is as in equation (2.56). If $\varphi \in \mathscr{S}(\mathbb{D})$ and $\varphi(\lambda) = z(\lambda)$ for all $\lambda \in \Lambda$, then Pick's lemma (Lemma 2.2) implies that P is positive semi-definite.

Conversely, if P is positive semi-definite on Λ, then by Moore's theorem (Theorem 2.5) there exist a Hilbert space $\mathcal{M}$ and $u\colon \Lambda \to \mathcal{M}$ such that

$$P(\lambda, \mu) = \langle u_\lambda, u_\mu \rangle_{\mathcal{M}} \qquad \text{for all } \lambda, \mu \in \Lambda,$$

or equivalently,

$$1 - \overline{z(\mu)}z(\lambda) = \langle (1 - \bar{\mu}\lambda)u_\lambda, u_\mu \rangle_{\mathcal{M}} \qquad \text{for all } \lambda, \mu \in \Lambda.$$

This last formula reshuffles to

$$\overline{z(\mu)}z(\lambda) + \langle u_\lambda, u_\mu \rangle_{\mathcal{M}} = 1 + \langle \lambda u_\lambda, \mu u_\mu \rangle_{\mathcal{M}} \qquad \text{for all } \lambda, \mu \in \Lambda,$$

which implies that

$$\left\langle \begin{pmatrix} z(\lambda) \\ u_\lambda \end{pmatrix}, \begin{pmatrix} z(\mu) \\ u_\mu \end{pmatrix} \right\rangle_{\mathbb{C} \oplus \mathcal{M}} = \left\langle \begin{pmatrix} 1 \\ \lambda u_\lambda \end{pmatrix}, \begin{pmatrix} 1 \\ \mu u_\mu \end{pmatrix} \right\rangle_{\mathbb{C} \oplus \mathcal{M}} \qquad \text{for all } \lambda, \mu \in \Lambda.$$

Consequently, by Lemma 2.18, there exists an isometry

$$V\colon \operatorname{dom} V \to \mathbb{C} \oplus \mathcal{M}, \tag{2.58}$$

where

$$\operatorname{dom} V = \left[\begin{pmatrix} 1 \\ \lambda u_\lambda \end{pmatrix} \colon \lambda \in \Lambda \right],$$

Operator Analysis on $\mathbb{D}$ 55

such that

$$V\begin{pmatrix} 1 \\ \lambda u_\lambda \end{pmatrix} = \begin{pmatrix} z(\lambda) \\ u_\lambda \end{pmatrix} \qquad \text{for all } \lambda \in \Lambda. \tag{2.59}$$

We wish to extend V to an isometry $W: \mathbb{C} \oplus \mathcal{M} \to \mathbb{C} \oplus \mathcal{M}$. As explained in Remark 2.31, this is possible if and only if

$$\dim\left((\mathbb{C} \oplus \mathcal{M}) \ominus \left[\begin{pmatrix} z(\lambda) \\ u_\lambda \end{pmatrix} : \lambda \in \Lambda\right]\right) \geq \dim\left((\mathbb{C} \oplus \mathcal{M}) \ominus \operatorname{dom} V\right). \tag{2.60}$$

To guarantee that this condition holds, if it does not already, choose a Hilbert space $\mathcal{N}$ such that

$$\dim \mathcal{N} \geq \max\{\aleph_0, \dim((\mathbb{C} \oplus \mathcal{M}) \ominus \operatorname{dom} V)\}.$$

Let $\mathcal{M}^\sim = \mathcal{M} \oplus \mathcal{N}$, define $u^\sim : \Lambda \to \mathcal{M}^\sim$ by

$$u_\lambda^\sim(\lambda) = u_\lambda \oplus 0 \qquad \text{for all } \lambda \in \Lambda,$$

and define $V^\sim : \operatorname{dom} V \oplus \mathcal{N} \to \mathbb{C} \oplus \mathcal{M}^\sim$ by

$$V^\sim(x \oplus y) = Vx \oplus 0 \qquad \text{for all } x \in \operatorname{dom} V, \ y \in \mathcal{N}.$$

With the constructions of the previous paragraph, equations (2.58), (2.59), and (2.60) hold if we replace $\mathcal{M}$ with $\mathcal{M}^\sim$, u_λ with $u_\lambda^\sim$, and V with $V^\sim$. Consequently, there exists an isometry

$$W = \begin{bmatrix} a & 1 \otimes \beta \\ \gamma \otimes 1 & D \end{bmatrix} : \mathbb{C} \oplus \mathcal{M}^\sim \to \mathbb{C} \oplus \mathcal{M}^\sim$$

such that

$$\begin{bmatrix} a & 1 \otimes \beta \\ \gamma \otimes 1 & D \end{bmatrix}\begin{pmatrix} 1 \\ \lambda u_\lambda^\sim \end{pmatrix} = \begin{pmatrix} z(\lambda) \\ u_\lambda^\sim \end{pmatrix} \qquad \text{for all } \lambda \in \Lambda.$$

But then

$$z(\lambda) = a + \lambda\langle u_\lambda^\sim, \beta\rangle_{\mathcal{M}^\sim} \quad \text{and} \quad u_\lambda^\sim = \gamma + \lambda D u_\lambda^\sim \qquad \text{for all } \lambda \in \Lambda.$$

Solving the second equation above for $u_\lambda^\sim$ and substituting into the first equation, we deduce that

$$z(\lambda) = a + \lambda\langle(1 - D\lambda)^{-1}\gamma, \beta\rangle_{\mathcal{M}^\sim} \qquad \text{for all } \lambda \in \Lambda. \tag{2.61}$$

Define φ by the formula

$$\varphi(\lambda) = a + \lambda\langle(1 - D\lambda)^{-1}\gamma, \beta\rangle_{\mathcal{M}} \qquad \text{for all } \lambda \in \mathbb{D}.$$

As W is an isometry, Theorem 2.36 implies that $\varphi \in \mathscr{S}(\mathbb{D})$ and equation (2.61) implies that $\varphi(\lambda) = z(\lambda)$ for all $\lambda \in \Lambda$. $\qquad\square$

56   Commutative Theory

2.7 The Müntz–Szász Interpolation Theorem

The classical Weierstrass approximation theorem asserts that the polynomials are uniformly dense in $C([0,1])$, the space of continuous complex-valued functions defined on the closed interval $[0,1]$. Equivalently, the monomials $1,t,t^2,\ldots$ have dense linear span in $C([0,1])$. We may choose a strictly increasing sequence $n_0 < n_1 < n_2 < \ldots$ of nonnegative integers and then ask the question: when do the monomials $t^{n_0}, t^{n_1}, t^{n_2}, \ldots$ have dense linear span in $C([0,1])$? The following beautiful result of Müntz [153] and Szász [196] answers this question.

Theorem 2.62. **(The Müntz–Szász theorem)** *If $n_1 < n_2 < n_3 \ldots$ is a strictly increasing sequence of nonnegative integers, then*

$$[t^{n_k}: k \geq 1] = C([0,1]) \quad \textit{if and only if} \quad n_1 = 0 \text{ and } \sum_{k=1}^{\infty} \frac{1}{1+n_k} = \infty.$$

Here we consider the L^2 version of Theorem 2.62.

Theorem 2.63. **(The Müntz–Szász theorem, L^2 version)** *If $n_1 < n_2 < n_3 < \ldots$ is a strictly increasing sequence of nonnegative integers then*

$$[t^{n_k}: k \geq 1] = L^2(0,1) \quad \textit{if and only if} \quad \sum_{k=1}^{\infty} \frac{1}{1+n_k} = \infty.$$

In this section we shall prove Theorem 2.63 and then use the Pick interpolation theorem to answer the following question.

Problem 2.64. **(The Müntz–Szász interpolation problem)**
Let $n_1 < n_2 < n_3 < \ldots$ be a strictly increasing sequence of nonnegative integers and let $c_1, c_2, c_3, \ldots$ be a sequence of complex numbers. When does there exist $X \in \mathcal{B}(L^2(0,1))$ satisfying $\|X\| \leq 1$ and

$$X(t^{n_k}) = c_k t^{n_k}$$

for all $k \geq 1$?

Our analysis will be is based on a transformation from the space $L^2(0,1)$ to the space H^2. Let $\Omega = \{z \in \mathbb{C}: \operatorname{Re} z > -\frac{1}{2}\}$ and notice that

$$t^z \in L^2(0,1) \quad \text{if and only if} \quad z \in \Omega.$$

Notice also that $z \in \Omega$ if and only if $\lambda \in \mathbb{D}$, where λ is defined by

$$\lambda = \frac{z}{1+z}. \tag{2.65}$$

Operator Analysis on $\mathbb{D}$ 57

We may, therefore, define a function v_z in H^2 by the formula

$$v_z = \frac{1}{1+z} s_{\bar{\lambda}}, \tag{2.66}$$

where s is the Szegő kernel function defined in equation (1.30). Now, if $z, w \in \Omega$, $\lambda = z/(1+z)$, and $\mu = w/(1+w)$, then

$$\begin{aligned}
\langle v_z, v_w \rangle_{\mathrm{H}^2} &= \left\langle \frac{1}{1+z} s_{\bar{\lambda}}, \frac{1}{1+w} s_{\bar{\mu}} \right\rangle_{\mathrm{H}^2} \\
&= \frac{1}{(1+z)(1+\bar{w})} \frac{1}{1-\lambda\bar{\mu}} \\
&= \frac{1}{(1+z)(1+\bar{w}) - z\bar{w}} \\
&= \frac{1}{1+z+\bar{w}} \\
&= \left\langle t^z, t^w \right\rangle_{L^2(0,1)}.
\end{aligned} \tag{2.67}$$

Therefore, by the lurking isometry lemma, Lemma 2.18, there exists an isometry $V \colon [t^z \colon z \in \Omega] \to \mathrm{H}^2$ such that

$$V(t^z) = v_z \quad \text{for all } z \in \Omega. \tag{2.68}$$

But $[t^z \colon z \in \Omega] = L^2(0,1)$ and $[v_z \colon z \in \Omega] = [s_{\bar{\lambda}} \colon \lambda \in \mathbb{D}] = \mathrm{H}^2$. So in fact,

$$V \text{ is a Hilbert space isomorphism.} \tag{2.69}$$

The following fact from classical function theory is a well-known result in the convergence theory of Blaschke products. See, for example, [93, section II.2] or [175, theorem 15.23] for a proof.

Lemma 2.70. *Let $\{a_n\}$ be a sequence of points in $\mathbb{D}$. There exists a non-zero function $f \in \mathrm{H}^2$ that vanishes[9] at each point a_n if and only if*

$$\sum_{n=1}^{\infty} (1 - |a_n|) < \infty.$$

This lemma enables us to say when a sequence of Szegő kernel functions has dense linear span.

Lemma 2.71. *If $\{a_n\}$ is a sequence of distinct points in $\mathbb{D}$, then*

$$[s_{a_n} \colon n \geq 1] = \mathrm{H}^2 \quad \text{if and only if} \quad \sum_{n=1}^{\infty} (1 - |a_n|) = \infty.$$

---

[9] If the points are not distinct, the statement about f is interpreted to mean that f vanishes at a point to order v, where v is the number of times that point appears in the sequence.

58   *Commutative Theory*

Proof. Fix a sequence of distinct points $\{a_n\}$ in $\mathbb{D}$ and assume that $\sum_{n=1}^{\infty}(1-|a_n|)<\infty$. By Lemma 2.70 there exists $f \in \mathrm{H}^2$ such that $f \neq 0$ and $f(a_n)=0$ for all n. As $f(a_n)=\langle f, s_{a_n}\rangle_{\mathrm{H}^2}$, it follows that $f \perp [s_{a_n}: n \geq 1]$. As $f \neq 0$, this implies that $[s_{a_n}: n \geq 1] \neq \mathrm{H}^2$.

Now assume that $[s_{a_n}: n \geq 1] \neq \mathrm{H}^2$. Choose $f \in \mathrm{H}^2$ with $f \neq 0$ and $f \perp [s_{a_n}: n \geq 1]$ (so that $f(a_n)=0$ for all n). If we let $z_1, z_2, \ldots$ be an enumeration of the zero set of f then, by Lemma 2.70, $\sum_{m=1}^{\infty}(1-|z_m|)<\infty$. Consequently,

$$\sum_{n=1}^{\infty}(1-|a_n|) \leq \sum_{m=1}^{\infty}(1-|z_m|)<\infty.$$

$\square$

We have assembled the ingredients for a proof of the L^2 version of the Müntz-Szász theorem.

Proof of Theorem 2.63. Fix a strictly increasing sequence of nonnegative integers $n_1<n_2<n_3<\ldots$. By equations (2.68) and (2.69),

$$[t^{n_k}: k \geq 1]=L^2(0,1) \iff [v_{n_k}: k \geq 1]=\mathrm{H}^2.$$

Hence by equations (2.65) and the assertion (2.66),

$$[t^{n_k}: k \geq 1]=L^2(0,1) \iff \left[\frac{1}{1+n_k}s_{\frac{n_k}{1+n_k}}: k \geq 1\right]=\mathrm{H}^2$$

$$\iff \left[s_{\frac{n_k}{1+n_k}}: k \geq 1\right]=\mathrm{H}^2.$$

Therefore, by Lemma 2.71,

$$[t^{n_k}: k \geq 1]=L^2(0,1) \iff \sum_{k=1}^{\infty}\frac{1}{1+n_k}=\infty. \quad \square$$

$\square$

We now turn to the analysis of the Müntz–Szász interpolation problem. Transformation of the question raised in Problem 2.64 by the Hilbert space isomorphism V defined above leads to the following answer.

Theorem 2.72. (The Müntz–Szász interpolation theorem)
Let $n_1<n_2<n_3<\ldots$ be a strictly increasing sequence of nonnegative integers and let $c_1, c_2, c_3, \ldots$ be a sequence of complex numbers. There exists $X \in \mathcal{B}(L^2(0,1))$ satisfying $\|X\| \leq 1$ and

$$X(t^{n_k})=c_k t^{n_k} \quad \text{for all } k \geq 1 \tag{2.73}$$

Operator Analysis on $\mathbb{D}$ 59

if and only if there exists $\varphi \in \mathscr{S}(\mathbb{D})$ satisfying

$$\varphi\left(\frac{n_k}{1+n_k}\right) = c_k \quad \text{for all } k \geq 1. \tag{2.74}$$

Proof. First assume that $X \in \mathcal{B}(L^2(0,1))$, $\|X\| \leq 1$ and the interpolation condition (2.73) holds. As $\|X\| \leq 1$ it follows that $1 - X^*X \geq 0$. Hence if we fix m indices $k_1 < k_2 < \ldots < k_m$, then

$$0 \leq \left\langle (1 - X^*X) \sum_{i=1}^{m} \alpha_i t^{n_{k_i}}, \sum_{i=1}^{m} \alpha_i t^{n_{k_i}} \right\rangle_{L^2(0,1)} \tag{2.75}$$

for all choices of m scalars $\alpha_1, \alpha_2, \ldots, \alpha_m$. Let M denote the right-hand side of equation (2.75). In view of equation (2.73),

$$
\begin{aligned}
M &= \left\langle \sum_{j=1}^{m} \alpha_j t^{n_{k_j}}, \sum_{i=1}^{m} \alpha_i t^{n_{k_i}} \right\rangle_{L^2(0,1)} - \left\langle X \sum_{j=1}^{m} \alpha_j t^{n_{k_j}}, X \sum_{i=1}^{m} \alpha_i t^{n_{k_i}} \right\rangle_{L^2(0,1)} \\
&= \left\langle \sum_{j=1}^{m} \alpha_j t^{n_{k_j}}, \sum_{i=1}^{m} \alpha_i t^{n_{k_i}} \right\rangle_{L^2(0,1)} - \left\langle \sum_{j=1}^{m} c_{k_j} \alpha_j t^{n_{k_j}}, \sum_{i=1}^{m} c_{k_i} \alpha_i t^{n_{k_i}} \right\rangle_{L^2(0,1)} \\
&= \sum_{i,j=1}^{m} (1 - \bar{c}_{k_i} c_{k_j}) \left\langle t^{n_{k_j}}, t^{n_{k_i}} \right\rangle_{L^2(0,1)} \alpha_j \bar{\alpha}_i \\
&= \sum_{i,j=1}^{m} (1 - \bar{c}_{k_i} c_{k_j}) \left\langle v_{n_{k_j}}, v_{n_{k_i}} \right\rangle_{H^2} \alpha_j \bar{\alpha}_i, \tag{2.76}
\end{aligned}
$$

the last step by equation (2.67). Let us define

$$x_k = s_{\frac{n_k}{1+n_k}} \in H^2 \qquad \text{for } k = 1, \ldots, m. \tag{2.77}$$

Then, by statements (2.76), (2.65), and (2.66),

$$
\begin{aligned}
M &= \sum_{i,j=1}^{m} (1 - \bar{c}_{k_i} c_{k_j}) \left\langle \frac{1}{1+n_{k_j}} x_{k_j}, \frac{1}{1+n_{k_i}} x_{k_i} \right\rangle_{H^2} \alpha_j \bar{\alpha}_i \\
&= \sum_{i,j=1}^{m} (1 - \bar{c}_{k_i} c_{k_j}) \left\langle x_{k_j}, x_{k_i} \right\rangle_{H^2} \frac{\alpha_j}{1+n_{k_j}} \frac{\bar{\alpha}_i}{1+n_{k_i}}.
\end{aligned}
$$

Therefore, letting $\beta_i = \frac{\alpha_i}{1+n_{k_i}}$ for $i = 1, \ldots, m$, we have, for all $\beta_1, \ldots, \beta_m \in \mathbb{C}$,

$$\sum_{i,j=1}^{m} (1 - \bar{c}_{k_i} c_{k_j}) \left\langle x_{k_j}, x_{k_i} \right\rangle_{H^2} \beta_j \bar{\beta}_i \geq 0,$$

60   *Commutative Theory*

which is to say that the $m \times m$ matrix A defined by

$$A = \left[(1 - \bar{c}_{k_i} c_{k_j}) \left\langle x_{k_j}, x_{k_i} \right\rangle_{H^2} \right] \tag{2.78}$$

is positive semi-definite.

To summarize, in the previous paragraph we showed that the matrix A defined by equation (2.78) is positive semi-definite whenever $m \geq 1$ and $k_1 < k_2 < \ldots < k_m$ is sequence of positive integers. Hence if $\Lambda = \{n_k : k \geq 1\}$ and a kernel P on Λ is defined by the formula

$$P(n_k, n_l) = (1 - \bar{c}_l c_k) \left\langle s_{\frac{n_k}{1+n_k}}, s_{\frac{n_l}{1+n_l}} \right\rangle_{H^2} \qquad \text{for all } k, l \geq 1,$$

then, in the light of equation (2.77), P is a positive semi-definite kernel on Λ. Now, P is just the Pick kernel on Λ corresponding to the data $n_k/(1+n_k) \mapsto c_k$. Consequently, Theorem 2.57 implies that there exists $\varphi \in \mathscr{S}(\mathbb{D})$ such that the interpolation condition (2.74) holds.

Now assume that $\varphi \in \mathscr{S}(\mathbb{D})$ and equation (2.74) holds. Observe that if we define

$$\varphi^\vee(\lambda) \stackrel{\text{def}}{=} \overline{\varphi(\bar{\lambda})},$$

then $\varphi^\vee \in \mathscr{S}(\mathbb{D})$, so that $\|M_{\varphi^\vee}^*\| \leq 1$. As V is a Hilbert space isomorphism, the operator $X = V^{-1} M_{\varphi^\vee}^* V$ is a contraction. There remains to show that equation (2.73) holds. But for $k \geq 1$,

$$X(t^{n_k}) = V^{-1} M_{\varphi^\vee}^* V (t^{n_k})$$

$$= \frac{1}{1+n_k} V^{-1} M_{\varphi^\vee}^* s_{\frac{n_k}{1+n_k}} \qquad \text{by equations (2.66), (2.68)}$$

$$= \frac{1}{1+n_k} V^{-1} \overline{\varphi^\vee \left(\frac{n_k}{1+n_k} \right)} s_{\frac{n_k}{1+n_k}} \qquad \text{by Lemma 1.65}$$

$$= \frac{1}{1+n_k} V^{-1} \varphi \left(\frac{n_k}{1+n_k} \right) s_{\frac{n_k}{1+n_k}}$$

$$= \frac{c_k}{1+n_k} V^{-1} s_{\frac{n_k}{1+n_k}} \qquad \text{by equation (2.74)}$$

$$= c_k t^{n_k} \qquad \text{by equations (2.66), (2.68).}$$

$$\square$$

Notice that, by Theorem 2.57, condition (2.74) can be reformulated to say that the matrix

$$\left[\frac{1 - \bar{c}_i c_j}{1 + n_i + n_j} \right]$$

is positive semi-definite.

$$Operator\ Analysis\ on\ \mathbb{D} \qquad 61$$

Similar arguments apply to prove an interpolation theorem for a sequence $\{t^{z_n}\}$ where $\operatorname{Re} z_n > -\frac{1}{2}$ for all n.

2.8 Positivity Arguments

In this section we shall derive von Neumann's inequality from the fundamental theorem for $\mathscr{S}(\mathbb{D})$ via a *positivity argument*. In a positivity argument one proves that an operator is positive by expressing it in terms of simpler operators that are known to be positive, using operations that preserve positivity.

To see how a positivity argument can be applied to prove von Neumann's inequality, observe that Theorem 1.47 asserts that

$$\|T\| \leq 1 \implies \|X\| \leq 1 \tag{2.79}$$

whenever $\mathcal{H}$ is a Hilbert space, $T \in \mathcal{B}(\mathcal{H})$, $\sigma(T) \subseteq \mathbb{D}$, $\varphi \in \mathscr{S}(\mathbb{D})$ and $X = \varphi(T)$. On the other hand, by the equivalence (1.13), the implication (2.79) is equivalent to

$$1 - T^*T \geq 0 \implies 1 - X^*X \geq 0. \tag{2.80}$$

We will prove the statement (2.80) by expressing $1 - X^*X$ in terms of $1 - T^*T$ by a sequence of operations that preserve positivity. To that end, the following lemma is useful.

Lemma 2.81. (The preservation of positivity) *Let $\mathcal{H}$ be a Hilbert space and let $A, B, C \in \mathcal{B}(\mathcal{H})$.*

(i) *If $A \geq 0$ and $B \geq 0$ then $A + B \geq 0$;*

(ii) *If $A \geq 0$ then $C^*AC \geq 0$;*

(iii) *if (A_n) is a sequence of positive operators on $\mathcal{H}$ and A_n converges to an operator A in norm, then $A \geq 0$.*

Proof. Suppose that $A \geq 0$ and $B \geq 0$; recall that this terminology was introduced in Definition 1.9. We have, for any $x \in \mathcal{H}$,

$$\langle (A+B)x, x \rangle = \langle Ax, x \rangle + \langle Bx, x \rangle \geq 0$$
$$\langle C^*ACx, x \rangle = \langle ACx, Cx \rangle \geq 0.$$

To prove the final statement, consider any $x \in \mathcal{H}$. Then $\langle A_n x, x \rangle \geq 0$ for each n, and $\langle A_n x, x \rangle \to \langle Ax, x \rangle$ as $n \to \infty$. Hence $\langle Ax, x \rangle \geq 0$. $\qquad\square$

62 *Commutative Theory*

2.8.1 Hereditary Functions and Functional Calculus

Another ingredient in positivity proofs is the *hereditary functional calculus*. We say that h is a *hereditary function on* $\mathbb{D}$ if h is a mapping from $\mathbb{D} \times \mathbb{D}$ to $\mathbb{C}$ and has the property that

$$(\lambda, \mu) \mapsto h(\lambda, \bar{\mu}) \in \mathbb{C}$$

is a holomorphic function on $\mathbb{D} \times \mathbb{D}$.

We denote by $\text{Her}(\mathbb{D})$ the set of hereditary functions on $\mathbb{D}$. The space $\text{Her}(\mathbb{D})$ is a complete metrizable locally convex topological vector space (a Fréchet space) when equipped with the topology of uniform convergence on compact subsets of $\mathbb{D} \times \mathbb{D}$.

If $\mathcal{H}$ is a Hilbert space, $T \in \mathcal{B}(\mathcal{H})$ with $\sigma(T) \subseteq \mathbb{D}$ and h is a hereditary function on $\mathbb{D}$, then we may define $h(T) \in \mathcal{B}(\mathcal{H})$ by the following procedure. Expand h into a power series

$$h(\lambda, \mu) = \sum_{m,n} c_{mn} \bar{\mu}^n \lambda^m \qquad \text{for all } \lambda, \mu \in \mathbb{D}, \tag{2.82}$$

and then define $h(T)$ by substituting T for λ and T^* for $\bar{\mu}$, taking care to group powers of T^* to the left of powers of T, that is,

$$h(T) = \sum_{m,n} c_{mn} (T^*)^n T^m. \tag{2.83}$$

An alternative way to define $h(T)$ is to use the double Riesz-Dunford integral,

$$h(T) = \frac{1}{(2\pi i)^2} \int_\gamma \int_\gamma h(\lambda, \bar{\mu})(\bar{\mu} - T^*)^{-1}(\lambda - T)^{-1} \, d\lambda \, d\bar{\mu}, \tag{2.84}$$

where γ is a contour in $\mathbb{D}$ of the form $\gamma(t) = re^{it}, 0 \leq t \leq 2\pi$, with r chosen so that $r < 1$ and $\sigma(T) \subseteq r\mathbb{D}$.

Lemma 2.85. *If $\mathcal{H}$ is a Hilbert space, $T \in \mathcal{B}(\mathcal{H})$ with $\sigma(T) \subseteq \mathbb{D}$, and $h \in \ker(\mathbb{D})$, then equations (2.83) and (2.84) define the same element $h(T)$ in $\mathcal{B}(\mathcal{H})$.*

Proof. Fix $h \in \text{Her}(\mathbb{D})$ and let equation (2.82) give the power series representation of $h(\lambda, \bar{\mu})$ on $\mathbb{D} \times \mathbb{D}$. Fix $T \in \mathcal{B}(\mathcal{H})$ with $\sigma(T) \subseteq \mathbb{D}$ and choose $r < 1$ such that $\sigma(T) \subseteq r\mathbb{D}$.

Let us take $h(T)$ to be defined by the integral formula (2.84), so that

$$h(T) = \frac{1}{(2\pi i)^2} \int_\gamma \int_\gamma h(\lambda, \bar{\mu})(\bar{\mu} - T^*)^{-1}(\lambda - T)^{-1} \, d\lambda \, d\bar{\mu}.$$

As the series (2.82) converges uniformly on compact subsets of $\mathbb{D} \times \mathbb{D}$, in particular the series converges uniformly on $r\mathbb{T}^2$. Therefore,

$$h(T) = \frac{1}{(2\pi i)^2} \int_\gamma \int_\gamma \left(\sum_{m,n} c_{mn} \bar{\mu}^n \lambda^m \right) (\bar{\mu} - T^*)^{-1} (\lambda - T)^{-1} \, d\lambda \, d\bar{\mu}$$

$$= \sum_{m,n} c_{mn} \frac{1}{(2\pi i)^2} \int_\gamma \int_\gamma \left(\bar{\mu}^n \lambda^m \right) (\bar{\mu} - T^*)^{-1} (\lambda - T)^{-1} \, d\lambda \, d\bar{\mu}$$

$$= \sum_{m,n} c_{mn} \left(\frac{1}{2\pi i} \int_\gamma \bar{\mu}^n (\bar{\mu} - T^*)^{-1} d\bar{\mu} \right) \left(\frac{1}{2\pi i} \int_\gamma \lambda^m (\lambda - T)^{-1} d\lambda \right)$$

$$= \sum_{m,n} c_{mn} \, (T^*)^n \, T^m,$$

in agreement with the alternative definition (2.83) of $h(T)$. $\qquad \square$

There is a natural involution $h \mapsto h^*$ on $\mathrm{Her}(\mathbb{D})$. By an *involution* on a complex vector space E we mean a self-map $x \mapsto x^*$ of E such that

(i) $(\lambda x + \mu y)^* = \bar{\lambda} x^* + \bar{\mu} y^*$ for all $\lambda, \mu \in \mathbb{C}$ and $x, y \in E$;
(ii) $(x^*)^* = x$ for all $x \in E$.

When E is in addition an algebra, we require also

(iii) for all $x, y \in E$, $(xy)^* = y^* x^*$.

We define an involution on $\mathrm{Her}(\mathbb{D})$ by

$$h^*(\lambda, \mu) = \overline{h(\mu, \lambda)} \qquad \text{for all } \lambda, \mu \in \mathbb{D}.$$

Lemma 2.86. (Properties of the hereditary functional calculus) *Let $\mathcal{H}$ be a Hilbert space and let $T \in \mathcal{B}(\mathcal{H})$ satisfy $\sigma(T) \subseteq \mathbb{D}$.*

(i) *If $f, g \in \mathrm{Her}(\mathbb{D})$ and $\alpha, \beta \in \mathbb{C}$, then*

$$(\alpha f + \beta g)(T) = \alpha f(T) + \beta g(T).$$

(ii) *If $\varphi \in \mathrm{Hol}(\mathbb{D})$ and $h(\lambda, \mu) = \varphi(\lambda)$ for all $\lambda, \mu \in \mathbb{D}$, then $h(T) = \varphi(T)$.*
(iii) *If $\psi \in \mathrm{Hol}(\mathbb{D})$ and $h(\lambda, \mu) = \overline{\psi(\mu)}$ for all $\lambda, \mu \in \mathbb{D}$, then $h(T) = \psi(T)^*$.*
(iv) *$h^*(T) = h(T)^*$.*
(v) *If $\varphi, \psi \in \mathrm{Hol}(\mathbb{D})$, $h \in \mathrm{Her}(\mathbb{D})$, and $g \in \mathrm{Her}(\mathbb{D})$ is defined by*

$$g(\lambda, \mu) = \overline{\psi(\mu)} \, h(\lambda, \mu) \, \varphi(\lambda) \qquad \text{for all } \lambda, \mu \in \mathbb{D},$$

then

$$g(T) = \psi(T)^* h(T) \varphi(T).$$

It is straightforward to prove these formulas from either of the two definitions (2.83) or (2.84). Alternatively, observe that, in the case that all the functions concerned are *polynomials*, the properties follow by simple algebra.

64 *Commutative Theory*

One can then deduce the corresponding statements for hereditary functions by continuity, in view of the following important result.

Proposition 2.87. *The hereditary functional calculus is continuous: if $T \in \mathcal{B}(\mathcal{H})$ satisfies $\sigma(T) \subseteq \mathbb{D}$ and if $\{h_n\}$ is a sequence in $\mathrm{Her}(\mathbb{D})$, $h \in \mathrm{Her}(\mathbb{D})$, and $h_n \to h$ uniformly on compact subsets of $\mathbb{D} \times \mathbb{D}$, then $h_n(T) \to h(T)$ in $\mathcal{B}(\mathcal{H})$ with respect to the operator norm.*

Proof. Choose a positive $r < 1$ such that $\sigma(T) \subseteq r\mathbb{D}$ and let $\gamma(t) = re^{it}$, $0 \le t \le 2\pi$. The hypotheses imply that h_n tends to h uniformly on $r(\mathbb{D}^-)^2$. Now

$$h_n(T) = \frac{1}{(2\pi i)^2} \int_\gamma \int_\gamma h_n(\lambda, \bar{\mu})(\bar{\mu} - T^*)^{-1}(\lambda - T)^{-1} \, d\lambda \, d\bar{\mu}.$$

It follows from Lemma 2.27, with T replaced by T/r, that

$$(\lambda, \mu) \mapsto (\bar{\mu} - T^*)^{-1}(\lambda - T)^{-1} \quad \text{is bounded on } r\mathbb{T} \times r\mathbb{T},$$

and therefore that

$$h_n(\lambda, \bar{\mu})(\bar{\mu} - T^*)^{-1}(\lambda - T)^{-1} \to h(\lambda, \bar{\mu})(\bar{\mu} - T^*)^{-1}(\lambda - T)^{-1}$$

in the operator norm, uniformly for $\lambda, \bar{\mu} \in r\mathbb{T}$. Hence

$$h_n(T) \to \frac{1}{(2\pi i)^2} \int_\gamma \int_\gamma h(\lambda, \bar{\mu})(\bar{\mu} - T^*)^{-1}(\lambda - T)^{-1} \, d\lambda \, d\bar{\mu}$$

$$= h(T)$$

as $n \to \infty$. $\qquad\qquad\qquad\qquad\qquad\qquad\qquad\qquad\qquad\qquad\qquad\square$

2.8.2 Positive Definite Functions Are Sums of Dyads

Another useful ingredient in positivity proofs is the following corollary of Moore's theorem. A *dyad* in $\mathrm{Her}(\mathbb{D})$ is a function of the form

$$(\lambda, \mu) \mapsto \overline{\psi(\mu)}\varphi(\lambda) \qquad \text{for all } \lambda, \mu \in \mathbb{D}$$

for some functions φ, ψ in $\mathrm{Hol}(\mathbb{D})$. Note that if $A \in \mathrm{Her}(\mathbb{D})$, then A is a kernel on $\mathbb{D}$, and so it makes sense to say that A is positive semi-definite.

Theorem 2.88. *If A is a positive semi-definite hereditary function on $\mathbb{D}$, then there exists a sequence $f_1, f_2, \ldots$ in $\mathrm{Hol}(\mathbb{D})$ such that*

$$A(\lambda, \mu) = \sum_{n=1}^{\infty} \overline{f_n(\mu)} f_n(\lambda) \quad \text{for all } \lambda, \mu \in \mathbb{D}, \tag{2.89}$$

the series in equation (2.89) converging uniformly on compact subsets of $\mathbb{D} \times \mathbb{D}$.

Operator Analysis on $\mathbb{D}$
65

To prove this statement we shall use Lemma 2.90. If $\mathcal{X}$ is a Banach space, $\Omega \subseteq \mathbb{C}^d$ is a domain, and $f \colon \Omega \to \mathcal{X}$ is a mapping, then f is said to be *weakly holomorphic* on Ω if the map $\lambda \mapsto x^*(f(\lambda))$ is holomorphic on Ω for all $x^* \in \mathcal{X}^*$.

Lemma 2.90. *A map* $f \colon \Omega \to \mathcal{X}$ *is weakly holomorphic on a domain* $\Omega \subset \mathbb{C}^d$ *if and only if* f *is holomorphic on* Ω *in the sense of Definition* 1.40.

Proof. It is clear that holomorphic functions are weakly holomorphic.

We first prove the converse for the case that $d = 1$. Consider a point $z \in \Omega$ and choose $r > 0$ such that the disc $D(z,r)$ with center z and radius r is contained in Ω. Next choose δ such that $0 < \delta < \frac{1}{2}r$. Consider any $x^* \in X^*$. By hypothesis $x^* \circ f$ is a $\mathbb{C}$-valued analytic function on Ω, and we may therefore apply Cauchy's integral formula to it. Hence if $0 < |s| < \delta$ and $0 < |t| < \delta$ and γ denotes the positively oriented boundary of $D(z,r)$, then

$$
x^* \left(\frac{f(z+s) - f(z)}{s} - \frac{f(z+t) - f(z)}{t} \right)
$$

$$
= \frac{x^* \circ f(z+s) - x^* \circ f(z)}{s} - \frac{x^* \circ f(z+t) - x^* \circ f(z)}{t}
$$

$$
= \frac{1}{2\pi i} \int_\gamma \frac{1}{s} \left(\frac{x^* \circ f(w)}{w - z - s} - \frac{x^* \circ f(w)}{w - z} \right) - \frac{1}{t} \left(\frac{x^* \circ f(w)}{w - z - t} - \frac{x^* \circ f(w)}{w - z} \right) dw
$$

$$
= \frac{s - t}{2\pi i} \int_\gamma \frac{(x^* \circ f)(w)\, dw}{(w - z)(w - z - s)(w - z - t)}.
$$

Hence there is a $c > 0$ such that

$$
\left| x^* \left(\frac{f(z+s) - f(z)}{s} - \frac{f(z+t) - f(z)}{t} \right) \right|
$$

$$
\leq \frac{|s - t|}{r} \sup_{w \in \gamma} \left| \frac{x^* \circ f(w)}{(w - z - s)(w - z - t)} \right|
$$

$$
\leq c|s - t| \, \|x^*\| \, \sup_\gamma \|f(w)\|. \tag{2.91}
$$

Taking the suprema of both sides over x^* of unit norm, we have

$$
\left\| \frac{f(z+s) - f(z)}{s} - \frac{f(z+t) - f(z)}{t} \right\| \leq c \sup_\gamma \|f(w)\| \cdot |s - t|. \tag{2.92}
$$

It follows that, for any sequence $S = (s_n)$ that tends to zero in $D(0,\delta)$,

$$
\left(\frac{f(z+s_n) - f(z)}{s_n} \right)_n \quad \text{is a Cauchy sequence in } X. \tag{2.93}
$$

By completeness of X, the sequence converges to a limit $l(S) \in X$. Now $l(S)$ is in fact independent of S. For suppose sequences S_1, S_2 tend to zero and

66 *Commutative Theory*

$l(S_1) \neq l(S_2)$. Then we may form the sequence with alternate terms from S_1 and S_2, and for this sequence the Cauchy sequence in the statement (2.93) will fail to converge. Thus $l(S)$ does not depend on S, which is to say that

$$\lim_{s \to 0} \frac{f(z+s) - f(z)}{s} \quad \text{exists.}$$

Hence f is complex-differentiable at z in the case that $d = 1$.

We shall next give a proof in the case that $d = 2$; extension to the case of general $d > 1$ is straightforward.

Let $f: \Omega \to X$ be weakly holomorphic, where Ω is a domain in $\mathbb{C}^2$. Let $z \in \Omega$. Choose $r > 0$ such that the polydisc $D(z,r)$ with center z and polyradius (r,r) is contained in Ω. Let δ satisfy $0 < \delta < \frac{1}{2}r$. By the case $d = 1$ of the lemma, the first-order partial derivatives of f exist, and by virtue of inequality (2.92), there exists $c > 0$ such that

$$\left\| \frac{\partial f}{\partial z^1}(z^1, z^2 + t^2) - \frac{f(z^1 + t^1, z^2 + t^2) - f(z^1, z^2 + t^2)}{t^1} \right\| \leq c\delta |t^1|,$$

$$\left\| \frac{\partial f}{\partial z^2}(z) - \frac{f(z^1, z^2 + t^2) - f(z)}{t^2} \right\| \leq c\delta |t^2| \quad (2.94)$$

for all non-zero t^1, t^2 of modulus less than δ. Moreover, the first-order partial derivatives of f are weakly holomorphic in Ω. Indeed, consider any $x^* \in X^*$. Since

$$x^* \circ \frac{\partial f}{\partial z^j} = \frac{\partial}{\partial z^j} x^* \circ f, \quad (2.95)$$

for $j = 1, 2$, the left-hand side is equal to a partial derivative of a scalar holomorphic function and hence is holomorphic.

In the identity

$$f(z+t) - f(z) - t^1 \frac{\partial f}{\partial z^1}(z) - t^2 \frac{\partial f}{\partial z^2}(z)$$

$$= f(z+t) - f(z^1, z^2 + t^2) - t^1 \frac{\partial f}{\partial z^1}(z^1, z^2 + t^2)$$

$$+ f(z^1, z^2 + t^2) - f(z) - t^2 \frac{\partial f}{\partial z^2}(z)$$

$$+ t^1 \left(\frac{\partial f}{\partial z^1}(z^1, z^2 + t^2) - \frac{\partial f}{\partial z^1}(z) \right)$$

Operator Analysis on $\mathbb{D}$ 67

take the norm of both sides and use the estimates (2.94) to obtain

$$\left\| f(z+t) - f(z) - t^1 \frac{\partial f}{\partial z^1}(z) - t^2 \frac{\partial f}{\partial z^2}(z) \right\|$$

$$\leq c\delta(|t^1|^2 + |t^2|^2) + |t^1| \left\| \frac{\partial f}{\partial z^1}(z^1, z^2 + t^2) - \frac{\partial f}{\partial z^1}(z) \right\| \qquad (2.96)$$

for small enough $\|t\|$. Again by the case $d = 1$ of the lemma, applied to the weakly holomorphic function $\frac{\partial f}{\partial z^1}(z^1, \cdot)$, the latter function is holomorphic (and hence locally Lipschitz) at z^2. That is,

$$\left\| \frac{\partial f}{\partial z^1}(z^1, z^2 + t^2) - \frac{\partial f}{\partial z^1}(z) \right\| = O(|t^2|) \quad \text{as } t^2 \to 0.$$

It follows from inequality (2.96) that

$$\left\| f(z+t) - f(z) - t^1 \frac{\partial f}{\partial z^1}(z) - t^2 \frac{\partial f}{\partial z^2}(z) \right\| = O(\|t\|^2) \quad \text{as } \|t\| \to 0.$$

Thus f is Fréchet-differentiable at z. $\qquad\square$

Remark 2.97. One can weaken the hypothesis significantly in Lemma 2.90. Let us say that a subset E of $\mathcal{X}^*$ is *norm-generating* if, for every $x \in \mathcal{X}$,

$$\|x\| = \sup_{x^* \in E \setminus \{0\}} \frac{|x^*(x)|}{\|x^*\|}.$$

In particular, any norm-dense subset of $\mathcal{X}^*$ is norm-generating, as is the unit sphere of the pre-dual of $\mathcal{X}$ if $\mathcal{X}$ is a dual space.

If the map $\lambda \mapsto x^*(f(\lambda))$ is holomorphic on Ω for every x^* in a norm-generating subset E of $\mathcal{X}^*$, then f is holomorphic.

The proof requires only a slight modification. The inequality (2.91) now holds for all x^* in E; divide through by $\|x^*\|$ and take the supremum over E to obtain the estimate (2.92) as before. Later, equation (2.95) holds for all $x^* \in E$, and so, again, the first-order partial derivatives of f are holomorphic. The proof then concludes as before.

Proof of Theorem 2.88. As A is positive semi-definite on $\mathbb{D}$, by Theorem 2.5, there exists a Hilbert space $\mathcal{M}$ and a function $u: \mathbb{D} \to \mathcal{M}$ such that $[\operatorname{ran} u] = \mathcal{M}$ and, for all $\lambda, \mu \in \mathbb{D}$,

$$A(\lambda, \mu) = \langle u_\lambda, u_\mu \rangle_{\mathcal{M}}. \qquad (2.98)$$

Commutative Theory

First we show that u is holomorphic. Fix v in the span of ran u, which is dense in $\mathcal{M}$. So v can be written as a finite sum $\sum c_j u_{\lambda_j}$. Then

$$\langle u_\lambda, v \rangle_{\mathcal{M}} = \left\langle u_\lambda, \sum c_j u_{\lambda_j} \right\rangle$$
$$= \sum \overline{c_j} A(\lambda, \lambda_j)$$

is holomorphic. Since ran u is a norm-generating set in $\mathcal{M}$, Lemma 2.90 and Remark 2.97 imply that u_λ is holomorphic on $\mathbb{D}$.

We next show that $\mathcal{M}$ is separable. Fix a sequence of distinct points $\{\lambda_n\}$ in $\mathbb{D}$ with $\lambda_n \to 0$. If $v \in \mathcal{M}$ is orthogonal to $\{u_{\lambda_n} : n \geq 1\}$, then the holomorphic function $g(\lambda) = \langle u_\lambda, v \rangle_{\mathcal{M}}$ vanishes on the set of uniqueness $\{\lambda_n : n \geq 1\}$. Therefore, $g(\lambda) = 0$ for all $\lambda \in \mathbb{D}$, that is, $v \perp \mathcal{M}$. This proves that $\{u_{\lambda_n} : n \geq 1\}$ has dense linear span in $[\text{ran} \, u]$. In particular, $\mathcal{M}$ is separable.

Now choose an orthonormal basis $\{e_n\}$ (either finite or countably infinite) for $\mathcal{M}$. By Parseval's equality, for every pair $\lambda, \mu \in \mathbb{D}$,

$$A(\lambda, \mu) = \langle u_\lambda, u_\mu \rangle_{\mathcal{M}} = \sum_n \langle u_\lambda, e_n \rangle_{\mathcal{M}} \overline{\langle u_\mu, e_n \rangle_{\mathcal{M}}} \tag{2.99}$$

and the series converges. Therefore, if we define f_n by $f_n(\lambda) = \langle u_\lambda, e_n \rangle_{\mathcal{M}}$, then $f_n \in \text{Hol}(\mathbb{D})$ and equation (2.89) holds.

Finally, we prove that the series (2.89) converges uniformly on compact subsets of $\mathbb{D} \times \mathbb{D}$. Fix a compact subset $K \subset \mathbb{D} \times \mathbb{D}$ and choose $r_1 < r < 1$ such that $K \subseteq r_1(\mathbb{D}^2)$. As A is bounded on $r\mathbb{D}^2$, there exists a constant M such that

$$A(re^{i\theta}, re^{i\theta}) = \sum_n |f_n(re^{i\theta})|^2 \leq M \qquad \text{for all } \theta \in \mathbb{R}.$$

For any $f \in \text{Hol}(\mathbb{D})$ let us denote by $(f)_r$ the function defined by $(f)_r(z) = f(rz)$ for $z \in \mathbb{D}$; then

$$\sum_n |(f_n)_r(e^{i\theta})|^2 \leq M \qquad \text{for all } \theta \in \mathbb{R}.$$

Integrate both sides of this inequality with respect to normalized Lebesgue measure on $\mathbb{T}$ to obtain

$$\sum_n \|(f_n)_r\|_{H^2}^2 \leq M.$$

Therefore, for every $\varepsilon > 0$, there exists N such that

$$\sum_{n \geq N} \|(f_n)_r\|_{H^2}^2 \leq \varepsilon. \tag{2.100}$$

As $r_1 < r < 1$, there exists $C > 0$ such that, for all $f \in \text{Hol}(\mathbb{D})$,

$$\sup\{|f(\lambda)| : |\lambda| \leq r_1\} \leq C \|(f)_r\|_{H^2}. \tag{2.101}$$

Therefore, for all $(\lambda, \mu) \in K$,

$$
\sum_{n \geq N} |f_n(\lambda)\overline{f_n(\mu)}| \leq \left\{\sum_{n \geq N} |f_n(\lambda)|^2\right\}^{\frac{1}{2}} \left\{\sum_{n \geq N} |f_n(\mu)|^2\right\}^{\frac{1}{2}}
$$

$$
\leq C^2 \sum_{n \geq N} \|(f_n)_r\|_{\mathrm{H}^2}^2 \qquad \text{by inequality (2.101)}
$$

$$
\leq C^2 \varepsilon \qquad \text{by inequality (2.100)}.
$$

Since $\varepsilon > 0$ is arbitrary, the series (2.89) converges uniformly on K. $\qquad\square$

2.8.3 Von Neumann's Inequality via Model Theory

In this subsection we shall use the fundamental theorem for $\mathscr{S}(\mathbb{D})$, the existence of models, and the positivity tools assembled in the previous subsections to prove the Schur form of von Neumann's inequality (Theorem 1.47): for any contraction T with spectrum in $\mathbb{D}$ and any function φ in the Schur class, $\varphi(T)$ is a contraction.

Proof. Let $\varphi \in \mathscr{S}(\mathbb{D})$, let $T \in \mathcal{B}(\mathcal{H})$ be a contraction and assume that $\sigma(T) \subseteq \mathbb{D}$. By the fundamental theorem for $\mathscr{S}(\mathbb{D})$ there exists a model $(\mathcal{M}, u)$ for φ. By Theorem 2.88 with $A(\lambda, \mu) = \langle u_\lambda, u_\mu \rangle_{\mathcal{M}}$, the model formula (2.8) becomes

$$
1 - \overline{\varphi(\mu)}\varphi(\lambda) = \sum_n \overline{f_n(\mu)}(1 - \bar{\mu}\lambda)f_n(\lambda) \tag{2.102}
$$

with uniform convergence on compact subsets of $\mathbb{D} \times \mathbb{D}$.

To complete the proof view each side of equation (2.102) as a hereditary function and substitute T into both sides using the hereditary calculus. The definition of the hereditary calculus (see equations (2.82) and (2.83)) implies that $(1 - \bar{\mu}\lambda)(T) = 1 - T^*T$. Therefore, by Property (v) in Lemma 2.86, for all $n \geq 1$,

$$
\left(\overline{f_n(\mu)}(1 - \bar{\mu}\lambda)f_n(\lambda)\right)(T) = f_n(T)^*(1 - T^*T)f_n(T).
$$

Hence by property (i) in Lemma 2.86, for all $N \geq 1$,

$$
\left(\sum_{n=1}^{N} \overline{f_n(\mu)}(1 - \bar{\mu}\lambda)f_n(\lambda)\right)(T) = \sum_{n=1}^{N} f_n(T)^*(1 - T^*T)f_n(T).
$$

Hence as

$$
\left(1 - \overline{\varphi(\mu)}\varphi(\lambda)\right)(T) = 1 - \varphi(T)^*\varphi(T),
$$

70 *Commutative Theory*

and
$$\sum_{n=1}^{N} \overline{f_n(\mu)}(1 - \bar{\mu}\lambda) f_n(\lambda) \;\to\; 1 - \overline{\varphi(\mu)}\varphi(\lambda) \quad \text{in} \quad \text{Her}(\mathbb{D}),$$

it follows by the continuity of the hereditary calculus, Proposition 2.87 that

$$1 - \varphi(T)^*\varphi(T) = \sum_{n=1}^{\infty} f_n(T)^*(1 - T^*T) f_n(T), \qquad (2.103)$$

where the series on the right hand side converges in the operator norm.

Since $\|T\| \le 1$, the fundamental fact (1.13) implies that $1 - T^*T \ge 0$. Hence, by Lemma 2.81(ii), $f_n(T)^*(1 - T^*T) f_n(T) \ge 0$ for each n. Therefore, by Lemma 2.81(i),

$$\sum_{n=1}^{N} f_n(T)^*(1 - T^*T) f_n(T) \ge 0$$

for each N. Since the right-hand side of equation (2.103) converges in the operator norm, Lemma 2.81(iii) now shows that $1 - \varphi(T)^*\varphi(T)$ is positive as well. Hence the fundamental fact implies that $\|\varphi(T)\| \le 1$. $\qquad \square$

2.9 Historical Notes

Lemma 2.5 is sometimes called the Aronszajn theorem. In 1935 E. H. Moore showed how to construct a Hilbert function space from a positive kernel [152], while in 1950 N. Aronszajn developed the theory of reproducing kernels in depth [35]. Kernels as positive definite functions had previously been studied by J. Mercer [151] in 1909.

Theorem 2.62 was proved by C. Müntz in [153]. O. Szász generalized the proof to complex powers, and also proved the L^2 version (Theorem 2.63) [196]. For a more modern proof of Theorem 2.62, see [175, theorem 15.26].

Lemma 2.90 on weakly holomorphic functions. Proofs of this result for the case that $d = 1$ can be found in many textbooks, for example, [176, 60]. The proof we give here (for $d = 1$) was posted on math.stackexchange.com by D. Ullrich. Unlike many, it does not depend on the Uniform Boundedness Principle or on other results proved by a Baire category argument. We prove the theorem for $d = 2$ since we were unable to find the case $d > 1$ in standard textbooks.

3
Further Development of Models on the Disc

The ideas developed in Chapter 2 for the study of Schur-class functions apply equally well, with modifications, to some other classes of functions. In this chapter we shall modify the model formula in two ways: first, to analyse *operator-valued* Schur-class functions and, second, to study the *corona problem*.

3.1 A Model Formula for $\mathscr{S}_{\mathcal{B}(\mathcal{H},\mathcal{K})}(\mathbb{D})$

For engineering applications, mappings on $\mathbb{D}$ that take values in $\mathcal{B}(\mathcal{H},\mathcal{K})$, not only in $\mathbb{C}$, are significant. In applications $\mathcal{H}, \mathcal{K}$ are typically Hilbert spaces of input and output signals, respectively, of a linear system. Even in a purely mathematical context, holomorphic operator-valued functions are increasingly important, as witness the notion of a complete spectral set, Definition 1.51, and Arveson's dilation theorem, Theorem 1.52.

Adjustments to the derivation of the fundamental theorem for $\mathscr{S}(\mathbb{D})$ yield the existence of models for $\mathscr{S}_{\mathcal{B}(\mathcal{H},\mathcal{K})}(\mathbb{D})$ and, as a consequence, allow one to generalize the entirety of the results from the previous chapter. In this section we sketch some of the details.

Recall from Section 1.11 that $\mathrm{H}^{\infty}_{\mathcal{B}(\mathcal{H},\mathcal{K})}(\mathbb{D})$ denotes the Banach space of bounded analytic $\mathcal{B}(\mathcal{H},\mathcal{K})$-valued functions on $\mathbb{D}$. To define a model formula in the setting of $\mathcal{B}(\mathcal{H},\mathcal{K})$-valued maps we need first to generalize Pick's lemma, Lemma 2.2. The following definition, which formalizes the concept of a positive semi-definite $\mathcal{B}(\mathcal{H})$-valued kernel, generalizes Definition 2.1.

Definition 3.1. *If Ω is a set and $\mathcal{H}$ is a Hilbert space then we say that A is a positive semi-definite $\mathcal{B}(\mathcal{H})$-valued kernel on Ω if*

$$A \colon \Omega \times \Omega \to \mathcal{B}(\mathcal{H})$$

72   *Commutative Theory*

is a mapping such that, for all $n \geq 1$ and for all choices of points $\lambda_1, \lambda_2, \ldots, \lambda_n$ in Ω, the $n \times n$ block matrix $[A(\lambda_j, \lambda_i)]$ is a positive operator acting on $\mathcal{H}^n$.

Here $\mathcal{H}^n$ is the orthogonal direct sum of n copies of $\mathcal{H}$; the block matrix $[A(\lambda_j, \lambda_i)]$ acts as a linear operator on $\mathcal{H}^n$ in an obvious way.

In generalizing the theory of $\mathscr{S}(\mathbb{D})$ we replace the ubiquitous scalar-valued function $1 - \overline{\varphi(\mu)}\varphi(\lambda)$ (where $\varphi \in \mathscr{S}(\mathbb{D})$) by the $\mathcal{B}(\mathcal{H})$-valued kernel

$$1 - \varphi(\mu)^*\varphi(\lambda) \qquad \text{for } \lambda, \mu \in \mathbb{D},$$

where $\varphi \in \mathscr{S}_{\mathcal{B}(\mathcal{H},\mathcal{K})}(\mathbb{D})$ and here 1 denotes the identity operator on $\mathcal{H}$.

Pick's lemma (Lemma 2.2) can now be generalized as follows.

Lemma 3.2. (Pick's lemma for operator-valued mappings)
If $\varphi \in \mathscr{S}_{\mathcal{B}(\mathcal{H},\mathcal{K})}(\mathbb{D})$, then the kernel A on $\mathbb{D}$ defined by the formula

$$A(\lambda, \mu) = \frac{1 - \varphi(\mu)^*\varphi(\lambda)}{1 - \bar{\mu}\lambda}$$

is a positive semi-definite $\mathcal{B}(\mathcal{H})$-valued kernel on $\mathbb{D}$.

The proof of this lemma is similar to the proof of the corresponding statement in the scalar case, Lemma 2.2, but it does require a technical enhancement, involving Hardy spaces of vector-valued functions. For any Hilbert space $\mathcal{H}$ we denote by $\mathrm{H}^2_{\mathcal{H}}$ the space of square-summable series

$$f(z) = \sum_{n=1}^{\infty} a_n z^n \tag{3.3}$$

in which the coefficients a_n belong to $\mathcal{H}$, and the square-summability hypothesis means that

$$\sum_{n=0}^{\infty} \|a_n\|^2_{\mathcal{H}} < \infty.$$

An inner product is defined on $\mathrm{H}^2_{\mathcal{H}}$ by

$$\left\langle \sum_n a_n z^n, \sum_n b_n z^n \right\rangle_{\mathrm{H}^2_{\mathcal{H}}} = \sum_n \langle a_n, b_n \rangle_{\mathcal{H}}. \tag{3.4}$$

This inner product is well defined, and with the obvious definitions of addition and scalar multiplication, it makes $\mathrm{H}^2_{\mathcal{H}}$ a Hilbert space. Again, the square-summability hypothesis ensures that the power series (3.3) converges locally uniformly on $\mathbb{D}$ and therefore defines f as an $\mathcal{H}$-valued analytic map on $\mathbb{D}$. It also ensures that the series (3.4) converges. The norm of f can be expressed in terms of the values of f by the formula

Further Development of Models on the Disc

$$\|f\|_{H_{\mathcal{H}}^2} = \left\{ \lim_{r \to 1-} \int_0^{2\pi} \|f(re^{i\theta})\|_{\mathcal{H}}^2 \frac{d\theta}{2\pi} \right\}^{\frac{1}{2}}.$$

It remains the case that the Szegő kernel is a reproducing kernel on $H_{\mathcal{H}}^2$. In the new context, this means the following. For any $x \in \mathcal{H}$ and $\lambda \in \mathbb{D}$, we denote by $s_\lambda \otimes x$ the element of $H_{\mathcal{H}}^2$ given by

$$(s_\lambda \otimes x)(z) = s_\lambda(z)x \quad \text{for all } z \in \mathbb{D}.$$

We use a $\otimes$ symbol here because (as explained in Section 3.4), $H_{\mathcal{H}}^2$ can be identified with the tensor product of the Hilbert spaces H^2 and $\mathcal{H}$. The reproducing property of s_λ is

$$\langle f, s_\lambda \otimes x \rangle_{H_{\mathcal{H}}^2} = \langle f(\lambda), x \rangle_{\mathcal{H}} \quad \text{for all } \lambda \in \mathbb{D}, x \in \mathcal{H} \text{ and } f \in H_{\mathcal{H}}^2.$$

It follows from the expression (3.5) that any map φ in $\mathscr{S}_{\mathcal{B}(\mathcal{H},\mathcal{K})}$ determines a contraction $M_\varphi \colon H_{\mathcal{H}}^2 \to H_{\mathcal{K}}^2$ by

$$(M_\varphi f)(\lambda) = \varphi(\lambda)f(\lambda) \quad \text{for } f \in H_{\mathcal{H}}^2, \lambda \in \mathbb{D}.$$

The generalization of Lemma 1.65 states that, for any $\lambda \in \mathbb{D}$ and $y \in \mathcal{K}$,

$$M_\varphi^*(s_\lambda \otimes y) = s_\lambda \otimes \varphi(\lambda)^* y \quad \text{in } H_{\mathcal{H}}^2. \tag{3.5}$$

With these ingredients it is simple to modify the proof of the scalar version of Pick's theorem, Lemma 2.2, to obtain Lemma 3.2.

Just as Moore's theorem (Theorem 2.5) gives a representation of positive semi-definite kernels on a set Ω in terms of a Hilbert space $\mathcal{M}$ and a function $u \colon \Omega \to \mathcal{M}$, there is a version of Moore's theorem for operator-valued kernels that gives a representation of positive semi-definite $\mathcal{B}(\mathcal{H})$-valued kernels on a set Ω in terms of a Hilbert space $\mathcal{M}$ and a function $u \colon \Omega \to \mathcal{B}(\mathcal{H},\mathcal{M})$.

Theorem 3.6. (Moore's theorem for operator-valued kernels) *If Ω is a set and $A \colon \Omega \times \Omega \to \mathcal{B}(\mathcal{H})$ is a mapping, then A is a positive semi-definite $\mathcal{B}(\mathcal{H})$-valued kernel on Ω if and only if there exist a Hilbert space $\mathcal{M}$ and a function $u \colon \Omega \to \mathcal{B}(\mathcal{H},\mathcal{M})$ satisfying*

$$A(\lambda,\mu) = u(\mu)^* u(\lambda) \quad \text{for all } \lambda, \mu \in \Omega. \tag{3.7}$$

Furthermore, if A is positive semi-definite, then we may choose $\mathcal{M}$ and u having the additional property that $\{u(\lambda)x \colon \lambda \in \Omega, x \in \mathcal{H}\}$ has dense linear span in $\mathcal{M}$.

Observe that Theorem 3.6 reduces to Theorem 2.5 when $\mathcal{H} = \mathbb{C}$. For if $u \colon \Omega \to \mathcal{B}(\mathbb{C},\mathcal{M})$ is a function we may write[1] $u(\lambda) = v(\lambda) \otimes 1$, where $v \colon \Omega \to \mathcal{M}$. We then have

---

[1] This notation is explained in equation (1.17).

74   *Commutative Theory*

$$u(\mu)^*u(\lambda) = (v(\mu) \otimes 1)^*(v(\lambda) \otimes 1) = \langle v(\lambda), v(\mu) \rangle_{\mathcal{M}},$$

which is equation (2.6). The proof of Theorem 3.6 is an easy modification of the proof of the scalar version, Theorem 2.5.

Definition 3.8. (**Models for** $\mathscr{S}_{\mathcal{B}(\mathcal{H},\mathcal{K})}(\mathbb{D})$) *Let* $\varphi\colon \mathbb{D} \to \mathcal{B}(\mathcal{H},\mathcal{K})$ *be a mapping. A* model *on* $\mathbb{D}$ *for* φ *is a pair* $(\mathcal{M},u)$ *where* $\mathcal{M}$ *is a Hilbert space,* $u\colon \mathbb{D} \to \mathcal{B}(\mathcal{H},\mathcal{M})$ *is a mapping and the following* model formula *holds:*

$$1 - \varphi(\mu)^*\varphi(\lambda) = u(\mu)^*(1 - \bar{\mu}\lambda)u(\lambda) \quad \textit{for all } \lambda,\mu \in \mathbb{D}. \tag{3.9}$$

It is an easy matter to adapt the proof of Theorem 2.9 to obtain the following generalization.

Theorem 3.10. (**The fundamental theorem for** $\mathscr{S}_{\mathcal{B}(\mathcal{H},\mathcal{K})}(\mathbb{D})$) *Every map* $\varphi \in \mathscr{S}_{\mathcal{B}(\mathcal{H},\mathcal{K})}(\mathbb{D})$ *has a model on* $\mathbb{D}$.

Proof. By Lemma 3.2, the $\mathcal{B}(\mathcal{H})$-valued kernel

$$A(\lambda,\mu) = \frac{1 - \varphi(\mu)^*\varphi(\lambda)}{1 - \bar{\mu}\lambda}$$

on $\mathbb{D}$ is positive semi-definite. Hence by Moore's theorem for operator-valued kernels, Lemma 3.6, there exist a Hilbert space $\mathcal{M}$ and a map $u\colon \mathbb{D} \to \mathcal{B}(\mathcal{H},\mathcal{M})$ such that

$$A(\lambda,\mu) = u(\mu)^*u(\lambda) \quad \text{for all } \lambda,\mu \in \mathbb{D}.$$

Hence

$$1 - \varphi(\mu)^*\varphi(\lambda) = (1 - \bar{\mu}\lambda)u(\mu)^*u(\lambda) \quad \text{for all } \lambda,\mu \in \mathbb{D}.$$

Thus $(\mathcal{M},u)$ is a model for φ. $\qquad\square$

3.2 Lurking Isometries Revisited

Lurking isometry arguments work equally well for operator-valued mappings.

Lemma 3.11. (**The lurking isometry lemma for operator-valued maps**) *Let* $\mathcal{L}$, $\mathcal{H}$, *and* $\mathcal{K}$ *be Hilbert spaces and let* Ω *be a set. If* $u\colon \Omega \to \mathcal{B}(\mathcal{L},\mathcal{H})$ *and* $v\colon \Omega \to \mathcal{B}(\mathcal{L},\mathcal{K})$ *are functions, then*

$$v(\mu)^*v(\lambda) = u(\mu)^*u(\lambda) \quad \textit{for all } \lambda,\mu \in \Omega \tag{3.12}$$

if and only if there exists a linear isometry

$$V\colon [u(\lambda)x\colon \lambda \in \Omega, x \in \mathcal{L}] \to \mathcal{K}$$

such that

$$Vu(\lambda) = v(\lambda) \quad \textit{for all } \lambda \in \Omega. \tag{3.13}$$

Further Development of Models on the Disc 75

Proof. Suppose that equation (3.12) holds. If we seek to define an operator V_0 on

$$\text{span}\{u(\lambda)x\colon \lambda \in \Omega, x \in \mathcal{L}\}$$

by the formula

$$V_0\left(\sum_i c_i u(\lambda_i)x_i\right) = \sum_i c_i v(\lambda_i)x_i,$$

then equation (3.12) guarantees that V_0 is both well defined and isometric. If we extend V_0 by continuity to an operator V defined on $[u(\lambda)x\colon \lambda \in \Omega, x \in \mathcal{L}]$, then $V\colon [u(\lambda)x\colon \lambda \in \Omega, x \in \mathcal{L}] \to \mathcal{K}$ is a linear isometry and equation (3.13) holds.

Conversely, assume that $V\colon [u(\lambda)x\colon \lambda \in \Omega, x \in \mathcal{L}] \to \mathcal{K}$ is isometric and equation (3.13) holds. Fix $\lambda, \mu \in \Omega$. Then for all $x, y \in \mathcal{L}$,

$$\begin{aligned}
\langle v(\mu)^* v(\lambda)x, y\rangle_{\mathcal{L}} &= \langle v(\lambda)x, v(\mu)y\rangle_{\mathcal{K}} \\
&= \langle Vu(\lambda)x, Vu(\mu)y\rangle_{\mathcal{K}} \\
&= \langle u(\lambda)x, u(\mu)y\rangle_{\mathcal{H}} \\
&= \langle u(\mu)^* u(\lambda)x, y\rangle_{\mathcal{L}},
\end{aligned}$$

which proves that equation (3.12) holds. $\qquad\square$

Just as was the case in Section 1.3, a lurking isometry argument implies that $\mathcal{B}(\mathcal{H}, \mathcal{M})$-valued models are automatically holomorphic, and as a consequence, the converse of the fundamental theorem for $\mathscr{S}_{\mathcal{B}(\mathcal{H},\mathcal{K})}(\mathbb{D})$ obtains. We formulate these facts without proof.

Proposition 3.14. (Automatic holomorphy of models) *Let* $\varphi\colon \mathbb{D} \to \mathcal{B}(\mathcal{H}, \mathcal{K})$ *be a function and let* $(\mathcal{M}, u)$ *be a model for* φ. *Then* u *is a holomorphic* $\mathcal{B}(\mathcal{H}, \mathcal{M})$-*valued function on* $\mathbb{D}$.

Proposition 3.15. (Converse to the fundamental theorem for $\mathscr{S}_{\mathcal{B}(\mathcal{H},\mathcal{K})}(\mathbb{D})$) *If* $\varphi\colon \mathbb{D} \to \mathcal{B}(\mathcal{H}, \mathcal{K})$ *is a mapping which has a model on* $\mathbb{D}$, *then* $\varphi \in \mathscr{S}_{\mathcal{B}(\mathcal{H},\mathcal{K})}(\mathbb{D})$.

3.3 The Network Realization Formula

In this section we prove the network realization formula (Theorem 1.58) from Chapter 1 by modifying the proof of the scalar case, as given in Section 2.5. We restate the theorem in a form parallel to Theorem 2.36.

Theorem 3.16. *Let* $\mathcal{H}, \mathcal{K}$ *be Hilbert spaces and let* $\varphi\colon \mathbb{D} \to \mathcal{B}(\mathcal{H}, \mathcal{K})$ *be a mapping. The following statements are equivalent.*

76 *Commutative Theory*

(i) φ *belongs to* $\mathscr{S}_{\mathcal{B}(\mathcal{H},\mathcal{K})}(\mathbb{D})$;

(ii) *there exist a Hilbert space* $\mathcal{M}$ *and a contraction*

$$V = \begin{bmatrix} A & B \\ C & D \end{bmatrix} : \mathcal{H} \oplus \mathcal{M} \to \mathcal{K} \oplus \mathcal{M} \tag{3.17}$$

such that, for all $\lambda \in \mathbb{D}$,

$$\varphi(\lambda) = A + B\lambda(1 - D\lambda)^{-1}C; \tag{3.18}$$

(iii) *there exist a Hilbert space* $\mathcal{M}$ *and an isometry* V *of the form* (3.17) *such that the formula* (3.18) *holds.*

Proof. Trivially, (iii) implies (ii).

(i)$\Rightarrow$(iii) Assume that $\varphi \in \mathscr{S}_{\mathcal{B}(\mathcal{H},\mathcal{K})}(\mathbb{D})$. By Theorem 3.10, φ has a model. However, for the purposes of the proof, we shall require a model that has an additional technical property.

Claim 3.19. There exists a model $(\mathcal{M}, u)$ for φ such that

$$\dim \left((\mathcal{H} \oplus \mathcal{M}) \ominus [x \oplus \lambda u(\lambda)x \colon \lambda \in \mathbb{D}, x \in \mathcal{H}] \right)$$
$$\leq \dim \left((\mathcal{K} \oplus \mathcal{M}) \ominus [\varphi(\lambda)x \oplus u(\lambda)x \colon \lambda \in \mathbb{D}, x \in \mathcal{H}] \right) \tag{3.20}$$

This claim can be proved by means of the trick that allowed us to assume that equation (2.60) holds in the proof of Theorem 2.57.

We reshuffle the model equation (3.9) to the form

$$\varphi(\mu)^*\varphi(\lambda) + u(\mu)^*u(\lambda) = 1 + (\mu u(\mu))^*(\lambda u(\lambda)) \quad \text{for all } \lambda, \mu \in \mathbb{D},$$

or equivalently,

$$\begin{bmatrix} \varphi(\mu) \\ u(\mu) \end{bmatrix}^* \begin{bmatrix} \varphi(\lambda) \\ u(\lambda) \end{bmatrix} = \begin{bmatrix} 1 \\ \mu u(\mu) \end{bmatrix}^* \begin{bmatrix} 1 \\ \lambda u(\lambda) \end{bmatrix} \quad \text{for all } \lambda, \mu \in \mathbb{D}. \tag{3.21}$$

Note that these block operators denote mappings

$$\begin{bmatrix} \varphi(\lambda) \\ u(\lambda) \end{bmatrix} : \mathcal{H} \to \mathcal{K} \oplus \mathcal{M}, \quad \begin{bmatrix} 1 \\ \lambda u(\lambda) \end{bmatrix} : \mathcal{H} \to \mathcal{H} \oplus \mathcal{M}.$$

By Lemma 3.11 there is an isometry

$$V_0 \colon [x \oplus \lambda u(\lambda)x \colon \lambda \in \mathbb{D}, \, x \in \mathcal{H}] \to [\varphi(\lambda)x \oplus u(\lambda)x \colon \lambda \in \mathbb{D}, x \in \mathcal{H}]$$

such that $V_0(x \oplus \lambda u_\lambda x) = \varphi(\lambda)x \oplus u_\lambda x$ for all $\lambda \in \mathbb{D}$ and $x \in \mathcal{H}$. Thus V_0 is an isometry from a subspace of $\mathcal{H} \oplus \mathcal{M}$ onto a subspace of $\mathcal{K} \oplus \mathcal{M}$, and by Claim 3.19, the orthogonal complement of the former subspace has dimension no greater than the dimension of the orthogonal complement of the latter

Further Development of Models on the Disc 77

subspace. Hence there exists an isometry $V\colon \mathcal{H}\oplus\mathcal{M}\to\mathcal{K}\oplus\mathcal{M}$ such that V_0 is a restriction of V.

Write V as a block matrix in the form

$$V = \begin{bmatrix} A & B \\ C & D \end{bmatrix}\colon \mathcal{H}\oplus\mathcal{M}\to\mathcal{K}\oplus\mathcal{M}.$$

Then

$$\begin{bmatrix} A & B \\ C & D \end{bmatrix}\begin{bmatrix} 1 \\ \lambda u(\lambda) \end{bmatrix} = \begin{bmatrix} \varphi(\lambda) \\ u(\lambda) \end{bmatrix} \qquad \text{for all } \lambda\in\mathbb{D}.$$

Thus

$$\varphi(\lambda) = A + B\lambda u(\lambda),$$
$$u(\lambda) = C + D\lambda u(\lambda).$$

Using the second equation to eliminate $u(\lambda)$ from the first equation, we deduce that

$$\varphi(\lambda) = A + B\lambda(1 - D\lambda)^{-1}C \qquad \text{for all } \lambda\in\mathbb{D},$$

which is equation (3.18).

(ii)$\Rightarrow$(i) Assume that V is a contraction given by equation (3.17) and φ is a mapping given by equation (3.18). If we define $u\colon \mathbb{D}\to\mathcal{B}(\mathcal{H},\mathcal{M})$ by

$$u(\lambda) = (1 - D\lambda)^{-1}C,$$

then u and φ are holomorphic on $\mathbb{D}$. Moreover, for $\lambda\in\mathbb{D}$,

$$V\begin{bmatrix} 1 \\ \lambda u(\lambda) \end{bmatrix} = \begin{bmatrix} A & B \\ C & D \end{bmatrix}\begin{bmatrix} 1 \\ \lambda u(\lambda) \end{bmatrix}$$
$$= \begin{bmatrix} A + B\lambda u(\lambda) \\ C + D\lambda u(\lambda) \end{bmatrix}$$
$$= \begin{bmatrix} \varphi(\lambda) \\ u(\lambda) \end{bmatrix}.$$

Since V is a contraction, $V^*V \leq 1$. Hence for all $\lambda\in\mathbb{D}$,

$$\begin{bmatrix} \varphi(\lambda)^* & u(\lambda)^* \end{bmatrix}\begin{bmatrix} \varphi(\lambda) \\ u(\lambda) \end{bmatrix} = \left(V\begin{bmatrix} 1 \\ \lambda u(\lambda) \end{bmatrix}\right)^*\left(V\begin{bmatrix} 1 \\ \lambda u(\lambda) \end{bmatrix}\right)$$
$$= \begin{bmatrix} 1 & \bar{\lambda}u(\lambda)^* \end{bmatrix}V^*V\begin{bmatrix} 1 \\ \lambda u(\lambda) \end{bmatrix}$$
$$\leq \begin{bmatrix} 1 & \bar{\lambda}u(\lambda)^* \end{bmatrix}\begin{bmatrix} 1 \\ \lambda u(\lambda) \end{bmatrix}. \qquad (3.22)$$

78   *Commutative Theory*

Thus

$$\varphi(\lambda)^*\varphi(\lambda) + u(\lambda)^*u(\lambda) \leq 1 + |\lambda|^2 u(\lambda)^*u(\lambda) \qquad \text{for all } \lambda \in \mathbb{D}. \qquad (3.23)$$

Since also $-u(\lambda)^*u(\lambda) \leq -|\lambda|^2 u(\lambda)^*u(\lambda)$, we may add the inequalities to deduce that $\varphi(\lambda)^*\varphi(\lambda) \leq 1$ for all $\lambda \in \mathbb{D}$. Thus $\varphi \in \mathscr{S}_{\mathcal{B}(\mathcal{H},\mathcal{K})}(\mathbb{D})$. $\qquad \square$

Remark 3.24. In the event that V in Theorem 3.16 is an isometry, the inequalities (3.22) and (3.23) hold with equality, so that

$$\varphi(\lambda)^*\varphi(\lambda) = 1 - (1 - |\lambda|^2)u(\lambda)^*u(\lambda) \qquad \text{for all } \lambda \in \mathbb{D}. \qquad (3.25)$$

It is tempting to infer that $\varphi(\lambda)$ is an isometry for $\lambda \in \mathbb{T}$, but this is incorrect since $(1 - D\lambda)^{-1}C$ need not be bounded as λ tends to a point on $\mathbb{T}$. Indeed, there is nothing to stop φ from being the zero function.

3.4 Tensor Products of Hilbert Spaces

There is a natural way to associate with Hilbert spaces $\mathcal{H}, \mathcal{K}$ a new Hilbert space, denoted by $\mathcal{H} \otimes \mathcal{K}$ and known as the tensor product of $\mathcal{H}$ and $\mathcal{K}$; this construction will be useful in particular for the extension of results about scalar functions to operator-valued functions. To give a simple example, if $\mathcal{H} = \mathbb{C}^n$, then $\mathcal{H} \otimes \mathcal{K}$ is expressible as the orthogonal direct sum of n copies of $\mathcal{K}$. Another example is the space $H^2_{\mathcal{H}}$, which we encountered in Section 3.1 and which can be regarded as the tensor product $H^2 \otimes \mathcal{H}$. For the general construction we follow Dixmier [75, chapter 1, section 2].

We shall assume the notion of the algebraic tensor product $V \otimes_a W$ of two vector spaces V and W (for example [97]). $V \otimes_a W$ is a vector space spanned by elements of the form $x \otimes y$, where $x \in V$, $y \in W$ and in which the relations

$$(\lambda_1 x_1 + \lambda_2 x_2) \otimes y = \lambda_1 x_1 \otimes y + \lambda_2 x_2 \otimes y,$$

$$x \otimes (\lambda_1 y_1 + \lambda_2 y_2) = \lambda_1 x \otimes y_1 + \lambda_2 x \otimes y_2$$

hold for all scalars λ_1, λ_2 and vectors $x, x_1, x_2 \in V$, $y, y_1, y_2 \in W$. In the case of Hilbert spaces $\mathcal{H}_1 \otimes_a \mathcal{H}_2$ is a pre-Hilbert space under an inner product $\langle \cdot, \cdot \rangle_{\mathcal{H}_1 \otimes \mathcal{H}_2}$ that satisfies

$$\langle x_1 \otimes y_1, x_2 \otimes y_2 \rangle_{\mathcal{H}_1 \otimes \mathcal{H}_2} = \langle x_1, x_2 \rangle_{\mathcal{H}_1} \langle y_1, y_2 \rangle_{\mathcal{H}_2} \qquad (3.26)$$

for all $x_1, x_2 \in \mathcal{H}_1$, $y_1, y_2 \in \mathcal{H}_2$.

Definition 3.27. *For Hilbert spaces $\mathcal{H}_1, \mathcal{H}_2$, the Hilbert tensor product $\mathcal{H}_1 \otimes \mathcal{H}_2$ of $\mathcal{H}_1$ and $\mathcal{H}_2$ is the completion of the pre-Hilbert space $\mathcal{H}_1 \otimes_a \mathcal{H}_2$ with respect to the inner product (3.26).*

Further Development of Models on the Disc 79

One can represent the general element of $\mathcal{H}_1 \otimes \mathcal{H}_2$ in terms of an orthonormal basis of either $\mathcal{H}_1$ or $\mathcal{H}_2$.

Proposition 3.28. *Let $\mathcal{H}_1, \mathcal{H}_2$ be Hilbert spaces with complete orthonormal systems $(e_i)_{i \in I}$ and $(f_j)_{j \in J}$ respectively. Then $(e_i \otimes f_j)_{i \in I, \, j \in J}$ is a complete orthonormal system in $\mathcal{H}_1 \otimes \mathcal{H}_2$.*

Proof. Clearly $(e_i \otimes f_j)$ is an orthonormal system. Let S denote the closed linear span of the system in $\mathcal{H}_1 \otimes \mathcal{H}_2$. For any $x \in \mathcal{H}_1$, $y \in \mathcal{H}_2$, one sees that $x \otimes f_j \in S$ for every $j \in J$, and therefore $x \otimes y \in S$. Hence $S \supseteq \mathcal{H}_1 \otimes_a \mathcal{H}_2$. Since $\mathcal{H}_1 \otimes_a \mathcal{H}_2$ is dense in $\mathcal{H}_1 \otimes \mathcal{H}_2$, it follows that $S = \mathcal{H}_1 \otimes \mathcal{H}_2$. The system $(e_i \otimes f_j)$ is therefore complete. $\qquad\square$

Corollary 3.29. *Let $\mathcal{H}_1, \mathcal{H}_2$ and (e_i) be as in Proposition 3.28. For every element h in $\mathcal{H}_1 \otimes \mathcal{H}_2$ there is a unique family $(y_i)_{i \in I}$ of vectors in $\mathcal{H}_2$ such that*

$$\sum_{i \in I} \|y_i\|^2 = \|h\|^2 \tag{3.30}$$

and

$$h = \sum_{i \in I} e_i \otimes y_i, \tag{3.31}$$

the sum converging in the norm of $\mathcal{H}_1 \otimes \mathcal{H}_2$. Conversely, every sum of the form (3.31) for which $\sum_{i \in I} \|y_i\|^2 < \infty$ converges in norm to an element of $\mathcal{H}_1 \otimes \mathcal{H}_2$.

Remark 3.32. If both $\mathcal{H}_1$ and $\mathcal{H}_2$ are infinite-dimensional, then a general element in $\mathcal{H}_1 \otimes \mathcal{H}_2$ can be expressed (in many ways) as a norm-convergent infinite sum of the form $\sum x_j \otimes y_j$. The algebraic tensor product $\mathcal{H}_1 \otimes_a \mathcal{H}_2$ can be identified with the set of all elements in $\mathcal{H}_1 \otimes \mathcal{H}_2$ that have a representation as a *finite* sum $\sum x_j \otimes y_j$.

Definition 3.4 identifies $H^2_{\mathcal{H}}$ with $\ell^2 \otimes \mathcal{H}$, which we can further identify with $H^2 \otimes \mathcal{H}$ using (z^n) as an orthonormal basis of H^2.

3.5 Tensor Products of Operators

Consider operators $T_j \colon \mathcal{H}_j \to \mathcal{K}_j$ for $j = 1, 2$ and Hilbert spaces $\mathcal{H}_1, \mathcal{H}_2$, $\mathcal{K}_1, \mathcal{K}_2$. The algebraic theory tells us that T_1, T_2 induce a linear map $T_1 \otimes T_2$ between the algebraic tensor products $\mathcal{H}_1 \otimes_a \mathcal{H}_2$ and $\mathcal{K}_1 \otimes_a \mathcal{K}_2$ such that

$$(T_1 \otimes T_2)(x_1 \otimes x_2) = (T_1 x_1) \otimes (T_2 x_2) \quad \text{for all } x_1 \in \mathcal{H}_1, \, x_2 \in \mathcal{H}_2.$$

Proposition 3.33. *If $T_j \colon \mathcal{H}_j \to \mathcal{K}_j$ are Hilbert space operators for $j = 1, 2$, then $T_1 \otimes T_2$ extends by continuity to be an operator from $\mathcal{H}_1 \otimes \mathcal{H}_2$ to $\mathcal{K}_1 \otimes \mathcal{K}_2$. Moreover $\|T_1 \otimes T_2\| = \|T_1\| \cdot \|T_2\|$.*

80 Commutative Theory

Proof. Let T_1, T_2 be contractions. Suppose first that $\mathcal{K}_2 = \mathcal{H}_2$ and T_2 is the identity operator. We can write the general element ξ of $\mathcal{H}_1 \otimes_a \mathcal{H}_2$ in the form $\xi = \sum_{j=1}^{n} x_1^j \otimes x_2^j$ for some n, where $x_2^1, \ldots, x_2^n$ are orthonormal in $\mathcal{H}_2$. Then

$$
\begin{aligned}
\|(T_1 \otimes 1)\xi\|^2 &= \left\| \sum_j (T_1 x_1^j) \otimes x_2^j \right\|^2 \\
&= \left\| \sum_j T_1 x_1^j \right\|^2 \\
&\leq \sum_j \|T_1 x_1^j\|^2 \\
&\leq \sum_j \|x_1^j\|^2 \\
&= \left\| \sum_j x_1^j \otimes x_2^j \right\|^2 .
\end{aligned}
$$

Thus $\|T_1 \otimes 1\| \leq 1$. Likewise $\|1 \otimes T_2\| \leq 1$, and therefore $(T_1 \otimes 1)(1 \otimes T_2) = T_1 \otimes T_2$ is a contraction.

Since $\mathcal{H}_1 \otimes_a \mathcal{H}_2$ is dense in $\mathcal{H}_1 \otimes \mathcal{H}_2$, we may now extend $T_1 \otimes T_2$ by continuity to obtain a contraction from $\mathcal{H}_1 \otimes \mathcal{H}_2$ to $\mathcal{K}_1 \otimes \mathcal{K}_2$.

Now consider arbitrary bounded operators T_1, T_2. By scaling we infer that $T_1 \otimes T_2$ extends to a bounded operator on $\mathcal{H}_1 \otimes \mathcal{H}_2$ that satisfies $\|T_1 \otimes T_2\| \leq \|T_1\| \cdot \|T_2\|$. The reverse inequality is easy—look at $T_1 \otimes T_2$ applied to the elementary tensors $x_n \otimes y_n$, where x_n and y_n are unit vectors satisfying $\|T_1 x_n\| \to \|T_1\|$ and $\|T_2 y_n\| \to \|T_2\|$. $\qquad\square$

Remark 3.34. Observe the similar properties $\|x \otimes y\| = \|x\| \, \|y\|$ and $\|T_1 \otimes T_2\| = \|T_1\| \, \|T_2\|$ for the tensors both of vectors in $\mathcal{H}_1 \otimes \mathcal{H}_2$ and of operators on $\mathcal{H}_1 \otimes \mathcal{H}_2$. These are two examples of *cross-norms*, which are norms defined on tensor products of normed spaces having such multiplicative properties. We discuss these a little more in Section 16.4.

Proposition 3.35. (Properties of the tensor product of operators) *The following relations hold for operators on Hilbert space.*

(i) $(T_1 \otimes T_2)^* = T_1^* \otimes T_2^*$;

(ii) $(S_1 \otimes S_2)(T_1 \otimes T_2) = (S_1 T_1) \otimes (S_2 T_2)$ *whenever both sides are defined;*

(iii) *if $T_1 \geq 0$ and $T_2 \geq 0$ then $T_1 \otimes T_2 \geq 0$.*

Further Development of Models on the Disc 81

Proof. (i) and (ii) are easy. For (iii), write the positive operator T_j as a product $A_j^* A_j$ for some operator A_j; then by (i) and (ii),

$$T_1 \otimes T_2 = (A_1 \otimes A_2)^* (A_1 \otimes A_2) \geq 0.$$

$\square$

With the aid of tensor products we can enhance the functional calculus to define $\varphi(T)$ for suitable operators T and *operator-valued* holomorphic functions φ. Consider a Hilbert space $\mathcal{H}$, an operator $T \in \mathcal{B}(\mathcal{H})$ and a domain U containing $\sigma(T)$. Suppppose that φ is a holomorphic function on U with values in $\mathcal{B}(\mathcal{K}_1, \mathcal{K}_2)$ for some Hilbert spaces $\mathcal{K}_1, \mathcal{K}_2$. Define an operator $\varphi(T) \colon \mathcal{K}_1 \otimes \mathcal{H} \to \mathcal{K}_2 \otimes \mathcal{H}$ by a slight modification of the Riesz–Dunford functional calculus described in Section 1.2. Choose a finite collection Γ of closed rectifiable curves in $U \setminus \sigma(T)$ that winds once around each point of $\sigma(T)$ and no times round each point in the complement of U. Then

$$\varphi(T) \stackrel{\text{def}}{=} \frac{1}{2\pi i} \int_\Gamma \varphi(w) \otimes (w - T)^{-1} \, dw. \tag{3.36}$$

We have merely inserted the symbol $\otimes$ into the formula (1.4). The integral exists as a Riemann integral and defines an operator $\varphi(T) \in \mathcal{B}(\mathcal{K}_1 \otimes \mathcal{H}, \mathcal{K}_2 \otimes \mathcal{H})$.

Similarly, we may enhance the hereditary functional calculus, as defined in equation (2.84). An *operator-valued hereditary function on* $\mathbb{D}$ is a map $h \colon \mathbb{D} \times \mathbb{D} \to \mathcal{B}(\mathcal{K})$, for some Hilbert space $\mathcal{K}$, such that $h(\lambda, \bar{\mu})$ is analytic in (λ, μ). The space of such functions will be denoted by $\mathrm{Her}_{\mathcal{B}(\mathcal{K})}(\mathbb{D})$. For such a function h and for any operator $T \in \mathcal{B}(\mathcal{H})$ with spectrum in $\mathbb{D}$ we define an operator $h(T)$ on $\mathcal{K} \otimes \mathcal{H}$ by the formula

$$h(T) \stackrel{\text{def}}{=} \frac{1}{(2\pi i)^2} \int_\gamma \int_\gamma h(\lambda, \bar{\mu}) \otimes (\bar{\mu} - T^*)^{-1} (\lambda - T)^{-1} \, d\bar{\mu} \, d\lambda, \tag{3.37}$$

where γ is a contour in $\mathbb{D}$ of the form $\gamma(t) = re^{it}, 0 \leq t \leq 2\pi$, and $r \in (0, 1)$ is chosen so that $\sigma(T) \subset r\mathbb{D}$.

Standard properties of the hereditary functional calculus (see Lemma 2.86, Proposition 2.87) continue to hold, with appropriate adjustments.

3.6 Realization of Rational Matrix Functions and the McMillan Degree

Consider a function $\varphi \colon \mathbb{D} \to \mathcal{B}(\mathcal{H}, \mathcal{K})$, which has a realization (3.18) with model space $\mathcal{M}$ and isometric V. If $\mathcal{H}, \mathcal{K}$ and $\mathcal{M}$ are finite-dimensional then $(1 - \lambda D)^{-1}$ is a rational matrix-valued function, with poles only where $1/\lambda$

82 *Commutative Theory*

is in the spectrum of D, and hence φ is a rational matrix function. Moreover, the map

$$u(\lambda) = (1 - \lambda D)^{-1} C$$

is a rational map from $\mathbb{D}$ to $\mathcal{M}$, and $(\mathcal{M}, u)$ is a model for φ. Being rational, u is analytic at all but finitely many points of $\mathbb{T}$. By equation (3.25), φ has isometric values at all but finitely many points on the unit circle. An analytic function on $\mathbb{D}$ that takes isometric values at almost every point of $\mathbb{T}$ with respect to Lebesgue measure is called an *operator-valued inner function*; in the case that $\mathcal{H}$ and $\mathcal{K}$ are finite-dimensional, it is called a *matrix-valued inner function*.

It is not entirely straightforward to define the degree of a matrix-valued rational function. The poles of such a function are naturally to be understood as the poles of its entries, but how are we to count the multiplicity of each pole (including ∞) correctly? If φ is a diagonal matrix, we simply add up the degrees of the diagonal entries. For a general rational matrix function φ, one proceeds as follows.

First, write $\varphi(z) = \frac{1}{q(z)} P(z)$, where q is the least common multiple of all the denominators of φ, and $P(z)$ is a polynomial matrix. Then put $P(z)$ into its Smith normal form, which means we write $P(z) = L(z)D(z)R(z)$, where L, D, and R are all matrices of polynomials, L and R are square matrices of determinant 1, and D is a diagonal matrix (or a diagonal matrix augmented with zeroes if it is not square) with the property that each diagonal entry of D divides the next one in $\mathbb{C}[z]$. Finally look at the diagonal matrix $\frac{1}{q(z)} D(z)$. The sum of the degrees of the entries of this diagonal matrix is called the *McMillan degree* of φ. It can be shown that the McMillan degree of a rational matrix-valued inner function φ exactly equals the dimension of the smallest model space $\mathcal{M}$ on which there is a network realization formula for φ. For a proof of this assertion, see, for example, [119, theorem 1.13.2] or [46, theorem 3.3]. The statement generalizes Proposition 2.51.

For a rational matrix-valued function that is not inner, as in the scalar case, one can still find a network realization formula on a finite-dimensional space $\mathcal{M}$; however, the price that must be paid for finite-dimensionality is that the matrix V can no longer be chosen to be isometric. It is still true that the McMillan degree equals the dimension of the smallest model space $\mathcal{M}$ on which there is a network realization formula for φ, with a matrix V that may not be an isometry [119, theorem 1.13.2]. For a thorough discussion of realizations of rational matrix functions we refer the reader to [41].

To summarize:

Theorem 3.38. *A matrix-valued function φ on $\mathbb{D}$ is a rational inner function if and only if it has a network realization formula as in Theorem 3.16 (with*

Further Development of Models on the Disc 83

an isometric V) on a finite-dimensional model space. If it does, the McMillan degree of φ equals the minimal dimension of model space on which φ has a realization.

3.7 Pick Interpolation Revisited

In this section we give a version of the Pick interpolation theorem, Theorem 1.81, for operator-valued maps.

Theorem 3.39. *Let $\mathcal{H}, \mathcal{K}$ be Hilbert spaces, let $\lambda_1, \ldots, \lambda_n$ be distinct points in $\mathbb{D}$, and let $Z_1, \ldots, Z_n \in \mathcal{B}(\mathcal{H}, \mathcal{K})$. There exists $\varphi \in \mathscr{S}_{\mathcal{B}(\mathcal{H},\mathcal{K})}(\mathbb{D})$ such that $\varphi(\lambda_i) = Z_i$ for $i = 1, \ldots, n$ if and only if the operator with matrix*

$$P = \left[\frac{1 - Z_i^* Z_j}{1 - \bar{\lambda}_i \lambda_j} \right]_{i,j=1}^{n}$$

on $\mathcal{H}^n$ is positive semi-definite.

Proof. Suppose that there exists $\varphi \in \mathscr{S}_{\mathcal{B}(\mathcal{H},\mathcal{K})}(\mathbb{D})$ such that $\varphi(\lambda_i) = Z_i$ for $i = 1, \ldots, n$. By Lemma 3.2 the $\mathcal{B}(\mathcal{H})$-valued kernel

$$A(\lambda, \mu) = \frac{1 - \varphi(\mu)^* \varphi(\lambda)}{1 - \bar{\mu}\lambda}$$

is positive semi-definite on $\mathbb{D}$. Hence, by restriction,

$$\left[A(\lambda_j, \lambda_i) \right]_{i,j=1}^{n} \geq 0,$$

which is to say that $P \geq 0$.

To prove the converse, suppose that $P \geq 0$. Apply Lemma 3.6 (Moore's theorem) with $\Omega = \{\lambda_1, \ldots, \lambda_n\}$ and

$$A(\lambda_j, \lambda_i) = \frac{1 - Z_i^* Z_j}{1 - \bar{\lambda}_i \lambda_j} \qquad \text{for } i, j = 1, \ldots, n$$

to deduce that there is a Hilbert space $\mathcal{M}$ and a map $u\colon \{\lambda_1, \ldots, \lambda_n\} \to \mathcal{B}(\mathcal{H}, \mathcal{M})$ satisfying

$$A(\lambda_j, \lambda_i) = u(\lambda_i)^* u(\lambda_j) \qquad \text{for } i, j = 1, \ldots, n.$$

Hence

$$1 - Z_i^* Z_j = (1 - \bar{\lambda}_i \lambda_j) u(\lambda_i)^* u(\lambda_j) \qquad \text{for } i, j = 1, \ldots, n.$$

After the standard reshuffle one can write this equation in the form

$$f(\lambda_i)^* f(\lambda_j) = g(\lambda_i)^* g(\lambda_j) \colon \mathcal{H} \to \mathcal{H} \qquad \text{for } i, j = 1. \ldots n,$$

84   *Commutative Theory*

where

$$f(\lambda_j) = \begin{bmatrix} 1 \\ \lambda_j u(\lambda_j) \end{bmatrix} : \mathcal{H} \to \mathcal{H} \oplus \mathcal{M},$$

$$g(\lambda_j) = \begin{bmatrix} Z_j \\ u(\lambda_j) \end{bmatrix} : \mathcal{H} \to \mathcal{K} \oplus \mathcal{M} \quad \text{for } j = 1,\ldots,n.$$

Now apply the enhanced lurking isometry lemma, Lemma 3.11, to infer that there exists an isometry

$$V : [f(\lambda_j)x : 1 \leq j \leq n, x \in \mathcal{H}] \to \mathcal{K} \oplus \mathcal{M}$$

such that

$$V f(\lambda_j) = g(\lambda_j) \qquad \text{for } j = 1,\ldots,n. \tag{3.40}$$

Extend V to a contraction

$$T = \begin{bmatrix} A & B \\ C & D \end{bmatrix} : \mathcal{H} \oplus \mathcal{M} \to \mathcal{K} \oplus \mathcal{M}$$

by defining T to be zero on

$$(\mathcal{H} \oplus \mathcal{M}) \ominus [f(\lambda_j)x : 1 \leq j \leq n, x \in \mathcal{H}].$$

Then, by equation (3.40),

$$\begin{bmatrix} A & B \\ C & D \end{bmatrix} \begin{pmatrix} x \\ \lambda_j u(\lambda_j)x \end{pmatrix} = \begin{pmatrix} Z_j x \\ u(\lambda_j)x \end{pmatrix} \qquad \text{for all } x \in \mathcal{H}, j = 1,\ldots,n.$$

Consequently

$$A + B\lambda_j u(\lambda_j) = Z_j,$$

$$C + D\lambda_j u(\lambda_j) = u(\lambda_j)$$

and therefore

$$A + B\lambda_j(1 - D\lambda_j)^{-1}C = Z_j \tag{3.41}$$

for $j = 1,\ldots,n$.

For $\lambda \in \mathbb{D}$ let

$$\varphi(\lambda) = A + B\lambda(1 - D\lambda)^{-1}C \in \mathcal{B}(\mathcal{H},\mathcal{K}).$$

By Theorem 3.16, $\varphi \in \mathscr{S}_{\mathcal{B}(\mathcal{H},\mathcal{K})}(\mathbb{D})$. By equation (3.41), $\varphi(\lambda_j) = Z_j$ for $j = 1,\ldots,n$. $\square$

Another use of models of operator-valued functions is to prove a generalization of von Neumann's inequality, Theorem 1.47.

Further Development of Models on the Disc 85

Theorem 3.42. (Von Neumann's inequality for operator-valued functions)
Let $\mathcal{H}, \mathcal{K}_1, \mathcal{K}_2$ *be Hilbert spaces. For any function* $\varphi \in \mathcal{S}_{\mathcal{B}(\mathcal{K}_1, \mathcal{K}_2)}(\mathbb{D})$
and any contraction T *on* $\mathcal{H}$ *having its spectrum in* $\mathbb{D}$, *the operator*
$\varphi(T) \in \mathcal{B}(\mathcal{K}_1 \otimes \mathcal{H}, \mathcal{K}_2 \otimes \mathcal{H})$ *is a contraction.*

Proof. By Theorem 3.10, φ has a model $(\mathcal{M}, u)$ in the sense of Definition 3.8,
so that u maps $\mathbb{D}$ to $\mathcal{B}(\mathcal{K}_1, \mathcal{M})$, and

$$1 - \varphi(\mu)^* \varphi(\lambda) = u(\mu)^* (1 - \bar{\mu}\lambda) u(\lambda) \quad \text{for all } \lambda, \mu \in \mathbb{D}.$$

By Proposition 3.14, u is automatically holomorphic on $\mathbb{D}$ and so both sides
of this equation are $\mathcal{B}(\mathcal{K}_1)$-valued hereditary functions on $\mathbb{D}$. Here $1 - \bar{\mu}\lambda$
is to be regarded as a $\mathcal{B}(\mathcal{M})$-valued hereditary function; let us write it
$1_{\mathcal{M}} - \bar{\mu}\lambda$ to emphasize this fact. Apply the enhanced hereditary functional
calculus, equation (3.37), to both sides, noting that $u(T)$ maps $\mathcal{K}_1 \otimes \mathcal{H}$ to
$\mathcal{M} \otimes \mathcal{H}$ and that $1_{\mathcal{M}} - \bar{\mu}\lambda$ applied to T yields the operator $1_{\mathcal{M}} \otimes (1_{\mathcal{H}} - T^*T)$
on $\mathcal{M} \otimes \mathcal{H}$. By the standard properties of the hereditary calculus,

$$1 - \varphi(T)^* \varphi(T) = u(T)^* \left(1_{\mathcal{M}} \otimes (1_{\mathcal{H}} - T^*T) \right) u(T)$$

$$\geq 0.$$

Hence $\varphi(T)$ is a contraction. $\qquad\square$

Observe that, if $\varphi \in \mathcal{S}_{\mathcal{B}(\mathbb{C}^n, \mathbb{C}^m)}(\mathbb{D})$, then for each $\lambda \in \mathbb{D}$, we may write $\varphi(\lambda)$
as an $m \times n$ matrix $[\varphi_{ij}(\lambda)]$. Each entry φ_{ij} belongs to the scalar Schur class
$\mathcal{S}(\mathbb{D})$, and so, if T is a contraction on some Hilbert space $\mathcal{H}$ and $\sigma(T) \subset \mathbb{D}$,
then the block matrix $[\varphi_{ij}(T)]$ is well defined and determines an operator from
$\mathcal{H}^n$ to $\mathcal{H}^m$. This operator is just $\varphi(T)$, as defined by the enhanced functional
calculus, equation (3.36). It is noteworthy that the matricial von Neumann
inequality, in conjunction with Arveson's dilation theorem, Theorem 1.52,
yields an alternative proof of the Sz.-Nagy dilation theorem, Theorem 1.50.
For the inequality shows that the closed unit disc $\mathbb{D}^-$ is a complete spectral
set for any contraction T, and so Arveson's result yields a dilation of T to a
normal operator having spectrum in $\partial\mathbb{D}$, which is to say, a unitary operator (to
be precise, one should apply von Neumann's inequality to rT, where $0 < r < 1$,
so that the algebraic spectrum of rT is contained in $\mathbb{D}$; then let $r \to 1$ to show
that $\mathbb{D}^-$ is a complete spectral set for T).

3.8 The Corona Problem

The commutative Banach algebra $H^\infty(\mathbb{D})$ possesses some obvious characters
(multiplicative linear functionals), to wit, the point evaluation functionals at

86   Commutative Theory

points of $\mathbb{D}$. Thus $\mathbb{D}$ is naturally embedded in the maximal ideal space of $H^\infty(\mathbb{D})$, which is a compact Hausdorff space. Is $\mathbb{D}$ dense in the maximal ideal space? This question can be re-formulated in a function-theoretic way: if finitely many H^∞ functions are jointly bounded below in modulus on $\mathbb{D}$, can they lie in a proper ideal of $H^\infty(\mathbb{D})$? Here is a more formal statement of the problem.

The corona problem. *Given functions* $\varphi_1, \ldots, \varphi_m \in H^\infty(\mathbb{D})$, *determine whether there exist functions* $\psi_1, \ldots, \psi_m \in H^\infty(\mathbb{D})$ *such that*

$$\varphi_1 \psi_1 + \cdots \varphi_m \psi_m = 1. \tag{3.43}$$

There is an obvious necessary condition for the existence of functions $\psi_1, \ldots, \psi_m$ satisfying equation (3.43): for any $\lambda \in \mathbb{D}$,

$$1 = |(\varphi_1 \psi_1 + \cdots + \varphi_m \psi_m)(\lambda)|$$
$$\leq (|\varphi_1(\lambda)|^2 + \cdots + |\varphi_m(\lambda)|^2)^{\frac{1}{2}} (|\psi_1(\lambda)|^2 + \cdots + |\psi_m(\lambda)|^2)^{\frac{1}{2}}.$$

Therefore, if we think of $\psi = (\psi_1, \ldots, \psi_m)$ as a $\mathbb{C}^m$-valued function defined on $\mathbb{D}$, so that

$$\|\psi\|_{\mathbb{D}} = \sup_{\lambda \in \mathbb{D}} (|\psi_1(\lambda)|^2 + \cdots + |\psi_m(\lambda)|^2)^{\frac{1}{2}},$$

then we have

$$1 \leq (|\varphi_1(\lambda)|^2 + \cdots + |\varphi_m(\lambda)|^2)^{\frac{1}{2}} \|\psi\|_{\mathbb{D}} \quad \text{for all } \lambda \in \mathbb{D}.$$

Defining $\delta(\varphi)$ by

$$\delta(\varphi) = \inf_{\lambda \in \mathbb{D}} (|\varphi_1(\lambda)|^2 + \cdots + |\varphi_m(\lambda)|^2)^{\frac{1}{2}}, \tag{3.44}$$

we are led to the condition

$$\delta(\varphi) \geq \|\psi\|^{-1}$$
$$> 0. \tag{3.45}$$

Thus a necessary condition for the solvability of the corona problem is that $\delta(\varphi) > 0$. The 'corona problem' was to show the converse; it was a famous open problem for many years. It was solved by Lennart Carleson [57] in 1962. Another proof of this theorem, due to Thomas Wolff, can be found in [134, appendix] or in [158, appendix 3].

Theorem 3.46. **(Carleson's corona theorem)** *If* $\varphi_1, \ldots, \varphi_m \in H^\infty$ *and if* $\delta(\varphi)$ *defined by equation (3.44) is positive, then there exist* $\psi_1, \ldots, \psi_m \in H^\infty$ *such that* $\varphi_1 \psi_1 + \cdots + \varphi_m \psi_m = 1$.

Further Development of Models on the Disc 87

Carleson also gave an estimate for the supremum norm of a solution $\psi = (\psi_1, \ldots, \psi_m)$ of equation (3.43) in terms of $\delta(\varphi)$. He showed that *if $\delta(\varphi)$ is positive then there exists $\psi \in (H^\infty)^m$ such that $\varphi_1 \psi_1 + \cdots + \varphi_m \psi_m = 1$ and*

$$\|\psi\|_{\mathbb{D}} \leq m! \, B^m \delta(\varphi)^{-Am},$$

for constants A, B.

It was then a matter of interest to find the smallest estimate for the norm of a solution ψ. Operator analysis can be used to solve this problem.

To this end, let us broaden the notion of model introduced in Definition 2.7 to hereditary functions on $\mathbb{D}$.

Definition 3.47. *A* model *of a function $h \in \mathrm{Her}(\mathbb{D})$ is a pair $(\mathcal{M}, u)$, where $\mathcal{M}$ is a Hilbert space and u is a map from $\mathbb{D}$ to $\mathcal{M}$ such that, for all $\lambda, \mu \in \mathbb{D}$,*

$$h(\lambda, \mu) = (1 - \overline{\mu}\lambda)\langle u_\lambda, u_\mu \rangle_{\mathcal{M}}. \tag{3.48}$$

Thus a model of φ in our earlier sense is the same as a model of $1 - \overline{\varphi(\mu)}\varphi(\lambda)$ in this broadened sense. There should be no conflict between the two terminologies.

Corresponding to data $\varphi_1, \ldots, \varphi_m \in H^\infty$ and any $\rho > 0$, we introduce the hereditary function

$$h_{\varphi,\rho}(\lambda, \mu) = \overline{\varphi_1(\mu)}\varphi_1(\lambda) + \cdots + \overline{\varphi_m(\mu)}\varphi_m(\lambda) - \rho \quad \text{for all } \lambda, \mu \in \mathbb{D}. \tag{3.49}$$

We shall then say that the data $(\varphi_1, \ldots, \varphi_m)$ have a *corona model with parameter ρ* if the hereditary function $h_{\varphi,\rho}$ has a model.

Theorem 3.50. *Let $\varphi_1, \ldots, \varphi_m \in H^\infty$ and let $\rho > 0$. There exist $\psi_1, \ldots, \psi_m \in H^\infty$ such that $\varphi_1 \psi_1 + \cdots + \varphi_m \psi_m = 1$ and $\|\psi\|_{\mathbb{D}} \leq \rho^{-\frac{1}{2}}$ if and only if φ has a corona model with parameter ρ.*

It is useful to identify an m-tuple $\psi = (\psi_1, \ldots, \psi_m)$ of H^∞ functions with the column vector function

$$\psi = \begin{bmatrix} \psi_1 \\ \vdots \\ \psi_m \end{bmatrix}.$$

In the notation introduced in Section 1.11, ψ is an element of the Banach space $H^\infty_{\mathcal{B}(\mathbb{C}, \mathbb{C}^m)}(\mathbb{D})$.

Proof. (Sufficiency). Suppose that $h_{\varphi,\rho}$ has a model $(\mathcal{M}, u)$, that is,

$$\varphi(\mu)^* \varphi(\lambda) - \rho = (1 - \overline{\mu}\lambda)\langle u_\lambda, u_\mu \rangle_{\mathcal{M}} \qquad \text{for all } \lambda, \mu \in \mathbb{D}.$$

88   *Commutative Theory*

Reshuffle this equation to obtain

$$\varphi(\mu)^*\varphi(\lambda) + \bar{\mu}\lambda\langle u_\lambda, u_\mu\rangle_{\mathcal{M}} = \rho + \langle u_\lambda, u_\mu\rangle_{\mathcal{M}} \qquad \text{for all } \lambda, \mu \in \mathbb{D}.$$

This is to say that the families of vectors

$$\begin{pmatrix} \varphi(\lambda) \\ \lambda u_\lambda \end{pmatrix}_{\lambda \in \mathbb{D}} \quad \text{in } \mathbb{C}^m \oplus \mathcal{M} \quad \text{and} \quad \begin{pmatrix} \sqrt{\rho} \\ u_\lambda \end{pmatrix}_{\lambda \in \mathbb{D}} \quad \text{in } \mathbb{C} \oplus \mathcal{M}$$

have the same gramians. Combining the lurking isometry and extension steps of the argument, we deduce that there is a contraction

$$\begin{bmatrix} A & B \\ C & D \end{bmatrix} : \mathbb{C}^m \oplus \mathcal{M} \to \mathbb{C} \oplus \mathcal{M}$$

such that

$$\begin{bmatrix} A & B \\ C & D \end{bmatrix} \begin{pmatrix} \varphi(\lambda) \\ \lambda u_\lambda \end{pmatrix} = \begin{pmatrix} \sqrt{\rho} \\ u_\lambda \end{pmatrix} \qquad \text{for all } \lambda \in \mathbb{D}.$$

Solve this equation to obtain

$$u_\lambda = (1 - D\lambda)^{-1} C\varphi(\lambda)$$

and

$$[A + B\lambda(1 - D\lambda)^{-1}C]\varphi(\lambda) = \sqrt{\rho} \qquad \text{for all } \lambda \in \mathbb{D}. \tag{3.51}$$

Let

$$\psi(\lambda) \overset{\text{def}}{=} \rho^{-\frac{1}{2}}[A + B\lambda(1 - D\lambda)^{-1}C]^T \qquad \text{for all } \lambda \in \mathbb{D}.$$

By Theorem 3.16, $\rho^{\frac{1}{2}}\psi$ is in the Schur class $\mathscr{S}_{\mathcal{B}(\mathbb{C},\mathbb{C}^m)}(\mathbb{D})$. Thus $\|\psi\|_{\mathbb{D}} \leq \rho^{-\frac{1}{2}}$ and, by equation (3.51), $\psi^T\varphi = 1$. We have proved sufficiency.

(Necessity). Suppose that there exists $\psi \in \mathrm{H}^\infty_{\mathcal{B}(\mathbb{C},\mathbb{C}^m)}(\mathbb{D})$ such that $\psi^T\varphi = 1$ and $\|\psi\|_{\mathbb{D}} \leq \rho^{-\frac{1}{2}}$. Then

$$\rho^{\frac{1}{2}}\psi^T \in \mathscr{S}_{\mathcal{B}(\mathbb{C}^m,\mathbb{C})}(\mathbb{D}),$$

and so, by Theorem 3.10, $\rho^{\frac{1}{2}}\psi^T$ has a model in the sense of Definition 3.8. That is to say, there is a Hilbert space $\mathcal{M}$ and a map $u \colon \mathbb{D} \to \mathcal{B}(\mathbb{C}^m, \mathcal{M})$ such that

$$1 - \rho\overline{\psi(\mu)}\psi(\lambda)^T = (1 - \bar{\mu}\lambda)u(\mu)^*u(\lambda) \qquad \text{for all } \lambda, \mu \in \mathbb{D}.$$

Pre- and post-multiply this equation by $\varphi(\mu)^*$ and $\varphi(\lambda)$, respectively, and use the equation $\psi^T\varphi = 1$ to deduce that

$$\varphi(\mu)^*\varphi(\lambda) - \rho = (1 - \bar{\mu}\lambda)v(\mu)^*v(\lambda) \qquad \text{for all } \lambda, \mu \in \mathbb{D}$$

where $v = u\varphi$. Thus $(\mathcal{M}, v)$ is a corona model for φ with parameter ρ. $\square$

Further Development of Models on the Disc 89

Because the corona problem and its variants have attracted so much attention among both function- and operator-theorists, we shall give two other formulations of the preceding theorem. The first of them illustrates one of the refrains of this book: substitution of *operators* into function-theoretic statements (such as the equation $\psi^T \varphi = 1$) often leads to more definitive results than working with scalar arguments (as in the inequality (3.45)).

Theorem 3.52. *Let* $\varphi_1, \dots, \varphi_m \in H^\infty$ *and let* $\rho > 0$. *The following statements are equivalent.*

(i) *There exist* $\psi_1, \dots, \psi_m \in H^\infty$ *such that* $\varphi_1 \psi_1 + \dots + \varphi_m \psi_m = 1$ *and*
$$\|\psi\|_{\mathbb{D}} \leq \rho^{-\frac{1}{2}};$$

(ii) *for every contraction T having spectrum contained in $\mathbb{D}$,*
$$\varphi_1(T)\varphi_1(T)^* + \dots + \varphi_m(T)\varphi_m(T)^* \geq \rho; \tag{3.53}$$

(iii) *the kernel*
$$(\lambda, \mu) \mapsto \left(\overline{\varphi_1(\mu)}\varphi_1(\lambda) + \dots + \overline{\varphi_m(\mu)}\varphi_m(\lambda) - \rho \right) \frac{1}{1 - \bar{\mu}\lambda} \tag{3.54}$$

is positive semi-definite on $\mathbb{D}$.

Proof. (i)$\Rightarrow$(ii) Suppose $\psi_1, \dots, \psi_m$ exist as described. By Theorem 3.50, φ has a corona model $(\mathcal{M}, u)$ with parameter ρ:
$$\sum_{j=1}^{m} \overline{\varphi_j(\mu)}\varphi_j(\lambda) - \rho = (1 - \bar{\mu}\lambda)\langle u_\lambda, u_\mu \rangle \quad \text{for all } \lambda, \mu \in \mathbb{D}.$$

By Theorem 2.88, the positive kernel $\langle u_\lambda, u_\mu \rangle$ is a sum of dyads. Thus there is a sequence (f_n) in $\mathrm{Hol}(\mathbb{D})$ such that
$$\sum_{j=1}^{m} \overline{\varphi_j(\mu)}\psi_j(\lambda) \quad \rho - \sum_{n=1}^{\infty} \overline{f_n(\mu)}(1 - \bar{\mu}\lambda)f_n(\lambda) \quad \text{for all } \lambda, \mu \in \mathbb{D}, \tag{3.55}$$

the right-hand side converging uniformly for (λ, μ) in compact subsets of $\mathbb{D} \times \mathbb{D}$.

Consider any contraction T such that $\sigma(T) \subset \mathbb{D}$. By equation (3.55) and the properties of the hereditary calculus,
$$\sum_{j=1}^{m} \varphi_j(T)^*\varphi_j(T) - \rho = \sum_{n=1}^{\infty} f_n(T)^*(1 - T^*T)f_n(T)$$
$$\geq 0.$$

(ii)$\Rightarrow$(iii) Suppose that the inequality (3.53) holds for every contraction T with spectrum contained in $\mathbb{D}$. We should like to apply this hypothesis to S, the unilateral shift operator on H^2 defined in Section 1.5, but we cannot do so

90   Commutative Theory

directly since $\sigma(S)$ is the *closed* unit disc. Instead choose $r \in (0,1)$ and $T = rS$. Then T is a contraction and $\sigma(T) \subset \mathbb{D}$, and therefore

$$\varphi_1(rS)\varphi_1(rS)^* + \cdots + \varphi_m(rS)\varphi_m(rS)^* - \rho \geq 0.$$

Let $(\varphi_j)_r$ be the analytic function on a neighborhood of $\mathbb{D}^-$ given by $(\varphi_j)_r(z) = \varphi_j(rz)$ for $z \in \mathbb{D}$. Then $(\varphi_j)_r(S) = \varphi_j(rS)$, and

$$(\varphi_1)_r(S)(\varphi_1)_r(S)^* + \cdots + (\varphi_m)_r(S)(\varphi_m)_r(S)^* - \rho \geq 0.$$

A self-adjoint bounded linear operator A on H^2 is positive if and only the kernel $(\lambda, \mu) \to \langle As_\mu, s_\lambda \rangle$ is positive on $\mathbb{D}$, where s_λ is the Szegő kernel of equation (1.30) (observe that the set of linear combinations $\sum c_j s_{\lambda_j}$ is dense in H^2). It follows that, for every $r \in (0,1)$, the kernel

$$(\lambda, \mu) \mapsto \left\langle \left((\varphi_1)_r(S)(\varphi_1)_r(S)^* + \cdots + (\varphi_m)_r(S)(\varphi_m)_r(S)^* - \rho\right) s_\mu, s_\lambda \right\rangle$$

on $\mathbb{D}$ is positive. By virtue of Lemma 1.65, the kernel

$$(\lambda, \mu) \mapsto \left(\varphi_1(r\lambda)\overline{\varphi_1(r\mu)} + \cdots + \varphi_m(r\lambda)\overline{\varphi_m(r\mu)} - \rho \right) \frac{1}{1 - \bar{\mu}\lambda}$$

is positive for all $r \in (0,1)$. On letting $r \to 1$ we deduce that the kernel remains positive when $r = 1$, which is to say that condition (iii) holds.

(iii)$\Rightarrow$(i) Suppose the kernel (3.54) is positive semi-definite; then, by Moore's theorem, it is expressible as $\langle u_\lambda, u_\mu \rangle_{\mathcal{M}}$ for some Hilbert space $\mathcal{M}$ and some map $u: \mathbb{D} \to \mathcal{M}$. That is to say, φ has a corona model with parameter ρ. Hence, by Theorem 3.50, there exist $\psi_1, \ldots, \psi_m \in \mathrm{H}^\infty$ such that

$$\varphi_1 \psi_1 + \cdots + \varphi_m \psi_m = 1 \text{ and } \|\psi\|_{\mathbb{D}} \leq \rho^{-\frac{1}{2}}. \qquad \square$$

The content of Theorem 3.50 is customarily expressed somewhat differently and called either the Toeplitz corona theorem or Leech's theorem. A *Toeplitz operator* is an operator T_φ, corresponding to a function $\varphi \in \mathrm{L}^\infty(\mathbb{T})$, acting on the Hardy space H^2 and defined by

$$T_\varphi f = P_+(\varphi f) \qquad \text{for all } f \in \mathrm{H}^2,$$

where $P_+: L^2 \to \mathrm{H}^2$ is the orthogonal projection operator. In the special case that $\varphi \in \mathrm{H}^\infty$, the P_+ is redundant and T_φ coincides with the multiplication operator M_φ defined in Section 1.12. Lemma 1.65 tells us that, for any $\varphi \in \mathrm{H}^\infty$ and $\lambda \in \mathbb{D}$,

$$T_\varphi^* s_\lambda = \overline{\varphi(\lambda)} s_\lambda,$$

where s_λ is the Szegő kernel.

Further Development of Models on the Disc 91

Theorem 3.56. **(Leech's theorem/Toeplitz corona theorem)**
Let $\varphi \in H^\infty_{\mathcal{B}(\mathbb{C},\mathbb{C}^m)}(\mathbb{D})$ *and let* $\rho > 0$. *There exists* $\psi \in H^\infty_{\mathcal{B}(\mathbb{C},\mathbb{C}^m)}(\mathbb{D})$ *such that* $\psi^T \varphi = 1$ *and* $\|\psi\|_{\mathbb{D}} \le \rho^{-\frac{1}{2}}$ *if and only if*

$$\sum_{j=1}^m T_{\varphi_j} T^*_{\varphi_j} \ge \rho. \tag{3.57}$$

Proof. We have

$$\sum_{j=1}^m T_{\varphi_j} T^*_{\varphi_j} - \rho \ge 0 \Leftrightarrow \left\langle \left(\sum T_{\varphi_j} T^*_{\varphi_j} - \rho \right) s_\mu, s_\lambda \right\rangle \text{ is a positive kernel}$$

$$\Leftrightarrow \sum \left\langle T^*_{\varphi_j} s_\mu, T^*_{\varphi_j} s_\lambda \right\rangle - \rho \langle s_\mu, s_\lambda \rangle \text{ is a positive kernel}$$

$$\Leftrightarrow \sum \left\langle \overline{\varphi_j(\mu)} s_\mu, \overline{\varphi_j(\lambda)} s_\lambda \right\rangle - \rho \langle s_\mu, s_\lambda \rangle \text{ is a positive kernel}$$

$$\Leftrightarrow (1 - \bar{\mu}\lambda)^{-1} \left(\sum \overline{\varphi_j(\mu)} \varphi_j(\lambda) - \rho \right) \text{ is a positive kernel}$$

$$\Leftrightarrow \sum \overline{\varphi_j(\mu)} \varphi_j(\lambda) - \rho = (1 - \bar{\mu}\lambda)\langle u_\lambda, u_\mu \rangle \text{ for all } \lambda, \mu \in \mathbb{D}$$

for some Hilbert space $\mathcal{M}$ and some map $u \colon \mathbb{D} \to \mathcal{M}$, that is, if and only if φ has a corona model with parameter ρ. $\square$

3.9 Historical Notes

Theorem 3.6, the Moore theorem for $\mathcal{B}(\mathcal{H})$-valued functions, is sometimes called the Kolmogorov decomposition of $A(\lambda, \mu)$.

Section 3.3, the network realization formula. A more algebraic approach to the results of this section is based on properties of the linear fractional transformation $\mathcal{F}_V$, which maps operators on $\mathcal{M}$ to elements of $\mathcal{B}(\mathcal{H}, \mathcal{K})$ by

$$\mathcal{F}_V(X) = A + BX(1 - DX)^{-1}C$$

for X such that $1 - DX$ is invertible, where $V = \begin{bmatrix} A & B \\ C & D \end{bmatrix} \colon \mathcal{H} \oplus \mathcal{M} \to \mathcal{K} \oplus \mathcal{M}$.
The identity

$$1 - \mathcal{F}_V(X)^* \mathcal{F}_V(X) = C^*(1 - X^*D^*)^{-1}(1 - X^*X)(1 - DX)^{-1}C$$

$$+ \begin{bmatrix} 1 & C^*(1 - X^*D^*)^{-1}X^* \end{bmatrix} (1 - V^*V) \begin{bmatrix} 1 \\ X(1 - DX)^{-1}C \end{bmatrix}.$$

can be verified by direct calculation.

92 *Commutative Theory*

Section 3.4, tensor products of Hilbert spaces. The algebraic tensor product of two vector spaces is constructed in many texts, for example [97]. One way to do it is to define $x \otimes y$ to be the linear functional $f \mapsto f(x, y)$ on the vector space of all bilinear functionals on $\mathcal{H} \times \mathcal{K}$.

For x, y in Hilbert spaces $\mathcal{H}, \mathcal{K}$, the expression $x \otimes y$ can denote either a rank-one operator from $\mathcal{K}$ to $\mathcal{H}$, as in equation (1.17) or an element of $\mathcal{H} \otimes \mathcal{K}$. If $\lambda, \mu \in \mathbb{C}$, then in the former case $(\lambda x) \otimes (\mu y) = \lambda \bar{\mu}(x \otimes y)$, while in the latter, $(\lambda x) \otimes (\mu y) = \lambda \mu(x \otimes y)$. A case can be made that the two objects denoted by $x \otimes y$ are nevertheless "the same": identify $y \in \mathcal{K}$ with the linear functional $\langle \cdot, y \rangle$ on $\mathcal{K}$, thereby identifying $\mathcal{K}$ with $\mathcal{K}^*$, and then identify $\mathcal{H} \otimes_a \mathcal{K}^*$ with a subset of $\mathcal{B}(\mathcal{K}, \mathcal{H})$ in the natural way. The fact that the first of these two identifications is a *conjugate* linear map accounts for the $\mu/\bar{\mu}$ anomaly. However, we prefer to regard the two entities $x \otimes y$ as two different objects and to rely on the context to distinguish them. The notation is long established, in both senses.

Section 3.8. Leech's theorem was originally proved by Robert Leech in 1972 [142]; however, due to unfavorable refereeing, it was not published until 2014. It is often called the *Toeplitz corona theorem* and often attributed to subsequent authors. See [124] for a historical discussion.

Even though Theorem 3.56 is sharp, one can ask what the best bounds on $\rho^{-1/2} = \|\psi\|$ are just in terms of the constant $\delta = \delta(\varphi)$. For a discussion see [158], which gives estimates due to A. Uchiyama and V. Tolokonnikov [202], and a more recent improvement by S. Treil [203]. They show that for δ small one has the estimates

$$A \, \delta^{-2} \log \log \frac{1}{\delta} \leq \|\psi\| \leq B \, \delta^{-2} \log \frac{1}{\delta}.$$

4

Operator Analysis on $\mathbb{D}^2$

Holomorphic functions of several complex variables differ greatly in their properties from functions of one complex variable, and the theories of the two species of function have quite different flavors. For introductory books to several complex variables, see, for example, [137, 118, 201]. Many theorems about holomorphic functions of one complex variable do not generalize easily (or at all) to several variables. However, in this chapter we show the power of the notion of model in the discovery and proof of generalizations of some single-variable theorems to two variables. We exploit the fact that functions in $\mathscr{S}(\mathbb{D}^2)$ also have models to extend the techniques in Chapter 2 to functions of two variables. Our development will follow that of Chapter 2 quite closely, but with one major difference. It will require a new idea to establish a fundamental theorem for $\mathscr{S}(\mathbb{D}^2)$.

In the previous chapter we used a positivity argument to derive von Neumann's inequality (Theorem 1.36) from the fundamental theorem for $\mathscr{S}(\mathbb{D})$ (Theorem 2.9). A powerful technique in operator analysis, the *duality construction*, allows one to go in the reverse direction, that is, to derive model formulas from von Neumann-like inequalities. In this chapter we shall first illustrate this technique by deriving a fundamental theorem for $\mathscr{S}(\mathbb{D}^2)$ using Andô's inequality.

Once the fundamental theorem for $\mathscr{S}(\mathbb{D}^2)$ is proved, only notational changes to the one-variable proofs presented in Chapter 2 are required to extend the results of Chapter 2 to the bidisc.

4.1 The Space of Hereditary Functions on $\mathbb{D}^2$

We use $\lambda = (\lambda^1, \lambda^2)$ and $\mu = (\mu^1, \mu^2)$ for the variables in $\mathbb{D}^2$. Following the definition of $\mathrm{Her}(\mathbb{D})$ in Section 2.8.1, we define $\mathrm{Her}(\mathbb{D}^2)$ to consist of the set of complex-valued functions defined on $\mathbb{D}^2 \times \mathbb{D}^2$ having the property that the map

93

94 *Commutative Theory*

$$(\lambda, \mu) \mapsto h(\lambda, \bar{\mu}) \in \mathbb{C}$$

is a holomorphic function on $\mathbb{D}^2 \times \mathbb{D}^2$. Equivalently, a function h on $\mathbb{D}^2 \times \mathbb{D}^2$ belongs to $\mathrm{Her}(\mathbb{D}^2)$ if, for each fixed $\mu \in \mathbb{D}^2$, $h(\lambda, \mu)$ is holomorphic in λ on $\mathbb{D}^2$, and for each fixed $\lambda \in \mathbb{D}^2$, $\overline{h(\lambda, \mu)}$ is holomorphic in μ on $\mathbb{D}^2$. We equip $\mathrm{Her}(\mathbb{D}^2)$ with the natural linear operations of pointwise addition and scalar multiplication and with the topology of uniform convergence on compact subsets of $\mathbb{D}^2 \times \mathbb{D}^2$. As in the case of the disc, $\mathrm{Her}(\mathbb{D}^2)$ is a complete metrizable locally convex topological vector space.

If $h \in \mathrm{Her}(\mathbb{D}^2)$ then h can be represented by a power series,

$$h(\lambda, \mu) = \sum_{m,n} a_{m,n} \, \bar{\mu}^n \lambda^m \qquad \text{for all } \lambda, \mu \in \mathbb{D}^2, \tag{4.1}$$

which converges uniformly on compact subsets of $\mathbb{D}^2 \times \mathbb{D}^2$. In equation (4.1) $m = (m_1, m_2)$ and $n = (n_1, n_2)$ are multi-indices that independently range through the set of pairs of nonnegative integers, and λ^m and $\bar{\mu}^n$ are defined by

$$\lambda^m = (\lambda^1)^{m_1} (\lambda^2)^{m_2} \qquad \text{and} \qquad \bar{\mu}^n = (\bar{\mu^1})^{n_1} (\bar{\mu^2})^{n_2}. \tag{4.2}$$

We define an involution (cf. Lemma 2.86) on $\mathrm{Her}(\mathbb{D}^2)$ by the formula

$$h^*(\lambda, \mu) = \overline{h(\mu, \lambda)} \qquad \text{for all } \lambda, \mu \in \mathbb{D}^2$$

and then define $\mathcal{R} \subseteq \mathrm{Her}(\mathbb{D}^2)$ by

$$\mathcal{R} = \{h \in \mathrm{Her}(\mathbb{D}^2) \colon h^* = h\}.$$

It is immediate that $\mathcal{R}$ is a real subspace of $\mathrm{Her}(\mathbb{D}^2)$. Furthermore, if $\mathrm{Her}(\mathbb{D}^2)^*$ denotes the dual space of $\mathrm{Her}(\mathbb{D}^2)$ over $\mathbb{C}$ and $\mathcal{R}^*$ denotes the dual space of $\mathcal{R}$ over $\mathbb{R}$, then there is a close relationship between $\mathrm{Her}(\mathbb{D}^2)^*$ and $\mathcal{R}^*$, described in the following statement.

Lemma 4.3. *Let E be a topological vector space over $\mathbb{C}$, let $*$ be a continuous involution on E and let $\mathcal{R}$ be the real subspace $\{x \colon x = x^*\}$ of E. For every continuous real linear functional L on $\mathcal{R}$ there exists a unique continuous complex linear functional $L^\sim$ on E such that $L = L^\sim | \mathcal{R}$.*

Proof. For $h \in E$, let

$$\mathrm{Re}\, h = \tfrac{1}{2}(h + h^*) \qquad \text{and} \qquad \mathrm{Im}\, h = \tfrac{1}{2i}(h - h^*).$$

Observe that $\mathrm{Re}\, h, \mathrm{Im}\, h \in \mathcal{R}$ and $h = \mathrm{Re}\, h + i\, \mathrm{Im}\, h$. Therefore, if $L^\sim \in E^*$, the dual of E over $\mathbb{C}$, and $L = L^\sim | \mathcal{R}$, then

$$L^\sim(h) = L^\sim(\mathrm{Re}\, h + i\, \mathrm{Im}\, h) = L^\sim(\mathrm{Re}\, h) + i L^\sim(\mathrm{Im}\, h) = L(\mathrm{Re}\, h) + i L(\mathrm{Im}\, h).$$

Operator Analysis on $\mathbb{D}^2$ 95

This calculation proves that there is at most one $L^{\sim}$ with the stated properties. It also suggests that, for a continuous real linear functional L on $\mathcal{R}$, we define $L^{\sim}$ by

$$L^{\sim}(h) = L(\operatorname{Re} h) + i L(\operatorname{Im} h) \qquad \text{for all } h \in E.$$

This formula does indeed define an element of E^* satisfying $L = L^{\sim}|\mathcal{R}$. $\quad\square$

We define a second set $\mathcal{P} \subseteq \operatorname{Her}(\mathbb{D}^2)$, called the *positive cone in* $\operatorname{Her}(\mathbb{D}^2)$, by

$$\mathcal{P} = \{h \in \operatorname{Her}(\mathbb{D}^2) \colon h \text{ is positive semi-definite on } \mathbb{D}^2\}.$$

It is easy to see that $\mathcal{P}$ is a closed cone[1] in $\mathcal{R}$. Also, the following analog of Theorem 2.88 obtains.

Theorem 4.4. *If A is a hereditary function on $\mathbb{D}^2$ and A is positive semi-definite, then there exists a sequence $f_1, f_2, \ldots$ in $\operatorname{Hol}(\mathbb{D}^2)$ such that, for all $\lambda, \mu \in \mathbb{D}^2$,*

$$A(\lambda, \mu) = \sum_{n=1}^{\infty} \overline{f_n(\mu)} \, f_n(\lambda), \tag{4.5}$$

where the series in equation (4.5) *converges uniformly on compact subsets of* $\mathbb{D}^2 \times \mathbb{D}^2$.

Proof. Follow the proof of theorem 2.88 with $\mathbb{D}^2$ in place of $\mathbb{D}$. $\quad\square$

4.2 The Hereditary Functional Calculus on $\mathbb{D}^2$

In this section we shall extend the hereditary functional calculus to commuting pairs of operators. First, though, let us consider the holomorphic functional calculus for commuting pairs. In Section 1.2 we briefly discussed the functional calculus for general domains in $\mathbb{C}^d$. On the bidisc, things are more straightforward. Suppose f is in $\operatorname{Hol}(\mathbb{D}^2)$. Then f has a power series expansion

$$f(\lambda) = \sum a_m \lambda^m = \sum_{m_1=0}^{\infty} \sum_{m_2=0}^{\infty} a_{m_1 m_2} (\lambda^1)^{m_1} (\lambda^2)^{m_2}. \tag{4.6}$$

(Here $m = (m_1, m_2)$ is a multi-index and on the far right of equation (4.6) the reader has to distinguish between the superscripts 1 and 2 used as indices, and m_1 and m_2 which are powers).

---

[1] $\mathcal{P}$ is a *cone* if (i) $\mathcal{P} + \mathcal{P} \subseteq \mathcal{P}$, (ii) $\mathcal{P} \cap (-\mathcal{P}) = \{0\}$ and (iii) $\alpha \mathcal{P} \subseteq \mathcal{P}$ whenever $\alpha \in \mathbb{R}$ and $\alpha \geq 0$.

96 *Commutative Theory*

If $T = (T^1, T^2)$ is a commuting pair of operators, then we would like to define

$$f(T) = \sum a_m T^m = \sum_{m_1=0}^{\infty} \sum_{m_2=0}^{\infty} a_{m_1 m_2} (T^1)^{m_1} (T^2)^{m_2} \qquad (4.7)$$

whenever the series (4.7) converges. This will certainly happen if both T^1 and T^2 have spectral radius less than one, since the series (4.6) converges absolutely on all of $\mathbb{D}^2$. So we can use the formula (4.7) as a *definition* of $f(T)$ whenever f is in $\text{Hol}(\mathbb{D}^2)$ and $\sigma(T) \subseteq \mathbb{D}^2$. Alternatively, as for the disc in equation (1.4), we may use the Riesz–Dunford integral, defining

$$f(T) = \frac{1}{(2\pi i)^2} \int_{\gamma} \int_{\gamma} f(\lambda)(\lambda^1 - T^1)^{-1}(\lambda^2 - T^2)^{-1} d\lambda^1 d\lambda^2 \qquad (4.8)$$

for a suitably chosen contour γ.

If $\mathcal{H}$ is a Hilbert space, $T = (T^1, T^2)$ is a commuting pair in $\mathcal{B}(\mathcal{H})$ with $\sigma(T) \subseteq \mathbb{D}^2$ (equivalently, $\sigma(T^1) \subseteq \mathbb{D}$ and $\sigma(T^2) \subseteq \mathbb{D}$), and $h \in \text{Her}(\mathbb{D}^2)$, then just as in the case of one variable, there are two equivalent ways that an operator $h(T)$ in $\mathcal{B}(\mathcal{H})$ may be defined.

(1) **Via power series.** For h having the power series expansion (4.1) define $h(T) \in \mathcal{B}(\mathcal{H})$ by the formula,

$$h(T) = \sum_{m,n} a_{m,n} (T^*)^n T^m, \qquad (4.9)$$

where, following equation (4.2), we adopt the notations

$$T^m = (T^1)^{m_1} (T^2)^{m_2} \quad \text{and} \quad (T^*)^n = ((T^1)^*)^{n_1} ((T^2)^*)^{n_2}.$$

(2) **Via the iterated Riesz–Dunford functional calculus.** As we assume that $\sigma(T) \subseteq \mathbb{D}^2$, there exists $r < 1$ such that $\sigma(T^1)$, $\sigma(T^2)$, $\sigma((T^1)^*)$, and $\sigma((T^2)^*)$ all lie in $r\mathbb{D}$. Define a path by $\gamma(t) = re^{it}$, $0 \le t \le 2\pi$ and define $h(T)$ by

$$h(T) = \frac{1}{(2\pi i)^4} \int_{\gamma} \int_{\gamma} \int_{\gamma} \int_{\gamma} h(\lambda, \mu)(\bar{\mu}^1 - (T^1)^*)^{-1}(\bar{\mu}^2 - (T^2)^*)^{-1}$$
$$(\lambda^1 - T^1)^{-1}(\lambda^2 - T^2)^{-1} d\lambda^1 d\lambda^2 d\bar{\mu}^1 \bar{\mu}^2. \qquad (4.11)$$

As a substitute for Lemmas 2.85 and 2.86 and Proposition 2.87 we have the following results, which are proved in much the same way.

Lemma 4.12. *If $\mathcal{H}$ is a Hilbert space, T is a commuting pair in $\mathcal{B}(\mathcal{H})$ with $\sigma(T) \subseteq \mathbb{D}^2$, and $h \in \text{Her}(\mathbb{D}^2)$, then equations (4.9) and (4.11) define the same element $h(T) \in \mathcal{B}(\mathcal{H})$.*

$$\text{Operator Analysis on } \mathbb{D}^2 \qquad\qquad 97$$

Lemma 4.13. (Properties of the hereditary functional calculus) *Let $\mathcal{H}$ be a Hilbert space and T be a commuting pair of operators on $\mathcal{H}$ satisfying $\sigma(T) \subseteq \mathbb{D}^2$.*

(i) *If $f, g \in \mathrm{Her}(\mathbb{D}^2)$ and $\alpha, \beta \in \mathbb{C}$, then*

$$(\alpha f + \beta g)(T) = \alpha f(T) + \beta g(T).$$

(ii) *If $\varphi \in \mathrm{Hol}(\mathbb{D}^2)$ and $h(\lambda, \mu) = \varphi(\lambda)$ for all $\lambda, \mu \in \mathbb{D}^2$, then $h(T) = \varphi(T)$.*

(iii) *If $\psi \in \mathrm{Hol}(\mathbb{D}^2)$ and $h(\lambda, \mu) = \overline{\psi(\mu)}$ for all $\lambda, \mu \in \mathbb{D}^2$, then $h(T) = \psi(T)^*$.*

(iv) *If $h \in \mathrm{Her}(\mathbb{D}^2)$ then $h^* \in \mathrm{Her}(\mathbb{D}^2)$ and $h^*(T) = h(T)^*$.*

(v) *If $\varphi, \psi \in \mathrm{Hol}(\mathbb{D}^2)$, $h \in \mathrm{Her}(\mathbb{D}^2)$ and $g \in \mathrm{Her}(\mathbb{D}^2)$ satisfies*

$$g(\lambda, \mu) = \overline{\psi(\mu)}\, h(\lambda, \mu)\, \varphi(\lambda), \qquad \text{for all } \lambda, \mu \in \mathbb{D}^2,$$

then

$$g(T) = \psi(T)^* h(T) \varphi(T).$$

Proposition 4.14. *The hereditary functional calculus is continuous, that is, if $\{h_n\}$ is a sequence in $\mathrm{Her}(\mathbb{D}^2)$, $h \in \mathrm{Her}(\mathbb{D}^2)$, and $h_n \to h$ uniformly on compact subsets of $\mathbb{D}^2 \times \mathbb{D}^2$, then $h_n(T) \to h(T)$ in $\mathcal{B}(\mathcal{H})$ with respect to the operator norm, whenever T is a commuting pair of operators with $\sigma(T) \subset \mathbb{D}^2$.*

Remark 4.15. (The hereditary calculus via nuclearity) There is a third way to define the hereditary functional calculus, which might be preferred by those who are familiar with theories of tensor products and nuclear spaces. Those who are not can safely skip this remark.

We work in the category $\mathcal{C}$ of *real* locally convex topological vector spaces and continuous real-linear maps. The *complete projective tensor product* of two locally convex topological vector spaces E and F is a complete locally convex topological vector space $E \hat{\otimes} F$ together with a continuous bilinear map $\iota \colon E \times F \to E \hat{\otimes} F$ having a certain universal property: for every complete locally convex topological vector space $\mathcal{X}$ and every continuous bilinear map $B \colon E \times F \to \mathcal{X}$, there exists a continuous linear map $L \colon E \hat{\otimes} F \to \mathcal{X}$ such that $L \circ \iota = B$.

For every pair E, F in $\mathcal{C}$ there does exist a complete projective tensor product $(E \hat{\otimes} F, \iota)$, and it is unique up to isomorphism in the category $\mathcal{C}$ [121, part III], [100, chapter II, theorem 4.7]. Moreover, for each B, the linear map L is uniquely determined (up to conjugation by an isomorphism).

We shall apply this result to the case $E = \mathrm{Hol}(\mathbb{D}^2)$, $F = \overline{\mathrm{Hol}(\mathbb{D}^2)}$ and $\mathcal{X} = \mathcal{B}(\mathcal{H})$ for some Hilbert space $\mathcal{H}$. For any function $\varphi \in \mathrm{Hol}(\mathbb{D}^2)$ and any commuting pair $T = (T_1, T_2)$ of operators on $\mathcal{H}$ having spectrum in $\mathbb{D}^2$,

98 Commutative Theory

the several-variable functional calculus defines an operator $\varphi(T) \in \mathcal{B}(\mathcal{H})$ with natural properties. We may therefore define a continuous bilinear map

$$B \colon \operatorname{Hol}(\mathbb{D}^2) \times \overline{\operatorname{Hol}(\mathbb{D}^2)} \to \mathcal{B}(\mathcal{H})$$

by

$$B(\varphi, \bar{\psi}) = \psi(T)^* \varphi(T) \quad \text{for all } \varphi, \psi \in \operatorname{Hol}(\mathbb{D}^2).$$

By the universal property, there exists a continuous real-linear map

$$L \colon \operatorname{Hol}(\mathbb{D}^2) \hat{\otimes} \operatorname{Hol}(\mathbb{D}^2) \to \mathcal{B}(\mathcal{H})$$

such that $L \circ \iota = B$, which is to say that

$$L \circ \iota(\varphi, \bar{\psi}) = \psi(T)^* \varphi(T) \quad \text{for all } \varphi, \psi \in \operatorname{Hol}(\mathbb{D}^2).$$

The projective tensor product

$$\operatorname{Hol}(\mathbb{D}^2) \hat{\otimes} \overline{\operatorname{Hol}(\mathbb{D}^2)} \cong \operatorname{Her}(\mathbb{D}^2), \tag{4.16}$$

where $\cong$ denotes isomorphism of real-linear topological vector spaces (this fact depends on the nuclearity of $\operatorname{Hol}(\mathbb{D}^2)$). Indeed, by [100, chapter II, corollary 4.15],

$$\operatorname{Hol}(\mathbb{D}^2) \hat{\otimes} \operatorname{Hol}(\mathbb{D}^2) \cong \operatorname{Hol}(\mathbb{D}^4).$$

Since $J \colon \operatorname{Hol}(\mathbb{D}^4) \to \operatorname{Her}(\mathbb{D}^2)$ given by $Jf(\lambda, \mu) = f(\lambda, \bar{\mu})$ is also a linear homeomorphism, the isomorphism (4.16) holds.

Under the identification (4.16), $\iota(\varphi, \bar{\psi})$ is the element $\overline{\psi(\mu)}\varphi(\lambda)$ of $\operatorname{Her}(\mathbb{D}^2)$.[2] Thus

$$L(\overline{\psi(\mu)}\varphi(\lambda)) = \psi(T)^* \varphi(T) \quad \text{for all } \varphi, \psi \in \operatorname{Hol}(\mathbb{D}^2).$$

We define $f(T)$ to be $L(f)$ for all $f \in \operatorname{Her}(\mathbb{D}^2)$. The properties (i) to (v) in Lemma 4.13 and Proposition 4.14 can then be deduced from the properties of the holomorphic functional calculus.

One can think of this definition as a refinement of the power series definition of the hereditary calculus given earlier. For the case of the bidisc it is more elaborate than is necessary, but the approach has one great merit: it can be implemented on an arbitrary domain in $\mathbb{C}^d$.

Remark 4.17. (A metamathematical observation) In working with the hereditary calculus, as for the functional calculus, one rarely uses the formulas that define it. Rather, one uses the properties of the hereditary calculus as outlined in Lemma 4.13 and Proposition 4.14.

---

[2] Here, and in the sequel, we shall allow ourselves the slight abuse of notation of letting $\overline{\psi(\mu)}\varphi(\lambda)$ denote the hereditary function whose value at (λ, μ) is $\overline{\psi(\mu)}\varphi(\lambda)$.

Operator Analysis on $\mathbb{D}^2$ 99

4.3 Models on $\mathbb{D}^2$

As we shall see, the following definition plays the same key role for the bidisc that Definition 2.7 played in the theory of the Schur class on $\mathbb{D}$, as laid out in Chapter 2.

Definition 4.18. (**Models on $\mathbb{D}^2$**) *Let $\varphi: \mathbb{D}^2 \to \mathbb{C}$ be a function. A model for φ is a pair $(\mathcal{M}, u)$ where $\mathcal{M} = (\mathcal{M}^1, \mathcal{M}^2)$ is a pair of Hilbert spaces, $u = (u^1, u^2)$ is a pair of functions with $u^1: \mathbb{D}^2 \to \mathcal{M}^1$ and $u^2: \mathbb{D}^2 \to \mathcal{M}^2$, and for which the following* model formula *holds for every λ and μ in $\mathbb{D}^2$:*

$$1 - \overline{\varphi(\mu)}\varphi(\lambda) = \left\langle (1 - \bar{\mu}^1\lambda^1)\, u_\lambda^1, u_\mu^1 \right\rangle_{\mathcal{M}^1} + \left\langle (1 - \bar{\mu}^2\lambda^2)\, u_\lambda^2, u_\mu^2 \right\rangle_{\mathcal{M}^2}. \quad (4.19)$$

In Definition 4.18 we identify $\mathcal{M}$ with the Hilbert space $\mathcal{M}^1 \oplus \mathcal{M}^2$, and refer to $\mathcal{M}$ as a *decomposed Hilbert space*. When $\lambda \in \mathbb{D}^2$ and $\mathcal{M} = \mathcal{M}^1 \oplus \mathcal{M}^2$ is a decomposed Hilbert space, we may view a point $\lambda \in \mathbb{C}^2$ as an operator acting on $\mathcal{M}$ by identifying λ with the 2×2 block operator

$$\begin{bmatrix} \lambda^1 & 0 \\ 0 & \lambda^2 \end{bmatrix} : \mathcal{M}^1 \oplus \mathcal{M}^2 \to \mathcal{M}^1 \oplus \mathcal{M}^2, \quad (4.20)$$

or equivalently,[3]

$$\lambda(u^1 \oplus u^2) = \lambda^1 u^1 \oplus \lambda^2 u^2 \qquad \text{for all } u^1 \in \mathcal{M}^1, u^2 \in \mathcal{M}^2. \quad (4.21)$$

Note that when λ is so interpreted, if we let

$$u_\lambda = u_\lambda^1 \oplus u_\lambda^2,$$

then the model formula (4.19) takes the form

$$1 - \overline{\varphi(\mu)}\varphi(\lambda) = \left\langle (1 - \mu^*\lambda)\, u_\lambda, u_\mu \right\rangle_{\mathcal{M}} \quad (4.22)$$

for all $\lambda, \mu \in \mathbb{D}^2$. Optically, this is almost identical to equation (2.8).

Remarks 4.23.

(i) The notation for λ suppresses the dependence of λ on the decomposition of $\mathcal{M}$.

(ii) λ cannot be treated as a scalar as in general $\lambda T \neq T\lambda$ when $T \in \mathcal{B}(\mathcal{M})$.

(iii) If P is the projection from $\mathcal{M}$ onto $\mathcal{M}^1$, then λ denotes $\lambda^1 P + \lambda^2(1 - P)$.

Low-hanging fruits in the theory of models on $\mathbb{D}^2$ are the analogs of automatic holomorphy (Proposition 2.21) and the converse to the fundamental theorem (Proposition 2.32).

---

[3] This is similar to viewing a point $\lambda \in \mathbb{C}$ as an element in $\mathcal{B}(\mathcal{M})$ via the formula $\lambda(u) = \lambda u$, $u \in \mathcal{M}$.

100 Commutative Theory

Proposition 4.24. (Automatic holomorphy of models on $\mathbb{D}^2$) *Let $\varphi\colon \mathbb{D}^2 \to \mathbb{C}$ be a function and suppose that there exists a model $(\mathcal{M}, u)$ for φ. Then $(\mathcal{M}, u)$ is holomorphic (i.e., u^1 and u^2 are holomorphic $\mathcal{M}$-valued functions on $\mathbb{D}^2$).*

Proof. Assume that $\varphi\colon \mathbb{D}^2 \to \mathbb{C}$ is a function for which $(\mathcal{M}, u)$ is a model. The formula (4.22) can be reshuffled to

$$\varphi(\lambda)\overline{\varphi(\mu)} + \langle u_\lambda, u_\mu \rangle = 1 + \langle \lambda u_\lambda, \mu u_\mu \rangle \qquad \text{for all } \lambda, \mu \in \mathbb{D}^2,$$

which in turn may be rewritten as

$$\left\langle \begin{pmatrix} \varphi(\lambda) \\ u_\lambda \end{pmatrix}, \begin{pmatrix} \varphi(\mu) \\ u_\mu \end{pmatrix} \right\rangle_{\mathbb{C} \oplus \mathcal{M}} = \left\langle \begin{pmatrix} 1 \\ \lambda u_\lambda \end{pmatrix}, \begin{pmatrix} 1 \\ \mu u_\mu \end{pmatrix} \right\rangle_{\mathbb{C} \oplus \mathcal{M}} \qquad \text{for all } \lambda, \mu \in \mathbb{D}^2.$$

Therefore, Lemma 2.18 implies that there exists a linear isometry

$$V\colon [1 \oplus \lambda u_\lambda\colon \lambda \in \mathbb{D}^2] \to \mathbb{C} \oplus \mathcal{M}$$

such that

$$V \begin{pmatrix} 1 \\ \lambda u_\lambda \end{pmatrix} = \begin{pmatrix} \varphi(\lambda) \\ u_\lambda \end{pmatrix} \qquad \text{for all } \lambda \in \mathbb{D}^2.$$

If we define $W \in \mathcal{B}(\mathbb{C} \oplus \mathcal{M})$ by setting $W = V$ on $[1 \oplus \lambda u_\lambda\colon \lambda \in \mathbb{D}^2]$ and $W = 0$ on the orthogonal complement of $[1 \oplus \lambda u_\lambda\colon \lambda \in \mathbb{D}^2]$ in $\mathbb{C} \oplus \mathcal{M}$, then W is a partial isometry (in particular, W is a contraction) such that

$$W \begin{pmatrix} 1 \\ \lambda u_\lambda \end{pmatrix} = \begin{pmatrix} \varphi(\lambda) \\ u_\lambda \end{pmatrix} \qquad \text{for all } \lambda \in \mathbb{D}^2.$$

If we represent W as a 2×2 block matrix,

$$W = \begin{bmatrix} a & 1 \otimes \beta \\ \gamma \otimes 1 & D \end{bmatrix} \colon \mathbb{C} \oplus \mathcal{M} \to \mathbb{C} \oplus \mathcal{M},$$

then

$$\begin{pmatrix} \varphi(\lambda) \\ u_\lambda \end{pmatrix} = W \begin{pmatrix} 1 \\ \lambda u_\lambda \end{pmatrix} = \begin{bmatrix} a & 1 \otimes \beta \\ \gamma \otimes 1 & D \end{bmatrix} \begin{pmatrix} 1 \\ \lambda u_\lambda \end{pmatrix} \qquad \text{for all } \lambda \in \mathbb{D}^2.$$

In particular,

$$u_\lambda = (\gamma \otimes 1)1 + D\lambda u_\lambda = \gamma + D\lambda u_\lambda \qquad \text{for all } \lambda \in \mathbb{D}^2.$$

As W is a contraction, so also is D. Consequently we may solve this last equation for u_λ to obtain the equation

$$u_\lambda = (1 - D\lambda)^{-1}\gamma \qquad \text{for all } \lambda \in \mathbb{D}^2.$$

$$\text{Operator Analysis on } \mathbb{D}^2 \qquad 101$$

To deduce that u is holomorphic on $\mathbb{D}^2$ we modify slightly the argument in Lemma 2.30; note that, now, λ and D need not commute. Consider $\lambda_0 \in \mathbb{D}^2$. By the resolvent identity, Proposition 2.28, for any $\lambda \in \mathbb{D}^2$,

$$
\begin{aligned}
u_\lambda - u_{\lambda_0} &= \big((1 - D\lambda)^{-1} - (1 - D\lambda_0)^{-1}\big)\gamma \\
&= (1 - D\lambda)^{-1} D(\lambda - \lambda_0)(1 - D\lambda_0)^{-1}\gamma. \qquad (4.25)
\end{aligned}
$$

Define a linear operator $L \colon \mathbb{C}^2 \to \mathcal{B}(\mathcal{M})$ by

$$
Lz = (1 - D\lambda_0)^{-1} Dz (1 - D\lambda_0)^{-1} \qquad \text{for } z \in \mathbb{C}^2.
$$

Then, by equation (4.25) and another application of the resolvent identity,

$$
\begin{aligned}
u_\lambda - u_{\lambda_0} - L(\lambda - \lambda_0) &= \big((1 - D\lambda)^{-1} - (1 - D\lambda_0)^{-1}\big) D(\lambda - \lambda_0)(1 - D\lambda_0)^{-1}\gamma \\
&= (1 - D\lambda)^{-1} D(\lambda - \lambda_0) D(\lambda - \lambda_0)(1 - D\lambda_0)^{-1}\gamma \\
&= O(\|\lambda - \lambda_0\|^2).
\end{aligned}
$$

Hence L is the Fréchet derivative of u at λ_0, and so u is holomorphic on $\mathbb{D}^2$. $\qquad\square$

Proposition 4.26. (Converse to the fundamental theorem for $\mathscr{S}(\mathbb{D}^2)$) *If $\varphi \colon \mathbb{D}^2 \to \mathbb{C}$ is a function and φ has a model, then $\varphi \in \mathscr{S}(\mathbb{D}^2)$.*

Proof. Assume that $\varphi \colon \mathbb{D}^2 \to \mathbb{C}$ has a model $(\mathcal{M}, u)$ on $\mathbb{D}^2$. We claim that φ is holomorphic on $\mathbb{D}^2$. If $\varphi(\mu) = 0$ for all $\mu \in \mathbb{D}^2$, then clearly, φ is holomorphic on $\mathbb{D}^2$. Otherwise, choose $\mu \in \mathbb{D}^2$ such that $\varphi(\mu) \neq 0$ and solve the model formula (4.22) for $\varphi(\lambda)$ to obtain

$$
\varphi(\lambda) = \overline{\varphi(\mu)}^{\,-1} \Big(1 - \big\langle (1 - \mu^* \lambda)\, u_\lambda, u_\mu \big\rangle_{\mathcal{M}}\Big) \qquad \text{for all } \lambda \in \mathbb{D}^2.
$$

As Proposition 4.24 implies that u is holomorphic, this equation implies that φ is holomorphic on $\mathbb{D}^2$.

To see that $\varphi \in \mathscr{S}(\mathbb{D}^2)$, notice that, by the model formula (4.22) with $\mu = \lambda$,

$$
1 - |\varphi(\lambda)|^2 = (1 - \|\lambda\|^2)\, \|u_\lambda\|_{\mathcal{M}}^2 \geq 0 \qquad \text{for all } \lambda \in \mathbb{D}^2.
$$

Hence $\|\varphi\|_{\mathbb{D}^2} \leq 1$. Since φ is holomorphic and $\|\varphi\|_{\mathbb{D}^2} \leq 1$, $\varphi \in \mathscr{S}(\mathbb{D}^2)$. $\qquad\square$

Remark 4.27. The last two proofs are identical to the proofs of Propositions 2.21 and 2.32 but for the replacement of $\mathbb{D}$ by $\mathbb{D}^2$, of $\bar{\mu}$ by μ^*, of $|\cdot|$ by $\|\cdot\|$ and the updating of references to foregoing propositions and equations. In future, when we encounter such simple modifications of known proofs, we shall feel at liberty to leave the details to the reader.

102 *Commutative Theory*

Motivated by Definition 4.18, we define the *model wedge*, $\mathcal{W}$, to be the subset of $\mathrm{Her}(\mathbb{D}^2)$ given by the formula

$$\mathcal{W} = \{(1 - \bar{\mu}^1\lambda^1)a^1 + (1 - \bar{\mu}^2\lambda^2)a^2 : a^1, a^2 \in \mathcal{P}\}. \tag{4.28}$$

Here the notation $(1 - \bar{\mu}^1\lambda^1)a^1$ is an abbreviation for the kernel

$$(\lambda, \mu) \mapsto (1 - \bar{\mu}^1\lambda^1)a^1(\lambda, \mu)$$

on $\mathbb{D}^2$. We shall need the properties described in the following three lemmas about the model wedge.

Lemma 4.29. $\mathcal{W}$ *is a closed wedge*[4] *in* $\mathcal{R}$.

Proof. Since $\mathcal{P}$ is a cone, $\mathcal{W}$ is a wedge. To see that $\mathcal{W}$ is closed assume that

$$h_n = (1 - \bar{\mu}^1\lambda^1)a_n^1 + (1 - \bar{\mu}^2\lambda^2)a_n^2, \qquad \text{where } a_n^1, a_n^2 \in \mathcal{P}, \tag{4.30}$$

is a sequence in $\mathcal{W}$ and $h_n \to h \in \mathcal{R}$ uniformly on compact subsets of $\mathbb{D}^2 \times \mathbb{D}^2$.

Consider any $r < 1$. Since $h_n \to h$ uniformly on compact subsets of $\mathbb{D}^2 \times \mathbb{D}^2$, there exists a constant M such that

$$\sup_{n} \sup_{\lambda \in r\mathbb{D}^2} h_n(\lambda, \lambda) \le M.$$

Therefore, if $n \ge 1$ and $\lambda \in r\mathbb{D}^2$,

$$(1 - r^2)\left(a_n^1(\lambda, \lambda) + a_n^2(\lambda, \lambda)\right) \le (1 - |\lambda^1|^2)a_n^1(\lambda, \lambda) + (1 - |\lambda^2|^2)a_n^2(\lambda, \lambda) \le M,$$

which implies that

$$\sup_{n} \sup_{\lambda \in r\mathbb{D}^2} a_n^1(\lambda, \lambda) \le \frac{M}{1 - r^2} \quad \text{and} \quad \sup_{n} \sup_{\lambda \in r\mathbb{D}^2} a_n^2(\lambda, \lambda) \le \frac{M}{1 - r^2}. \tag{4.31}$$

Now, for any positive semi-definite kernel p on $\mathbb{D}^2$,

$$|p(\lambda, \mu)|^2 \le p(\lambda, \lambda)p(\mu, \mu) \qquad \text{for all } \lambda, \mu \in \mathbb{D}^2.$$

Therefore, as $a_n^1, a_n^2 \in \mathcal{P}$, the inequalities (4.31) imply that

$$\sup_{n} \sup_{\lambda, \mu \in r\mathbb{D}^2} |a_n^1(\lambda, \mu)| \le \frac{M}{1 - r^2} \quad \text{and} \quad \sup_{n} \sup_{\lambda, \mu \in r\mathbb{D}^2} |a_n^2(\lambda, \mu)| \le \frac{M}{1 - r^2},$$

so that $\{a_n^1\}$ and $\{a_n^2\}$ are uniformly bounded on $r\mathbb{D}^2 \times r\mathbb{D}^2$.

To summarize, in the previous paragraph we showed that $\{a_n^1\}$ and $\{a_n^2\}$ are uniformly locally bounded on $\mathbb{D}^2 \times \mathbb{D}^2$, and so, by Montel's theorem, there

---

[4] $\mathcal{W}$ is a *wedge* if $\mathcal{W} + \mathcal{W} \subseteq \mathcal{W}$ and $\alpha\mathcal{W} \subseteq \mathcal{W}$ whenever $\alpha \in \mathbb{R}$ and $\alpha \ge 0$. So a cone $\mathcal{P}$ is a wedge with the extra requirement that $\mathcal{P} \cap (-\mathcal{P}) = \{0\}$.

Operator Analysis on $\mathbb{D}^2$ 103

exists an increasing sequence of indices $n_1, n_2, \ldots$ and functions $a^1, a^2 \in \mathcal{R}$ such that

$$a^1_{n_k} \to a^1 \text{ and } a^2_{n_k} \to a^2 \text{ in } \mathrm{Her}(\mathbb{D}^2) \text{ as } k \to \infty.$$

Hence equation (4.30) implies that

$$h = (1 - \bar{\mu}^1 \lambda^1) a^1 + (1 - \bar{\mu}^2 \lambda^2) a^2.$$

But $\mathcal{P}$ is closed, so that $a^1, a^2 \in \mathcal{P}$. Therefore $h \in \mathcal{W}_{\mathcal{B}(\mathcal{X})}$. $\square$

Lemma 4.32. $\mathcal{P}$ *is contained in* $\mathcal{W}$*. In particular, if* $f \in \mathrm{Hol}(\mathbb{D}^2)$*, then* $\overline{f(\mu)} f(\lambda) \in \mathcal{W}$*.*

Proof. Fix $p \in \mathcal{P}$. Since $(\bar{\mu}^1)^n p(\lambda, \mu)(\lambda^1)^n \in \mathcal{P}$ for each $n \geq 0$, it follows from the fact that $\mathcal{P}$ is a cone that, for each $m \geq 0$,

$$\sum_{n=0}^{m} (\bar{\mu}^1)^n p(\lambda, \mu)(\lambda^1)^n \in \mathcal{P}.$$

Hence since $\mathcal{P}$ is closed,

$$\frac{p(\lambda, \mu)}{1 - \bar{\mu}^1 \lambda^1} = \lim_{m \to \infty} \sum_{n=0}^{m} (\bar{\mu}^1)^n p(\lambda, \mu)(\lambda^1)^n \in \mathcal{P}.$$

Therefore, as $0 \in \mathcal{P}$ as well,

$$p(\lambda, \mu) = (1 - \bar{\mu}^1 \lambda^1) \left(\frac{p(\lambda, \mu)}{1 - \bar{\mu}^1 \lambda^1} \right) + (1 - \bar{\mu}^2 \lambda^2) \cdot 0$$

is an element of $\mathcal{W}$. $\square$

Lemma 4.33. *If* $\varphi \colon \mathbb{D}^2 \to \mathbb{C}$ *is a function, then* φ *has a model if and only if* $1 - \overline{\varphi(\mu)} \varphi(\lambda) \in \mathcal{W}$*.*

Proof. First assume that $(\mathcal{M}, u)$ is a model for φ. If $a^1(\lambda, \mu)$ is defined to be $\langle u^1_\lambda, u^1_\mu \rangle_{\mathcal{M}}$ and $a^2(\lambda, \mu)$ is defined to be $\langle u^2_\lambda, u^2_\mu \rangle_{\mathcal{M}}$, then equation (4.19) becomes

$$1 - \overline{\varphi(\mu)} \varphi(\lambda) = (1 - \bar{\mu}^1 \lambda^1) a^1(\lambda, \mu) + (1 - \bar{\mu}^2 \lambda^2) a^2(\lambda, \mu). \tag{4.34}$$

Now, Moore's theorem (Theorem 2.5) implies that a^1 and a^2 are positive semi-definite on $\mathbb{D}^2$ and Proposition 4.24 implies that a^1 and a^2 are elements of $\mathrm{Her}(\mathbb{D}^2)$. Therefore $a^1, a^2 \in \mathcal{P}$. Hence equation (4.34) implies that $1 - \overline{\varphi(\mu)} \varphi(\lambda) \in \mathcal{W}$.

Conversely, assume that $1 - \overline{\varphi(\mu)} \varphi(\lambda) \in \mathcal{W}$, that is, there exist $a^1, a^2 \in \mathcal{P}$ such that equation (4.34) holds. By Moore's theorem there exist Hilbert spaces

104 *Commutative Theory*

$\mathcal{M}^1$ and $\mathcal{M}^2$ and functions $u^1\colon \mathbb{D}^2 \to \mathcal{M}^1$ and $u^2\colon \mathbb{D}^2 \to \mathcal{M}^2$ such that, for all $\lambda, \mu \in \mathbb{D}^2$,

$$a^1(\lambda,\mu) = \left\langle u_\lambda^1, u_\mu^1 \right\rangle_{\mathcal{M}^1} \quad \text{and} \quad a^2(\lambda,\mu) = \left\langle u_\lambda^2, u_\mu^2 \right\rangle_{\mathcal{M}^2}.$$

Substituting these formulas for a^1 and a^2 into equation (4.34) we deduce equation (4.19), so that $(\mathcal{M},u)$ is a model for φ. $\qquad\square$

4.4 Models on $\mathbb{D}^2$ via the Duality Construction

In this section we shall prove a fundamental theorem for $\mathscr{S}(\mathbb{D}^2)$. The proof illustrates a powerful line of reasoning in operator analysis, the *duality construction*. The duality construction starts from a linear functional L on the space $\mathrm{Her}(\mathbb{D}^2)$ that is assumed to act positively on the model wedge and from it constructs a Hilbert space H_L and a commuting pair of operators $T = (T^1, T^2)$ acting on H_L.

Theorem 4.35. (The fundamental theorem for $\mathscr{S}(\mathbb{D}^2)$) *Every function in* $\mathscr{S}(\mathbb{D}^2)$ *has a model on* $\mathbb{D}^2$.

Proof. Assume that $\varphi \in \mathscr{S}(\mathbb{D}^2)$. By Lemma 4.33 it suffices to show that

$$1 - \overline{\varphi(\mu)}\varphi(\lambda) \in \mathcal{W}. \tag{4.36}$$

As Lemma 4.29 guarantees that $\mathcal{W}$ is a closed wedge, the assertion (4.36) will follow from the Hahn–Banach theorem if we can show that

$$L(1 - \overline{\varphi(\mu)}\varphi(\lambda)) \geq 0 \tag{4.37}$$

whenever $L \in \mathcal{R}^*$ and

$$L(h) \geq 0 \qquad \text{for all } h \in \mathcal{W}. \tag{4.38}$$

Accordingly, fix $L \in \mathcal{R}^*$ satisfying the inequality (4.38).

Let $L^\sim$ be as in Lemma 4.3 and define a sesquilinear form on $\mathrm{Hol}(\mathbb{D}^2)$ by the formula

$$\langle f, g \rangle_0 = L^\sim(\overline{g(\mu)}f(\lambda)) \qquad \text{for all } f, g \in \mathrm{Hol}(\mathbb{D}^2).$$

Since $\overline{f(\mu)}f(\lambda) \in \mathcal{R}$ whenever $f \in \mathrm{Hol}(\mathbb{D}^2)$, and Lemma 4.3 implies that $L = L^\sim | \mathcal{R}$,

$$\langle f, f \rangle_0 = L(\overline{f(\mu)}f(\lambda)) \qquad \text{for all } f \in \mathrm{Hol}(\mathbb{D}^2). \tag{4.39}$$

$$\textit{Operator Analysis on } \mathbb{D}^2 \qquad\qquad 105$$

But Lemma 4.32 implies that $\overline{f(\mu)} f(\lambda) \in \mathcal{W}$ whenever $f \in \mathrm{Hol}(\mathbb{D}^2)$. Hence inequality (4.38) guarantees that

$$\langle f, f \rangle_0 \geq 0 \quad \text{for all } f \in \mathrm{Hol}(\mathbb{D}^2),$$

that is, $\langle \cdot, \cdot \rangle_0$ is a positive semi-definite sesquilinear form on $\mathrm{Hol}(\mathbb{D}^2)$. Consequently, if

$$\mathcal{N} \stackrel{\text{def}}{=} \{ f \in \mathrm{Hol}(\mathbb{D}^2) \colon \langle f, f \rangle_0 = 0 \},$$

then $\mathcal{N}$ is a subspace of $\mathrm{Hol}(\mathbb{D}^2)$ and $\langle \cdot, \cdot \rangle_0$ induces an inner product $\langle \cdot, \cdot \rangle_L$ on $\mathrm{Hol}(\mathbb{D}^2)/\mathcal{N}$ via the formula

$$\langle f + \mathcal{N}, g + \mathcal{N} \rangle_L = \langle f, g \rangle_0. \tag{4.40}$$

Let H_L denote the completion of $\mathrm{Hol}(\mathbb{D}^2)/\mathcal{N}$ with respect to this inner product. Using equation (4.39), we have

$$\| f + \mathcal{N} \|_L^2 = L(\overline{f(\mu)} f(\lambda)) \quad \text{for all } f \in \mathrm{Hol}(\mathbb{D}^2). \tag{4.41}$$

From equation (4.41) it follows that

$$\| f + \mathcal{N} \|_L^2 - \| \lambda^1 f + \mathcal{N} \|_L^2 = L(\overline{f(\mu)} f(\lambda)) - L\left(\overline{\mu^1 f(\mu)} \lambda^1 f(\lambda) \right)$$
$$= L\big((1 - \bar{\mu}^1 \lambda^1) \overline{f(\mu)} f(\lambda) \big)$$

for all $f \in \mathrm{Hol}(\mathbb{D}^2)$. But an examination of the definition (4.28) reveals that

$$(1 - \bar{\mu}^1 \lambda^1) \overline{f(\mu)} f(\lambda) \in \mathcal{W}.$$

Therefore, inequality (4.38) implies that, for all $f \in \mathrm{Hol}(\mathbb{D}^2)$,

$$\| \lambda^1 f + \mathcal{N} \|_L^2 \quad \leq \quad \| f + \mathcal{N} \|_L^2.$$

This inequality ensures that we may define a contractive operator $T^1 \in \mathcal{B}(H_L)$ that satisfies

$$T^1(f + \mathcal{N}) = \lambda^1 f + \mathcal{N} \quad \text{for all } f \in \mathrm{Hol}(\mathbb{D}^2).$$

In similar fashion we may define a contractive operator $T^2 \in \mathcal{B}(H_L)$ that satisfies

$$T^2(f + \mathcal{N}) = \lambda^2 f + \mathcal{N} \quad \text{for all } f \in \mathrm{Hol}(\mathbb{D}^2).$$

Clearly T^1 and T^2 commute. Now, as T^1 and T^2 are multiplication by λ^1 and multiplication by λ^2 it is reasonable to expect that $\varphi(T)$ is multiplication by $\varphi(\lambda)$. However, as $\sigma(T)$ is not necessarily in $\mathbb{D}^2$, $\varphi(T)$ is not necessarily well defined. But T^1 and T^2 are contractions, and so $\sigma(T) \in (\mathbb{D}^2)^-$. Consequently, if we fix $r < 1$ (at the end of the argument we shall let $r \to 1$), $\sigma(rT) \subseteq \mathbb{D}^2$ and $\varphi(rT)$ is well defined.

We make the following claim.

106 Commutative Theory

Claim 4.42. $\varphi(rT)(1+\mathcal{N}) = \varphi(r\lambda)+\mathcal{N}$.

To prove this claim, choose a sequence of polynomials $p_n(\lambda)$ such that p_n converges uniformly to $\varphi(\lambda)$ on compact subsets of $\mathbb{D}^2$. By algebra,

$$p_n(rT)(1+\mathcal{N}) = p_n(r\lambda)+\mathcal{N}$$

for all n. But by the continuity of the functional calculus, $p_n(rT) \to \varphi(rT)$ in $\mathcal{B}(H_L)$, and the continuity of L and equation (4.41) imply that

$$p_n(r\lambda)+\mathcal{N} \to \varphi(r\lambda)+\mathcal{N}.$$

This establishes the claim.

Now rT is a pair of commuting contractions and $\sigma(rT) \subseteq \mathbb{D}^2$. Therefore, by Andô's inequality (Theorem 1.55),

$$\|\varphi(rT)\| \le 1.$$

Hence, in the light of Claim 4.42 and the definition (4.40), if we write $\varphi_r(\lambda) = \varphi(r\lambda)$,

$$L(1 - \overline{\varphi_r(\mu)}\varphi_r(\lambda)) = \big\langle (1 - \varphi(rT)^*\varphi(rT))(1+\mathcal{N}), 1+\mathcal{N} \big\rangle_L \ge 0$$

Since L is continuous and $1 - \overline{\varphi_r(\mu)}\varphi_r(\lambda) \to 1 - \overline{\varphi(\mu)}\varphi(\lambda)$ in $\mathcal{R}$ as $r \to 1$, it follows that inequality (4.37) holds. $\qquad\square$

4.5 The Network Realization Formula for $\mathbb{D}^2$

In this section we generalize Theorem 2.36 to the bidisc. The proof will follow that of Theorem 2.36 via a lurking isometry argument, but with two significant differences. First, λ now needs to be interpreted as an operator as in equation (4.20), that is, the operator defined on $\mathcal{M} = \mathcal{M}^1 \oplus \mathcal{M}^2$ by the formula

$$\lambda(u^1 \oplus u^2) = \lambda^1 u^1 \oplus \lambda^2 u^2 \qquad \text{for all } u^1 \in \mathcal{M}^1, u^2 \in \mathcal{M}^2.$$

Second, the vectors $\begin{bmatrix} 1 \\ \lambda u_\lambda \end{bmatrix}$ no longer necessarily have dense linear span as they did in the proof of Theorem 2.36, a fact that complicates the extension step. This issue is handled by the following simple lemma, which is adapted from an idea in the proof of Theorem 2.57.

Lemma 4.43. *If $\varphi\colon \mathbb{D}^2 \to \mathbb{C}$ is a function and $(\mathcal{M}, u)$ is a model for φ, then there exists a model $(\mathcal{M}^\sim, u^\sim)$ for φ such that*

$$\dim [1 \oplus \lambda u_\lambda^\sim \colon \lambda \in \mathbb{D}^2]^\perp = \dim [\varphi(\lambda) \oplus u_\lambda^\sim \colon \lambda \in \mathbb{D}^2]^\perp,$$

where the orthogonal projections are taken in $\mathbb{C} \oplus \mathcal{M}^\sim$.

Operator Analysis on $\mathbb{D}^2$ 107

Proof. Let $\alpha = \max\{\aleph_0, \dim \mathcal{M}\}$, choose a Hilbert space $\mathcal{N}$ with $\dim = \alpha$, and let $(\mathcal{M}^\sim, u^\sim)$ denote the trivial extension of $(\mathcal{M}, u)$ by $\mathcal{N}$ (Remark (2.10) (ii)). Then

$$\dim\,[1 \oplus \lambda u_\lambda^\sim \colon \lambda \in \mathbb{D}^2]^\perp = \alpha = \dim\,[\varphi(\lambda) \oplus u_\lambda^\sim \colon \lambda \in \mathbb{D}^2]^\perp.$$

$\square$

Theorem 4.44. (The network realization formula for $\mathscr{S}(\mathbb{D}^2)$**)** *A function* φ *on* $\mathbb{D}^2$ *belongs to* $\mathscr{S}(\mathbb{D}^2)$ *if and only if there exist a decomposed Hilbert space* $\mathcal{M} = \mathcal{M}^1 \oplus \mathcal{M}^2$ *and a unitary operator*

$$V = \begin{bmatrix} a & 1 \otimes \beta \\ \gamma \otimes 1 & D \end{bmatrix} \colon \mathbb{C} \oplus \mathcal{M} \to \mathbb{C} \oplus \mathcal{M}, \tag{4.45}$$

such that

$$\varphi(\lambda) = a + \langle \lambda(1 - D\lambda)^{-1}\gamma, \beta \rangle \qquad \text{for all } \lambda \in \mathbb{D}, \tag{4.46}$$

where λ *in the right hand side of equation* (4.46) *is interpreted as in equation* (4.20).

Proof. Consider $\varphi \in \mathscr{S}(\mathbb{D}^2)$. Theorem 4.35 implies that φ has a model $(\mathcal{M}, u)$, which, by Lemma 4.43, we may assume to satisfy

$$\dim\,[1 \oplus \lambda u_\lambda \colon \lambda \in \mathbb{D}^2]^\perp = \dim\,[\varphi(\lambda) \oplus u_\lambda \colon \lambda \in \mathbb{D}^2]^\perp. \tag{4.47}$$

As in the proof of Theorem 2.36, the model formula (4.22) can be reshuffled to

$$\varphi(\lambda)\overline{\varphi(\mu)} + \langle u_\lambda, u_\mu \rangle = 1 + \langle \lambda u_\lambda, \mu u_\mu \rangle,$$

which in turn may be rewritten as

$$\left\langle \begin{pmatrix} \varphi(\lambda) \\ u_\lambda \end{pmatrix}, \begin{pmatrix} \varphi(\mu) \\ u_\mu \end{pmatrix} \right\rangle_{\mathbb{C} \oplus \mathcal{M}} = \left\langle \begin{pmatrix} 1 \\ \lambda u_\lambda \end{pmatrix}, \begin{pmatrix} 1 \\ \mu u_\mu \end{pmatrix} \right\rangle_{\mathbb{C} \oplus \mathcal{M}}.$$

Therefore, Lemma 2.18 implies that there exists a linear isometry

$$V \colon [1 \oplus \lambda u_\lambda \colon \lambda \in \mathbb{D}^2] \to [\varphi(\lambda) \oplus u_\lambda \colon \lambda \in \mathbb{D}^2]$$

satisfying

$$V \begin{pmatrix} 1 \\ \lambda u_\lambda \end{pmatrix} = \begin{pmatrix} \varphi(\lambda) \\ u_\lambda \end{pmatrix}. \tag{4.48}$$

Now, by the relation (4.47), we may construct a Hilbert space isomorphism L that maps $[1 \oplus \lambda u_\lambda \colon \lambda \in \mathbb{D}^2]^\perp$ onto $[\varphi(\lambda) \oplus u_\lambda \colon \lambda \in \mathbb{D}^2]^\perp$. Therefore, we may

108   *Commutative Theory*

define a unitary $W \in \mathcal{B}(\mathbb{C} \oplus \mathcal{M})$ by setting $W = V$ on $[1 \oplus \lambda u_\lambda \colon \lambda \in \mathbb{D}^2]$ and $W = L$ on $[1 \oplus \lambda u_\lambda \colon \lambda \in \mathbb{D}^2]^\perp$. Furthermore, in light of equation (4.48),

$$W \begin{pmatrix} 1 \\ \lambda u_\lambda \end{pmatrix} = \begin{pmatrix} \varphi(\lambda) \\ u_\lambda \end{pmatrix} \qquad \text{for all } \lambda \in \mathbb{D}^2.$$

If we represent W as a 2×2 block matrix,

$$W = \begin{bmatrix} a & 1 \otimes \beta \\ \gamma \otimes 1 & D \end{bmatrix} \colon \mathbb{C} \oplus \mathcal{M} \to \mathbb{C} \oplus \mathcal{M},$$

then

$$\begin{pmatrix} \varphi(\lambda) \\ u_\lambda \end{pmatrix} = W \begin{pmatrix} 1 \\ \lambda u_\lambda \end{pmatrix} = \begin{bmatrix} a & 1 \otimes \beta \\ \gamma \otimes 1 & D \end{bmatrix} \begin{pmatrix} 1 \\ \lambda u_\lambda \end{pmatrix} \qquad \text{for all } \lambda \in \mathbb{D}^2.$$

This relation gives rise to the equations

$$\varphi(\lambda) = a + \langle \lambda u_\lambda, \beta \rangle$$

and

$$u_\lambda = \gamma + D \lambda u_\lambda \qquad \text{for all } \lambda \in \mathbb{D}^2.$$

Use the second equation to eliminate u_λ from the first equation to yield the desired realization formula (4.46).

Now assume that V is an isometry given by equation (4.45) and φ is given by equation (4.46). If we define $u \colon \mathbb{D}^2 \to \mathcal{M}$ by

$$u_\lambda = (1 - D\lambda)^{-1} \gamma \qquad \text{for all } \lambda \in \mathbb{D}^2,$$

then

$$\begin{aligned}
V \begin{pmatrix} 1 \\ \lambda u_\lambda \end{pmatrix} &= \begin{bmatrix} a & 1 \otimes \beta \\ \gamma \otimes 1 & D \end{bmatrix} \begin{pmatrix} 1 \\ \lambda u_\lambda \end{pmatrix} \\
&= \begin{pmatrix} a + \langle \lambda u_\lambda, \beta \rangle \\ \gamma + D\lambda u_\lambda \end{pmatrix} \\
&= \begin{pmatrix} \varphi(\lambda) \\ u_\lambda \end{pmatrix} \qquad \text{for all } \lambda \in \mathbb{D}^2.
\end{aligned}$$

Consequently, as V is an isometry, Lemma 2.18 implies that

$$\left\langle \begin{pmatrix} \varphi(\lambda) \\ u_\lambda \end{pmatrix}, \begin{pmatrix} \varphi(\mu) \\ u_\mu \end{pmatrix} \right\rangle_{\mathbb{C} \oplus \mathcal{M}} = \left\langle \begin{pmatrix} 1 \\ \lambda u_\lambda \end{pmatrix}, \begin{pmatrix} 1 \\ \mu u_\mu \end{pmatrix} \right\rangle_{\mathbb{C} \oplus \mathcal{M}} \qquad \text{for all } \lambda, \mu \in \mathbb{D}^2.$$

But this last formula reshuffles to the formula (4.22), so that φ has a model. Therefore, by the converse to the fundamental theorem (Proposition 4.26), $\varphi \in \mathscr{S}(\mathbb{D}^2)$. $\qquad \square$

$$\textit{Operator Analysis on } \mathbb{D}^2 \qquad\qquad 109$$

4.6 Nevanlinna–Pick Interpolation on $\mathbb{D}^2$

In this section we generalize the Pick interpolation theorem (Theorem 1.81) to the bidisc.

Theorem 4.49. (Pick's interpolation theorem on $\mathbb{D}^2$) *Let $\lambda_1,\dots,\lambda_n$ be distinct points in $\mathbb{D}^2$ and let $z_1,\dots,z_n \in \mathbb{C}$. There exists $\varphi \in \mathscr{S}(\mathbb{D}^2)$ such that $\varphi(\lambda_i) = z_i$ for $i = 1,\dots,n$ if and only if there exist a pair of $n \times n$ positive semi-definite matrices $A^1 = [a_{ij}^1]$ and $A^2 = [a_{ij}^2]$ such that*

$$1 - \bar{z}_i z_j = (1 - \bar{\lambda}_i^1 \lambda_j^1)a_{ij}^1 + (1 - \bar{\lambda}_i^2 \lambda_j^2)a_{ij}^2 \tag{4.50}$$

for $i,j = 1,\dots,n$.

Proof. First assume that $\varphi \in \mathscr{S}(\mathbb{D}^2)$, and $\varphi(\lambda_i) = z_i$ for $i = 1,\dots,n$. By Theorem 4.35, φ has a model $(\mathcal{M},u)$. But then, setting $\lambda = \lambda_j$ and $\mu = \lambda_i$ in equation (4.19), we have

$$1 - \bar{z}_i z_j = 1 - \overline{\varphi(\lambda_i)}\varphi(\lambda_j)$$
$$= (1 - \bar{\lambda}_i^1 \lambda_j^1)\langle u_j^1, u_i^1 \rangle_{\mathcal{M}^1} + (1 - \bar{\lambda}_i^2 \lambda_j^2)\langle u_j^2, u_i^2 \rangle_{\mathcal{M}^2}.$$

If we set $A^1 = [\langle u_j^1, u_i^1 \rangle_{\mathcal{M}^1}]$ and $A^2 = [\langle u_j^2, u_i^2 \rangle_{\mathcal{M}^2}]$, then equation (4.50) holds, and by Theorem 2.5, A^1 and A^2 are positive semi-definite.

Now assume that A^1 and A^2 are positive semi-definite and equation (4.50) holds. Use Theorem 2.5 to express

$$A^1 = \left[\langle u_j^1, u_i^1 \rangle_{\mathcal{M}^1}\right] \quad \text{and} \quad A^2 = \left[\langle u_j^2, u_i^2 \rangle_{\mathcal{M}^2}\right] \tag{4.51}$$

so that equation (4.50) becomes

$$1 - \bar{z}_i z_j = (1 - \bar{\lambda}_i^1 \lambda_j^1)\langle u_j^1, u_i^1 \rangle_{\mathcal{M}^1} + (1 - \bar{\lambda}_i^2 \lambda_j^2)\langle u_j^2, u_i^2 \rangle_{\mathcal{M}^2}. \tag{4.52}$$

Now, reshuffle equation (4.52) to

$$\bar{z}_i z_j + \langle u_j^1, u_i^1 \rangle_{\mathcal{M}^1} + \langle u_j^2, u_i^2 \rangle_{\mathcal{M}^2} = 1 + \bar{\lambda}_i^1 \lambda_j^1 \langle u_j^1, u_i^1 \rangle_{\mathcal{M}^1} + \bar{\lambda}_i^2 \lambda_j^2 \langle u_j^2, u_i^2 \rangle_{\mathcal{M}^2},$$

or equivalently,

$$\left\langle \begin{pmatrix} z_j \\ u_j^1 \\ u_j^2 \end{pmatrix}, \begin{pmatrix} z_i \\ u_i^1 \\ u_i^2 \end{pmatrix} \right\rangle_{\mathbb{C} \oplus \mathcal{M}^1 \oplus \mathcal{M}^2} = \left\langle \begin{pmatrix} 1 \\ \lambda_j^1 u_j^1 \\ \lambda_j^2 u_j^2 \end{pmatrix}, \begin{pmatrix} 1 \\ \lambda_i^1 u_i^1 \\ \lambda_i^2 u_i^2 \end{pmatrix} \right\rangle_{\mathbb{C} \oplus \mathcal{M}^1 \oplus \mathcal{M}^2}.$$

Letting $\mathcal{M} = \mathcal{M}^1 \oplus \mathcal{M}^2$, $u_j = u_j^1 \oplus u_j^2$, and interpreting λ_j as in equation (4.20), we may write the last equation in the simpler form

$$\left\langle \begin{pmatrix} z_j \\ u_j \end{pmatrix}, \begin{pmatrix} z_i \\ u_i \end{pmatrix} \right\rangle_{\mathbb{C} \oplus \mathcal{M}} = \left\langle \begin{pmatrix} 1 \\ \lambda_j u_j \end{pmatrix}, \begin{pmatrix} 1 \\ \lambda_i u_i \end{pmatrix} \right\rangle_{\mathbb{C} \oplus \mathcal{M}}. \tag{4.53}$$

110   *Commutative Theory*

Now, equation (4.53) and Lemma 2.18 together imply that there exists an isometry

$$V_0 \colon \left[\begin{pmatrix} 1 \\ \lambda_j u_j \end{pmatrix} \colon j = 1, \ldots, n \right] \to \left[\begin{pmatrix} z_j \\ u_j \end{pmatrix} \colon j = 1, \ldots, n \right]$$

satisfying

$$V_0 \begin{pmatrix} 1 \\ \lambda_j u_j \end{pmatrix} = \begin{pmatrix} z_j \\ u_j \end{pmatrix}, \qquad j = 1, \ldots, n.$$

Since V_0 is an isometry, $\dim(\mathrm{dom}\, V_0) = \dim(\mathrm{ran}\, V_0)$. Hence $\dim(\mathrm{dom}\, V_0^{\perp}) = \dim(\mathrm{ran}\, V_0^{\perp})$ and we may extend V_0 to a unitary $V \in \mathcal{B}(\mathbb{C} \oplus \mathcal{M})$ by choosing a Hilbert space isomorphism $L \colon \mathrm{dom}\, V_0^{\perp} \to \mathrm{ran}\, V_0^{\perp}$ and then letting $V = V_0$ on $\mathrm{dom}\, V_0$ and $V = L$ on $\mathrm{dom}\, V_0^{\perp}$. If we decompose

$$V = \begin{bmatrix} a & 1 \otimes \beta \\ \gamma \otimes 1 & D \end{bmatrix} \colon \mathbb{C} \oplus \mathcal{M} \to \mathbb{C} \oplus \mathcal{M},$$

then, as

$$\begin{bmatrix} a & 1 \otimes \beta \\ \gamma \otimes 1 & D \end{bmatrix} \begin{pmatrix} 1 \\ \lambda_j u_j \end{pmatrix} = \begin{pmatrix} z_j \\ u_j \end{pmatrix} \qquad \text{for } j = 1, \ldots, n,$$

we have

$$z_j = a + \langle \lambda_j u_j, \beta \rangle \quad \text{and} \quad u_j = \gamma + D \lambda_j u_j \qquad \text{for } j = 1, \ldots, n.$$

These equations imply that

$$z_j = a + \langle \lambda_j (1 - D\lambda_j)^{-1} \gamma, \beta \rangle, \qquad j = 1, \ldots, n. \tag{4.54}$$

The proof of Theorem 4.49 can now be completed if φ is defined by the formula

$$\varphi(\lambda) = a + \lambda \langle (1 - D\lambda)^{-1} \gamma, \beta \rangle \qquad \text{for } \lambda \in \mathbb{D},$$

since Theorem 4.44 implies that $\varphi \in \mathscr{S}(\mathbb{D}^2)$ and equation (4.54) implies that $\varphi(\lambda_j) = z_j$ for $j = 1, \ldots, n$. $\qquad \square$

There is an analog of Theorem 2.57 on the bidisc as well. We may replace the set of nodes $\{\lambda_1, \lambda_2, \ldots, \lambda_n\}$ with a general set $\Lambda \subseteq \mathbb{D}^2$, replace the targets $\{z_1, z_2, \ldots, z_n\}$ with a function $z \colon \Lambda \to \mathbb{C}$, and replace the condition that there exist a pair of $n \times n$ matrices a^1, a^2 such that equation (4.50) holds with the condition that there exist a pair of positive semi-definite kernels $P = (P^1, P^2)$ on Λ such that, for all $\lambda, \mu \in \Lambda$,

$$1 - \overline{z(\mu)} z(\lambda) = (1 - \bar{\mu}^1 \lambda^1) P^1(\lambda, \mu) + (1 - \bar{\mu}^2 \lambda^2) P^2(\lambda, \mu). \tag{4.55}$$

Operator Analysis on $\mathbb{D}^2$ 111

Theorem 4.56. *Let data* $\Lambda \subseteq \mathbb{D}^2$ *and* $z\colon \Lambda \to \mathbb{C}$ *be given. There exists* $\varphi \in \mathscr{S}(\mathbb{D}^2)$ *such that* $\varphi(\lambda) = z(\lambda)$ *for all* $\lambda \in \Lambda$ *if and only if there exists an ordered pair* $P = (P^1, P^2)$ *of positive semi-definite kernels on* Λ *such that equation* (4.55) *holds.*

Proof. Assume that $\Lambda \subseteq \mathbb{D}^2$, $z\colon \Lambda \to \mathbb{C}$ and $\varphi \in \mathscr{S}(\mathbb{D}^2)$ satisfies $\varphi(\lambda) = z(\lambda)$ for all $\lambda \in \Lambda$. By Theorem 4.35, φ has a model $(\mathcal{M}, u)$. If we define P^1 and P^2 on $\Lambda \times \Lambda$ by the formulas

$$P^1(\lambda, \mu) = \langle u^1_\lambda, u^1_\mu \rangle_{\mathcal{M}^1} \quad \text{and} \quad P^2(\lambda, \mu) = \langle u^2_\lambda, u^2_\mu \rangle_{\mathcal{M}^2} \qquad \text{for all } \lambda, \mu \in \Lambda,$$

then Moore's theorem (Theorem 2.5) implies that P^1 and P^2 are positive semi-definite, and equation (4.19) implies that equation (4.55) holds.

Conversely, assume that P^1 and P^2 are positive semi-definite on Λ and equation (4.55) holds. By Moore's theorem (Theorem 2.5) there exist Hilbert spaces $\mathcal{M}^1$ and $\mathcal{M}^2$ and functions $u^1\colon \Lambda \to \mathcal{M}^1$ and $u^2\colon \Lambda \to \mathcal{M}^2$ such that, for all $\lambda, \mu \in \Lambda$,

$$P^1(\lambda, \mu) = \langle u^1_\lambda, u^1_\mu \rangle_{\mathcal{M}^1} \quad \text{and} \quad P^2(\lambda, \mu) = \langle u^1_\lambda, u^1_\mu \rangle_{\mathcal{M}^1}.$$

Letting $\mathcal{M} = \mathcal{M}^1 \oplus \mathcal{M}^2$, $u = u^1 \oplus u^2$ and interpreting λ as in equation (4.20), we may rewrite equation (4.55) as

$$1 - \overline{z(\mu)}z(\lambda) = \langle (1 - \bar{\mu}\lambda)u_\lambda, u_\mu \rangle_{\mathcal{M}} \quad \text{for all } \lambda, \mu \in \Lambda,$$

which in turn may be rewritten

$$\left\langle \begin{pmatrix} z(\lambda) \\ u_\lambda \end{pmatrix}, \begin{pmatrix} z(\mu) \\ u_\mu \end{pmatrix} \right\rangle_{\mathbb{C} \oplus \mathcal{M}} = \left\langle \begin{pmatrix} 1 \\ \lambda u_\lambda \end{pmatrix}, \begin{pmatrix} 1 \\ \mu u_\mu \end{pmatrix} \right\rangle_{\mathbb{C} \oplus \mathcal{M}} \qquad \text{for all } \lambda, \mu \in \Lambda. \tag{4.57}$$

As in the proof of Theorem 2.57, by replacing $\mathcal{M}$ with $\mathcal{M}^\sim = \mathcal{M} \oplus \mathcal{N}$ and u_λ with $u^\sim_\lambda = u_\lambda \oplus 0$ we may construct from equation (4.57) an isometry

$$W = \begin{bmatrix} a & 1 \otimes \beta \\ \gamma \otimes 1 & D \end{bmatrix} \colon \mathbb{C} \oplus \mathcal{M}^\sim \to \mathbb{C} \oplus \mathcal{M}^\sim$$

such that, for all $\lambda \in \Lambda$,

$$\begin{bmatrix} a & 1 \otimes \beta \\ \gamma \otimes 1 & D \end{bmatrix} \begin{pmatrix} 1 \\ \lambda u^\sim_\lambda \end{pmatrix} = \begin{pmatrix} z(\lambda) \\ u^\sim_\lambda \end{pmatrix}.$$

But then,

$$z(\lambda) = a + \lambda \langle u^\sim_\lambda, \beta \rangle_{\mathcal{M}^\sim} \quad \text{and} \quad u^\sim_\lambda = \gamma + D\lambda u^\sim_\lambda,$$

which implies that, for all $\lambda \in \Lambda$,

$$z(\lambda) = a + \lambda \langle (1 - D\lambda)^{-1}\gamma, \beta \rangle_{\mathcal{M}^\sim}. \tag{4.58}$$

112 *Commutative Theory*

The proof of Theorem 4.56 can now be completed by definition of φ according to the formula

$$\varphi(\lambda) = a + \lambda\big\langle(1 - D\lambda)^{-1}\gamma, \beta\big\rangle_{\mathcal{M}} \qquad \text{for all } \lambda \in \mathbb{D}^2.$$

As W is an isometry, Theorem 4.44 implies that $\varphi \in \mathscr{S}(\mathbb{D}^2)$, and equation (4.58) implies that $\varphi(\lambda) = z(\lambda)$ for all $\lambda \in \Lambda$. $\qquad\square$

4.7 Toeplitz Corona for the Bidisc

It is not currently known whether the corona theorem holds for $H^\infty(\mathbb{D}^2)$. That is to say, for given functions $\varphi_1, \ldots, \varphi_m$ in $H^\infty(\mathbb{D}^2)$, it is unknown whether the condition

$$\inf_{\lambda \in \mathbb{D}^2} |\varphi_1(\lambda)|^2 + \cdots + |\varphi_m(\lambda)|^2 > 0$$

suffices to ensure that there exist functions $\psi_1, \ldots, \psi_m \in H^\infty(\mathbb{D}^2)$ such that $\psi_1\varphi_1 + \cdots + \psi_m\varphi_m = 1$. There *is* a Toeplitz corona theorem for the bidisc, however. The hypotheses are essentially the same as for one variable (Theorems 3.50, 3.52, and 3.56), with λ and μ interpreted as in equation (4.20).

We adapt the terminology introduced in Section 3.8 to the bidisc.

Definition 4.59. *A* model *of a function $h \in \mathrm{Her}(\mathbb{D}^2)$ is a pair $(\mathcal{M}, u)$ where $\mathcal{M} = \mathcal{M}^1 \oplus \mathcal{M}^2$ is a decomposed separable Hilbert space and $u = u^1 \oplus u^2$ is a map from $\mathbb{D}^2$ to $\mathcal{M}$ such that, for all $\lambda, \mu \in \mathbb{D}^2$,*

$$h(\lambda, \mu) = (1 - \overline{\mu^1}\lambda^1)\big\langle u_\lambda^1, u_\mu^1\big\rangle_{\mathcal{M}^1} + (1 - \overline{\mu^2}\lambda^2)\big\langle u_\lambda^2, u_\mu^2\big\rangle_{\mathcal{M}^2}$$

$$= \big\langle(1 - \bar{\mu}\lambda)u_\lambda, u_\mu\big\rangle_{\mathcal{M}}.$$

For $\rho > 0$, an m-tuple $\varphi = (\varphi_1, \ldots, \varphi_m)$ of functions in $H^\infty(\mathbb{D}^2)$ has a corona model with parameter ρ *if the hereditary function*

$$\overline{\varphi_1(\mu)}\varphi_1(\lambda) + \cdots + \overline{\varphi_m(\mu)}\varphi_m(\lambda) - \rho \qquad (4.60)$$

has a model $(\mathcal{M}, u)$.

Theorem 4.61. *Let φ be in $H^\infty_{\mathcal{B}(\mathbb{C}, \mathbb{C}^m)}(\mathbb{D}^2)$ and let $\rho > 0$. There exists ψ in $H^\infty_{\mathcal{B}(\mathbb{C}, \mathbb{C}^m)}(\mathbb{D}^2)$ of norm less than or equal to $\rho^{-\frac{1}{2}}$ such that*

$$\psi^T \varphi = 1$$

if and only if φ has a corona model with parameter ρ.

The proof of Theorem 4.61 is very similar to that of Theorem 3.50. One invokes Theorem 4.93 in place of Theorem 3.10, and instead of Theorem

Operator Analysis on $\mathbb{D}^2$ 113

3.16 one uses the corresponding network realization formula for matrix-valued functions in the Schur class of the bidisc, which can be proved in much the same way as Theorem 3.16.

Recall from Theorem 3.52 that, in one variable, an m-tuple φ has a corona model if and only if the kernel (4.60) (or more precisely, the kernel (3.49)) multiplied by the Szegő kernel is positive semi-definite on $\mathbb{D}$. The analogous statement for two variables does not hold, but there *is* a similar criterion involving a *family* of kernels on $\mathbb{D}^2$.

Definition 4.62. *A kernel* $k(\lambda,\mu)$ *on* $\mathbb{D}^2$ *is said to be* admissible *if the kernels*

$$(1 - \bar{\mu}^1 \lambda^1) k(\lambda,\mu) \ \text{and} \ (1 - \bar{\mu}^2 \lambda^2) k(\lambda,\mu)$$

are positive semi-definite on $\mathbb{D}^2$.

If k is an admissible kernel on $\mathbb{D}^2$, then the operations of multiplication by the co-ordinate functions constitute a pair of commuting contractions on the reproducing kernel Hilbert space with reproducing kernel k (recall Theorem 2.14). The analog of Theorem 3.52 for the bidisc is the following.

Theorem 4.63. *Let* $\varphi_1,\ldots,\varphi_m \in H^\infty(\mathbb{D}^2)$ *and let* $\rho > 0$. *The following statements are equivalent.*

(i) *There exist* $\psi_1,\ldots,\psi_m \in H^\infty(\mathbb{D}^2)$ *such that* $\varphi_1\psi_1 + \cdots + \varphi_m\psi_m = 1$ *and*
$$\|\psi\|_{\mathbb{D}^2} \le \rho^{-\frac{1}{2}};$$

(ii) *for every commuting pair of contractions* T *having spectrum contained in* $\mathbb{D}^2$,
$$\varphi_1(T)\varphi_1(T)^* + \cdots + \varphi_m(T)\varphi_m(T)^* \ge \rho;$$

(iii) *for every admissible kernel* k *on* $\mathbb{D}^2$, *the kernel*
$$(\lambda,\mu) \mapsto \left(\overline{\varphi_1(\mu)}\varphi_1(\lambda) + \cdots + \overline{\varphi_m(\mu)}\varphi_m(\lambda) - \rho \right) k(\lambda,\mu)$$

is positive semi-definite on $\mathbb{D}^2$.

The implications (i)$\Rightarrow$(ii)$\Rightarrow$(iii) are shown much as in the proof of Theorem 3.52. Proof that (iii)$\Rightarrow$(i) requires the development of the theory of admissible kernels. See for example [11, theorem 11.5].

4.8 Operator-Valued Functions on $\mathbb{D}^2$

The elements of operator analysis on $\mathbb{D}^2$ (the functional calculus, models, hereditary functions) can be extended to *operator-valued* holomorphic maps on $\mathbb{D}^2$, just as they were for the disc in Section 3.5.

114   *Commutative Theory*

Let us begin with the holomorphic functional calculus. Consider Hilbert spaces $\mathcal{X}, \mathcal{Y}$ and a holomorphic map $f\colon \mathbb{D}^2 \to \mathcal{B}(\mathcal{X}, \mathcal{Y})$; then f is given by the locally uniformly convergent power series (4.6), but where now each each Taylor coefficient $a_{m_1 m_2}$ belongs to $\mathcal{B}(\mathcal{X}, \mathcal{Y})$. Consider also a commuting pair $T = (T^1, T^2)$ of operators on a Hilbert space $\mathcal{H}$ satisfying $\sigma(T) \subset \mathbb{D}^2$; we wish to define $f(T)$ as an operator from $\mathcal{X} \otimes \mathcal{H}$ to $\mathcal{Y} \otimes \mathcal{H}$. We may do this either by the power-series method or by a variant of the Riesz-Dunford integral, equation (1.4). Let

$$f(T) = \sum a_m \otimes T^m = \sum_{m_1=0}^{\infty} \sum_{m_2=0}^{\infty} a_{m_1 m_2} \otimes (T^1)^{m_1} (T^2)^{m_2}.$$

The series converges in the operator norm and defines a functional calculus. Alternatively, choose $r \in (0,1)$ such that $\sigma(T) \subset r\mathbb{D}^2$, let $\gamma(t) = re^{it}$ for $0 \leq t \leq 2\pi$ and define

$$f(T) = \frac{1}{(2\pi i)^2} \int_\gamma \int_\gamma f(\lambda^1, \lambda^2) \otimes (\lambda^1 - T^1)^{-1} (\lambda^2 - T^2)^{-1} d\lambda^1 \, d\lambda^2.$$

The definitions agree, and yield the expected properties (compare Section 1.2).

Lemma 4.64. (Properties of the enhanced holomorphic functional calculus on $\mathbb{D}^2$) *Let $\mathcal{H}, \mathcal{X}, \mathcal{Y}$, and $\mathcal{Z}$ be Hilbert spaces and let $T = (T^1, T^2)$ be a commuting pair of operators on $\mathcal{H}$ that satisfies $\sigma(T) \subseteq \mathbb{D}^2$.*

(i) *If $f, g \in \mathrm{Hol}_{\mathcal{B}(\mathcal{X}, \mathcal{Y})}(\mathbb{D}^2)$ and $\alpha, \beta \in \mathbb{C}$, then*

$$(\alpha f + \beta g)(T) = \alpha f(T) + \beta g(T).$$

(ii) *If $f \in \mathrm{Hol}_{\mathcal{B}(\mathcal{X}, \mathcal{Y})}(\mathbb{D}^2)$ and $g \in \mathrm{Hol}_{\mathcal{B}(\mathcal{Y}, \mathcal{Z})}(\mathbb{D}^2)$ then*

$$(gf)(T) = g(T)f(T) \qquad in \ \mathcal{B}(\mathcal{X} \otimes \mathcal{H}, \mathcal{Z} \otimes \mathcal{H}).$$

(iii) *The holomorphic functional calculus map $f \mapsto f(T)$ is continuous with respect to the topology on $\mathrm{Hol}_{\mathcal{B}(\mathcal{X}, \mathcal{Y})}(\mathbb{D}^2)$ of uniform convergence on compact sets and the operator norm on $\mathcal{B}(\mathcal{X} \otimes \mathcal{H}, \mathcal{Y} \otimes \mathcal{H})$.*

Likewise, the hereditary functional calculus on $\mathbb{D}^2$ can be enhanced to encompass operator-valued hereditary functions. Let $\mathcal{X}$ be a Hilbert space. A *hereditary $\mathcal{B}(\mathcal{X})$-valued function* on $\mathbb{D}^2$ is a map $h\colon \mathbb{D}^2 \to \mathcal{B}(\mathcal{X})$ such that the map

$$(\lambda, \mu) \mapsto h(\lambda, \bar{\mu})$$

is holomorphic on $\mathbb{D}^2$. The set of all such functions is denoted by $\mathrm{Her}_{\mathcal{B}(\mathcal{X})}(\mathbb{D}^2)$ and comprises a complex vector space, which is a Fréchet space with respect to the topology of uniform convergence in the operator norm on compact subsets of $\mathbb{D}^2$.

Operator Analysis on $\mathbb{D}^2$

115

Consider $h \in \mathrm{Her}_{\mathcal{B}(\mathcal{X})}(\mathbb{D}^2)$ and T as in the foregoing paragraph. To define $h(T) \in \mathcal{B}(\mathcal{X} \otimes \mathcal{H})$ it suffices to insert the symbol $\otimes$ into the defining equation (4.11). Define $h(T)$ by

$$h(T) = \frac{1}{(2\pi i)^4} \int_\gamma \int_\gamma \int_\gamma \int_\gamma h(\lambda, \bar{\mu}) \otimes (\bar{\mu}^1 - (T^1)^*)^{-1} (\bar{\mu}^2 - (T^2)^*)^{-1}$$

$$(\lambda^1 - T^1)^{-1} (\lambda^2 - T^2)^{-1} d\bar{\mu}^1 \, d\bar{\mu}^2 \, d\lambda^1 \, d\lambda^2.$$

Alternatively, $h(T)$ can be defined by a power series as in equation (4.9). If $h(\lambda, \bar{\mu}) = \sum_{n,m} a_{nm} \lambda^n \bar{\mu}^m$, where m, n are multi-indices and each $a_{nm} \in \mathcal{B}(\mathcal{X})$, then we may define

$$h(T) = \sum_{n,m} a_{nm} \otimes (T^*)^m T^n.$$

The enhanced hereditary functional calculus continues to behave well (compare Lemma 2.86 and Proposition 2.87). We define an involution $h \mapsto h^*$ on $\mathrm{Her}_{\mathcal{B}(\mathcal{X})}(\mathbb{D}^2)$ by $h^*(\lambda, \mu) = h(\mu, \lambda)^*$ for $\lambda, \mu \in \mathbb{D}^2$.

Lemma 4.65. (Properties of the enhanced hereditary functional calculus on $\mathbb{D}^2$) *Let* $\mathcal{H}, \mathcal{X}, \mathcal{Y}$ *be Hilbert spaces and let* $T = (T^1, T^2)$ *be a commuting pair of operators on* $\mathcal{H}$ *that satisfies* $\sigma(T) \subseteq \mathbb{D}^2$.

(i) *If* $f, g \in \mathrm{Her}_{\mathcal{B}(\mathcal{X})}(\mathbb{D}^2)$ *and* $\alpha, \beta \in \mathbb{C}$, *then*

$$(\alpha f + \beta g)(T) = \alpha f(T) + \beta g(T).$$

(ii) *If* $\varphi \in \mathrm{Hol}_{\mathcal{B}(\mathcal{X})}(\mathbb{D}^2)$ *and* $h(\lambda, \mu) = \varphi(\lambda)$ *for all* $\lambda, \mu \in \mathbb{D}^2$, *then* $h(T) = \varphi(T)$.

(iii) *If* $\psi \in \mathrm{Hol}_{\mathcal{B}(\mathcal{X})}(\mathbb{D}^2)$ *and* $h(\lambda, \mu) = \psi(\mu)^*$ *for all* $\lambda, \mu \in \mathbb{D}^2$, *then* $h(T) = \psi(T)^*$.

(iv) *If* $h \in \mathrm{Her}_{\mathcal{B}(\mathcal{X})}(\mathbb{D}^2)$ *then* $h^*(T) = h(T)^*$.

(v) *If* $\varphi, \psi \in \mathrm{Hol}_{\mathcal{B}(\mathcal{X},\mathcal{Y})}(\mathbb{D}^2)$, $h \in \mathrm{Her}_{\mathcal{B}(\mathcal{Y})}(\mathbb{D}^2)$, *and* $g \in \mathrm{Her}_{\mathcal{B}(\mathcal{X})}(\mathbb{D}^2)$ *is defined by*

$$g(\lambda, \mu) = \psi(\mu)^* h(\lambda, \mu) \, \varphi(\lambda) \qquad \textit{for all } \lambda, \mu \in \mathbb{D}^2,$$

then

$$g(T) = \psi(T)^* h(T) \varphi(T).$$

(vi) *The hereditary functional calculus map* $h \mapsto h(T)$ *is continuous with respect to the topology on* $\mathrm{Her}_{\mathcal{B}(\mathcal{X})}(\mathbb{D}^2)$ *of uniform convergence on compact sets and the operator norm on* $\mathcal{B}(\mathcal{X} \otimes \mathcal{H})$.

116   *Commutative Theory*

4.9 Models of Operator-Valued Functions on $\mathbb{D}^2$

The following definition plays the role in the theory of $\mathcal{B}(\mathcal{X},\mathcal{Y})$-valued Schur functions on $\mathbb{D}^2$ that Definition 4.18 and the model formula (4.22) played in the theory of the Schur class on $\mathbb{D}^2$. In this section we shall assume for convenience that the Hilbert space $\mathcal{X}$ is separable (in many applications, $\mathcal{X}$ is even finite-dimensional, in which case some of the arguments can be greatly simplified).

Definition 4.66. (**Models on** $\mathbb{D}^2$**-enhanced**) *Let* $\varphi\colon \mathbb{D}^2 \to \mathcal{B}(\mathcal{X},\mathcal{Y})$ *be a function. A* model on $\mathbb{D}^2$ *for* φ *is a pair* $(\mathcal{M},u)$ *where*

(i) $\mathcal{M} = \mathcal{M}^1 \oplus \mathcal{M}^2$ *is a decomposed Hilbert space,*

(ii) $u = u^1 \oplus u^2$ *where* u^1 *and* u^2 *are maps from* $\mathbb{D}^2$ *to* $\mathcal{B}(\mathcal{X},\mathcal{M}^1)$ *and* $\mathcal{B}(\mathcal{X},\mathcal{M}^2)$ *repectively, and*

(iii) *the following model formula holds for every* λ *and* μ *in* $\mathbb{D}^2$:

$$1 - \varphi(\mu)^*\varphi(\lambda) = (1 - \bar{\mu}^1\lambda^1)u^1(\mu)^*u^1(\lambda) + (1 - \bar{\mu}^2\lambda^2)u^2(\mu)^*u^2(\lambda). \tag{4.67}$$

Note that if λ in $\mathbb{D}^2$ is interpreted as an operator on $\mathcal{M}$, as in the formula (4.21), then equation (4.67) can be written in the form

$$1 - \varphi(\mu)^*\varphi(\lambda) = u(\mu)^*(1 - \mu^*\lambda)u(\lambda). \tag{4.68}$$

Proposition 4.69. (**Automatic holomorphy of models on** $\mathbb{D}^2$)
Let $\varphi\colon \mathbb{D}^2 \to \mathcal{B}(\mathcal{X},\mathcal{Y})$ *be a function and suppose that* $(\mathcal{M},u)$ *is a model for* φ. *Then* $(\mathcal{M},u)$ *is holomorphic, that is,* u *is a holomorphic* $\mathcal{B}(\mathcal{X},\mathcal{M})$*-valued function on* $\mathbb{D}^2$.

Proof. Assume that $\varphi\colon \mathbb{D}^2 \to \mathcal{B}(\mathcal{X},\mathcal{Y})$ is a function for which $(\mathcal{M},u)$ is a model. The formula (4.68) can be reshuffled to

$$\varphi(\mu)^*\varphi(\lambda) + u(\mu)^*u(\lambda) = 1 + (\mu u(\mu))^*(\lambda u(\lambda)) \qquad \text{for all } \lambda,\mu \in \mathbb{D}^2,$$

which in turn may be rewritten as

$$\begin{bmatrix} \varphi(\mu) \\ u(\mu) \end{bmatrix}^* \begin{bmatrix} \varphi(\lambda) \\ u(\lambda) \end{bmatrix} = \begin{bmatrix} 1 \\ \mu u(\mu) \end{bmatrix}^* \begin{bmatrix} 1 \\ \lambda u(\lambda) \end{bmatrix} \qquad \text{for all } \lambda,\mu \in \mathbb{D}^2.$$

Therefore, if we define $\mathcal{L} \subseteq \mathcal{X} \oplus \mathcal{M}$ by

$$\mathcal{L} = \left[\begin{bmatrix} 1 \\ \lambda u(\lambda) \end{bmatrix} x \colon \lambda \in \mathbb{D}^2, x \in \mathcal{X} \right],$$

then Lemma 3.11 implies that there exists a linear isometry

$$V\colon \mathcal{L} \to \mathcal{Y} \oplus \mathcal{M}$$

Operator Analysis on $\mathbb{D}^2$ 117

such that

$$V \begin{bmatrix} 1 \\ \lambda u(\lambda) \end{bmatrix} = \begin{bmatrix} \varphi(\lambda) \\ u(\lambda) \end{bmatrix} \qquad \text{for all } \lambda \in \mathbb{D}^2. \tag{4.70}$$

If we define $W \in \mathcal{B}(\mathcal{X} \oplus \mathcal{M}, \mathcal{Y} \oplus \mathcal{M})$ by setting $W = V$ on $\mathcal{L}$ and $W = 0$ on the orthogonal complement of $\mathcal{L}$ in $\mathcal{X} \oplus \mathcal{M}$, then W is a partial isometry (in particular, W is a contraction) and equation (4.70) implies that

$$W \begin{bmatrix} 1 \\ \lambda u(\lambda) \end{bmatrix} = \begin{bmatrix} \varphi(\lambda) \\ u(\lambda) \end{bmatrix} \qquad \text{for all } \lambda \in \mathbb{D}^2.$$

If we represent W as a 2×2 block matrix,

$$W = \begin{bmatrix} A & B \\ C & D \end{bmatrix} : \mathcal{X} \oplus \mathcal{M} \to \mathcal{Y} \oplus \mathcal{M},$$

then

$$\begin{bmatrix} \varphi(\lambda) \\ u(\lambda) \end{bmatrix} = W \begin{bmatrix} 1 \\ \lambda u(\lambda) \end{bmatrix} = \begin{bmatrix} A & B \\ C & D \end{bmatrix} \begin{bmatrix} 1 \\ \lambda u(\lambda) \end{bmatrix} \qquad \text{for all } \lambda \in \mathbb{D}^2. \tag{4.71}$$

Therefore,

$$u(\lambda) = C + \lambda D u(\lambda) \qquad \text{for all } \lambda \in \mathbb{D}^2.$$

As W is a contraction, so also is D. Consequently, we may solve this last equation for $u(\lambda)$ to obtain the formula

$$u(\lambda) = (1 - D\lambda)^{-1} C \qquad \text{for all } \lambda \in \mathbb{D}^2,$$

which implies that u is holomorphic on $\mathbb{D}^2$. $\qquad\square$

Proposition 4.72. (Converse to the fundamental theorem for $\mathscr{S}_{\mathcal{B}(\mathcal{X},\mathcal{Y})}(\mathbb{D}^2)$)
If $\psi. \ \mathbb{D}^2 \ \to \ \mathcal{B}(\mathcal{X},\mathcal{Y})$ *is a function and* φ *has a model on* $\mathbb{D}^2$, *then* $\varphi \in \mathscr{S}_{\mathcal{B}(\mathcal{X},\mathcal{Y})}(\mathbb{D}^2)$.

Proof. Assume that $\varphi \colon \mathbb{D}^2 \to \mathcal{B}(\mathcal{X},\mathcal{Y})$ has a model $(\mathcal{M},u)$ on $\mathbb{D}^2$. We first claim that φ is holomorphic on $\mathbb{D}^2$. To prove this claim, observe that if W is constructed as in the proof of Proposition 4.69, then equation (4.71) implies that

$$\varphi(\lambda) = A + \lambda B u(\lambda) \qquad \text{for all } \lambda \in \mathbb{D}^2.$$

Consequently, as Proposition 4.69 implies that u is holomorphic on $\mathbb{D}^2$, so also φ is holomorphic on $\mathbb{D}^2$. To see that $\varphi \in \mathscr{S}_{\mathcal{B}(\mathcal{X},\mathcal{Y})}(\mathbb{D}^2)$, notice that, by the model formula (4.68) with $\mu = \lambda$,

$$1 - \varphi(\lambda)^* \varphi(\lambda) = u(\lambda)^* (1 - \lambda^* \lambda) u(\lambda) \geq 0 \qquad \text{for all } \lambda \in \mathbb{D}^2.$$

118   *Commutative Theory*

Hence $\|\varphi(\lambda)\| \le 1$ for all $\lambda \in \mathbb{D}^2$. Since φ is holomorphic and $\|\varphi(\lambda)\| \le 1$ for all $\lambda \in \mathbb{D}^2$, $\varphi \in \mathscr{S}_{\mathcal{B}(\mathcal{X},\mathcal{Y})}(\mathbb{D}^2)$. $\qquad\square$

We now assemble the ingredients for a fundamental theorem for $\mathcal{B}(\mathcal{X},\mathcal{Y})$-valued Schur functions on $\mathbb{D}^2$. We will follow the development in Section 4.3 quite closely. As in Section 4.8, $\text{Her}_{\mathcal{B}(\mathcal{X})}(\mathbb{D}^2)$ carries the topology of uniform convergence on compact subsets of $\mathbb{D}^2$, that is, $\text{Her}_{\mathcal{B}(\mathcal{X})}(\mathbb{D}^2)$ is topologized by the countable collection of semi-norms,

$$\rho_n(A) = \max_{\|\lambda\|,\|\mu\| \le 1 - 1/n} \|A(\lambda,\mu)\| \qquad \text{for } n = 1,2,\dots,$$

where $\|\lambda\| = \max\{|\lambda^1|, |\lambda^2|\}$. Thus $\text{Her}_{\mathcal{B}(\mathcal{X})}(\mathbb{D}^2)$ is a locally convex topological vector space (so that the Hahn–Banach theorem applies) as well as a Fréchet space (so that sets are closed if and only if they are sequentially closed).

Lemma 4.73. *Let $\mathcal{X}$ be a separable Hilbert space and assume that $\{A_n\}$ is a sequence in $\text{Her}_{\mathcal{B}(\mathcal{X})}(\mathbb{D}^2)$ that is uniformly bounded on compact subsets of $\mathbb{D}^2 \times \mathbb{D}^2$. There exists a subsequence $\{n_k\}$ and $A \in \text{Her}_{\mathcal{B}(\mathcal{X})}(\mathbb{D}^2)$ such that for each $x, y \in \mathcal{X}$*

$$\langle A_{n_k}(\lambda,\mu)x, y\rangle \to \langle A(\lambda,\mu)x, y\rangle \quad \text{as } k \to \infty \qquad (4.74)$$

uniformly on compact subsets of $\mathbb{D}^2 \times \mathbb{D}^2$. If, in addition, $A_n \in \mathcal{P}_{\mathcal{B}(\mathcal{X})}$ for all n, then $A \in \mathcal{P}_{\mathcal{B}(\mathcal{X})}$.

Proof. We first observe that if $x, y \in \mathcal{X}$, then the hypotheses guarantee that the functions

$$(\lambda,\mu) \mapsto \langle A_n(\lambda,\mu)x, y\rangle$$

constitute a locally bounded sequence in $\text{Her}(\mathbb{D}^2)$. Consequently, by Montel's theorem, for every $x, y \in \mathcal{X}$ there exist $f \in \text{Her}(\mathbb{D}^2)$ and a sequence $\{n_k\}$ of natural numbers such that

$$\langle A_{n_k}(\lambda,\mu)x, y\rangle \to f \text{ in } \text{Her}(\mathbb{D}^2) \text{ as } k \to \infty. \qquad (4.75)$$

As $\mathcal{X}$ is assumed to be separable, we may enumerate a sequence $\{(x_m, y_m)\}_{m=1}^{\infty}$ in $\mathcal{X} \times \mathcal{X}$ such that

$$\{(x_m, y_m): m \ge 1\} \text{ is dense in } \mathcal{X} \times \mathcal{X}. \qquad (4.76)$$

Invoke the statement (4.75) with $x = x_1$ and $y = y_1$ to obtain $f_1 \in \text{Her}(\mathbb{D}^2)$ and $\{n_k^1\}$ such that

$$\left\langle A_{n_k^1}(\lambda,\mu)x_1, y_1\right\rangle \to f_1 \text{ in } \text{Her}(\mathbb{D}^2) \text{ as } k \to \infty. \qquad (4.77)$$

Operator Analysis on $\mathbb{D}^2$ 119

Now, invoke the statement (4.75) again, with $\{A_n\}$ replaced with $\{A_{n_k^1}\}$, $x = x_2$, and $y = y_2$ to obtain $f_2 \in \mathrm{Her}(\mathbb{D}^2)$ and a subsequence $\{n_k^2\}$ of $\{n_k^1\}$ such that

$$\langle A_{n_k^2}(\lambda, \mu) x_2, y_2 \rangle \to f_2 \ \text{ in } \ \mathrm{Her}(\mathbb{D}^2) \text{ as } k \to \infty.$$

As $\{n_k^2\}_{k \geq 1}$ is a subsequence of $\{n_k^1\}_{k \geq 1}$, the statement (4.77) implies that

$$\langle A_{n_k^2}(\lambda, \mu) x_1, y_1 \rangle \to f_1 \ \text{ in } \ \mathrm{Her}(\mathbb{D}^2) \text{ as } k \to \infty.$$

Continuing inductively, we obtain for each $m \geq 1$ a function $f_m \in \mathrm{Her}(\mathbb{D}^2)$ and a sequence $\{n_k^m\}_{k \geq 1}$ that satisfies

$$\{n_k^{m+1}\}_{k \geq 1} \text{ is a subsequence of } \{n_k^m\}_{k \geq 1}$$

and

$$\langle A_{n_k^m}(\lambda, \mu) x_i, y_i \rangle \to f_i \ \text{ in } \mathrm{Her}(\mathbb{D}^2) \text{ as } k \to \infty \text{ for } i = 1, \dots, m.$$

It follows that if we let $n_k = n_k^k$ for $k = 1, 2, \dots$, then

$$\langle A_{n_k}(\lambda, \mu) x_m, y_m \rangle \to f_m \ \text{ in } \mathrm{Her}(\mathbb{D}^2) \text{ as } k \to \infty \text{ for all } m \geq 1. \quad (4.78)$$

Claim 4.79. For each $x, y \in \mathcal{X}$, $\{\langle A_{n_k}(\lambda, \mu) x, y \rangle\}_{k \geq 1}$ is a Cauchy sequence in $\mathrm{Her}(\mathbb{D}^2)$.

Proof. Fix $x, y \in \mathcal{X}$, $r < 1$, and $\varepsilon > 0$. As

$$\sup_{k \geq 1} \sup_{\lambda, \mu \in r\mathbb{D}^2} \|A_{n_k}(\lambda, \mu)\| < \infty,$$

the property (4.76) and the Cauchy-Schwarz inequality imply that there exists m such that for all $k \geq 1$,

$$\sup_{\lambda, \mu \in r\mathbb{D}^2} \left(|\langle A_{n_k}(\lambda, \mu)(x - x_m), y_m \rangle| + |\langle A_{n_k}(\lambda, \mu) x, (y - y_m) \rangle| \right) < \varepsilon/3.$$

$$(4.80)$$

Also, by assertion (4.78), there exists l such that if $k, j \geq l$,

$$\sup_{\lambda, \mu \in r\mathbb{D}^2} |\langle A_{n_k}(\lambda, \mu) x_m, y_m \rangle - \langle A_{n_j}(\lambda, \mu) x_m, y_m \rangle| < \varepsilon/3. \quad (4.81)$$

But

$$\begin{aligned}
\langle A_{n_k}&(\lambda, \mu) x, y \rangle - \langle A_{n_j}(\lambda, \mu) x, y \rangle \\
&= \langle A_{n_k}(\lambda, \mu) x_m, y_m \rangle - \langle A_{n_j}(\lambda, \mu) x_m, y_m \rangle \\
&\quad + \langle A_{n_k}(\lambda, \mu)(x - x_m), y_m \rangle + \langle A_{n_k}(\lambda, \mu) x, y - y_m \rangle \\
&\quad - \left(\langle A_{n_j}(\lambda, \mu)(x - x_m), y_m \rangle + \langle A_{n_j}(\lambda, \mu) x, y - y_m \rangle \right).
\end{aligned}$$

120　　　　　　　　　　　　*Commutative Theory*

Therefore, if $k, j \geq l$ and $\lambda, \mu \in r\mathbb{D}^2$, as inequality (4.81) implies that the first term in this last expression is less than $\varepsilon/3$ and inequality (4.80) implies that the second and third terms are less than $\varepsilon/3$, it follows from the triangle inequality that

$$\sup_{\lambda, \mu \in r\mathbb{D}^2} \left| \langle A_{n_k}(\lambda, \mu)x, y \rangle - \langle A_{n_j}(\lambda, \mu)x, y \rangle \right| \leq \varepsilon.$$

This proves the claim. □

As $\mathrm{Her}(\mathbb{D}^2)$ is complete, Claim 4.79 implies that, for each $x, y \in \mathcal{X}$, there exists $F_{x,y} \in \mathrm{Her}(\mathbb{D}^2)$ such that

$$F_{x,y}(\lambda, \mu) = \lim_{k \to \infty} \langle A_{n_k}(\lambda, \mu)x, y \rangle, \qquad (4.82)$$

where the convergence is uniform on compact subsets of $\mathbb{D}^2 \times \mathbb{D}^2$. Therefore, the convergence assertion (4.74), the first conclusion in Lemma 4.73, is a consequence of the following claim.

Claim 4.83. There exists $A \in \mathrm{Her}_{\mathcal{B}(\mathcal{X})}(\mathbb{D}^2)$ such that,

$$\text{for all } x, y \in \mathcal{X} \text{ and } \lambda, \mu \in \mathbb{D}^2, \quad F_{x,y}(\lambda, \mu) = \langle A(\lambda, \mu)x, y \rangle. \qquad (4.84)$$

To prove Claim 4.83, observe first that, as A_{n_k} is uniformly bounded on compact subsets of $\mathbb{D}^2 \times \mathbb{D}^2$, for fixed λ and μ, equation (4.82) implies that the map

$$(x, y) \mapsto F_{x,y}(\lambda, \mu) \qquad \text{for } x, y \in \mathcal{X}$$

is a bounded sesquilinear form on $\mathcal{X}$. Therefore, the Riesz representation theorem implies that, for each $\lambda, \mu \in \mathbb{D}^2$, there exists $A(\lambda, \mu) \in \mathcal{B}(\mathcal{X})$ such that

$$F_{x,y}(\lambda, \mu) = \langle A(\lambda, \mu)x, y \rangle \quad \text{for all } x, y \in \mathcal{X},$$

that is, the statement (4.84) holds. To see that $A \in \mathrm{Her}_{\mathcal{B}(\mathcal{X})}(\mathbb{D}^2)$, it suffices to show that A is holomorphic in λ for each fixed μ and holomorphic in $\bar{\mu}$ for each fixed λ. If, for each $x, y \in \mathcal{X}$, we define $L_{x,y} \in \mathcal{B}(\mathcal{X})^*$ by the formula

$$L_{x,y}(T) = \langle Tx, y \rangle \qquad \text{for } T \in \mathcal{B}(\mathcal{X})$$

and then define $E \subseteq \mathcal{B}(\mathcal{X})^*$ by

$$E = \{L_{x,y} : x, y \in \mathcal{X}\}$$

then it is easy to see that E is a norm-generating subset of $\mathcal{B}(\mathcal{X})^*$ (cf. Remark 2.97). The fact that $F_{x,y} \in \mathrm{Her}(\mathbb{D}^2)$, together with the statement (4.84), implies that, for each $\mu \in \mathbb{D}^2$, the map $\lambda \mapsto L(A(\lambda, \mu))$ is holomorphic for every $L \in E$. Therefore, by Theorem 2.90 and Remark 2.97, A is holomorphic in λ for each fixed $\mu \in \mathbb{D}^2$. Likewise, A is holomorphic in $\bar{\mu}$ for each fixed λ.

Operator Analysis on $\mathbb{D}^2$ 121

To prove the second assertion of the lemma, assume that $A_n \in \mathcal{P}_{\mathcal{B}(\mathcal{X})}$ for all n so that in particular,

$$A_{n_k} \text{ is positive semi-definite for all } k. \tag{4.85}$$

We wish to show that $A \in \mathcal{P}_{\mathcal{B}}(\mathcal{X})$, or equivalently, as $A \in \mathrm{Her}_{\mathcal{B}(\mathcal{X})}(\mathbb{D}^2)$, that A is positive semi-definite. Accordingly, choose n points $\lambda_1, \lambda_2, \ldots, \lambda_n$ in $\mathbb{D}^2$ and n vectors in $\mathcal{X}$. Using the properties (4.74) and (4.85) we see that

$$\sum_{i,j=1}^{n} \langle A(\lambda_j, \lambda_i) x_j, x_i \rangle = \lim_{k \to \infty} \sum_{i,j=1}^{n} \langle A_{n_k}(\lambda_j, \lambda_i) x_j, x_i \rangle \geq 0.$$

Therefore, A is positive semi-definite. $\qquad\square$

Let $\mathcal{R}_{\mathcal{B}(\mathcal{X})}$ denote the set of *symmetric* elements of $\mathrm{Her}_{\mathcal{B}(\mathcal{X})}(\mathbb{D}^2)$, that is, the elements h that satisfy

$$h(\mu, \lambda) = h(\lambda, \mu)^* \qquad \text{for all } \lambda, \mu \in \mathbb{D}^2.$$

It is easy to see that $\mathcal{R}_{\mathcal{B}(\mathcal{X})}$ is a closed real subspace of $\mathrm{Her}_{\mathcal{B}(\mathcal{X})}(\mathbb{D}^2)$. We also define a subset of $\mathrm{Her}_{\mathcal{B}(\mathcal{X})}(\mathbb{D}^2)$ by

$$\mathcal{P}_{\mathcal{B}(\mathcal{X})} = \{h \in \mathrm{Her}_{\mathcal{B}(\mathcal{X})}(\mathbb{D}^2) \colon h \text{ is positive semi-definite}\}.$$

Evidently, $\mathcal{P}_{\mathcal{B}(\mathcal{X})}$ is a closed cone in $\mathcal{R}_{\mathcal{B}(\mathcal{X})}$.

In light of Definition 4.66, we modify the definition of the *model wedge* as follows. $\mathcal{W}_{\mathcal{B}(\mathcal{X})}$ is the subset of $\mathrm{Her}_{\mathcal{B}(\mathcal{X})}(\mathbb{D}^2)$ given by the formula

$$\mathcal{W}_{\mathcal{B}(\mathcal{X})} = \{(1 - \bar{\mu}^1 \lambda^1) A^1 + (1 - \bar{\mu}^2 \lambda^2) A^2 \colon A^1, A^2 \in \mathcal{P}_{\mathcal{B}(\mathcal{X})}\}. \tag{4.86}$$

Here, the notation $(1 - \bar{\mu}^1 \lambda^1) A^1$ is an abbreviation for the kernel

$$(\lambda, \mu) \mapsto (1 - \bar{\mu}^1 \lambda^1) A^1(\lambda, \mu)$$

on $\mathbb{D}^2$. As before, the model wedge has the properties described in the following three lemmas.

Lemma 4.87. *The model wedge* $\mathcal{W}_{\mathcal{B}(\mathcal{X})}$ *is closed in* $\mathcal{R}_{\mathcal{B}(\mathcal{X})}$.

Proof. Clearly, $\mathcal{P}_{\mathcal{B}(\mathcal{X})} \subseteq \mathcal{R}_{\mathcal{B}(\mathcal{X})}$. Also, since $\mathcal{P}_{\mathcal{B}(\mathcal{X})}$ is a cone, $\mathcal{W}_{\mathcal{B}(\mathcal{X})}$ is a wedge. To see that $\mathcal{W}_{\mathcal{B}(\mathcal{X})}$ is closed, consider a convergent sequence (h_n) in $\mathcal{W}_{\mathcal{B}(\mathcal{X})}$ with limit h, where

$$h_n = (1 - \bar{\mu}^1 \lambda^1) A_n^1 + (1 - \bar{\mu}^2 \lambda^2) A_n^2 \qquad \text{for some } A_n^1, A_n^2 \in \mathcal{P}_{\mathcal{B}(\mathcal{X})}. \tag{4.88}$$

Fix $r < 1$. Since $h_n \to h$ uniformly on compact subsets of $\mathbb{D}^2 \times \mathbb{D}^2$, there exists a constant M such that

$$\sup_n \sup_{\lambda \in r\mathbb{D}^2} \|h_n(\lambda, \lambda)\| \leq M.$$

122   *Commutative Theory*

Therefore, by equation (4.88), if $n \geq 1$ and $\lambda \in r\mathbb{D}^2$,

$$(1 - r^2)\left(A_n^1(\lambda, \lambda) + A_n^2(\lambda, \lambda)\right) \leq (1 - |\lambda^1|^2)A_n^1(\lambda, \lambda) + (1 - |\lambda^2|^2)A_n^2(\lambda, \lambda)$$
$$= h_n(\lambda, \lambda)$$
$$\leq M.$$

Therefore,

$$\sup_n \sup_{\lambda \in r\mathbb{D}^2} \|A_n^1(\lambda, \lambda)\| \leq \frac{M}{1 - r^2} \quad \text{and} \quad \sup_n \sup_{\lambda \in r\mathbb{D}^2} \|A_n^2(\lambda, \lambda)\| \leq \frac{M}{1 - r^2}.$$
$$(4.89)$$

But if A is a positive semi-definite $\mathcal{B}(\mathcal{X})$-valued kernel on $\mathbb{D}^2$,

$$\|A(\lambda, \mu)\|^2 \leq \|A(\lambda, \lambda)\|\|A(\mu, \mu)\| \qquad \text{for all } \lambda, \mu \in \mathbb{D}^2.$$

Therefore, as $A_n^1, A_n^2 \in \mathcal{P}_{\mathcal{B}(\mathcal{X})}$, the inequalities (4.89) imply that

$$\sup_n \sup_{\lambda, \mu \in r\mathbb{D}^2} \|A_n^1(\lambda, \mu)\| \leq \frac{M}{1 - r^2} \quad \text{and} \quad \sup_n \sup_{\lambda, \mu \in r\mathbb{D}^2} \|A_n^2(\lambda, \mu)\| \leq \frac{M}{1 - r^2}.$$

To summarize, the previous paragraph shows that for each $r < 1$, $\{A_n^1\}$ and $\{A_n^2\}$ are uniformly bounded on $r\mathbb{D}^2 \times r\mathbb{D}^2$. Therefore, by Lemma 4.73, there exist an increasing sequence of indices $n_1, n_2, \ldots$ and functions $A^1, A^2 \in \mathcal{P}_{\mathcal{B}(\mathcal{X})}$ such that, for all $x, y \in \mathcal{X}$,

$$\langle A_{n_k}^1 x, y \rangle \to \langle A^1 x, y \rangle \text{ and } \langle A_{n_k}^2 x, y \rangle \to \langle A^2 x, y \rangle \text{ in } \mathrm{Her}(\mathbb{D}^2) \text{ as } k \to \infty.$$

Apply equation (4.88) with $n = n_k$ and let $k \to \infty$ to deduce that, for all $x, y \in \mathcal{X}$ and $\lambda, \mu \in \mathbb{D}^2$,

$$\langle h(\lambda, \mu)x, y \rangle = (1 - \bar{\mu}^1 \lambda^1)\langle A^1(\lambda, \mu)x, y \rangle + (1 - \bar{\mu}^2 \lambda^2)\langle A^2(\lambda, \mu)x, y \rangle.$$

Therefore, as x and y are arbitrary,

$$h = (1 - \bar{\mu}^1 \lambda^1)A^1 + (1 - \bar{\mu}^2 \lambda^2)A^2.$$

Thus $h \in \mathcal{W}_{\mathcal{B}(\mathcal{X})}$. $\square$

Lemma 4.90. *$\mathcal{P}_{\mathcal{B}(\mathcal{X})}$ is contained in $\mathcal{W}_{\mathcal{B}(\mathcal{X})}$. In particular, if $f \in \mathrm{Hol}_{\mathcal{B}(\mathcal{X}, \mathbb{C})}$ $(\mathbb{D}^2)$, then $f(\mu)^* f(\lambda) \in \mathcal{W}_{\mathcal{B}(\mathcal{X})}$.*

Proof. Fix $A \in \mathcal{P}_{\mathcal{B}(\mathcal{X})}$. Since $(\bar{\mu}^1)^n A(\lambda, \mu)(\lambda^1)^n \in \mathcal{P}_{\mathcal{B}(\mathcal{X})}$ for each $n \geq 0$, it follows from the fact that $\mathcal{P}_{\mathcal{B}(\mathcal{X})}$ is a cone that, for each $m \geq 0$,

$$\sum_{n=0}^{m} (\bar{\mu}^1)^n A(\lambda, \mu)(\lambda^1)^n \in \mathcal{P}_{\mathcal{B}(\mathcal{X})}.$$

Hence, since $\mathcal{P}_{\mathcal{B}(\mathcal{X})}$ is closed,

$$\frac{A(\lambda,\mu)}{1-\bar{\mu}^1\lambda^1} = \lim_{m\to\infty}\sum_{n=0}^{m}(\bar{\mu}^1)^n A(\lambda,\mu)(\lambda^1)^n \in \mathcal{P}_{\mathcal{B}(\mathcal{X})}.$$

Therefore, as

$$A(\lambda,\mu) = (1-\bar{\mu}^1\lambda^1)\frac{A(\lambda,\mu)}{1-\bar{\mu}^1\lambda^1},$$

it follows from the choice

$$A^1 = \frac{A(\lambda,\mu)}{1-\bar{\mu}^1\lambda^1} \quad \text{and} \quad A^2 = 0$$

in equation (4.86) that $A \in \mathcal{W}_{\mathcal{B}(\mathcal{X})}$. $\qquad\square$

Lemma 4.91. *If $\varphi\colon \mathbb{D}^2 \to \mathcal{B}(\mathcal{X},\mathcal{Y})$ is a function, then φ has a model on $\mathbb{D}^2$ if and only if $1 - \varphi(\mu)^*\varphi(\lambda) \in \mathcal{W}_{\mathcal{B}(\mathcal{X})}$.*

Proof. First assume that $(\mathcal{M},u)$ is a model for φ. If we let

$$A^1(\lambda,\mu) = u^1(\mu)^*u^1(\lambda) \text{ and } A^2(\lambda,\mu) = u^2(\mu)^*u^2(\lambda),$$

then equation (4.67) becomes

$$1 - \varphi(\mu)^*\varphi(\lambda) = (1-\bar{\mu}^1\lambda^1)A^1(\lambda,\mu) + (1-\bar{\mu}^2\lambda^2)A^2(\lambda,\mu). \qquad (4.92)$$

Now, Moore's theorem (Theorem 3.6) implies that A^1 and A^2 are positive semi-definite on $\mathbb{D}^2$, and Proposition 4.69 implies that A^1 and A^2 are elements of $\mathrm{Her}_{\mathcal{B}(\mathcal{X})}(\mathbb{D}^2)$. Therefore, $A^1, A^2 \in \mathcal{P}_{\mathcal{B}(\mathcal{X})}$. Hence equation (4.92) implies that $1 - \varphi(\mu)^*\varphi(\lambda) \in \mathcal{W}_{\mathcal{B}(\mathcal{X})}$.

Conversely, assume that $1 - \varphi(\mu)^*\varphi(\lambda) \in \mathcal{W}_{\mathcal{B}(\mathcal{X})}$, that is, there exist $A^1, A^2 \subset \mathcal{P}_{\mathcal{B}(\mathcal{X})}$ such that equation (4.92) holds. By Moore's theorem again there exist Hilbert spaces $\mathcal{M}^1$ and $\mathcal{M}^2$ and functions $u^1\colon \mathbb{D}^2 \to \mathcal{M}^1$ and $u^2\colon \mathbb{D}^2 \to \mathcal{M}^2$ such that, for all $\lambda,\mu \in \mathbb{D}^2$,

$$A^1(\lambda,\mu) = u^1(\mu)^*u^1(\lambda) \quad \text{and} \quad A^2(\lambda,\mu) = u^2(\mu)^*u^2(\lambda).$$

Substituting these formulas for A^1 and A^2 into equation (4.92) we deduce equation (4.67), so that $(\mathcal{M},u)$ is a model for φ. $\qquad\square$

Finally, we are primed to prove a fundamental theorem for $\mathcal{B}(\mathcal{X},\mathcal{Y})$-valued Schur functions on $\mathbb{D}^2$.

Theorem 4.93. (The fundamental theorem for $\mathscr{S}_{\mathcal{B}(\mathcal{X},\mathcal{Y})}(\mathbb{D}^2)$) *Let $\mathcal{X}$ and $\mathcal{Y}$ be Hilbert spaces, and assume $\mathcal{X}$ is separable. Then every function in the Schur class $\mathscr{S}_{\mathcal{B}(\mathcal{X},\mathcal{Y})}(\mathbb{D}^2)$ has a model on $\mathbb{D}^2$.*

124 *Commutative Theory*

Proof. Assume that $\varphi \in \mathscr{S}_{\mathcal{B}(\mathcal{X},\mathcal{Y})}(\mathbb{D}^2)$. By Lemma 4.91 it suffices to show that

$$1 - \varphi(\mu)^*\varphi(\lambda) \in \mathcal{W}_{\mathcal{B}(\mathcal{X})}. \tag{4.94}$$

As Lemma 4.87 guarantees that $\mathcal{W}_{\mathcal{B}(\mathcal{X})}$ is a closed wedge, the assertion (4.94) will follow from the Hahn–Banach theorem if we can show that

$$L(1 - \varphi(\mu)^*\varphi(\lambda)) \geq 0 \tag{4.95}$$

whenever $L \in \mathcal{R}^*_{\mathcal{B}(\mathcal{X})}$ and

$$L(h) \geq 0 \qquad \text{for all } h \in \mathcal{W}_{\mathcal{B}(\mathcal{X})}. \tag{4.96}$$

Accordingly, fix $L \in \mathcal{R}^*_{\mathcal{B}(\mathcal{X})}$ satisfying the inequality (4.96).

For the remainder of the proof we shall denote by V, W the vector spaces of $\mathcal{B}(\mathcal{X},\mathbb{C})$-valued and $\mathcal{B}(\mathcal{X},\mathcal{Y})$-valued holomorphic functions on $\mathbb{D}^2$, respectively, that is,

$$V = \text{Hol}_{\mathcal{B}(\mathcal{X},\mathbb{C})}(\mathbb{D}^2), \quad W = \text{Hol}_{\mathcal{B}(\mathcal{X},\mathcal{Y})}(\mathbb{D}^2).$$

Let $L^\sim$ be as in Lemma 4.3. Noting that if $f, g \in V$, then $g(\mu)^* f(\lambda)$ lies in $\text{Her}_{\mathcal{B}(\mathcal{X})}(\mathbb{D}^2)$, we may define a sesquilinear form $\langle \cdot, \cdot \rangle_0$ on V by the formula

$$\langle f, g \rangle_0 = L^\sim(g(\mu)^* f(\lambda)) \quad \text{for all} \quad f, g \in V.$$

Since $f(\mu)^* f(\lambda) \in \mathcal{R}_{\mathcal{B}(\mathcal{X})}$ whenever $f \in V$, and $L = L^\sim | \mathcal{R}_{\mathcal{B}(\mathcal{X})}$ by Lemma 4.3,

$$\langle f, f \rangle_0 = L(f(\mu)^* f(\lambda)) \quad \text{for all } f \in V. \tag{4.97}$$

But Lemma 4.90 implies that $f(\mu)^* f(\lambda) \in \mathcal{W}_{\mathcal{B}(\mathcal{X})}$ whenever $f \in V$. Hence inequality (4.38) tells us that

$$\langle f, f \rangle_0 \geq 0 \quad \text{for all } f \in V,$$

that is, $\langle \cdot, \cdot \rangle_0$ is a positive semi-definite sesquilinear form on V. Consequently, if

$$\mathcal{N} \overset{\text{def}}{=} \{f \in V \colon \langle f, f \rangle_0 = 0\},$$

then $\mathcal{N}$ is a subspace of V and $\langle \cdot, \cdot \rangle_0$ induces an inner product $\langle \cdot, \cdot \rangle_L$ on $V/\mathcal{N}$ via the formula

$$\langle f + \mathcal{N}, g + \mathcal{N} \rangle_L = \langle f, g \rangle_0.$$

Let H_L denote the completion of $V/\mathcal{N}$ with respect to the inner product $\langle \cdot, \cdot \rangle_L$. Using equation (4.97), we have

$$\|f + \mathcal{N}\|_L^2 = L(f(\mu)^* f(\lambda)) \quad \text{for all } f \in V. \tag{4.98}$$

Operator Analysis on $\mathbb{D}^2$ 125

Similarly we define a positive semi-definite sesquilinear form on W by

$$\langle f,g \rangle_1 = L^\sim(g(\mu)^* f(\lambda)) \qquad \text{for all } f,g \in W,$$

and define $\mathcal{N}'$ to be the subspace $\{f : \langle f,f \rangle_1 = 0\}$ of W. Let $\langle \cdot, \cdot \rangle_{L'}$ be the inner product induced on $W/\mathcal{N}'$ and its completion $H_{L'}$ by the pre-inner product $\langle \cdot, \cdot \rangle_1$. Thus, for $\psi \in W$,

$$\| \psi + \mathcal{N}' \|_{L'}^2 = L(\psi(\mu)^* \psi(\lambda)). \qquad (4.99)$$

From equation (4.98) it follows that if $f \in V$,

$$\| f + \mathcal{N} \|_L^2 - \| \lambda^1 f + \mathcal{N} \|_L^2 = L\big(f(\mu)^* f(\lambda)\big) - L\big((\mu^1 f(\mu))^*(\lambda^1 f(\lambda))\big)$$
$$= L\big((1 - \bar{\mu}^1 \lambda^1) f(\mu)^* f(\lambda)\big).$$

But an examination of the definition (4.86) reveals that

$$(1 - \bar{\mu}^1 \lambda^1) f(\mu)^* f(\lambda) \in W_{\mathcal{B}(\mathcal{X})}.$$

Therefore, inequality (4.96) implies that, for all $f \in V$,

$$\| \lambda^1 f + \mathcal{N} \|_L^2 \leq \| f + \mathcal{N} \|_L^2.$$

As $V/\mathcal{N}$ is dense in H_L, this inequality ensures that we may define a contractive operator $T^1 \in \mathcal{B}(H_L)$ that satisfies

$$T^1(f + \mathcal{N}) = \lambda^1 f + \mathcal{N} \qquad \text{for all } f \in V.$$

In similar fashion we may define a contractive operator $T^2 \in \mathcal{B}(H_L)$ that satisfies

$$T^2(f + \mathcal{N}) = \lambda^2 f + \mathcal{N} \qquad \text{for all } f \in V.$$

Choose orthonormal bases $\{e'_j\}_{j \in J}$ and $\{e_i\}_{i \in I}$ of $\mathcal{X}, \mathcal{Y}$ respectively, for some index sets J, I.

Let us adapt Claim 4.42 to the present, more general, setting.

Claim 4.100. For any function ψ, holomorphic in a neighborhood of the closed bidisc and taking values in $\mathcal{B}(\mathcal{X}, \mathcal{Y})$ and for any family $(g_j)_{j \in J}$ of functions in V such that $\sum_{j \in J} L(g_j(\mu)^* g_j(\lambda)) < \infty$,

$$\psi(T)\left(\sum_{j \in J} e'_j \otimes (g_j + \mathcal{N}) \right) = \sum_{i \in I} \left(e_i \otimes \left(\sum_{j \in J} \psi_{ij} g_j + \mathcal{N} \right) \right), \qquad (4.101)$$

where ψ_{ij} is the scalar holomorphic function on $\mathbb{D}^2$ given by

$$\psi_{ij}(\lambda) = \left\langle \psi(\lambda) e'_j, e_i \right\rangle \qquad \text{for all } j \in J, i \in I \text{ and } \lambda \in \mathbb{D}^2.$$

Moreover, all the sums in equation (4.101) converge in norm.

126   *Commutative Theory*

Proof. Note first that, since T^1, T^2 are commuting contractions on H_L, the spectrum $\sigma(T)$ lies in the closed bidisc, and hence $\psi(T)$ is a well-defined operator from $\mathcal{X} \otimes H_L$ to $\mathcal{Y} \otimes H_L$.

The hypothesis on the g_j implies that $\sum_J \|g_j + \mathcal{N}\|_L^2 < \infty$, and therefore, by Corollary 3.29, $\sum_J e'_j \otimes (g_j + \mathcal{N})$ is a sum of pairwise orthogonal vectors that converges in norm to an element of $\mathcal{X} \otimes H_L$.

Because of the linearity and continuity of the functional calculus, it is enough to prove the claim in the case that ψ is a monomial, $\psi(\lambda) = a\lambda^m$ for some multi-index $m = (m_1, m_2)$ and some $a \in \mathcal{B}(\mathcal{X}, \mathcal{Y})$. In this case, for any $i \in I$, $j \in J$, we have $\psi_{ij}(\lambda) = a_{ij}\lambda^m$, where $a_{ij} = \langle ae'_j, e_i \rangle_{\mathcal{Y}}$. We also have $\psi(T) = a \otimes T^m$. Thus

$$\psi(T) \sum_{j \in J} e'_j \otimes (g_j + \mathcal{N}) = \sum_{j \in J} \psi(T) \left(e'_j \otimes (g_j + \mathcal{N}) \right)$$

$$= \sum_{j \in J} a \otimes T^m \left(e'_j \otimes (g_j + \mathcal{N}) \right)$$

$$= \sum_{j \in J} \left(ae'_j \otimes T^m (g_j + \mathcal{N}) \right)$$

$$= \sum_{j \in J} \left(ae'_j \otimes \lambda^m (g_j + \mathcal{N}) \right) \qquad \text{by Claim 4.42,}$$

the sum on the right-hand side being convergent in the norm of $\mathcal{Y} \otimes H_L$.

By Corollary 3.29, there are uniquely determined elements f_i in H_L such that

$$\sum_{j \in J} \left(\left(\sum_{i \in I} a_{ij} e_i \right) \otimes \lambda^m (g_j + \mathcal{N}) \right) = \sum_{i \in I} e_i \otimes f_i, \qquad (4.102)$$

all the sums converging in norm. Moreover,

$$f_i = \sum_{j \in J} a_{ij} \lambda^m (g_j + \mathcal{N}). \qquad (4.103)$$

Indeed, the sum on the right-hand side converges in the norm of H_L (for any $i \in I$) by virtue of the Cauchy–Schwartz inequality, since both $\sum_j |a_{ij}|^2 < \infty$ and

$$\sum_j \|\lambda^m (g_j + \mathcal{N})\|^2 \le \sum_j \|g_j + \mathcal{N}\|^2 < \infty.$$

Take inner products of both sides of equation (4.102) with $e_i \otimes h$, for any $i \in I$ and $h \in H_L$, to deduce equation (4.103). Thus

$$\psi(T)\sum_{j\in J}e'_j\otimes(g_j+\mathcal{N}) = \sum_{j\in J}\left(\left(\sum_{i\in I}a_{ij}e_i\right)\otimes\lambda^m(g_j+\mathcal{N})\right)$$

$$= \sum_{i\in I}e_i\otimes f_i$$

$$= \sum_{i\in I}\left(e_i\otimes\sum_{j\in J}a_{ij}\lambda^m(g_j+\mathcal{N})\right),$$

which is Claim 4.100. $\qquad\square$

We resume the proof of Theorem 4.93. Just for this proof we introduce some notations. The symbol $\mathbf{1}$ denotes the constant scalar function with value 1 on $\mathbb{D}^2$. For any $x\in\mathcal{X}$ we denote by x^* the element $\langle\cdot,x\rangle_\mathcal{X}$ of $\mathcal{B}(\mathcal{X},\mathbb{C})$ (we have previously used the notation $1\otimes x$ for this object, but here we wish to avoid confusion with other uses of the tensor product symbol). For any operator A and any scalar function g on $\mathbb{D}^2$ we denote by $g\cdot A$ the map $\lambda\mapsto g(\lambda)A$ on $\mathbb{D}^2$. Thus $\mathbf{1}\cdot x^*\in V$ for $x\in\mathcal{X}$.

We shall find an element $\gamma\in\mathcal{X}\otimes H_L$ such that, roughly speaking, $\psi(T)\gamma=\psi$ for every map ψ of the type in Claim 4.100. Since $\psi(T)\gamma\in\mathcal{Y}\otimes H_L$ and $\psi\in W$ we shall need an identification of the last two spaces.

Lemma 4.104. *The linear map* $\iota\colon\mathcal{Y}\otimes_a V\to W$ *defined by*

$$\iota(y\otimes h)(\lambda)=yh(\lambda)\qquad\text{for all }\lambda\in\mathbb{D}^2\text{ and }h\in V$$

preserves the semi-inner products induced by $\mathcal{Y}\otimes H_L$ *and* $H_{L'}$ *respectively.*

Proof. For any $y_1,y_2\in\mathcal{Y}$ and $h_1,h_2\in V$,

$$\langle\iota(y_1\otimes h_1),\iota(y_2\otimes h_2)\rangle_{L'} = \langle y_1h_1,y_2h_2\rangle_{L'}$$

$$= L^\sim\big(h_2(\mu)^*y_2^*y_1h_1(\lambda)\big)$$

$$- \langle y_1,y_2\rangle_\mathcal{Y}\,L^\sim\big(h_2(\mu)^*h_1(\lambda)\big)$$

$$= \langle y_1,y_2\rangle_\mathcal{Y}\langle h_1,h_2\rangle_L$$

$$= \langle y_1\otimes h_1,y_2\otimes h_2\rangle_{\mathcal{Y}\otimes H_L}.$$

$\qquad\square$

It follows from the lemma that $\iota(\mathcal{Y}\otimes_a\mathcal{N})\subseteq\mathcal{N}'$, and so ι induces an isometry from $\mathcal{Y}\otimes H_L$ to $H_{L'}$. We have, for any $i\in I$ and $j\in J$, any $f\in\mathrm{Hol}(\mathbb{D}^2)$ and $\lambda\in\mathbb{D}^2$,

$$\iota(e_i\otimes(f\cdot e'_j{}^*))(\lambda)=f(\lambda)e_ie'_j{}^*.$$

128   *Commutative Theory*

We claim that

$$\gamma = \sum_{j \in J} e'_j \otimes (\mathbf{1} \cdot e'_j{}^* + \mathcal{N}) \tag{4.105}$$

defines an element of $\mathcal{X} \otimes H_L$. Indeed, by Corollary 3.29, elements of $\mathcal{X} \otimes H_L$ are exactly norm-convergent sums $\sum_{j \in J} e'_j \otimes h_j$, where $(h_j)_{j \in J}$ is a family of elements of H_L such that $\sum_j \|h_j\|_L^2 < \infty$. Take $h_j = \mathbf{1} \cdot e'_j{}^* + \mathcal{N}$, so that $h_j + \mathcal{N} \in V/\mathcal{N} \subseteq H_L$. For any finite subset J_0 of J,

$$\sum_{J_0} \|h_j\|_L^2 = \sum_{J_0} L(\mathbf{1} \cdot e'_j e'_j{}^*) = L\left(\mathbf{1} \cdot \sum_{J_0} e'_j e'_j{}^*\right) = L(\mathbf{1} \cdot P_{J_0})$$

where P_{J_0} denotes the orthogonal projection from $\mathcal{X}$ to the span of $\{e'_j : j \in J_0\}$. Since $P_{J_0} \leq 1_{\mathcal{X}}$, the kernel $\mathbf{1} \cdot (1_{\mathcal{X}} - P_{J_0})$ is positive semi-definite on $\mathbb{D}^2$, and hence, since $L \geq 0$ on $\mathcal{P}_{\mathcal{B}(\mathcal{X})}$, we have

$$L(\mathbf{1} \cdot P_{J_0}) \quad \leq \quad L(\mathbf{1} \cdot 1_{\mathcal{X}}).$$

Thus

$$\sum_{J_0} \|h_j\|_L^2 \leq L(\mathbf{1} \cdot 1_{\mathcal{X}}) \quad \text{for all finite subsets } J_0 \subseteq J.$$

That is to say, $\sum_J \|h_j\|_L^2 \leq L(\mathbf{1} \cdot 1_{\mathcal{X}})$. Hence the formula (4.105) does define an element γ of $\mathcal{X} \otimes H_L$ as claimed, the sum converges in the norm of $\mathcal{X} \otimes H_L$ and

$$\|\gamma\|_{\mathcal{X} \otimes H_L}^2 = \sum_J \|h_j\|_L^2 \leq L(\mathbf{1} \cdot 1_{\mathcal{X}}). \tag{4.106}$$

Moreover, for ψ as above, Claim 4.100 tells us that

$$\psi(T)\gamma = \sum_{i \in I} e_i \otimes \left(\sum_{j \in J} \psi_{ij} \cdot e'_j{}^* + \mathcal{N}\right).$$

Combine this relation with equation (4.105) to obtain

$$\iota \psi(T)\gamma = \sum_{i \in I} \sum_{j \in J} \psi_{ij} e_i e'_j{}^* + \mathcal{N}'$$

$$= \psi + \mathcal{N}_1.$$

Consider $r \in (0, 1)$. The map $\varphi_r(\lambda) = \varphi(r\lambda)$ belongs to $\mathscr{S}_{\mathcal{B}(\mathcal{X}, \mathcal{Y})}(\mathbb{D}^2)$ and is holomorphic in a neighborhood of the closed bidisc. As rT is a pair of commuting contractions and $\sigma(rT) \subseteq \mathbb{D}^2$, Andô's dilation theorem (Theorem 1.54) implies that

$$\|\varphi(rT)\| \leq 1.$$

Operator Analysis on $\mathbb{D}^2$ 129

Moreover, we may apply the above calculations with $\psi = \varphi_r$ to assert that

$$\iota\varphi(rT)\gamma = \varphi_r + \mathcal{N}'.$$

Hence

$$\|\varphi_r + \mathcal{N}'\|_{L'}^2 = \|\iota\varphi(rT)\gamma\|_{L'}^2$$
$$\leq \|\gamma\|_{\mathcal{X} \otimes H_L}^2$$
$$\leq L(1 \cdot 1_{\mathcal{X}}) \qquad \text{by inequality (4.106)}.$$

Consequently, by the definition (4.99),

$$L(1 - \varphi(r\mu)^*\varphi(r\lambda)) \geq 0. \qquad (4.107)$$

Since L is continuous, on letting $r \to 1$ we obtain the desired relation (4.95). It follows that $1_{\mathcal{X}} - \varphi(\mu)^*\varphi(\lambda) \in \mathcal{W}_{\mathcal{B}(\mathcal{X})}$, and so φ has a model. $\square$

Andô's dilation theorem and the fundamental theorem 4.93 are equivalent in the informal sense that, given an independent proof of either, one can comparatively easily deduce the other. In the foregoing proof we used Andô's dilation theorem to prove the existence of models of functions in $\mathscr{S}_{\mathcal{B}(\mathcal{X},\mathcal{Y})}(\mathbb{D}^2)$. Conversely, from the statement of the fundamental theorem one can derive Andô's inequality for matrix-valued functions, and thereby Andô's dilation theorem. A proof of the fundamental theorem that does not use Andô's theorem is given in [52].

Corollary 4.108. *If* T *is a pair of commuting contractions and* $\sigma_{\mathrm{alg}}(T) \subseteq \mathbb{D}^2$, *then* $(\mathbb{D}^2)^-$ *is a complete spectral set for* T.

Proof. Let $\varphi \in \mathscr{S}_{\mathcal{B}(\mathcal{X})}(\mathbb{D}^2)$ where $\mathcal{X}$ is a finite-dimensional Hilbert space (and therefore separable, as we assume in this section). We need to show that $\|\varphi(T)\| \leq 1$. To do this, substitute T into the model formula for φ, just as we did in Section 2.8.3. $\square$

Andô's dilation theorem (Theorem 1.54) can then be deduced as a special case of Arveson's dilation theorem (Theorem 1.52).

4.10 Historical Notes

There is extensive literature on the function theory of polydiscs. W. Rudin's 1969 monograph [174] already contains many references.

Section 4.2. The hereditary functional calculus was introduced (in a more general context) and its basic properties demonstrated in [6]. There is a fourth method of defining the hereditary functional calculus: one can reduce it to

130 Commutative Theory

the Taylor functional calculus [32]. For any operator T on $\mathcal{H}$, one defines operators R_T, L_T on $\mathcal{B}(\mathcal{H})$ to be the operations of multiplication on the right and left, respectively, by T. If T_1, T_2 are commuting operators on $\mathcal{H}$ with spectra contained in $\mathbb{D}$, then $X \stackrel{\text{def}}{=} (R_{T_1}, R_{T_2}, L_{T_1^*}, L_{T_2^*})$ is a commuting 4-tuple of operators on $\mathcal{B}(\mathcal{H})$ with joint spectrum contained in $\mathbb{D}^4$. For $h \in \text{Her}(\mathbb{D}^2)$, the function $(\lambda, \mu) \mapsto h(\lambda, \bar{\mu})$ is analytic on $\mathbb{D}^4$ and so can be evaluated at X by any of the known functional calculi for commuting operators on a Banach space. The result will be an operator Y on $\mathcal{B}(\mathcal{H})$; one can then define $h(T)$ to be the image of the identity operator $1_{\mathcal{B}(\mathcal{H})}$ under Y. In this way the properties of the hereditary functional calculus can be deduced from those of the Taylor calculus.

Section 4.3. The existence of models for functions in the Schur class of the bidisc was proved in [7], which paper also demonstrated the power of models in the function theory of the bidisc (e.g., by the proof of a Pick theorem for the bidisc). The 2-dimensional systems theory approach to the network realization formula was pioneered by A. Kummert [138].

Theorem 4.49 was proved in [5]. Theorems 4.61 and 4.63 were proved in [10] and [45].

5
Carathéodory–Julia Theory on the Disc and the Bidisc

In this chapter we shall show how models can be used to discover a rich generalization of the classical Carathéodory–Julia theory to the bidisc.

5.1 The One-Variable Results

The Carathéodory–Julia theorem describes the geometric behavior of an analytic function φ defined on a smoothly bounded simply connected domain D near a point τ on the boundary of D at which $|\varphi|$ attains its supremum. It dates back to Gaston Julia's 1920 paper [123] and Carathéodory's 1929 paper [56]. By the Riemann mapping theorem, we may assume that $D = \mathbb{D}$, $\tau \in \mathbb{T}$, and after normalization, $\varphi \in \mathscr{S}(\mathbb{D})$.

One perspective on the results is that an a priori very weak regularity condition on φ at τ is equivalent to a much stronger regularity condition at τ. We adopt the following two definitions.

Definition 5.1. (**Julia quotient**) *For $\varphi \in \mathscr{S}(\mathbb{D})$, the Julia quotient J_φ of φ is the function on $\mathbb{D}$ given by the formula*

$$J_\varphi(\lambda) = \frac{1 - |\varphi(\lambda)|}{1 - |\lambda|} \qquad \text{for } \lambda \in \mathbb{D}. \tag{5.2}$$

Definition 5.3. *If $S \subseteq \mathbb{D}$ and $\tau \in \mathbb{T}$, we say that S approaches τ nontangentially and write $S \xrightarrow{\text{nt}} \tau$ if $\tau \in S^-$ and there exists a constant c such that*

$$|\lambda - \tau| \le c(1 - |\lambda|) \tag{5.4}$$

for all $\lambda \in S$. If $\{\lambda_n\}_{n \ge 1}$ is a sequence in $\mathbb{D}$, we write $\lambda_n \xrightarrow{\text{nt}} \tau$ if $\lambda_n \to \tau$ and $\{\lambda_n : n \ge 1\}$ approaches τ non-tangentially. We say that a property holds non-tangentially at τ if it holds on every set $S \subseteq \mathbb{D}$ that approaches τ nontangentially.

131

132 *Commutative Theory*

Now fix $\varphi \in \mathscr{S}(\mathbb{D})$ and $\tau \in \mathbb{T}$, and consider the following increasingly stringent conditions on the behavior of φ near the point τ.

Definition 5.5. (I) τ *is a* carapoint *for φ, that is,*

$$\liminf_{\substack{\lambda \to \tau \\ \lambda \in \mathbb{D}}} J_\varphi(\lambda) < \infty,$$

or equivalently, there exists a sequence $\{\lambda_n\}$ in $\mathbb{D}$ such that $\lambda_n \to \tau$ and

$$\sup_n J_\varphi(\lambda_n) < \infty.$$

(II) $J_\varphi(\lambda)$ *is non-tangentially bounded at τ, that is,*

$$\sup_{\lambda \in S} |J_\varphi(\lambda)| < \infty$$

whenever $S \subseteq \mathbb{D}$ and $S \overset{\text{nt}}{\to} \tau$.

(III) φ *is directionally differentiable at τ, that is, there exists a complex number ω such that, whenever $\delta \in \mathbb{C}$ and $\tau + t\delta \in \mathbb{D}$ for sufficiently small positive t,*

$$\lim_{t \to 0+} \frac{\varphi(\tau + t\delta) - \omega}{t} \text{ exists} \tag{5.6}$$

and, in addition, $|\omega| = 1$.

(IV) φ *is non-tangentially differentiable at τ, that is, there exist complex numbers ω and η such that*

$$\lim_{\substack{\lambda \to \tau \\ \lambda \in S}} \frac{\varphi(\lambda) - \omega - \eta(\lambda - \tau)}{|\lambda - \tau|} = 0 \tag{5.7}$$

whenever $S \subseteq \mathbb{D}$ and $S \overset{\text{nt}}{\to} \tau$, and, in addition, $|\omega| = 1$.

(V) φ *is non-tangentially continuously differentiable at τ, that is, condition (IV) holds and, in addition,*

$$\lim_{\substack{\lambda \to \tau \\ \lambda \in \mathbb{D}}} \varphi'(\lambda) = \eta \tag{5.8}$$

whenever $S \subseteq \mathbb{D}$ and $S \overset{\text{nt}}{\to} \tau$.

Theorem 5.9. (Carathéodory–Julia theorem) *Let $\tau \in \mathbb{T}$ and let $\varphi \in \mathscr{S}(\mathbb{D})$. Then conditions (I)–(V) are equivalent.*

Carathéodory gave the first proof of Theorem 5.9 in 1929 using geometric methods [56]. Since then, treatments of various aspects of the theorem in one

Carathéodory–Julia Theory on the Disc and the Bidisc 133

variable from numerous perspectives have been developed. Notable contributors are Harry Dym [85] and Donald Sarason [181], who gave Hilbert space proofs that (I) implies (V). Their approach involved studying de Branges–Rovnyak spaces (the model spaces that arise as taut models for a function in the Schur class) as reproducing kernel Hilbert spaces.

5.2 The Model Approach to Regularity on $\mathbb{D}$: B-points and C-points

It is surprising that the regularity conditions (I)–(V) for a function φ in $\mathscr{S}(\mathbb{D})$ correspond precisely to simple geometric properties of a model $(\mathcal{M}, u)$ for φ. We begin with the following elegant relation between the Julia quotient of φ and the norm of u.

Lemma 5.10. *Let $\tau \in \mathbb{T}$, $\varphi \in \mathscr{S}(\mathbb{D})$ and let $(\mathcal{M}, u)$ be a model for φ. Then, for all $\lambda \in \mathbb{D}$,*

$$\tfrac{1}{2} J_\varphi(\lambda) \leq \|u_\lambda\|^2 \leq 2 J_\varphi(\lambda).$$

Proof. If $\lambda \in \mathbb{D}$, then

$$1 - |\lambda| \leq 1 - |\lambda|^2 \leq 2(1 - |\lambda|).$$

Likewise, since $\varphi \in \mathscr{S}(\mathbb{D})$, if $\lambda \in \mathbb{D}$, then

$$1 - |\varphi(\lambda)| \leq 1 - |\varphi(\lambda)|^2 \leq 2(1 - |\varphi(\lambda)|).$$

As a consequence of these inequalities it follows that

$$\frac{1}{2} \frac{1 - |\varphi(\lambda)|}{1 - |\lambda|} \leq \frac{1 - |\varphi(\lambda)|^2}{1 - |\lambda|^2} \leq 2 \frac{1 - |\varphi(\lambda)|}{1 - |\lambda|}$$

Hence the proposition follows from the definition (5.2) and by the fact that the model formula (2.8) implies that

$$\|u_\lambda\|^2 = \frac{1 - |\varphi(\lambda)|^2}{1 - |\lambda|^2} \qquad \text{for all } \lambda \in \mathbb{D}.$$

$\square$

As a corollary to this lemma, we are able in Theorem 5.12 below to interpret the regularity conditions (I) and (II) in terms of models.

Definition 5.11. *Let $\mathcal{M}$ be a Hilbert space and assume that $u: \mathbb{D} \to \mathcal{M}$ is a map. We say that $\tau \in \mathbb{T}$ is a weak B-point for u if there exists a sequence $\{\lambda_n\}$ in $\mathbb{D}$ such that $\lambda_n \to \tau$ and*

134 *Commutative Theory*

$$\sup_n \|u_{\lambda_n}\| < \infty.$$

We say that τ is a strong B *point for u if u is nontangentially bounded at τ, that is,*

$$\sup_{\lambda \in S} \|u_\lambda\| < \infty$$

whenever $S \subseteq \mathbb{D}$ and $S \overset{\mathrm{nt}}{\to} \tau$.

Evidently, with the language of this definition, we have the following theorem.

Theorem 5.12. (The B-point theorem on $\mathbb{D}$) *If $\tau \in \mathbb{T}$, $\varphi \in \mathscr{S}(\mathbb{D})$ and $(\mathcal{M}, u)$ is a model for φ, then τ is a carapoint for φ if and only if τ is a weak B-point for u, and J_φ is nontangentially bounded at τ if and only if τ is a strong B point for u.*

Just as there is a simple characterization of the non-tangential boundedness of the Julia quotient in terms of models, so is there a simple characterization of non-tangential differentiability in terms of models.

Definition 5.13. *Let $\mathcal{M}$ be a Hilbert space and let $u \colon \mathbb{D} \to \mathcal{M}$ be a map. We say that $\tau \in \mathbb{T}$ is a* C-point *for u if u extends by weak continuity to be nontangentially continuous at τ, that is, if there exists a vector $u_\tau \in \mathcal{M}$ such that the map $u^\sim \colon \mathbb{D} \cup \{\tau\} \to \mathcal{M}$ defined on $\mathbb{D} \cup \{\tau\}$ by the formula*

$$u_\lambda^{\sim} = \begin{cases} u_\lambda & \text{if } \lambda \in \mathbb{D} \\ u_\tau & \text{if } \lambda = \tau \end{cases} \tag{5.14}$$

is weakly continuous on $S \cup \{\tau\}$ whenever $S \subseteq \mathbb{D}$ and $S \overset{\mathrm{nt}}{\to} \tau$.

In Definition 5.13 to say that $u^\sim$ is weakly continuous on S means that the map $u^\sim | S \cup \{\tau\} \colon S \cup \{\tau\} \to \mathcal{M}$ is continuous when $\mathcal{M}$ is endowed with the weak topology.

Lemma 5.15. *Let $\mathcal{M}$ be a Hilbert space and let $u \colon \mathbb{D} \to \mathcal{M}$ be a map. If $\tau \in \mathbb{T}$ is a* C-point *for u, then τ is a strong B-point for u.*

Proof. Consider any set $S \subseteq \mathbb{D}$ such that $S \overset{\mathrm{nt}}{\to} \tau$. By hypothesis there exists $u_\tau \in \mathcal{M}$ such that the map $u^\sim$ defined by equation (5.14) is continuous with respect to the weak topology on $\mathcal{M}$. Let $\bar{S}$ be the closure of S in $\mathbb{C}$; thus $\bar{S}$ is compact. Notice that, by inequality (5.4), the only point in $\bar{S}$ that is on the unit circle is τ. So $u^\sim$ is weakly continuous on $\bar{S}$, and therefore $u^\sim(\bar{S})$ is weakly compact in $\mathcal{M}$. Thus $u(S)$ is weakly bounded in $\mathcal{M}$. By the fact that weakly bounded sets are bounded in any locally convex space [176, theorem 3.18] (or by the uniform boundedness principle), $\sup_{\lambda \in S} \|u_\lambda\| < \infty$, as desired. $\square$

Carathéodory–Julia Theory on the Disc and the Bidisc 135

Lemma 5.16. *If $\varphi \in \mathscr{S}(\mathbb{D})$, $(\mathcal{M}, u)$ is a model for φ and τ is a C-point for u, then there exists a unique $\omega \in \mathbb{T}$ such that $\varphi(\lambda) \to \omega$ as $\lambda \overset{nt}{\to} \tau$. Furthermore,*

$$1 - \overline{\omega}\varphi(\lambda) = (1 - \bar{\tau}\lambda)\langle u_\lambda, u_\tau \rangle_{\mathcal{M}} \tag{5.17}$$

for all $\lambda \in \mathbb{D}$.

Proof. We first prove the second assertion of the lemma. As $(\mathcal{M}, u)$ is a model for φ,

$$1 - \overline{\varphi(\mu)}\varphi(\lambda) = (1 - \bar{\mu}\lambda)\langle u_\lambda, u_\mu \rangle_{\mathcal{M}} \quad \text{for all } \lambda, \mu \in \mathbb{D}. \tag{5.18}$$

Choose a sequence $\{\mu_n\}$ in $\mathbb{D}$ such that $\mu_n \overset{nt}{\to} \tau$. Since τ is a C-point for u, $u_{\mu_n} \to u_\tau$ weakly in $\mathcal{M}$. Also, as Lemma 5.15 guarantees that τ is a strong B-point for u, $\{J_\varphi(\mu_n)\}$ is bounded, $|\varphi(\mu_n)| \to 1$. Hence by replacing $\{\mu_n\}$ with a subsequence we may assume that there exists $\omega \in \mathbb{T}$ such that $\varphi(\mu_n) \to \omega$. Setting $\mu = \mu_n$ in equation (5.18) and letting $n \to \infty$ yields equation (5.17).

To prove the first assertion of the lemma we argue by contradiction. Suppose $\{\nu_n\}$ is a sequence in $\mathbb{D}$ such that $\nu_n \overset{nt}{\to} \tau$ and $\varphi(\nu_n) \nrightarrow \omega$. But then by replacing $\{\nu_n\}$ with a subsequence, we may assume that $\nu_n \to \eta$ where $\eta \neq \omega$. Setting $\mu = \nu$ in equation (5.18) and letting $n \to \infty$ leads to the formula

$$1 - \overline{\eta}\varphi(\lambda) = (1 - \bar{\tau}\lambda)\langle u_\lambda, u_\tau \rangle_{\mathcal{M}} \quad \text{for all } \lambda \in \mathbb{D},$$

which, together with equation (5.17), implies that $(\bar{\omega} - \bar{\eta})\varphi(\lambda) = 0$ for all $\lambda \in \mathbb{D}$. Since $\omega \neq \eta$, this implies that $\varphi(\lambda) = 0$ for all $\lambda \in \mathbb{D}$, contradicting the fact that $|\varphi(\mu_n)| \to 1$. $\square$

Theorem 5.19. (The C-point theorem on $\mathbb{D}$) *Assume that $\tau \in \mathbb{T}$, $\varphi \in \mathscr{S}(\mathbb{D})$ and $(\mathcal{M}, u)$ is a model for φ. Then τ is a C-point for u if and only if φ satisfies regularity condition* (IV), *that is, φ is non-tangentially differentiable at τ. Moreover, in the notation of Definitions 5.5 and 5.13, $\eta = \bar{\tau}\omega \|u_\tau\|^2$.*

Proof. Let $\tau \in \mathbb{T}$, $\varphi \in \mathscr{S}(\mathbb{D})$, and let $(\mathcal{M}, u)$ be a model for φ. First assume that τ is a C-point for u. By Lemma 5.16 there exists $\omega \in \mathbb{T}$ such that equation (5.17) holds, or equivalently,

$$\omega - \varphi(\lambda) = \bar{\tau}\omega(\tau - \lambda)\langle u_\lambda, u_\tau \rangle_{\mathcal{M}} \quad \text{for all } \lambda \in \mathbb{D}.$$

Consequently,

$$\varphi(\lambda) = \omega + \bar{\tau}\omega(\lambda - \tau)\langle u_\lambda, u_\tau \rangle_{\mathcal{M}}$$
$$= \omega + \bar{\tau}\omega(\lambda - \tau)\langle u_\tau + (u_\lambda - u_\tau), u_\tau \rangle_{\mathcal{M}}$$
$$= \omega + \bar{\tau}\omega\|u_\tau\|^2(\lambda - \tau) + \bar{\tau}\omega(\lambda - \tau)\langle u_\lambda - u_\tau, u_\tau \rangle_{\mathcal{M}}.$$

Commutative Theory

Therefore,

$$\frac{\varphi(\lambda) - \omega - \bar{\tau}\omega(\lambda - \tau)\|u_\tau\|^2}{\lambda - \tau} = -\bar{\tau}\omega \langle u_\lambda - u_\tau, u_\tau \rangle_{\mathcal{M}}$$

But Definition 5.13 implies that

$$\lim_{\substack{\lambda \to \tau \\ \lambda \in S}} \langle u_\lambda - u_\tau, u_\tau \rangle_{\mathcal{M}} = 0$$

whenever $S \subseteq \mathbb{D}$ and $S \overset{\text{nt}}{\to} \tau$. Therefore, if we let $\eta = \bar{\tau}\omega\|u_\tau\|^2$, then equation (5.7) holds.

Conversely, assume that φ satisfies regularity condition (IV). We rewrite the condition in the form

$$\varphi(\lambda) - \omega = \eta(\lambda - \tau) + o(|\lambda - \tau|),$$

which in turn implies that

$$1 - \bar{\omega}\varphi(\lambda) = \tau\bar{\omega}\eta(1 - \bar{\tau}\lambda) + o(|\lambda - \tau|). \tag{5.20}$$

In these formulas $o(|\lambda - \tau|)$ represents a quantity with the property that

$$\lim_{\substack{\lambda \to \tau \\ \lambda \in S}} \frac{o(|\lambda - \tau|)}{|\lambda - \tau|} = 0$$

whenever $S \subseteq \mathbb{D}$ and $S \overset{\text{nt}}{\to} \tau$.

Now, as regularity condition (IV) implies regularity condition (II), Theorem 5.12 implies that τ is a strong B-point for u. Therefore, if $\{\mu_n\}$ is a sequence in $\mathbb{D}$ such that $\mu_n \overset{\text{nt}}{\to} \tau$,

$$\sup_n \|u_{\mu_n}\| < \infty.$$

Since ball $\mathcal{M}$ is compact in the weak topology, it follows that τ will be a C-point for u if we can establish the following claim.

Claim 5.21. Let $\{\mu_n\}$ and $\{\nu_n\}$ be sequences in $\mathbb{D}$ with $\mu_n \overset{\text{nt}}{\to} \tau$ and $\nu_n \overset{\text{nt}}{\to} \tau$. If $u_{\lambda_n} \to x$ weakly in $\mathcal{M}$ and $u_{\nu_n} \to y$ weakly in $\mathcal{M}$, then $x = y$.

To prove this claim, first note that equation (5.20) implies that $\varphi(\mu_n) \to \omega$. We may deduce, by setting $\mu = \mu_n$ in the model formula (5.18) and then letting $n \to \infty$, that

$$1 - \bar{\omega}\varphi(\lambda) = (1 - \bar{\tau}\lambda) \langle u_\lambda, x \rangle_{\mathcal{M}}. \tag{5.22}$$

Combining equations (5.20) and (5.22) we find that

$$\langle u_\lambda, x \rangle_{\mathcal{M}} = \tau\bar{\omega}\eta + \frac{o(|\lambda - \tau|)}{|\lambda - \tau|}. \tag{5.23}$$

Carathéodory–Julia Theory on the Disc and the Bidisc 137

Likewise, using the sequence $\{v_n\}$ we derive the formula

$$\langle u_\lambda, y \rangle_{\mathcal{M}} = \tau \bar{\omega} \eta + \frac{o(|\lambda - \tau|)}{|\lambda - \tau|}. \tag{5.24}$$

Setting $\lambda = \mu_n$ in equation (5.23), setting $\lambda = v_n$ in equation (5.24) and letting $n \to \infty$ we infer that

$$\langle x, x \rangle = \langle y, x \rangle = \langle y, y \rangle.$$

Hence $y = x$. $\square$

5.3 A Proof of the Carathéodory–Julia Theorem on $\mathbb{D}$ via Models

The following theorem gives a surprisingly simple characterization of B-points.

Proposition 5.25. (**The range criterion for** B-**points on** $\mathbb{D}$) *Assume that* $\tau \in \mathbb{T}$, $\varphi \in \mathscr{S}(\mathbb{D})$, $(\mathcal{M}, u)$ *is a model for* φ, *and* (a, β, γ, D) *is a realization of* $(\mathcal{M}, u)$. *Then the following conditions are equivalent.*

(i) τ *is a weak* B-*point for* u.
(ii) $\gamma \in \mathrm{ran}(1 - D\tau)$.
(iii) τ *is a strong* B-*point for* u.

Proof. First assume that τ is a weak B-point for u. We may choose a sequence $\{\lambda_n\}$ in $\mathbb{D}$ satisfying $\sup_n \|u_{\lambda_n}\| < \infty$ and

$$\lambda_n \to \tau. \tag{5.26}$$

Since $\sup_n \|u_{\lambda_n}\| < \infty$ and ball $\mathcal{H}$ is weakly compact, by passing to a subsequence if necessary, we may assume that, in addition to the condition (5.17), there exists $x \in \mathcal{M}$ such that

$$u_{\lambda_n} \to x \text{ weakly in } \mathcal{M}. \tag{5.27}$$

Now, equation (2.26) implies that

$$(1 - D\lambda_n) u_{\lambda_n} = \gamma \quad \text{for all } n.$$

Hence equations (5.26) and (5.27) imply that

$$(1 - D\tau)x = \gamma.$$

In particular, $\gamma \in \mathrm{ran}(1 - D\tau)$.

138 *Commutative Theory*

Now assume that $\gamma \in \operatorname{ran}(1 - D\tau)$. We wish to show that τ is a strong B-point for u. Accordingly, fix $S \subseteq \mathbb{D}$ satisfying $S \overset{\mathrm{nt}}{\to} \tau$ and choose c satisfying inequality (5.4) for all $\lambda \in S$. If we choose $x \in \mathcal{M}$ such that $(1 - D\tau)x = \gamma$, then

$$\begin{aligned}
u_\lambda &= (1 - D\lambda)^{-1}\gamma \\
&= (1 - D\lambda)^{-1}(1 - D\tau)x \\
&= (1 - D\lambda)^{-1}\big(1 - D\lambda + D(\lambda - \tau)\big)x \\
&= x + (1 - D\lambda)^{-1}D(\lambda - \tau)x.
\end{aligned}$$

As D is a contraction,

$$\|(1 - D\lambda)^{-1}\| \le (1 - |\lambda|)^{-1}.$$

Hence if $\lambda \in S$,

$$\begin{aligned}
\|u_\lambda\| &= \|x + (1 - D\lambda)^{-1}D(\lambda - \tau)x\| \\
&\le \|x\| + \|(1 - D\lambda)^{-1}\| \, |\lambda - \tau| \, \|x\| \\
&\le \|x\| + (1 - |\lambda|)^{-1} c(1 - |\lambda|) \, \|x\| \\
&= (1 + c)\|x\|,
\end{aligned}$$

that is, $\sup_\lambda \|u_\lambda\| < \infty$.

To summarize, we have shown that (i) implies (ii) and (ii) implies (iii). As trivially, (iii) implies (i), the proof of the theorem is complete. $\qquad\square$

Lemma 5.28. *Suppose that $(\mathcal{M}, u)$ is a taut model for φ, τ is a B-point for u and $(\alpha, \beta, \gamma, D)$ is a realization of $(\mathcal{M}, u)$. Then $\ker(1 - D\tau) = \{0\}$.*

Proof. Let $\mathcal{K} = \ker(1 - D\tau)$. Since D is a contraction and τ is unimodular, $\mathcal{K}$ is a reducing subspace for D. By Proposition 5.25, $\gamma \in \operatorname{ran}(1 - D\tau)$. Hence γ is perpendicular to $\mathcal{K}$ and so therefore is $u_\lambda = (1 - D\lambda)^{-1}\gamma$ for every $\lambda \in \mathbb{D}$. As the model is taut, this means that $\mathcal{K} = \{0\}$. $\qquad\square$

Lemma 5.29. *If τ is a C-point for u then the function $u^\sim$ of Definition 5.13 is strongly continuous on every set S in $\mathbb{D}$ that approaches τ non-tangentially.*

Proof. Suppose $\lambda \to \tau$ from within S. We know $u_\lambda \to u_\tau$ weakly; we must show it converges in norm, which is the same as proving

$$\|u_\lambda\|^2 - \operatorname{Re}\langle u_\lambda, u_\tau \rangle \to 0.$$

We know from Lemma 5.16 that

$$1 - \bar{\omega}\varphi(\lambda) = (1 - \bar{\tau}\lambda)\langle u_\lambda, u_\tau \rangle.$$

Carathéodory–Julia Theory on the Disc and the Bidisc 139

Subtract and add twice the real part of this expression to obtain

$$(1 - |\lambda|^2)(\|u_\lambda\|^2 - \mathrm{Re}\langle u_\lambda, u_\tau \rangle)$$

$$= 1 - |\varphi(\lambda)|^2 - (1 - |\lambda|^2)\mathrm{Re}\,\langle u_\lambda, u_\tau \rangle$$

$$= 1 - |\varphi(\lambda)|^2 - 2\mathrm{Re}(1 - \bar{\omega}\varphi(\lambda)) + 2\mathrm{Re}[(1 - \bar{\tau}\lambda)\langle u_\lambda, u_\tau \rangle]$$

$$\quad - \mathrm{Re}[(1 - |\lambda|^2)\langle u_\lambda, u_\tau \rangle]$$

$$= -|1 - \bar{\omega}\varphi(\lambda)|^2 + \mathrm{Re}[(1 - 2\bar{\tau}\lambda + |\lambda|^2)\langle u_\lambda, u_\tau \rangle]$$

$$= -|1 - \bar{\omega}\varphi(\lambda)|^2 + |\tau - \lambda|^2\|u_\tau\|^2 + |\tau - \lambda|^2 \mathrm{Re}\langle u_\lambda - u_\tau, u_\tau \rangle.$$

Thus

$$\left| \|u_\lambda\|^2 - \mathrm{Re}\langle u_\lambda, u_\tau \rangle \right| \leq \frac{|1 - \bar{\omega}\varphi(\lambda)|^2}{1 - |\lambda|^2} + \frac{|\tau - \lambda|^2}{1 - |\lambda|^2}\|u_\tau\|^2$$

$$+ \frac{|\tau - \lambda|^2}{1 - |\lambda|^2}|\langle u_\lambda - u_\tau, u_\tau \rangle|.$$

Since $|\tau - \lambda|$ is comparable to $1 - |\lambda|$ for $\lambda \in S$, each of the three terms on the right-hand side is the product of a bounded factor and one tending to zero. Thus the right-hand side tends to zero as $\lambda \to \tau$ in S. $\square$

Proposition 5.30. *Assume that $\tau \in \mathbb{T}$, $\varphi \in \mathscr{S}(\mathbb{D})$ and $(\mathcal{M}, u)$ is a model for φ. If τ is a B-point for u then τ is a C-point for u.*

Proof. Let (a, β, γ, D) be a taut realization of $(\mathcal{M}, u)$. By Theorem 5.12 there exists a vector $u_\tau \in \mathcal{M}$ such that $(1 - D\tau)u_\tau = \gamma$. Fix a sequence $\{\lambda_n\}$ in $\mathbb{D}$ that approaches τ nontangentially. In particular, as τ is a B point, $\{u_{\lambda_n}\}$ is bounded. We claim that $u_{\lambda_n} \to u_\tau$ weakly in $\mathcal{M}$. For if not, as $\{u_{\lambda_n}\}$ is bounded, there exists $x \in \mathcal{M}$ such that $x \neq u_\tau$ and a subsequence $\{n_k\}$ such that $u_{\lambda_{n_k}} \to x$ weakly in $\mathcal{M}$. But then, as $(1 - D\lambda_{n_k})u_{\lambda_{n_k}} = \gamma$ for all n, $(1 - D\tau)x = \gamma$. By Lemma 5.28 this forces $x = u_\tau$, a contradiction. $\square$

The following sequence of steps completes the proof of the Carathéodory–Julia theorem, Theorem 5.9, by model theory.

1. Suppose $\tau \in \mathbb{T}$, $\varphi \in \mathscr{S}(\mathbb{D})$ and let $(\mathcal{M}, u)$ be a model for φ.
2. It is trivial that, in Definition 5.5, (V) $\implies$ (IV) $\implies$ (III) $\implies$ (II) $\implies$ (I). It therefore suffices to show that

$$(I) \implies (II), \quad (II) \implies (IV) \quad \text{and} \quad (IV) \implies (V).$$

3. To prove that (I) $\implies$ (II), suppose (I) holds. By the B-point theorem, Theorem 5.12, τ is a weak B-point for u. Therefore, by the range condition, Proposition 5.25, τ is a strong B-point for u. Hence, by the B-point theorem, (II) holds.

140   *Commutative Theory*

4. For the implication (II) $\implies$ (IV), suppose that (II) holds. By the B-point theorem, τ is a strong B-point for u. Therefore, by Proposition 5.30, τ is a C-point for u. Hence, by the C-point theorem, (IV) holds.

5. To prove that (IV) $\implies$ (V), consider any $S \subseteq \mathbb{D}$ such that $S \overset{\text{nt}}{\to} \tau$. We have

$$\varphi(\lambda) = \omega + \eta(\lambda - \tau) + o(|\lambda - \tau|)$$

for any $\lambda \in S$; we want to show that $\varphi'(\lambda) \to \eta$ as $\lambda \to \tau$, $\lambda \in S$. Since

$$(1 - D\lambda)u_\lambda = \gamma = (1 - D\tau)u_\tau$$
$$= (1 - D\lambda)u_\tau + D(\lambda - \tau)u_\tau,$$

we have

$$u_\lambda = u_\tau + (1 - D\lambda)^{-1} D(\lambda - \tau)u_\tau. \tag{5.31}$$

Since τ is a C-point, we deduce from Lemma 5.29 that

$$(1 - D\lambda)^{-1} D(\lambda - \tau)u_\tau \to 0 \tag{5.32}$$

strongly as $\lambda \to \tau$ in S. Differentiating the equation

$$1 - \bar{\omega}\varphi(\lambda) = (1 - \lambda\bar{\tau})\langle u_\lambda, u_\tau \rangle$$

with respect to λ and using the fact that $\eta = \omega\bar{\tau}\|u_\tau\|^2$, we find that

$$\varphi'(\lambda) = \omega\bar{\tau}\langle u_\lambda, u_\tau \rangle - \omega(1 - \lambda\bar{\tau})\left\langle \frac{d}{d\lambda}u_\lambda, u_\tau \right\rangle$$
$$= \eta + \omega\bar{\tau}\langle u_\lambda - u_\tau, u_\tau \rangle + \omega\bar{\tau}(\lambda - \tau)\left\langle \frac{d}{d\lambda}u_\lambda, u_\tau \right\rangle.$$

The second term in the right-hand side tends to 0 as $\lambda \to \tau$, so it is enough to show that

$$(\lambda - \tau)\frac{d}{d\lambda}u_\lambda$$

tends weakly to 0.

Differentiating equation (5.31) with respect to λ and multiplying by $(\lambda - \tau)$, we get

$$(\lambda - \tau)\frac{d}{d\lambda}u_\lambda = [(1 - D\lambda)^{-1} D(\lambda - \tau) + 1](1 - D\lambda)^{-1} D(\lambda - \tau)u_\tau.$$

The norm of the expression in brackets is bounded by

$$\frac{|\lambda - \tau|}{1 - |\lambda|} + 1,$$

Carathéodory–Julia Theory on the Disc and the Bidisc 141

which is bounded on S. Therefore by the limit relation (5.32) we can conclude that

$$(\lambda - \tau)\frac{d}{d\lambda}u_\lambda \to 0$$

in norm. □

5.4 Pick Interpolation on the Boundary

Let E be a finite or infinite subset of the unit circle and suppose we are given a unimodular function ψ defined on E. When can ψ be extended to a function $\varphi \in \mathscr{S}(\mathbb{D})$, in the sense that each point of E is a B-point of φ and the non-tangential limit of φ agrees with ψ on E? The condition should be that the Pick matrix be positive. But the diagonal is no longer well-defined. The answer is: it can be done if and only if there exist diagonal entries that make the Pick matrix positive. The proof of the following theorem starts like that of Theorem 2.57, but one has to be careful to ensure that φ does indeed have the right boundary values.

Theorem 5.33. *Let $E \subset \mathbb{T}$, and let $\psi \colon E \to \mathbb{T}$ be a function. There exists a function φ in the Schur class such that every point of E is a B-point of φ and*

$$\lim_{\lambda \overset{nt}{\to} \tau} \varphi(\lambda) = \psi(\tau)$$

if and only if there exists a function $\Delta \colon E \to \mathbb{R}^+$ such that

$$P(\tau,\sigma) \overset{def}{=} \begin{cases} \dfrac{1 - \psi(\tau)\overline{\psi(\sigma)}}{1 - \tau\overline{\sigma}} & \text{if } \tau \neq \sigma \\[2mm] \Delta(\tau) & \text{if } \tau = \sigma \end{cases}$$

is a positive semidefinite kernel on E.

Proof. (Necessity) Suppose such a φ exists. Let $(\mathcal{M},u)$ be a model for φ. By Definition 5.13, for each $\tau \in E$ there exists a vector u_τ in $\mathcal{M}$ such that $u^\sim$ defined by equation (5.14) is nontangentially weakly continuous at τ. Define $\Delta \colon E \to \mathbb{R}$ by $\Delta(\tau) = \|u_\tau\|^2$. Then $P(\tau,\sigma) = \langle u_\tau, u_\sigma \rangle$, since

$$\langle u_\tau, u_\sigma \rangle = \lim_{t \uparrow 1} \lim_{s \uparrow 1} \langle u_{t\tau}, u_{s\sigma} \rangle$$

$$= \lim_{t \uparrow 1} \lim_{s \uparrow 1} \frac{1 - \varphi(t\tau)\overline{\varphi(s\sigma)}}{1 - ts\tau\overline{\sigma}}$$

$$= \frac{1 - \psi(\tau)\overline{\psi(\sigma)}}{1 - \tau\overline{\sigma}}.$$

Thus P is a gramian, and so it is positive semi-definite.

142   *Commutative Theory*

(Sufficiency) Suppose Δ exists. Since P is positive, we can find vectors u_τ in some Hilbert space $\mathcal{M}$ such that

$$P(\tau,\sigma) = \langle u_\tau, u_\sigma \rangle_\mathcal{M}$$

and $\|u_\tau\|^2 = \Delta(\tau)$ for all $\sigma, \tau \in E$. Define V in the usual way by

$$V \begin{pmatrix} 1 \\ \tau u_\tau \end{pmatrix} = \begin{pmatrix} \psi(\tau) \\ u_\tau \end{pmatrix} \qquad \text{for all } \tau \in E,$$

and observe that V is an isometry on the span of the vectors on which it is defined. Extend it to a partial isometry on $\mathbb{C} \oplus \mathcal{M}$, and write it in $ABCD$ form:

$$V = \begin{bmatrix} a & 1 \otimes \beta \\ \gamma \otimes 1 & D \end{bmatrix} = \begin{bmatrix} A & B \\ C & D \end{bmatrix} : \mathbb{C} \oplus \mathcal{M} \to \mathbb{C} \oplus \mathcal{M}.$$

Note that, by construction,

$$\psi(\tau) = A + B\tau(1 - D\tau)^{-1}C \qquad \text{for all } \tau \in E.$$

Let φ be the function in $\mathscr{S}(\mathbb{D})$ which has the realization

$$\varphi(\lambda) = A + B\lambda(1 - D\lambda)^{-1}C$$

and let

$$\mathcal{K} = \bigvee \{\ker(1 - \tau D): \tau \in E\}.$$

Notice that since V is a contraction, so is D, and therefore any eigenvector of D with eigenvalue τ of unit modulus is also an eigenvector of D^* with eigenvalue $\bar{\tau}$. Since $B^*B + D^*D \leq 1$, and $CC^* + DD^* \leq 1$, it follows that any vector in $\mathcal{K}$ is in the kernel of both B and C^*. Let P be projection onto $\mathcal{K}^\perp$. Since $BP^\perp = 0 = P^\perp C$, and $DP^\perp = P^\perp D$, we can take the equations

$$A + B\lambda u_\lambda = \varphi(\lambda)$$

$$C + D\lambda u_\lambda = u_\lambda$$

and replace B by BP and multiply the second equation by P to get

$$A + B\lambda P u_\lambda = \varphi(\lambda)$$

$$C + D\lambda P u_\lambda = P u_\lambda.$$

In other words, without loss of generality we can assume that the model is taut and $\mathcal{K} = 0$ (by replacing u_λ with $P u_\lambda$). The advantage of tautness is that even if $1 - \tau D$ is not invertible, it has no kernel, and therefore it has a left inverse. Thus we have

$$(1 - \lambda D)^{-1}C = u_\lambda \qquad \text{for all } \lambda \in \mathbb{D} \cup E.$$

Carathéodory–Julia Theory on the Disc and the Bidisc 143

For τ in E and $0 < t < 1$, we have

$$u_{t\tau} - u_\tau = (1 - t\tau D)^{-1}[(1 - \tau D) - (1 - t\tau D)](1 - \tau D)^{-1}C$$
$$= -(1 - t)(1 - t\tau D)^{-1}[\tau D]\, u_\tau.$$

As

$$\|(1 - t)(1 - t\tau D)^{-1}\| \leq 1,$$

we conclude that

$$\|u_{t\tau} - u_\tau\| \leq \|u_\tau\|.$$

Therefore τ is a weak B-point for φ, and hence, by Proposition 5.30, τ is a C-point for φ, and u_λ tends weakly to u_τ as λ tends nontangentially to τ. Therefore

$$\psi(\tau) = A + B\tau u_\tau$$
$$= A + \lim_{\lambda \overset{\mathrm{nt}}{\to} \tau} B\lambda u_\lambda$$
$$= \lim_{\lambda \overset{\mathrm{nt}}{\to} \tau} \varphi(\lambda),$$

as desired. $\qquad\square$

Remark 5.34. You can *always* find a Δ to make P positive if E is a finite set. Given such a Δ, you can then solve the Pick problem with the extra condition that the angular derivative $|\varphi'(\tau)| \leq \Delta(\tau)$. In [182], Sarason gave necessary and sufficient conditions for the solvability of the Pick problem on a finite subset of $\mathbb{T}$, where both the function value and the angular derivative are specified at each interpolation point.

5.5 Regularity, B-points and C-points on the Bidisc

Many of the ideas in the proof of the Carathéodory–Julia theorem presented in the previous sections extend to the bidisc by modification of the notation. There are, however, a few new wrinkles.

5.5.1 The Regularity Conditions

We assume that $\varphi \in \mathscr{S}(\mathbb{D}^2)$ and $\tau \in \mathbb{T}^2$. In one variable we may interpret the Julia quotient geometrically as the ratio between the distance from $\varphi(\lambda)$ to the boundary of $\mathbb{D}$ and the distance from λ to the boundary of $\mathbb{D}$, that is,

$$J_\varphi(\lambda) = \frac{1 - |\varphi(\lambda)|}{1 - |\lambda|} = \frac{\mathrm{dist}(\varphi(\lambda), \partial\mathbb{D})}{\mathrm{dist}(\lambda, \partial\mathbb{D})}.$$

144   *Commutative Theory*

Furthermore, if $\lambda = (\lambda^1, \lambda^2) \in \mathbb{D}^2$, then

$$\mathrm{dist}(\lambda, \partial(\mathbb{D}^2)) = \min\{1 - |\lambda^1|, 1 - |\lambda^2|\}.$$

Therefore, the following definition takes the place of Definition 5.1.

Definition 5.35. (**Julia quotient on** $\mathbb{D}^2$) *For* $\varphi \in \mathscr{S}(\mathbb{D}^2)$, *we define* J_φ, *the Julia quotient of* φ, *to be the function on* $\mathbb{D}^2$ *given by the formula,*

$$J_\varphi(\lambda) = \frac{1 - |\varphi(\lambda)|}{\min\{1 - |\lambda^1|, 1 - |\lambda^2|\}}. \tag{5.36}$$

In similar fashion, we replace Definition 5.3 with the following.

Definition 5.37. (**Nontangential convergence in** $\mathbb{D}^2$) *If* $S \subseteq \mathbb{D}^2$ *and* $\tau \in \mathbb{T}^2$, *we say that* S *approaches* τ *non-tangentially and write* $S \overset{\mathrm{nt}}{\to} \tau$ *if* $\tau \in S^-$ *and there exists a constant* c *such that*

$$\max\{|\lambda^1 - \tau^1|, |\lambda^2 - \tau^2|\} \le c \min\{1 - |\lambda^1|, 1 - |\lambda^2|\} \tag{5.38}$$

for all $\lambda \in S$. *If* $\{\lambda_n\}_{n \ge 1}$ *is a sequence in* $\mathbb{D}^2$, *we write* $\lambda_n \overset{\mathrm{nt}}{\to} \tau$ *if* $\lambda_n \to \tau$ *and* $\{\lambda_n : n \ge 1\}$ *approaches* τ *non-tangentially. We say that a property holds non-tangentially at* τ *if it holds on every set* $S \subseteq \mathbb{D}^2$ *that approaches* τ *non-tangentially.*

The regularity conditions can then be formulated as follows.

Regularity conditions on $\mathbb{D}^2$.

(I) τ is a *carapoint for* φ if

$$\liminf_{\substack{\lambda \to \tau \\ \lambda \in \mathbb{D}}} J_\varphi(\lambda) < \infty,$$

or equivalently, if there exists a sequence $\{\lambda_n\}$ in $\mathbb{D}^2$ such that $\lambda_n \to \tau$ and

$$\sup_n J_\varphi(\lambda_n) < \infty.$$

(II) $J_\varphi(\lambda)$ is *non-tangentially bounded at* τ if

$$\sup_{\lambda \in S} |J_\varphi(\lambda)| < \infty$$

whenever $S \subseteq \mathbb{D}$ and $S \overset{\mathrm{nt}}{\to} \tau$.

(III) φ is *directionally differentiable at* τ if there exists a complex number ω of unit modulus such that, whenever $\delta \in \mathbb{C}^2$ is such that $\tau + t\delta \in \mathbb{D}^2$ for sufficiently small positive t, then

$$\lim_{t \to 0+} \frac{\varphi(\tau + t\delta) - \omega}{t} \quad \text{exists.} \tag{5.39}$$

Carathéodory–Julia Theory on the Disc and the Bidisc 145

(IV) φ is *non-tangentially differentiable at* τ if there exist $\omega \in \mathbb{C}$ of unit modulus and $\eta \in \mathbb{C}^2$ such that

$$\lim_{\substack{\lambda \to \tau \\ \lambda \in S}} \frac{\varphi(\lambda) - \omega - \eta \cdot (\lambda - \tau)}{|\lambda - \tau|} = 0 \tag{5.40}$$

whenever $S \subseteq \mathbb{D}$ and $S \overset{\text{nt}}{\to} \tau$. Here

$$\eta \cdot (\lambda - \tau) \overset{\text{def}}{=} \eta^1(\lambda^1 - \tau^1) + \eta^2(\lambda^2 - \tau^2).$$

(V) φ is *non-tangentially continuously differentiable at* τ, that is, (IV) holds and in addition

$$\lim_{\substack{\lambda \to \tau \\ \lambda \in \mathbb{D}}} \nabla \varphi(\lambda) = \eta$$

whenever $S \subseteq \mathbb{D}$ and $S \overset{\text{nt}}{\to} \tau$.

5.5.2 B- and C-points on the Bidisc

B-points and C-points work in much the same way on the bidisc as they do on the disc. Note first that the definitions of B- and C-points involve a Hilbert-space-valued function defined on $\mathbb{D}$. In the new setting we modify the definitions to apply to functions defined on $\mathbb{D}^2$.

Definition 5.41. *Let* $\mathcal{M}$ *be a Hilbert space and assume that* $u \colon \mathbb{D}^2 \to \mathcal{M}$ *is a function. We say that* $\tau \in \mathbb{T}^2$ *is a* weak B-point *for* u *if there exists a sequence* $\{\lambda_n\}$ *in* $\mathbb{D}$ *such that* $\lambda_n \to \tau$ *and*

$$\sup_n \|u_{\lambda_n}\| < \infty.$$

We say that τ *is a* strong B-point *for* u *if* u *is nontangentially bounded at* τ, *that is,*

$$\sup_{\lambda \in S} \|u_\lambda\| < \infty$$

whenever $S \subseteq \mathbb{D}^2$ *and* $S \overset{\text{nt}}{\to} \tau$.

Definition 5.42. *Let* $\mathcal{M}$ *be a Hilbert space and assume that* $u \colon \mathbb{D}^2 \to \mathcal{M}$ *is a function. We say that* $\tau \in \mathbb{T}^2$ *is a* C-point *for* u *if* u *extends by weak continuity to be non-tangentially continuous at* τ, *that is, there exists a vector* $u_\tau \in \mathcal{M}$ *such that the function* $u^{\sim} \colon \mathbb{D}^2 \cup \{\tau\} \to \mathcal{M}$ *defined on* $\mathbb{D}^2 \cup \{\tau\}$ *by the formula*

$$u_\lambda^{\sim} = \begin{cases} u_\lambda & \text{if } \lambda \in \mathbb{D}^2 \\ u_\tau & \text{if } \lambda = \tau \end{cases}$$

is weakly continuous on $S \cup \{\tau\}$ *whenever* $S \subseteq \mathbb{D}^2$ *and* $S \overset{\text{nt}}{\to} \tau$.

146　　　　　　　　　　　*Commutative Theory*

With these simple modifications to Definitions 5.11 and 5.13, the following adaptations of the B-point theorem (Theorem 5.12) and the C-point theorem (Theorem 5.19) to the bidisc are true.

Theorem 5.43. (The B-point theorem on $\mathbb{D}^2$) *If $\tau \in \mathbb{T}^2$, $\varphi \in \mathscr{S}(\mathbb{D}^2)$ and $(\mathcal{M}, u)$ is a model for φ, then the following conditions are equivalent.*

(i) *τ is a carapoint for φ;*
(ii) *τ is a weak B-point for u and J_φ is nontangentially bounded at τ;*
(iii) *τ is a strong B-point for u.*

Theorem 5.44. (The C-point theorem on $\mathbb{D}^2$) *Assume that $\tau \in \mathbb{T}^2$, $\varphi \in \mathscr{S}(\mathbb{D}^2)$ and $(\mathcal{M}, u)$ is a model for φ. Then τ is a C-point for u if and only if φ satisfies Regularity condition (IV), that is, φ is non-tangentially differentiable at τ.*

Theorems 5.43 and 5.44 are surprising because models for φ are in general not unique, so can contain some noise. Nonetheless, the property of having a B-point or C-point does not depend on the choice of model for a given function.

5.6 The Missing Link

The proof of the Carathéodory–Julia theorem in one variable presented in Section 5.3 had three ingredients:

1. repeated applications of the B- and C-point theorems;
2. the fact that if τ is a weak B-point, then τ is a strong B-point (the range criterion for B-points, Proposition 5.25);
3. the fact that if τ is a strong B-point, then τ is a C-point.

As pointed out in the previous section, the B- and C-point theorems survive to the bidisc. The range criterion for B-points is true on the bidisc as well. To formulate the range criterion for the bidisc, simply replace $\mathbb{D}$ with $\mathbb{D}^2$ and $\mathbb{T}$ with $\mathbb{T}^2$ in the statement of Proposition 5.25. To prove the range criterion for the bidisc, follow the proof of Proposition 5.25, replacing the two appearances of $\mathbb{D}$ in the proof with $\mathbb{D}^2$ and the appearances of $|\lambda|$ and $|\lambda - \tau|$ in the estimates at the end of the proof with

$$\max\{|\lambda^1|, |\lambda^2|\} \quad \text{and} \quad \max\{|\lambda^1 - \tau^1|, |\lambda^2 - \tau^2|\},$$

respectively.

The third ingredient, that B-points are C-points does not carry over to the bidisc. Rather, instead of B-points guaranteeing non-tangential differentiability

Carathéodory–Julia Theory on the Disc and the Bidisc 147

(Regularity condition IV), B-points guarantee only non-tangential directional differentiability (Regularity condition III). In other words, the limit in (III) is not necessarily linear in δ.

The missing link is Lemma 5.28. It is possible that the set of limits of sequences u_{λ_n} as $\lambda_n \to \tau$ non-tangentially is larger than a single vector.

Theorem 5.45. (Carathéodory–Julia theorem for the bidisc) *Let $\varphi \in \mathscr{S}(\mathbb{D}^2)$ and let $\tau \in \mathbb{T}^2$. Regularity conditions* (I)–(III) *are equivalent, and conditions* (IV) *and* (V) *are equivalent.*

For a proof see [23].

Example 5.46. The function

$$\varphi(\lambda) = \frac{2\lambda^1 \lambda^2 - \lambda_1 - \lambda_2}{2 - \lambda_1 - \lambda_2}$$

belongs to $\mathscr{S}(\mathbb{D}^2)$ and has a B-point at $(1,1)$ that is not a C-point.

5.7 Historical Notes

Julia proved in [123] that if $\varphi \in \mathscr{S}(\mathbb{D})$ and is regular at a boundary point τ, then there exists c such that

$$\left| \frac{\varphi(\lambda) - \varphi(\tau)}{\lambda - \tau} \right|^2 \leq c \, \frac{1 - |\varphi(\lambda)|^2}{1 - |\varphi(\tau)|^2}. \tag{5.47}$$

He noted that regularity could be weakened. Carathéodory proved [56] that condition (I) in Definition 5.5 is sufficient, with c equal to the lim inf in the Julia quotient. Dym [85] and Sarason [181] gave Hilbert space proofs of Theorem 5.9. Sarason pointed out that Julia's inequality (5.47), which is a boundary version of Schwarz's inequality, is a consequence of Cauchy–Schwarz: $|\langle u_\lambda, u_\tau \rangle|^2 \leq \|u_\lambda\|^2 \|u_\tau\|^2$. He quipped that this was an example of the conservation of Schwarz.

K. Włodarczyk [213], F. Jafari [120], and M. Abate [1] obtained partial analogs of the Julia–Carathéodory theorem for the polydisc. Theorems 5.43 and 5.44 are proved in [23]. Further results about B-points and C-points on the bidisc are given in [22, 25, 147, 205].

6
Herglotz and Nevanlinna Representations in Several Variables

The techniques discussed in the previous chapters are restricted neither to the study of Schur functions nor to the disc and the bidisc. In this chapter we analyze some classes of functions on other domains.

6.1 Overview

We adopt the notation $\mathbb{H}$ for the upper half plane,

$$\mathbb{H} = \{z \in \mathbb{C}: \operatorname{Im} z > 0\}.$$

For U a domain in $\mathbb{C}^d$, we define $\mathscr{H}(U)$, the *Herglotz class on* U, by

$$\mathscr{H}(U) = \{h \in \operatorname{Hol}(U): \operatorname{Re} h(\lambda) \geq 0 \text{ for all } \lambda \in U\},$$

and define $\mathscr{P}(U)$, the *Pick class on* U, by

$$\mathscr{P}(U) = \{f \in \operatorname{Hol}(U): \operatorname{Im} f(\lambda) \geq 0 \text{ for all } \lambda \in U\}.$$

The classes $\mathscr{H}(\mathbb{D})$ and $\mathscr{P}(\mathbb{H})$ are the subjects of three famous theorems. The first of these, due to Herglotz [114], dates back to 1911.

Theorem 6.1. (Herglotz representation theorem, function theory version) *A function h belongs to $\mathscr{H}(\mathbb{D})$ if and only if there exist $b \in \mathbb{R}$ and a finite non-negative Borel measure μ on $\mathbb{T}$ such that*

$$h(\lambda) = ib + \int_{\mathbb{T}} \frac{1 + \tau\lambda}{1 - \tau\lambda} \, d\mu(\tau) \quad \text{for all } \lambda \in \mathbb{D}. \tag{6.2}$$

The other two theorems, due to Nevanlinna [155], were published in 1922.

Herglotz and Nevanlinna Representations in Several Variables 149

Theorem 6.3. (Nevanlinna's first representation theorem) *A function* f *belongs to* $\mathscr{P}(\mathbb{H})$ *if and only if there exist* $a \in \mathbb{R}$, $b \geq 0$, *and a finite non-negative Borel measure* μ *on* $\mathbb{R}$ *such that*

$$f(z) = a + bz + \int_{\mathbb{R}} \frac{1 + tz}{t - z} \, d\mu(t) \quad \text{for all } z \in \mathbb{H}. \tag{6.4}$$

Theorem 6.5. (Nevanlinna's second representation theorem) *A function* f *satisfies* $f \in \mathscr{P}(\mathbb{H})$ *and*

$$\liminf_{y \to \infty} y|h(iy)| < \infty \tag{6.6}$$

if and only if there exists a finite non-negative Borel measure μ *on* $\mathbb{R}$ *such that*

$$f(z) = \int_{\mathbb{R}} \frac{1}{t - z} \, d\mu(t) \quad \text{for all } z \in \mathbb{H}. \tag{6.7}$$

In this chapter we shall show how the model theory methods of Chapter 4 can be adapted to generalize the Herglotz representation theorem to $\mathscr{H}(\mathbb{D}^2)$ and the Nevanlinna representation theorems to $\mathscr{P}(\mathbb{H}^2)$.

6.2 The Herglotz Representation on $\mathbb{D}^2$

A key to extending the Herglotz representation to $\mathbb{D}^2$ is to reformulate the one-variable theorem in operator-theoretic terms. Notice that, for μ as in Theorem 6.1, we may define a unitary operator $U \in \mathcal{B}(L^2(\mu))$ by the formula

$$(Uf)(\tau) = \tau f(\tau), \qquad \text{for } f \in L^2(\mu) \text{ and } \mu\text{-almost all } \tau \in \mathbb{T}.$$

Furthermore, if $\gamma \in L^2(\mu)$ denotes the constant function $\gamma(\tau) = 1$ for all $\tau \in \mathbb{T}$, then

$$\left\langle \frac{1 + U\lambda}{1 - U\lambda} \gamma, \gamma \right\rangle_{L^2(\mu)} = \int_{\mathbb{T}} \frac{1 + \tau\lambda}{1 - \tau\lambda} \, d\mu(\tau) \quad \text{for all } \lambda \in \mathbb{D}. \tag{6.8}$$

Thus equation (6.2) can be reformulated to read

$$h(\lambda) = ib + \left\langle \frac{1 + U\lambda}{1 - U\lambda} \gamma, \gamma \right\rangle_{L^2(\mu)} \qquad \text{for all } \lambda \in \mathbb{D}, \tag{6.9}$$

where U is a unitary operator acting on $L^2(\mu)$, and γ is a vector in $L^2(\mu)$. Conversely, assume that $\mathcal{M}$ is a Hilbert space, $U \in \mathcal{B}(\mathcal{M})$ is unitary, and $\gamma \in \mathcal{M}$. If E denotes the spectral measure of U and $d\mu = \langle dE\gamma, \gamma \rangle$, then equation (6.8) holds (with the inner product formed in $\mathcal{M}$ rather than $L^2(\mu)$). These remarks establish that Theorem 6.1 is equivalent to the following theorem.

150   Commutative Theory

Theorem 6.10. (**Herglotz representation theorem, operator theory version**) *A function h belongs to $\mathcal{H}(\mathbb{D})$ if and only if there exist $b \in \mathbb{R}$, a Hilbert space $\mathcal{M}$, a unitary operator $U \in \mathcal{B}(\mathcal{M})$, and a vector $\gamma \in \mathcal{M}$ such that*

$$h(\lambda) = ib + \left\langle \frac{1+U\lambda}{1-U\lambda} \gamma, \gamma \right\rangle_{\mathcal{M}} \quad \text{for all } \lambda \in \mathbb{D}. \tag{6.11}$$

In this formulation, Herglotz's theorem immediately generalizes by a mere change in the interpretation of λ.

Theorem 6.12. (**Herglotz representation theorem on $\mathbb{D}^2$**) *A function h belongs to $\mathcal{H}(\mathbb{D}^2)$ if and only if there exist $b \in \mathbb{R}$, a decomposed Hilbert space $\mathcal{M} = \mathcal{M}_1 \oplus \mathcal{M}_2$, a unitary operator $U \in \mathcal{B}(\mathcal{M})$, and $\gamma \in \mathcal{M}$ such that*

$$h(\lambda) = ib + \left\langle \frac{1+U\lambda}{1-U\lambda} \gamma, \gamma \right\rangle_{\mathcal{M}} \quad \text{for all } \lambda \in \mathbb{D}^2. \tag{6.13}$$

In this formula λ is interpreted as in equation (4.20), that is, λ acts as the direct sum of multiplication by λ^1 on $\mathcal{M}_1$ and by λ^2 on $\mathcal{M}_2$.

Proof. First assume that h is as in equation (6.13). As U is unitary, in particular $1 - U\lambda$ is invertible for all $\lambda \in \mathbb{D}^2$. Therefore, $h \in \text{Hol}(\mathbb{D}^2)$. There remains to show that $\text{Re}\, h(\lambda) \geq 0$ for all $\lambda \in \mathbb{D}^2$. But if $\mathcal{H}$ is a Hilbert space, $T \in \mathcal{B}(\mathcal{H})$, and $u \in \mathcal{H}$,

$$\text{Re}\ \langle Tu,u\rangle_{\mathcal{H}} = \tfrac{1}{2}\left(\langle Tu,u\rangle_{\mathcal{H}} + \overline{\langle Tu,u\rangle}_{\mathcal{H}} \right)$$
$$= \tfrac{1}{2}\left(\langle Tu,u\rangle_{\mathcal{H}} + \langle u,Tu\rangle_{\mathcal{H}} \right)$$
$$= \tfrac{1}{2}\left(\langle Tu,u\rangle_{\mathcal{H}} + \langle T^*u,u\rangle_{\mathcal{H}} \right)$$
$$= \left\langle \tfrac{1}{2}(T + T^*)u,u \right\rangle_{\mathcal{H}}$$
$$= \langle (\text{Re}\,T)u,u\rangle_{\mathcal{H}},$$

the last step being the *definition* of the operator $\text{Re}\,T$. Also, if $\lambda \in \mathbb{D}^2$,

$$\text{Re}\ \frac{1+U\lambda}{1-U\lambda} = \frac{1}{2}\left[\frac{1+U\lambda}{1-U\lambda} + \left(\frac{1+U\lambda}{1-U\lambda} \right)^* \right]$$
$$= \frac{1}{2} \frac{1}{(1-U\lambda)^*}\left[(1-U\lambda)^*(1+U\lambda) + (1+U\lambda)^*(1-U\lambda) \right]\frac{1}{1-U\lambda}$$
$$= \frac{1}{2} \frac{1}{(1-U\lambda)^*}\left[2 - 2\lambda^*\lambda \right]\frac{1}{1-U\lambda}$$
$$= \frac{1}{(1-U\lambda)^*}\left[1 - \lambda^*\lambda \right]\frac{1}{1-U\lambda}.$$

Therefore, if $\lambda \in \mathbb{D}^2$,

$$
\begin{aligned}
\mathrm{Re}\,h(\lambda) &= \mathrm{Re}(ib) + \mathrm{Re}\left\langle \frac{1+U\lambda}{1-U\lambda}\,\gamma,\gamma \right\rangle_{\mathcal{M}} \\
&= \left\langle \mathrm{Re}\left(\frac{1+U\lambda}{1-U\lambda}\right)\gamma,\gamma \right\rangle_{\mathcal{M}} \\
&= \left\langle \frac{1}{(1-U\lambda)^*}[1-\lambda^*\lambda]\frac{1}{1-U\lambda}\,\gamma,\gamma \right\rangle_{\mathcal{M}} \\
&= \left\langle [1-\lambda^*\lambda]\frac{1}{1-U\lambda}\,\gamma, \frac{1}{1-U\lambda}\gamma \right\rangle_{\mathcal{M}} \\
&\geq 0.
\end{aligned}
$$

This completes the proof that $h \in \mathcal{H}(\mathbb{D}^2)$ under the assumption that h has the form (6.13).

Conversely, assume that $h \in \mathcal{H}(\mathbb{D}^2)$. We first handle the case where

$$
h(0) = 1. \tag{6.14}
$$

For this case we show that there exist a Hilbert space $\mathcal{M}$, a unitary U on $\mathcal{M}$, and a vector $\gamma \in \mathcal{M}$ with $\|\gamma\| = 1$ and such that equation (6.11) holds with $b = 0$. We make use of the conformal map $z \mapsto w$ from $\mathbb{D}$ onto the right half plane, $-i\,\mathbb{H}$, where z and w are related by the formulas

$$
w = \frac{1+z}{1-z} \quad \text{and} \quad z = \frac{w-1}{w+1}.
$$

Evidently,

$$
h \in \mathcal{H}(\mathbb{D}^2) \iff \frac{h-1}{h+1} \in \mathcal{S}(\mathbb{D}^2).
$$

Therefore, by Theorem 4.35, $(h-1)/(h+1)$ has a model on $\mathbb{D}^2$, that is, there exist a decomposed Hilbert space $\mathcal{M}$ and a function $u \colon \mathbb{D}^2 \to \mathcal{M}$ such that

$$
1 - \frac{\overline{h(\mu)}-1}{\overline{h(\mu)}+1}\,\frac{h(\lambda)-1}{h(\lambda)+1} = \left\langle (1-\mu^*\lambda)u_\lambda, u_\mu \right\rangle_{\mathcal{M}} \quad \text{for all } \lambda,\mu \in \mathbb{D}^2. \tag{6.15}
$$

Multiply both sides of equation (6.15) by $(\overline{h(\mu)}+1)(h(\lambda)+1)$ and set $v_\lambda = (h(\lambda)+1)u_\lambda$ to obtain

$$
\begin{aligned}
(\overline{h(\mu)}+1)(h(\lambda)+1) &- (\overline{h(\mu)}-1)(h(\lambda)-1) \\
&= \left\langle (1-\mu^*\lambda)v_\lambda, v_\mu \right\rangle_{\mathcal{M}} \quad \text{for all } \lambda,\mu \in \mathbb{D}^2. \tag{6.16}
\end{aligned}
$$

152 Commutative Theory

Reshuffle equation (6.16) to

$$\left\langle \begin{bmatrix} h(\lambda)-1 \\ v_\lambda \end{bmatrix}, \begin{bmatrix} h(\mu)-1 \\ v_\mu \end{bmatrix} \right\rangle = \left\langle \begin{bmatrix} h(\lambda)+1 \\ \lambda v_\lambda \end{bmatrix}, \begin{bmatrix} h(\mu)+1 \\ \mu v_\mu \end{bmatrix} \right\rangle \quad \text{for all } \lambda, \mu \in \mathbb{D}^2. \tag{6.17}$$

By a trivial extension of $(\mathcal{M}, u)$ if necessary, it follows as in Remark 2.31 that there exists a unitary $W \in \mathcal{B}(\mathbb{C} \oplus \mathcal{M})$ such that

$$W \begin{bmatrix} h(\lambda)+1 \\ \lambda v_\lambda \end{bmatrix} = \begin{bmatrix} h(\lambda)-1 \\ v_\lambda \end{bmatrix} \quad \text{for all } \lambda \in \mathbb{D}^2.$$

If we write W as a block matrix

$$W = \begin{bmatrix} a & 1 \otimes \beta \\ \gamma \otimes 1 & D \end{bmatrix} : \mathbb{C} \oplus \mathcal{M} \to \mathbb{C} \oplus \mathcal{M},$$

we obtain the equations

$$h(\lambda) - 1 = a(h(\lambda) + 1) + \langle \lambda v_\lambda, \beta \rangle \tag{6.18}$$

and

$$v_\lambda = (h(\lambda) + 1)\gamma + D\lambda v_\lambda. \tag{6.19}$$

Now, setting $\lambda = 0$ in equation (6.18) and recalling our assumption (6.14) we deduce that $a = 0$, so that equation (6.18) becomes

$$h(\lambda) - 1 = \langle \lambda v_\lambda, \beta \rangle. \tag{6.20}$$

In addition, the fact that W is unitary implies the relations

$$\|\beta\| = \|\gamma\| = 1, \quad D\beta = D^*\gamma = 0, \quad D^*D + \beta \otimes \beta = DD^* + \gamma \otimes \gamma = 1. \tag{6.21}$$

It follows that the operator $U \in \mathcal{B}(\mathcal{M})$ defined by $U = \gamma \otimes \beta + D$ satisfies

$$U \text{ is unitary.} \tag{6.22}$$

Furthermore,

$$\begin{aligned} v_\lambda &= (h(\lambda) + 1)\gamma + D\lambda v_\lambda && \text{by equation (6.19)} \\ &= \big[(h(\lambda) - 1) + 2\big]\gamma + D\lambda v_\lambda \\ &= \big[\langle \lambda v_\lambda, \beta \rangle + 2\big]\gamma + D\lambda v_\lambda && \text{by equation (6.20)} \\ &= (\gamma \otimes \beta + D)\lambda v_\lambda + 2\gamma \\ &= U\lambda v_\lambda + 2\gamma, \end{aligned}$$

so that

$$v_\lambda = 2(1 - U\lambda)^{-1}\gamma. \tag{6.23}$$

Herglotz and Nevanlinna Representations in Several Variables 153

Now, by equation (6.16),

$$2(\overline{h(\mu)} + h(\lambda))$$
$$= (\overline{h(\mu)} + 1)(h(\lambda) + 1) - (\overline{h(\mu)} - 1)(h(\lambda) - 1)$$
$$= \langle (1 - \mu^* \lambda) v_\lambda, v_\mu \rangle_{\mathcal{M}} \qquad \text{by equation (6.16)}$$
$$= \langle v_\lambda, v_\mu \rangle_{\mathcal{M}} - \langle \lambda v_\lambda, \mu v_\mu \rangle_{\mathcal{M}}$$
$$= \langle U\lambda v_\lambda + 2\gamma, U\mu v_\mu + 2\gamma \rangle_{\mathcal{M}} - \langle \lambda v_\lambda, \mu v_\mu \rangle_{\mathcal{M}} \qquad \text{by equation (6.23)}$$
$$= 2\langle \gamma, U\mu v_\mu \rangle_{\mathcal{M}} + 2\langle U\lambda v_\lambda, \gamma \rangle_{\mathcal{M}} + 4.$$

Letting $\mu = 0$, we find that

$$2(1 + h(\lambda)) = 2\langle U\lambda v_\lambda, \gamma \rangle_{\mathcal{M}} + 4,$$

or equivalently,

$$h(\lambda) = \langle U\lambda v_\lambda, \gamma \rangle_{\mathcal{M}} + 1 \quad \text{for all } \lambda \in \mathbb{D}^2.$$

Hence

$$h(\lambda) = \langle U\lambda v_\lambda, \gamma \rangle_{\mathcal{M}} + 1$$
$$= \langle 2U\lambda(1 - U\lambda)^{-1}\gamma, \gamma \rangle_{\mathcal{M}} + \langle \gamma, \gamma \rangle_{\mathcal{M}} \quad \text{by equations (6.23) and (6.21)}$$
$$= \langle [2U\lambda(1 - U\lambda)^{-1} + 1]\gamma, \gamma \rangle_{\mathcal{M}}$$
$$= \left\langle \frac{1 + U\lambda}{1 - U\lambda} \gamma, \gamma \right\rangle_{\mathcal{M}} \qquad \text{for all } \lambda \in \mathbb{D}^2,$$

so that equation (6.11) holds with $b = 0$.

We now consider the general case when $h \in \mathscr{H}(\mathbb{D}^2)$, but we do not assume that $h(0) = 1$. Let $h(0) = a + ib$. If $a = 0$, the maximum principle implies that $h(\lambda) = ib$ for all $\lambda \in \mathbb{D}^2$, in which case equation (6.11) holds when we choose $\mathcal{M}$ and U arbitrarily and choose $\gamma = 0$. Otherwise, $a > 0$. If we set $g = (h - ib)/a$, then $g \subset \mathscr{H}(\mathbb{D}^2)$ and $g(0) = 1$. Hence by the case just proved, there exist $\mathcal{M}$, unitary $U \in \mathcal{B}(\mathcal{M})$, and $\beta \in \mathcal{M}$ such that $\|\beta\| = 1$ and

$$g(\lambda) = \left\langle \frac{1 + U\lambda}{1 - U\lambda} \beta, \beta \right\rangle_{\mathcal{M}} \qquad \text{for all } \lambda \in \mathbb{D}^2.$$

But then equation (6.11) holds with $\gamma = a^{1/2}\beta$. $\qquad \square$

6.3 Nevanlinna Representations on $\mathbb{H}$ via Operator Theory

Nevanlinna proved Theorem 6.3 by a change of variable in the integral formula (6.2) of Theorem 6.1, using the conformal map $\lambda \mapsto z$ from $\mathbb{D}$ onto $\mathbb{H}$, where λ and z are related by the formulas

154   *Commutative Theory*

$$z = i\frac{1+\lambda}{1-\lambda} \quad \text{and} \quad \lambda = \frac{z-i}{z+i}. \tag{6.24}$$

In this section we shall implement this simple idea of Nevanlinna in a slightly modified setting: we shall use operator theory to transform formula (6.2) in Theorem 6.1.

First assume that $f \in \mathscr{P}(\mathbb{H})$. If we define $h \in \text{Hol}(\mathbb{D})$ by

$$h(\lambda) = -if(z), \tag{6.25}$$

where λ and z are related as in equations (6.24), then $h \in \mathscr{H}(\mathbb{D})$. Hence, by Theorem 6.10, there exist $b \in \mathbb{R}$, a Hilbert space $\mathcal{M}$, a unitary operator $U \in \mathcal{B}(\mathcal{M})$, and a vector $\gamma \in \mathcal{M}$ such that equation (6.11) holds, that is,

$$h(\lambda) = ib + \left\langle \frac{1+U\lambda}{1-U\lambda} \gamma, \gamma \right\rangle_{\mathcal{M}} \quad \text{for all } \lambda \in \mathbb{D}. \tag{6.26}$$

We wish to re-express equation (6.26) in notation more intrinsic to the function f. To this end, let $a = -b$, $L = U^{-1}$ and $v = L\gamma$. For $z \in \mathbb{H}$ we define $M(z) \in \mathcal{B}(\mathcal{M})$ by the formula,

$$M(z) = i\left(L - \frac{z-i}{z+i}\right)^{-1}\left(L + \frac{z-i}{z+i}\right). \tag{6.27}$$

In the notations of the previous paragraph,

$$f(z) = ih(\lambda)$$

$$= i\left(ib + \left\langle \frac{1+U\lambda}{1-U\lambda} \gamma, \gamma \right\rangle_{\mathcal{M}}\right)$$

$$= -b + i\left\langle U\frac{U^{-1}+\lambda}{U^{-1}-\lambda} U^{-1}\gamma, \gamma \right\rangle_{\mathcal{M}}$$

$$= -b + i\left\langle \frac{L+\lambda}{L-\lambda} L\gamma, L\gamma \right\rangle_{\mathcal{M}}$$

$$= a + \left\langle i\frac{L+\lambda}{L-\lambda} v, v \right\rangle_{\mathcal{M}}$$

$$= a + \left\langle i\left(L - \frac{z-i}{z+i}\right)^{-1}\left(L + \frac{z-i}{z+i}\right) v, v \right\rangle_{\mathcal{M}}$$

$$= a + \langle M(z)v, v \rangle_{\mathcal{M}}.$$

Conversely, if $a \in \mathbb{R}$, $\mathcal{M}$ is a Hilbert space, $v \in \mathcal{M}$, L is a unitary on $\mathcal{M}$ and M is defined by equation (6.27), then the formula

$$f(z) = a + \langle M(z)v, v \rangle_{\mathcal{M}} \tag{6.28}$$

Herglotz and Nevanlinna Representations in Several Variables   155

defines an element in $\mathscr{P}(\mathbb{H})$. Summarizing, we have proven the following lemma.

Lemma 6.29. *A function f belongs to $\mathscr{P}(\mathbb{H})$ if and only if there exist $a \in \mathbb{R}$, a Hilbert space $\mathcal{M}$, $v \in \mathcal{M}$, and a unitary $L \in \mathcal{B}(\mathcal{M})$ such that equation (6.28) holds, where M is defined by equation (6.27).*

To obtain Nevanlinna's classical formulas one transforms the representation in this lemma by expressing L as the Cayley transform of a self-adjoint operator. Let $\mathcal{N} = \ker(1 - L)$. Then L has the 2×2 block form

$$L = \begin{bmatrix} 1 & 0 \\ 0 & L_0 \end{bmatrix} \tag{6.30}$$

with respect to the decomposition $\mathcal{M} = \mathcal{N} \oplus \mathcal{N}^{\perp}$, where $L_0 = L|\mathcal{N}^{\perp}$ is a unitary operator satisfying $\ker(1 - L_0) = \{0\}$. Consequently, it follows that there is a densely defined self-adjoint operator A on $\mathcal{N}^{\perp}$ satisfying

$$A(1 - L_0) = i(1 + L_0) \tag{6.31}$$

(see, for example, [172, section 121]).

Note that we have the formulas

$$(z+i)L_0 - (z-i) = i(1+L_0) - z(1-L_0) = (A-z)(1-L_0)$$

and

$$i\big((z+i)L_0 + (z-i)\big) = (1-L_0) + i(1+L_0)z = (1+Az)(1-L_0),$$

so that

$$\frac{i\big((z+i)L_0 + (z-i)\big)}{(z+i)L_0 - (z-i)} = \frac{1+Az}{A-z}. \tag{6.32}$$

The following computation derives a formula for $M(z)$ in terms of A in 2×2 block form with respect to the decomposition $\mathcal{M} = \mathcal{N} \oplus \mathcal{N}^{\perp}$.

$$
\begin{aligned}
M(z) &= i \left(\begin{bmatrix} 1 & 0 \\ 0 & L_0 \end{bmatrix} - \begin{bmatrix} \frac{z-i}{z+i} & 0 \\ 0 & \frac{z-i}{z+i} \end{bmatrix} \right)^{1} \left(\begin{bmatrix} 1 & 0 \\ 0 & L_0 \end{bmatrix} + \begin{bmatrix} \frac{z-i}{z+i} & 0 \\ 0 & \frac{z-i}{z+i} \end{bmatrix} \right) \\
&= i \begin{bmatrix} 1 - \frac{z-i}{z+i} & 0 \\ 0 & L_0 - \frac{z-i}{z+i} \end{bmatrix}^{-1} \begin{bmatrix} 1 + \frac{z-i}{z+i} & 0 \\ 0 & L_0 + \frac{z-i}{z+i} \end{bmatrix} \\
&= i \begin{bmatrix} 2i & 0 \\ 0 & (z+i)L_0 - (z-i) \end{bmatrix}^{-1} \begin{bmatrix} 2z & 0 \\ 0 & (z+i)L_0 + (z-i) \end{bmatrix} \\
&= \begin{bmatrix} z & 0 \\ 0 & \frac{1+Az}{A-z} \end{bmatrix}.
\end{aligned}
$$

156　　　　　　　　　　　*Commutative Theory*

As a consequence of this formula for $M(z)$ in terms of A we obtain the following equivalent form of Lemma 6.29.

Theorem 6.33. (**Nevanlinna's first representation theorem, operator theory version**) *A function f belongs to $\mathscr{P}(\mathbb{H})$ if and only if there exist $a \in \mathbb{R}$, a decomposed Hilbert space $\mathcal{M} = \mathcal{N} \oplus \mathcal{N}^{\perp}$, $v \in \mathcal{M}$, and a densely defined self-adjoint operator A on $\mathcal{N}^{\perp}$ such that*

$$f(z) = a + \left\langle \begin{bmatrix} z & 0 \\ 0 & \frac{1+Az}{A-z} \end{bmatrix} v, v \right\rangle. \tag{6.34}$$

Once Theorem 6.33 is established, it is a simple matter to unravel Nevanlinna's classical theorems. If we decompose $v = v_0 \oplus v_1$, where $v_0 \in \mathcal{N}$ and $v_1 \in \mathcal{N}^{\perp}$, and let $A = \int t dE$ be the spectral representation of A, then the formula (6.34) becomes

$$f(z) = a + \|v_0\|^2 z + \int_{t \in \mathbb{R}} \frac{1+tz}{t-z} \langle dE(t)v_1, v_1 \rangle, \tag{6.35}$$

which is Nevanlinna's formula (6.4) with $b = \|v_0\|^2$ and $\mu = \langle E(\cdot)v_1, v_1 \rangle$.

6.4 The Nevanlinna Representations on $\mathbb{H}^2$

To extend Nevanlinna's representation theorems to two variables we shall follow the strategy of the previous section. We shall transform Theorems 6.10 and 6.12 using operator theory. The result is Theorem 6.46, a precise analog of Theorem 6.33. Much of the calculation is the same as before, but with one significant difference: λ and z can no longer be treated as scalars when they appear in expressions involving operators.

Let $f \in \mathscr{P}(\mathbb{H}^2)$. If we define $h \in \text{Hol}(\mathbb{D}^2)$ by

$$h(\lambda) = -if(z), \tag{6.36}$$

where

$$z^1 = i\frac{1+\lambda^1}{1-\lambda^1} \quad \text{and} \quad z^2 = i\frac{1+\lambda^2}{1-\lambda^2}, \tag{6.37}$$

then $h \in \mathscr{H}(\mathbb{D}^2)$. Hence, by Theorem 6.12, there exist $b \in \mathbb{R}$, a decomposed Hilbert space $\mathcal{M} = \mathcal{M}_1 \oplus \mathcal{M}_2$, a unitary operator $U \in \mathcal{B}(\mathcal{M})$, and $\gamma \in \mathcal{M}$ such that equation (6.13) holds.[1]

---

[1] To enhance legibility in this chapter, we shall use superscripts to denote co-ordinates of scalar variables, as in (6.37), but subscripts for the decomposition of spaces, as in $\mathcal{M} = \mathcal{M}_1 \oplus \mathcal{M}_2$, and operators, as in (6.51).

Herglotz and Nevanlinna Representations in Several Variables   157

As in the previous section we re-express equation (6.13) in notation more appropriate to the function f. Let $a = -b$, $L = U^{-1}$ and $v = L\gamma$. Also, for $z = (z^1, z^2) \in \mathbb{H}^2$, interpret z as the block operator $\begin{bmatrix} z^1 & 0 \\ 0 & z^2 \end{bmatrix}$ acting on $\mathcal{M} = \mathcal{M}_1 \oplus \mathcal{M}_2$. Note that, with this convention, equations (6.37) imply that

$$\lambda = \begin{bmatrix} \lambda^1 & 0 \\ 0 & \lambda^2 \end{bmatrix} = \begin{bmatrix} \frac{z^1-i}{z^1+i} & 0 \\ 0 & \frac{z^2-i}{z^2+i} \end{bmatrix} = \frac{z-i}{z+i}.$$

Finally, for $z \in \mathbb{H}^2$, define $M(z) \in \mathcal{B}(\mathcal{M})$ by the formula

$$M(z) = i \left(L - \frac{z-i}{z+i} \right)^{-1} \left(L + \frac{z-i}{z+i} \right). \tag{6.38}$$

While optically, equation (6.38) is the same as equation (6.27), in the former z and L do not necessarily commute, as they do in equation (6.27). Nevertheless, we compute exactly as before to derive the representation formula

$$f(z) = a + \langle M(z)v, v \rangle_{\mathcal{M}}. \tag{6.39}$$

Continuing as in the previous section we reformulate equation (6.39) by expressing L as the Cayley transform of a self-adjoint operator. Here, as z and L do not necessarily commute, it is not possible to derive equation (6.32), and we must proceed with care. Noting that

$$L + \frac{z-i}{z+i} = (z+i)^{-1} \big((z+i)L + (z-i) \big)$$

and

$$\left(L - \frac{z-i}{z+i} \right)^{-1} = \big((z+i)L - (z-i) \big)^{-1} (z+i),$$

we infer from equation (6.38) that

$$M(z) - i \left((z+i)L - (z-i) \right)^{-1} \left((z+i)L + z - i \right),$$

which we rewrite in the form

$$M(z) = \big(z(L-1) + i(L+1) \big)^{-1} \big(iz(1+L) + 1 - L \big). \tag{6.40}$$

As before, let $\mathcal{N} = \ker(1 - L)$ and define $L_0 = L|\mathcal{N}^{\perp}$, so that L_0 is a unitary operator satisfying $\ker(1 - L_0) = \{0\}$ and equation (6.30) holds with respect to the decomposition $\mathcal{M} = \mathcal{N} \oplus \mathcal{N}^{\perp}$. Again densely define a self-adjoint operator A, the Cayley transform of L_0, on $\mathcal{N}^{\perp}$ by equation (6.31):

$$A(1 - L_0) = i(1 + L_0). \tag{6.41}$$

158 Commutative Theory

The following computation derives a formula for $M(z)$ in terms of A in 2×2 block form with respect to the decomposition $\mathcal{M} = \mathcal{N} \oplus \mathcal{N}^\perp$.

$$M(z) \stackrel{\text{(i)}}{=} \left(z \left(\begin{bmatrix} 1 & 0 \\ 0 & L_0 \end{bmatrix} - \begin{bmatrix} 1 & 0 \\ 0 & 1 \end{bmatrix} \right) + i \left(\begin{bmatrix} 1 & 0 \\ 0 & L_0 \end{bmatrix} + \begin{bmatrix} 1 & 0 \\ 0 & 1 \end{bmatrix} \right) \right)^{-1}$$

$$\times \left(iz \left(\begin{bmatrix} 1 & 0 \\ 0 & 1 \end{bmatrix} + \begin{bmatrix} 1 & 0 \\ 0 & L_0 \end{bmatrix} \right) + \begin{bmatrix} 1 & 0 \\ 0 & 1 \end{bmatrix} - \begin{bmatrix} 1 & 0 \\ 0 & L_0 \end{bmatrix} \right)$$

$$= \left(z \begin{bmatrix} 0 & 0 \\ 0 & L_0 - 1 \end{bmatrix} + \begin{bmatrix} 2i & 0 \\ 0 & i(L_0 + 1) \end{bmatrix} \right)^{-1}$$

$$\times \left(z \begin{bmatrix} 2i & 0 \\ 0 & i(1 + L_0) \end{bmatrix} + \begin{bmatrix} 0 & 0 \\ 0 & 1 - L_0 \end{bmatrix} \right)$$

$$\stackrel{\text{(ii)}}{=} \left(\left(z \begin{bmatrix} 0 & 0 \\ 0 & -1 \end{bmatrix} + \begin{bmatrix} 1 & 0 \\ 0 & A \end{bmatrix} \right) \begin{bmatrix} 2i & 0 \\ 0 & 1 - L_0 \end{bmatrix} \right)^{-1}$$

$$\times \left(z \begin{bmatrix} 1 & 0 \\ 0 & A \end{bmatrix} + \begin{bmatrix} 0 & 0 \\ 0 & 1 \end{bmatrix} \right) \begin{bmatrix} 2i & 0 \\ 0 & 1 - L_0 \end{bmatrix}$$

$$\stackrel{\text{(iii)}}{=} \begin{bmatrix} -\frac{1}{2}i & 0 \\ 0 & \frac{1}{2}(1 - iA) \end{bmatrix} \left(\begin{bmatrix} 1 & 0 \\ 0 & A \end{bmatrix} - z \begin{bmatrix} 0 & 0 \\ 0 & 1 \end{bmatrix} \right)^{-1}$$

$$\times \left(z \begin{bmatrix} 1 & 0 \\ 0 & A \end{bmatrix} + \begin{bmatrix} 0 & 0 \\ 0 & 1 \end{bmatrix} \right) \begin{bmatrix} -\frac{1}{2}i & 0 \\ 0 & \frac{1}{2}(1 - iA) \end{bmatrix}^{-1}$$

$$\stackrel{\text{(iv)}}{=} \begin{bmatrix} -i & 0 \\ 0 & 1 - iA \end{bmatrix} \left(\begin{bmatrix} 1 & 0 \\ 0 & A \end{bmatrix} - z \begin{bmatrix} 0 & 0 \\ 0 & 1 \end{bmatrix} \right)^{-1}$$

$$\times \left(z \begin{bmatrix} 1 & 0 \\ 0 & A \end{bmatrix} + \begin{bmatrix} 0 & 0 \\ 0 & 1 \end{bmatrix} \right) \begin{bmatrix} -i & 0 \\ 0 & 1 - iA \end{bmatrix}^{-1}.$$

In this calculation the Roman numerals over the equality signs indicate the following justifications.

(i) Use equations (6.40) and (6.41).

(ii) By equation (6.41).

(iii) Equation (6.41) can be re-arranged to $(A + i)(1 - L_0) = 2i$, so that

$$1 - L_0 = 2(1 - iA)^{-1}.$$

(iv) In the previous expression multiply the first factor by 2 and the last factor by $\frac{1}{2}$.

The calculation just carried out motivates the following definition.

Herglotz and Nevanlinna Representations in Several Variables 159

Definition 6.42. *A* $\mathcal{B}(\mathcal{M})$-*valued Nevanlinna resolvent of type 4 is a function*

$$M : \mathbb{H}^2 \to \mathcal{B}(\mathcal{M})$$

that has the form

$$M(z) = \begin{bmatrix} -i & 0 \\ 0 & 1 - iA \end{bmatrix} \left(\begin{bmatrix} 1 & 0 \\ 0 & A \end{bmatrix} - z \begin{bmatrix} 0 & 0 \\ 0 & 1 \end{bmatrix} \right)^{-1}$$
$$\times \left(z \begin{bmatrix} 1 & 0 \\ 0 & A \end{bmatrix} + \begin{bmatrix} 0 & 0 \\ 0 & 1 \end{bmatrix} \right) \begin{bmatrix} -i & 0 \\ 0 & 1 - iA \end{bmatrix}^{-1}, \qquad (6.43)$$

where

(i) *the* 2×2 *block structure in formula* (6.43) *is with respect to an orthogonal decomposition* $\mathcal{M} = \mathcal{N}_1 \oplus \mathcal{N}_2$ *of* $\mathcal{M}$;
(ii) *A is a densely defined self-adjoint operator on* $\mathcal{N}_2$; *and*
(iii) *for* $z \in \mathbb{H}^2$, *z acts on* $\mathcal{M}$ *by the formula* $z(m_1 \oplus m_2) = z^1 m_1 \oplus z^2 m_2$ *relative to a second orthogonal decomposition* $\mathcal{M}_1 \oplus \mathcal{M}_2$ *of* $\mathcal{M}$.

It is important for the reader to remember that our notation for the operator z suppresses its dependence on the orthogonal decomposition $\mathcal{M}_1 \oplus \mathcal{M}_2$ of $\mathcal{M}$, which is not explicit in the formula (6.43).

A careful inspection of equation (6.43) shows that $M(z)$ is a well-defined bounded linear operator on the whole of $\mathcal{M}$ for every $z \in \mathbb{H}^2$ (for details see [26, proposition 3.1]).

The use of the term *resolvent* calls for some justification. The resolvent operator $(A - z)^{-1}$ of a self-adjoint operator A on $\mathcal{M}$ plays an important role in operator theory. One of the original proofs of the spectral theorem for self-adjoint operators, that of Stone [191] in 1932, was based on the following properties of the resolvent, together with Nevanlinna's representation theorems.

(i) $z \mapsto (A - z)^{-1}$ is an analytic $\mathcal{B}(\mathcal{M})$-valued function on the upper halfplane $\mathbb{H}$;
(ii) $\mathrm{Im}(A - z)^{-1} \geq 0$ for all $z \in \mathbb{H}$;
(iii) the growth estimate $\|(A - z)^{-1}\| \leq 1/\mathrm{Im}\, z$ holds for $z \in \mathbb{H}$;
(iv) the "resolvent identity"

$$(A - z)^{-1} - (A - w)^{-1} = (z - w)(A - z)^{-1}(A - w)^{-1}$$

holds for all $z, w \in \mathbb{H}$.

$M(z)$, as defined in equation (6.43), has fully analogous properties.

(i) M is an analytic $\mathcal{B}(\mathcal{M})$-valued function on $\mathbb{H}^2$;
(ii) $\mathrm{Im}\, M(z) \geq 0$ for all $z \in \mathbb{H}^2$;

160 Commutative Theory

(iii) for all $z \in \mathbb{H}^2$,

$$\|M(z)\| \le (1 + \sqrt{10}\|z\|_1)\left(1 + \frac{1 + \sqrt{2}\|z\|_1}{\min\{\operatorname{Im} z^1, \operatorname{Im} z^2\}}\right), \qquad (6.44)$$

where $\|z\|_1 = |z^1| + |z^2|$;

(iv) M satisfies the resolvent identity

$$M(z) - M(w) = \begin{bmatrix} -i & 0 \\ 0 & 1 - iA \end{bmatrix}\left(\begin{bmatrix} 1 & 0 \\ 0 & A \end{bmatrix} - z\begin{bmatrix} 0 & 1 \\ 0 & 0 \end{bmatrix}\right)^{-1}(z - w)$$

$$\times \left(\begin{bmatrix} 1 & 0 \\ 0 & A \end{bmatrix} - \begin{bmatrix} 0 & 0 \\ 0 & 1 \end{bmatrix}w\right)^{-1}\begin{bmatrix} i & 0 \\ 0 & 1 + iA \end{bmatrix} \qquad (6.45)$$

for all $z, w \in \mathbb{H}^2$.

Proofs of these statements can be found in [26]; in [26], see Proposition 3.1 for (i) and (iii), Corollary 3.6 for (ii), and Proposition 10.2 for (iv).

Our calculations in this section lead to the following statement.

Theorem 6.46. *A function f belongs to $\mathscr{P}(\mathbb{H}^2)$ if and only if there exist $a \in \mathbb{R}$, a Hilbert space $\mathcal{M}$, $v \in \mathcal{M}$, and a $\mathcal{B}(\mathcal{M})$-valued Nevanlinna resolvent M of type 4 such that*

$$f(z) = a + \langle M(z)v, v \rangle_{\mathcal{M}} \quad \text{for all } z \in \mathbb{H}^2. \qquad (6.47)$$

6.5 A Classification Scheme for Nevanlinna Representations in Two Variables

There is a way to classify two-variable Nevanlinna representations using the Hilbert space geometry associated with formula (6.47). In one variable Nevanlinna's two representations identify two classes of functions, $\mathscr{P}(\mathbb{H})$ and a proper subset comprising the functions expressible as the Cauchy transform of a positive measure. In two variables, *four* classes arise naturally. Let us first examine how these four classes manifest themselves in Nevanlinna's context.

6.5.1 How It Works in One Variable

Recall that a one-variable analog of Theorem 6.65, couched in operator-theoretic terms, is Theorem 6.33, which asserts that $f \in \mathscr{P}(\mathbb{H})$ if and only if there exist $a \in \mathbb{R}$, a decomposed Hilbert space $\mathcal{M} = \mathcal{N} \oplus \mathcal{N}^\perp$, $v \in \mathcal{M}$, and a densely defined self-adjoint A on $\mathcal{M}$ such that

Herglotz and Nevanlinna Representations in Several Variables 161

$$f(z) = a + \langle M(z)v, v \rangle_{\mathcal{M}} \qquad \text{for all } z \in \mathbb{H}, \tag{6.48}$$

where

$$M(z) = \begin{bmatrix} z & 0 \\ 0 & \frac{1+Az}{A-z} \end{bmatrix} \qquad \text{for all } z \in \mathbb{H}. \tag{6.49}$$

The operator theory associated with these formulas identifies the following four decreasing classes of functions, which we label successively *types* 4, 3, 2, *and* 1.

Type 4. We refer to the function M defined in equation (6.49) as a *Nevanlinna resolvent of type* 4 and to the formula (6.48) as a *Nevanlinna representation of type* 4.

Type 3. A representation of the form (6.48), with the additional assumption that $v \perp \mathcal{N}$, is said to be *of type* 3. In this case equation (6.48) holds with M having the form

$$M(z) = \frac{1+Az}{A-z} \qquad \text{for all } z \in \mathbb{H} \tag{6.50}$$

for some densely defined self-adjoint operator A on $\mathcal{M}$. We refer to the function M defined in equation (6.50) as a *Nevanlinna resolvent of type* 3. When M is a Nevanlinna resolvent of type 3, we refer to the formula (6.48) as a *Nevanlinna representation of type* 3.

Type 2. This is the case of a representation (6.50) of type 3 but with the additional assumption that $v \in \operatorname{dom} A$. First note the identity

$$\frac{1+Az}{A-z} = (1-iA)(A-z)^{-1}(1+iA) - A.$$

Also, if $v \in \operatorname{dom} A$, then $\langle Av, v \rangle$ and $(1+iA)v$ are well defined. Consequently, if equation (6.48) is a type 3 Nevanlinna representation for f with $v \in \operatorname{dom} A$, then

$$\begin{aligned}
f(z) &= a + \langle M(z)v, v \rangle \\
&= a + \left\langle \frac{1+Az}{A-z}v, v \right\rangle \\
&= a + \left\langle \left((1-iA)(A-z)^{-1}(1+iA) - A \right)v, v \right\rangle \\
&= a - \langle Av, v \rangle + \left\langle (A-z)^{-1}(1+iA)v, (1+iA)v \right\rangle \\
&= a^{\sim} + \left\langle (A-z)^{-1}v^{\sim}, v^{\sim} \right\rangle,
\end{aligned}$$

where $a^{\sim} = a - \langle Av, v \rangle$ and $v^{\sim} = (1+iA)v$. Consequently, we say that M is a *Nevanlinna resolvent of type* 2 if M has the form

$$M(z) = (A-z)^{-1} \qquad \text{for all } z \in \mathbb{H}.$$

162   *Commutative Theory*

When M is a Nevanlinna resolvent of type 2, we refer to the formula (6.48) as a *Nevanlinna representation of type* 2.

Type 1. We do not define type 1 Nevanlinna resolvents. However, we say that equation (6.48) is a *Nevanlinna representation of type* 1 *of* f if M is type 2 and $a = 0$.

6.5.2 Type 3 Nevanlinna Resolvents and Representations in Two Variables

Every function in $\mathscr{P}(\mathbb{H}^2)$ has a Nevanlinna representation of type 4 (Definition 6.42 and Theorem 6.65). As in the one-variable type 3 case, assume that, for some $f \in \mathscr{P}(\mathbb{H}^2)$ with representation (6.47), the vector v is orthogonal to $\mathcal{N}_1$ and hence can be written $\begin{pmatrix} 0 & v \end{pmatrix}^T$. Then $M(z)$ can be expressed in the following simpler form. Let P_j be the orthogonal projection operator on $\mathcal{M}$ with range $\mathcal{M}_j$ for $j = 1, 2$. Let

$$P_j \sim \begin{bmatrix} X_j & B_j \\ B_j^* & Y_j \end{bmatrix}, \tag{6.51}$$

so that Y_j is the compression of P_j to $\mathcal{N}_j$.

Definition 6.52. *For any pair* $T = (T_1, T_2)$ *of operators and any* $z \in \mathbb{C}^2$, *let us write*

$$z_T \overset{\text{def}}{=} z^1 T_1 + z^2 T_2. \tag{6.53}$$

In this notation the operator z is expressible as

$$z = z^1 P_1 + z^2 P_2 \sim \begin{bmatrix} z_X & z_B \\ z_{B^*} & z_Y \end{bmatrix}.$$

On substituting this expression into the definition (6.43) of $M(z)$, we obtain the formula

$$M(z) = \begin{bmatrix} * & * \\ * & (1 - iA)(A - z_Y)^{-1}(1 + z_Y A)(1 - iA)^{-1} \end{bmatrix}.$$

Hence the representation (6.47) becomes

$$f(z) = a + \langle (1 - iA)(A - z_Y)^{-1}(1 + z_Y A)(1 - iA)^{-1}v, v \rangle \qquad \text{for all } z \in \mathbb{H}^2.$$

Definition 6.54. *A* $\mathcal{B}(\mathcal{M})$-*valued Nevanlinna resolvent of type* 3 *is a function* $M \colon \mathbb{H}^2 \to \mathcal{B}(\mathcal{M})$ *that has the form*

$$M(z) = (1 - iA)(A - z_Y)^{-1}(1 + z_Y A)(1 - iA)^{-1}, \tag{6.55}$$

Herglotz and Nevanlinna Representations in Several Variables   163

where

(i) *A is a densely defined self-adjoint operator on $\mathcal{M}$, and*

(ii) *$Y = (Y_1, Y_2)$ is a pair of positive self-adjoint contractions on $\mathcal{M}$ satisfying $Y_1 + Y_2 = 1$.*

Definition 6.56. *If $f : \mathbb{H}^2 \to \mathbb{C}$ is a function, $a \in \mathbb{R}$, $\mathcal{M}$ is a Hilbert space, $v \in \mathcal{M}$, M is a type 3 $\mathcal{B}(\mathcal{M})$-valued Nevanlinna resolvent, and equation (6.47) holds, then we say that equation (6.47)* is a *type 3 Nevanlinna representation for f.*

In the language of these definitions, the calculations leading up to Definition 6.54 imply the following proposition.

Proposition 6.57. *If $f : \mathbb{H}^2 \to \mathbb{C}$ is a function, then f has a type 3 Nevanlinna representation if and only if f has a type 4 Nevanlinna representation with the property that v is orthogonal to $\mathcal{N}_1$.*

6.5.3 Type 2 Nevanlinna Resolvents and Representations

Type 2 representations arise from type 3 representations in the case that the vector v in equation (6.47) is in the domain of A.

Lemma 6.58. *Assume that $f : \mathbb{H}^2 \to \mathbb{C}$ is a function, $a \in \mathbb{R}$, $\mathcal{M}$ is a Hilbert space, $v \in \mathcal{M}$, M is a type 3 $\mathcal{B}(\mathcal{M})$-valued Nevanlinna resolvent, and equation (6.47) holds. If v is in the domain of A and $u = (1 + iA)v$, then*

$$f(z) = a + \left\langle \left((A - z_Y)^{-1} - \frac{A}{1 + A^2} \right) u, u \right\rangle_{\mathcal{M}}$$

for all $z \in \mathbb{H}^2$.

Proof. Observe that

$$(A - z_Y)^{-1}(1 + z_Y A)(1 + A^2)^{-1} - (A - z_Y)^{-1}\left(1 + A^2 - (A - z_Y)A \right)(1 + A^2)^{-1}$$

$$= (A - z_Y)^{-1} - \frac{A}{1 + A^2}.$$

Consequently, if M is a type 3 Nevanlinna resolvent, then

$$M(z) = (1 - iA)(A - z_Y)^{-1}(1 + z_Y A)(1 - iA)^{-1}$$

$$= (1 - iA)(A - z_Y)^{-1}(1 + z_Y A)(1 + A^2)^{-1}(1 + iA)$$

$$= (1 - iA)\left((A - z_Y)^{-1} - \frac{A}{1 + A^2} \right)(1 + iA).$$

164 *Commutative Theory*

Therefore, if equation (6.47) is a type 3 Nevanlinna representation for f and $v \in \text{dom}\, A$ with $u = (1 + iA)v$, we have

$$f(z) = a + \left\langle (1 - iA)\left((A - z_Y)^{-1} - \frac{A}{1 + A^2} \right)(1 + iA)\,v, v \right\rangle_{\mathcal{M}}$$
$$= \left\langle \left((A - z_Y)^{-1} - \frac{A}{1 + A^2} \right)(1 + iA)v, (1 + iA)v \right\rangle_{\mathcal{M}}$$
$$= \left\langle \left((A - z_Y)^{-1} - \frac{A}{1 + A^2} \right)u, u \right\rangle_{\mathcal{M}}.$$

$\square$

Lemma 6.59. *Assume that* $f \colon \mathbb{H}^2 \to \mathbb{C}$ *is a function,* $a \in \mathbb{R}$, $\mathcal{M}$ *is a Hilbert space,* $v \in \mathcal{M}$, M *is a type* 3 $\mathcal{B}(\mathcal{M})$*-valued Nevanlinna resolvent, and equation* (6.47) *holds. If* v *is in the domain of* A *and* $u = (1 + iA)v$, *then there exist* $a^{\sim} \in \mathbb{R}$ *and* $v^{\sim} \in \mathcal{M}$ *such that*

$$f(z) = a^{\sim} + \left\langle (A - z_Y)^{-1}\, v^{\sim}, v^{\sim} \right\rangle_{\mathcal{M}}$$

for all $z \in \mathbb{H}^2$.

Lemma 6.59 suggests the following definitions.

Definition 6.60. *A* $\mathcal{B}(\mathcal{M})$-valued Nevanlinna resolvent of type 2 *is a function* $M \colon \mathbb{H}^2 \to \mathcal{B}(\mathcal{M})$ *that has the form*

$$M(z) = (A - z_Y)^{-1}, \tag{6.61}$$

where

(i) *A is a densely defined self-adjoint operator on* $\mathcal{M}$, *and*

(ii) $Y = (Y_1, Y_2)$ *is a pair of positive self-adjoint contractions on* $\mathcal{M}$ *satisfying* $Y_1 + Y_2 = 1$.

Definition 6.62. *If* $f \colon \mathbb{H}^2 \to \mathbb{C}$ *is a function,* $a \in \mathbb{R}$, $\mathcal{M}$ *is a Hilbert space,* $v \in \mathcal{M}$, M *is a type* 2 $\mathcal{B}(\mathcal{M})$-valued Nevanlinna resolvent, and equation (6.47) *holds, then we say that equation* (6.47) *is a* type 2 Nevanlinna representation *for* f.

From the way we have introduced the four types, the following statement is immediate.

Proposition 6.63. *Let* $f \colon \mathbb{H}^2 \to \mathbb{C}$ *be a function. If* f *has a type* 2 *Nevanlinna representation then* f *has a type* 3 *Nevanlinna representation. If* f *has a type* 3 *Nevanlinna representation with* $v \in \text{dom}\, A$, *then* f *has a type* 2 *Nevanlinna representation.*

Herglotz and Nevanlinna Representations in Several Variables 165

6.5.4 Type 1 Representations

There are no new resolvents of type 1. However, noting that in one variable in the representation formula in Nevanlinna's second representation, $a = 0$, we make the following definition.

Definition 6.64. *If $f: \mathbb{H}^2 \to \mathbb{C}$ is a function, $\mathcal{M}$ is a Hilbert space, $v \in \mathcal{M}$, M is a type 2 $\mathcal{B}(\mathcal{M})$-valued Nevanlinna resolvent, and equation (6.47) holds with $a = 0$, then we say that equation (6.47) is a* type 1 Nevanlinna representation *for f.*

6.5.5 Growth Conditions at ∞

There are simple necessary and sufficient function-theoretic conditions for a function $f \in \mathscr{P}(\mathbb{H}^2)$ to have a Nevanlinna representation of a given type. Proofs of the results in this subsection can be found in section 6 of [26]. In the sequel s is a positive real variable.

Theorem 6.65. *If $f \in \mathscr{P}(\mathbb{H}^2)$, then f has a representation of type 3 if and only if*

$$\liminf_{s \to \infty} \frac{1}{s} \operatorname{Im} f(is, is) = 0.$$

Theorem 6.66. *If $f \in \mathscr{P}(\mathbb{H}^2)$, then f has a representation of type 2 if and only if*

$$\liminf_{s \to \infty} s \operatorname{Im} f(is, is) < \infty.$$

Theorem 6.67. *If $f \in \mathscr{P}(\mathbb{H}^2)$, then f has a representation of type 1 if and only if*

$$\liminf_{s \to \infty} s \, |f(is, is)| < \infty.$$

Observe that the last theorem is a direct generalization of Nevanlinna's second representation theorem, Theorem 6.5, to two variables.

6.6 The Type of a Function

The elements of $\mathscr{P}(\mathbb{H}^2)$ can be classified according to the types of representation they admit.

Definition 6.68. *Let $f \in \mathscr{P}(\mathbb{H}^2)$. If f has a type 1 Nevanlinna representation then we say that f is* type 1. *Otherwise, if $r \in \{2, 3, 4\}$, we say that f is* type r *if f has a type r Nevanlinna representation but does not have a type $r - 1$ Nevanlinna representation.*

166 *Commutative Theory*

Definition 6.68 defines the type of a function f in terms of the underlying operator theory of the Nevanlinna representation of f. In Section 6.5.5, we gave intrinsic characterizations in terms of the growth of f. One can also use the function theory of carapoints developed in Chapter 5.

For $f \in \mathscr{P}(\mathbb{H}^2)$ we may transform f to the function $\varphi \in \mathscr{S}(\mathbb{D}^2)$ defined by

$$f(z) = i\frac{1+\varphi(\lambda)}{1-\varphi(\lambda)},$$

where

$$z^1 = i\frac{1+\lambda^1}{1-\lambda^1} \quad \text{and} \quad z^2 = i\frac{1+\lambda^2}{1-\lambda^2}.$$

We may then ordain that (∞,∞) *is a carapoint for* f if $(1,1)$ is a carapoint for φ and, in that case, define $f(\infty,\infty) \in \mathbb{C} \cup \{\infty\}$ by

$$f(\infty,\infty) = i\frac{1+\varphi(1,1)}{1-\varphi(1,1)}.$$

Here, if $\varphi(1,1) = 1$, then $h(\infty,\infty) = \infty$.

For the next theorem see [26, section 8].

Theorem 6.69. *If $f \in \mathscr{P}(\mathbb{H}^2)$ then the following statements obtain.*

(i) *f is type 1 if and only if (∞,∞) is a carapoint of f and $f(\infty,\infty) = 0$.*

(ii) *f is type 2 if and only if (∞,∞) is a carapoint of f and $f(\infty,\infty) \in \mathbb{R} \setminus \{0\}$.*

(iii) *f is type 3 if and only if (∞,∞) is not a carapoint of f.*

(iv) *f is type 4 if and only if (∞,∞) is a carapoint of f and $f(\infty,\infty) = \infty$.*

6.6.1 Examples

The four types of Nevanlinna representation described in this chapter enable us to write down concrete examples of functions in $\mathscr{P}(\mathbb{H}^2)$ of the four types.

Example 6.70. Let

$$h_1(z) = -\frac{1}{z^1+z^2} \quad \text{for all } z^1, z^2 \in \mathbb{H}.$$

We have $h_1(is,is) = \frac{1}{2}i/s$, and hence

$$\liminf_{s\to\infty} s|h_1(is,is)| = \frac{1}{2}.$$

Thus, by Theorem 6.67, h_1 has a representation of type 1, and therefore f has type 1. One representation of h_1 of type 1 is

$$h_1(z) = \langle (0 - z_Y)^{-1}v, v\rangle_{\mathbb{C}},$$

where $\mathcal{M} = \mathbb{C}$, $Y = (\frac{1}{2}, \frac{1}{2})$ and $v = 1/\sqrt{2}$, with $A = 0$.

Herglotz and Nevanlinna Representations in Several Variables   167

Example 6.71. Let

$$h_2(z) = 1 - \frac{1}{z^1 + z^2} = 1 + \langle (0 - z_Y)^{-1} v, v \rangle_\mathbb{C},$$

where Y, v are as in Example 6.70. Since $h_2(\infty, \infty) = 1$, h_2 does not have a representation of type 1, and again by Theorem 6.65, h_2 has type 2.

Example 6.72. Let h_3 be the constant function equal to i on $\mathbb{H}^2$. Certainly $h_3 \in \mathscr{P}(\mathbb{H}^2)$, and it follows from Theorems 6.65 and 6.66 that h_3 has type 3. A type 3 representation of h_3 is

$$h_3(z) = \langle M(z)v, v \rangle_{L^2(\mathbb{R})}, \tag{6.73}$$

where $M(z)$ is the Nevanlinna resolvent of type 3 defined by equation (6.55) corresponding to the choices $\mathcal{M} = L^2(\mathbb{R})$, $A =$ the operation of multiplication by the independent variable t, $Y_1 = Y_2 = \frac{1}{2}$, and $v(t) = 1/\sqrt{\pi(1 + t^2)}$.

Although the above is a valid example of a function of type 3, one might ask for a function having more function-theoretic interest than a constant. Consider the function

$$H_3(z) = \begin{cases} \dfrac{1}{1 + z^1 z^2} \left(z^1 - z^2 + \dfrac{iz^2(1 + (z^1)^2)}{\sqrt{z^1 z^2}} \right) & \text{if } z^1 z^2 \neq -1 \\[2ex] \frac{1}{2}(z^1 + z^2) & \text{if } z^1 z^2 = -1 \end{cases} \tag{6.74}$$

where we take the branch of the square root that is analytic in $\mathbb{C} \setminus [0, \infty)$ with range Π. It is shown in [26, examples 2.11 and 5.6] that $H_3 \in \mathscr{P}(\mathbb{H}^2)$ and that H_3 has the type 3 representation

$$H_3(z) = \langle M(z)v, v \rangle_{L^2(\mathbb{R})}, \tag{6.75}$$

where $M(z)$ is the Nevanlinna resolvent of type 3 defined by equation (6.55) corresponding to the choices $\mathcal{M} = L^2(\mathbb{R})$; $A =$ the operation of multiplication by the independent variable t; $Y = (P, Q)$, where P, Q are the orthogonal projection operators onto the subspaces of even and odd functions, respectively, in $L^2(\mathbb{R})$; and $v(t) = 1/\sqrt{\pi(1 + t^2)}$.

Thus equation (6.75) is a type 3 representation of the function H_3 given by equation (6.74). This function is *constant* and equal to i on the diagonal $z_1 = z_2$. Hence it is false that

$$\liminf_{s \to \infty} \operatorname{Im} H_3(is, is) < \infty.$$

Therefore, by Theorem 6.66, H_3 has no representation of type 2. Thus H_3 has type 3.

Example 6.76. The function

$$h_4(z) = \frac{z^1 z^2}{z^1 + z^2} = -\left(-\frac{1}{z^1} - \frac{1}{z^2} \right)^{-1}$$

168   *Commutative Theory*

clearly belongs to $\mathscr{P}(\mathbb{H}^2)$. Since

$$\frac{1}{s}\operatorname{Im}h_4(is,is) = \tfrac{1}{2},$$

it follows from Theorem 6.65 that h_4 has no representation of type 3. Thus h_4 has type 4.

A type 4 representation of h_4 is

$$h_4(z) = \langle M(z)v,v\rangle_{\mathbb{C}^2} \quad \text{for all } z \in \mathbb{H}^2,$$

where

$$M(z) = \frac{1}{z^1 + z^2}\begin{bmatrix} 2z^1z^2 & i(z^1 - z^2) \\ -i(z^1 - z^2) & -2 \end{bmatrix} \tag{6.77}$$

is the Nevanlinna resolvent of type 4, as in Definition 6.42, corresponding to

$$\mathcal{M} = \mathbb{C}^2, \quad \mathcal{N}_1 = \mathbb{C}\begin{pmatrix} 1 \\ 0 \end{pmatrix}, \quad \mathcal{N}_2 = \mathbb{C}\begin{pmatrix} 0 \\ 1 \end{pmatrix}, \quad A = 0 \text{ on } \mathcal{N}_2,$$

$$\mathcal{M}_1 = \mathbb{C}\begin{pmatrix} 1 \\ 1 \end{pmatrix}, \quad \mathcal{M}_2 = \mathbb{C}\begin{pmatrix} 1 \\ -1 \end{pmatrix}, \text{ and } v = \frac{1}{\sqrt{2}}\begin{pmatrix} 1 \\ 0 \end{pmatrix}.$$

Another function of type 4 is $\sqrt{z^1z^2}$.

6.7 Historical Notes

Rolf Nevanlinna proved his two representation theorems for $\mathscr{P}(\mathbb{H})$ in the course of solving the important problem of the uniqueness of solutions of the Stieltjes moment problem [155]. His purely function-theoretic results had an unforeseen application to operator theory, when Stone used them to construct the spectral measure of a self-adjoint operator and thereby prove the spectral theorem for such operators. A modern version of this approach can be found in [140].

Nevanlinna's representation theorems are extended to n variables in [26]. Representations are found not for the class $\mathscr{P}(\mathbb{H}^n)$ but for the strictly (when $n > 2$) smaller *Löwner class*. An integral representation for functions in $\mathscr{P}(\mathbb{H}^n)$ has recently been given by Luger and Nedic [146].

7

Model Theory on the Symmetrized Bidisc

The *symmetrized bidisc* is the domain G in $\mathbb{C}^2$ defined by the formula

$$G = \{(\lambda^1 + \lambda^2, \lambda^1 \lambda^2): \lambda^1, \lambda^2 \in \mathbb{D}\}.$$

One reason to study the function theory of this domain is the following classical result. Let $\pi \colon \mathbb{D}^2 \to G$ be defined by

$$\pi(\lambda) = (\lambda^1 + \lambda^2, \lambda^1 \lambda^2) \qquad \text{for all } \lambda \in \mathbb{D}^2.$$

We say that $\psi \in \text{Hol}(\mathbb{D}^2)$ is *symmetric* if $\psi(\lambda^1, \lambda^2) = \psi(\lambda^2, \lambda^1)$ for all $\lambda \in \mathbb{D}^2$. Clearly, if $\varphi \in \text{Hol}(G)$, then $\varphi \circ \pi$ is a symmetric function in $\text{Hol}(\mathbb{D}^2)$.

Theorem 7.1. (The Waring–Lagrange theorem for $\text{Hol}(\mathbb{D}^2)$**)** *If ψ is a symmetric function in* $\text{Hol}(\mathbb{D}^2)$, *then there exists $\varphi \in \text{Hol}(G)$ such that* $\psi = \varphi \circ \pi$.

Thus function theory on G is equivalent to the theory of *symmetric* functions on $\mathbb{D}^2$.

G is also interesting from the point of view of complex geometry, especially for the theory of invariant metrics [122, chapter 7]. Although G is closely related to the bidisc, its geometry is in many ways richer. For example, its distinguished boundary is topologically a Möbius band, a fact that endows the space of holomorphic maps from $\mathbb{D}$ to G with a subtle structure. Another feature of interest is its connection to the *spectral Nevanlinna–Pick problem*, which is a special case of a problem posed in the 1980s by control engineers.

In this chapter we apply operator-theoretic methods to the function theory of G. We show how to symmetrize the model formula for the bidisc to obtain a model formula for the symmetrized bidisc. As a consequence, we deduce a von Neumann inequality for G (via a positivity argument) and a realization formula for G (via a lurking isometry argument). We discuss the spectral Nevanlinna–Pick problem in Section 7.7.

169

170   *Commutative Theory*

7.1 Adding Symmetry to the Fundamental Theorem for $\mathbb{D}^2$

Let the "transposition" $\sigma\colon \mathbb{D}^2 \to \mathbb{D}^2$ be defined by

$$\sigma(\lambda^1, \lambda^2) = (\lambda^2, \lambda^1) \qquad \text{for all } \lambda \in \mathbb{D}^2.$$

It will be convenient to use the notation λ^σ instead of $\sigma(\lambda)$, so that $(\lambda^1, \lambda^2)^\sigma \stackrel{\text{def}}{=} (\lambda^2, \lambda^1)$.

We shall say that a function h on $\mathbb{D}^2 \times \mathbb{D}^2$ is *doubly symmetric* if

$$h(\sigma(\lambda), \mu) = h(\lambda, \mu) = h(\lambda, \sigma(\mu)) \quad \text{for all } \lambda, \mu \in \mathbb{D}^2.$$

As we did for the disc in Definition 3.47, we extend slightly the notion of model for functions on $\mathbb{D}^2$ introduced in Definition 4.18. We shall say that a function h on $\mathbb{D}^2 \times \mathbb{D}^2$ has a *model on the bidisc* if there exist a Hilbert space $\mathcal{M} = \mathcal{M}_1 \oplus \mathcal{M}_2$ and a map $u = u^1 \oplus u^2\colon \mathbb{D}^2 \to \mathcal{M}$ such that

$$h(\lambda, \bar{\mu}) = (1 - \bar{\mu}^1 \lambda^1)\langle u_\lambda^1, u_\mu^1 \rangle + (1 - \bar{\mu}^2 \lambda^2)\langle u_\lambda^2, u_\mu^2 \rangle \qquad \text{for all } \lambda, \mu \in \mathbb{D}^2. \tag{7.2}$$

Thus a function φ on $\mathbb{D}^2$ has a model in the sense of Definition 4.18 if and only if the hereditary function $1 - \overline{\varphi(\mu)}\varphi(\lambda)$ has a model in this broadened sense.

Lemma 7.3. *If h is a doubly symmetric function on $\mathbb{D}^2 \times \mathbb{D}^2$ and h has a model on the bidisc, then there exists a Hilbert space $\mathcal{M}$, a unitary operator U on $\mathcal{M}$, and a function $w\colon G \to \mathcal{M}$ such that, for all $\lambda, \mu \in \mathbb{D}^2$,*

$$h(\lambda, \bar{\mu}) = \Big\langle \Xi(s, \bar{t}) \, w_s, w_t \Big\rangle_{\mathcal{M}}, \tag{7.4}$$

where $s = (\lambda^1 + \lambda^2, \lambda^1 \lambda^2)$, $t = (\mu^1 + \mu^2, \mu^1 \mu^2)$, and

$$\Xi(s, \bar{t}) = \left[(1 - \bar{t}^2 s^2) - \tfrac{1}{2}(\bar{t}^1 - s^1 \bar{t}^2) U - \tfrac{1}{2}(\bar{s}^1 - t^1 \bar{s}^2) U^* \right]. \tag{7.5}$$

Proof. We break the proof up into a sequence of follow-your-nose steps.

Step 1. Assume that h is doubly symmetric and has the model (7.2) for some Hilbert space $\mathcal{M} = \mathcal{M}_1 \oplus \mathcal{M}_2$ and some map $u = u^1 \oplus u^2\colon \mathbb{D}^2 \to \mathcal{M}$.

Step 2. Replacing λ with λ^σ and μ with μ^σ in equation (7.2), we obtain

$$h(\lambda, \bar{\mu}) = (1 - \bar{\mu}^2 \lambda^2)\langle u_{\lambda^\sigma}^1, u_{\mu^\sigma}^1 \rangle_{\mathcal{M}_1} + (1 - \bar{\mu}^1 \lambda^1)\langle u_{\lambda^\sigma}^2, u_{\mu^\sigma}^2 \rangle_{\mathcal{M}_2}. \tag{7.6}$$

Step 3. Addition of equations (7.2) and (7.6) yields

$$2h(\lambda, \bar{\mu}) = (1 - \bar{\mu}^1 \lambda^1)\left\langle \begin{bmatrix} u_\lambda^1 \\ u_{\lambda^\sigma}^2 \end{bmatrix}, \begin{bmatrix} u_\mu^1 \\ u_{\mu^\sigma}^2 \end{bmatrix} \right\rangle_{\mathcal{M}} + (1 - \bar{\mu}^2 \lambda^2)\left\langle \begin{bmatrix} u_{\lambda^\sigma}^1 \\ u_\lambda^2 \end{bmatrix}, \begin{bmatrix} u_{\mu^\sigma}^1 \\ u_\mu^2 \end{bmatrix} \right\rangle_{\mathcal{M}}. \tag{7.7}$$

Model Theory on the Symmetrized Bidisc

Step 4. Define

$$v_\lambda = \begin{bmatrix} u_\lambda^1 \\ u_{\lambda^\sigma}^2 \end{bmatrix} \in \mathcal{M},$$

so that equation (7.7) becomes

$$2\,h(\lambda,\bar\mu) = (1 - \bar\mu^1\lambda^1)\langle v_\lambda, v_\mu\rangle_\mathcal{M} + (1 - \bar\mu^2\lambda^2)\langle v_{\lambda^\sigma}, v_{\mu^\sigma}\rangle_\mathcal{M}. \tag{7.8}$$

Step 5. Now, $h(\lambda^\sigma, \bar\mu) = h(\lambda, \mu)$. We may therefore deduce from equation (7.8) that

$$(1 - \bar\mu^1\lambda^1)\langle v_\lambda, v_\mu\rangle + (1 - \bar\mu^2\lambda^2)\langle v_{\lambda^\sigma}, v_{\mu^\sigma}\rangle = (1 - \bar\mu^1\lambda^2)\langle v_{\lambda^\sigma}, v_\mu\rangle$$
$$+ (1 - \bar\mu^2\lambda^1)\langle v_\lambda, v_{\mu^\sigma}\rangle.$$

Step 6. (The reshuffle) Reshuffle the last formula to

$$\langle v_\lambda, v_\mu\rangle + \langle v_{\lambda^\sigma}, v_{\mu^\sigma}\rangle - \langle v_{\lambda^\sigma}, v_\mu\rangle - \langle v_\lambda, v_{\mu^\sigma}\rangle$$
$$= \bar\mu^1\lambda^1\langle v_\lambda, v_\mu\rangle + \bar\mu^2\lambda^2\langle v_{\lambda^\sigma}, v_{\mu^\sigma}\rangle - \bar\mu^1\lambda^2\langle v_{\lambda^\sigma}, v_\mu\rangle - \bar\mu^2\lambda^1\langle v_\lambda, v_{\mu^\sigma}\rangle,$$

and then factor the left and right sides of this new formula to obtain

$$\langle v_\lambda - v_{\lambda^\sigma}, v_\mu - v_{\mu^\sigma}\rangle = \langle \lambda^1 v_\lambda - \lambda^2 v_{\lambda^\sigma}, \mu^1 v_\mu - \mu^2 v_{\mu^\sigma}\rangle.$$

Step 7. (A lurking isometry) Enlarge, if necessary, the model space $\mathcal{M}$ (as in the proof of Theorem 2.57 or Remark 2.31) to ensure that the spaces

$$[v_\lambda - v_{\lambda_\sigma} : \lambda \in \mathbb{D}^2] \quad \text{and} \quad [\lambda^1 v_\lambda - \lambda^2 v_{\lambda^\sigma} : \lambda \in \mathbb{D}^2]$$

have equal codimension in $\mathcal{M}$. By the lurking isometry lemma together with the equality of codimensions, there is a unitary operator U on $\mathcal{M}$ such that

$$U(v_\lambda - v_{\lambda^\sigma}) = \lambda^1 v_\lambda - \lambda^2 v_{\lambda^\sigma} \qquad \text{for all } \lambda \in \mathbb{D}^2. \tag{7.9}$$

Step 8. Reshuffle equation (7.9) to

$$(U - \lambda^1)v_\lambda = (U - \lambda^2)v_{\lambda^\sigma}$$

or equivalently,

$$(U - \lambda^2)^{-1} v_\lambda = (U - \lambda^1)^{-1} v_{\lambda^\sigma} \quad \text{for all } \lambda \in \mathbb{D}^2.$$

This equation implies that if we define w_λ by either of the formulas

$$w_\lambda = (U - \lambda^2)^{-1} v_\lambda \quad \text{or} \quad w_\lambda = (U - \lambda^1)^{-1} v_{\lambda^\sigma},$$

then

$$v_\lambda = (U - \lambda^2)w_\lambda \quad \text{and} \quad v_{\lambda^\sigma} = (U - \lambda^1)w_\lambda \quad \text{for all } \lambda \in \mathbb{D}^2. \tag{7.10}$$

172   *Commutative Theory*

Step 9. If we substitute the formulas (7.10) into equation (7.8) we obtain

$$2\,h(\lambda,\bar{\mu}) = (1 - \bar{\mu}^1\lambda^1)\langle (U - \lambda^2)w_\lambda, (U - \mu^2)w_\mu\rangle$$
$$+ (1 - \bar{\mu}^2\lambda^2)\langle (U - \lambda^1)w_\lambda, (U - \mu^1)w_\mu\rangle$$
$$= (1 - \bar{\mu}^1\lambda^1)\langle (U - \mu^2)^*(U - \lambda^2)w_\lambda, w_\mu\rangle$$
$$+ (1 - \bar{\mu}^2\lambda^2)\langle (U - \mu^1)^*(U - \lambda^1)w_\lambda, w_\mu\rangle$$
$$= \langle \Omega\, w_\lambda, w_\mu\rangle \quad \text{for all } \lambda, \mu \in \mathbb{D}^2, \tag{7.11}$$

where

$$\Omega = (1 - \bar{\mu}^1\lambda^1)(U - \mu^2)^*(U - \lambda^2) + (1 - \bar{\mu}^2\lambda^2)(U - \mu^1)^*(U - \lambda^1).$$

Step 10. (An identity) For any $\lambda, \mu \in \mathbb{D}^2$ and any unitary operator U,

$$(1 - \bar{\mu}^1\lambda^1)(U - \mu^2)^*(U - \lambda^2) + (1 - \bar{\mu}^2\lambda^2)(U - \mu^1)^*(U - \lambda^1)$$
$$= 2(1 - \bar{t}^2 s^2) - \left(\bar{t}^1 - s^1\bar{t}^2\right) U - \left(\bar{s}^1 - \bar{t}^1 s^2\right) U^*, \tag{7.12}$$

where

$$s^1 = \lambda^1 + \lambda^2 \qquad s^2 = \lambda^1\lambda^2$$

and

$$t^1 = \mu^1 + \mu^2 \qquad t^2 = \mu^1\mu^2.$$

The left-hand side of equation (7.12) is Ω from the previous step. Now

$$(U - \mu^2)^*(U - \lambda^2) = 1 + \bar{\mu}^2\lambda^2 - \bar{\mu}^2 U - \lambda^2 U^*$$

and

$$(U - \mu^1)^*(U - \lambda^1) = 1 + \bar{\mu}^1\lambda^1 - \bar{\mu}^1 U - \lambda^1 U^*.$$

Therefore,

$$\Omega = (1 - \bar{\mu}^1\lambda^1)(1 + \bar{\mu}^2\lambda^2 - \bar{\mu}^2 U - \lambda^2 U^*)$$
$$+ (1 - \bar{\mu}^2\lambda^2)(1 + \bar{\mu}^1\lambda^1 - \bar{\mu}^1 U - \lambda^1 U^*)$$
$$= (1 - \bar{\mu}^1\lambda^1)(1 + \bar{\mu}^2\lambda^2) + (1 - \bar{\mu}^2\lambda^2)(1 + \bar{\mu}^1\lambda^1)$$
$$- \left((1 - \bar{\mu}^1\lambda^1)\bar{\mu}^2 + (1 - \bar{\mu}^2\lambda^2)\bar{\mu}^1\right) U$$
$$- \left((1 - \bar{\mu}^1\lambda^1)\lambda^2 + (1 - \bar{\mu}^2\lambda^2)\lambda^1\right) U^*$$
$$= 2(1 - \bar{\mu}^1\bar{\mu}^2\lambda^1\lambda^2)$$
$$- \left(\overline{\mu^1 + \mu^2} - (\lambda^1 + \lambda^2)\,\overline{\mu^1\mu^2}\right) U$$
$$- \left(\overline{\lambda^1 + \lambda^2} - (\mu^1 + \mu^2)\,\overline{\lambda^1\lambda^2}\right) U^*,$$

$$\text{Model Theory on the Symmetrized Bidisc} \qquad 173$$

so that, in the symmetric variables s,t, the formula for Ω becomes

$$\Omega = 2(1 - \bar{t}^2 s^2) - \left(\bar{t}^1 - s^1 \bar{t}^2\right) U - \left(\bar{s}^1 - t^1 \bar{s}^2\right) U^*, \qquad (7.13)$$

which is the identity (7.12).

Step 11. To complete the proof of Lemma 7.3 let $\Xi = \frac{1}{2}\Omega$ and note that equation (7.13) implies equation (7.5) and equation (7.11) implies equation (7.4). $\qquad\square$

7.2 How to Define Models on the Symmetrized Bidisc

In this section we prove a fundamental theorem for the symmetrized bidisc.

First seek to express $\Xi(s,\bar{t})$, defined in equation (7.5), as a difference of positive terms, that is, in the form

$$\Xi(s,\bar{t}) = \left(a(t) + b(t)U\right)^* \left(a(s) + b(s)U\right) - \left(c(t) + d(t)U\right)^* \left(c(s) + d(s)U\right),$$

where a, b, c, $d \in \mathrm{Hol}(G)$. This is a simple algebraic exercise, which leads to the formula

$$\Xi(s,\bar{t}) = \left(1 - \tfrac{1}{2}t^1 U\right)^* \left(1 - \tfrac{1}{2}s^1 U\right) - \left(\tfrac{1}{2}t^1 - t^2 U\right)^* \left(\tfrac{1}{2}s^1 - s^2 U\right). \quad (7.14)$$

If we substitute this formula for Ξ into equation (7.4) we obtain

$$h(\lambda,\bar{\mu}) = \left\langle \Xi(s,\bar{t})w_s, w_t \right\rangle_{\mathcal{M}}$$

$$= \left\langle \left[\left(1 - \tfrac{1}{2}t^1 U\right)^* \left(1 - \tfrac{1}{2}s^1 U\right) - \left(\tfrac{1}{2}t^1 - t^2 U\right)^* \left(\tfrac{1}{2}s^1 - s^2 U\right) \right] w_s, w_t \right\rangle_{\mathcal{M}}$$

$$= \left\langle \left[1 - \left(\frac{\tfrac{1}{2}t^1 - t^2 U}{1 - \tfrac{1}{2}t^1 U}\right)^* \left(\frac{\tfrac{1}{2}s^1 - s^2 U}{1 - \tfrac{1}{2}s^1 U}\right) \right] \right.$$

$$\left. \times \left(1 - \tfrac{1}{2}s^1 U\right) w_s, \left(1 - \tfrac{1}{2}t^1 U\right) w_t \right\rangle_{\mathcal{M}}.$$

Therefore, if we define the operator s_U on $\mathcal{M}$, for $s \in G$ and a unitary operator U on $\mathcal{M}$, by the formula

$$s_U = \frac{\tfrac{1}{2}s^1 - s^2 U}{1 - \tfrac{1}{2}s^1 U},$$

and we set

$$u_s = \left(1 - \tfrac{1}{2}s^1 U\right) w_s,$$

then

$$h(\lambda,\bar{\mu}) = \left\langle (1 - t_U^* s_U) u_s, u_t \right\rangle_{\mathcal{M}}. \qquad (7.15)$$

This suggests the following definition.

174   *Commutative Theory*

Definition 7.16. *Let $\varphi\colon G \to \mathbb{C}$ be a function. A G-model for φ is a triple $(\mathcal{M}, U, u)$ where $\mathcal{M}$ is a Hilbert space, U is a unitary operator acting on $\mathcal{M}$, $u\colon G \to \mathcal{M}$, and*

$$1 - \overline{\varphi(t)}\varphi(s) = \left\langle (1 - t_U^* s_U)\, u_s, u_t \right\rangle_{\mathcal{M}} \quad \text{for all } s, t \in G. \tag{7.17}$$

Theorem 7.18. (**Fundamental theorem for the symmetrized bidisc**) *Every function in $\mathscr{S}(G)$ has a G-model.*

Proof. Let $\varphi \in \mathscr{S}(G)$. Then $\varphi \circ \pi$ is a symmetric function in $\mathscr{S}(\mathbb{D}^2)$. Hence, by the fundamental theorem for the bidisc (Theorem 4.35), the function

$$h(\lambda, \bar{\mu}) = 1 - \overline{\varphi(\mu^1 + \mu^2, \mu^1 \mu^2)}\varphi(\lambda^1 + \lambda^2, \lambda^1 \lambda^2)$$

has a model representation on the bidisc. It is clearly doubly symmetric. Therefore, by Lemma 7.3 and the discussion leading up to equation (7.15), φ has a G-model. $\qquad\square$

Because of the presence of a unitary operator in the model formula, it is possible to rewrite this fundamental theorem for G in a form that is amenable to positivity arguments. For $\omega \in \mathbb{T}$, we may view ω as the unitary operator $U = [\omega]$ acting on $\mathcal{M} = \mathbb{C}$. For this U we have

$$s_U = \frac{\frac{1}{2}s^1 - s^2 U}{1 - \frac{1}{2}s^1 U} = \frac{\frac{1}{2}s^1 - s^2[\omega]}{1 - \frac{1}{2}s^1[\omega]}.$$

It follows that if we define $C_\omega\colon G \to \mathbb{C}$ by

$$C_\omega(s) = \frac{\frac{1}{2}s^1 - s^2\omega}{1 - \frac{1}{2}s^1\omega} \qquad \text{for } s \in G, \tag{7.19}$$

then

$$s_{[\omega]} = [C_\omega(s)].$$

Lemma 7.20. *For every $\omega \in \mathbb{T}$ and $s \in G$,*

$$|C_\omega(s)| < 1.$$

Proof. Invoke the identity (7.12) with U the operator of multiplication by ω on $\mathbb{C}$ and with $\mu = \lambda$ (so that also $s = t$) to obtain

$$(1 - |\lambda^1|^2)\, |\omega - \lambda^2|^2 + (1 - |\lambda^2|^2)\, |\omega - \lambda^1|^2$$
$$= 2(1 - |s^2|)^2 - (\bar{s}^1 - s^1\bar{s}^2)\omega - (s^1 - \bar{s}^1 s^2)\bar{\omega}.$$

Model Theory on the Symmetrized Bidisc   175

The right-hand side here is $\Xi(s,\bar{s})$ (with $U = \omega$), and so, by the identity (7.14),

$$(1 - |\lambda^1|^2)\,|\omega - \lambda^2|^2 + (1 - |\lambda^2|^2)\,|\omega - \lambda^1|^2$$

$$= \left(1 - \tfrac{1}{2}s^1\omega\right)^* \left(1 - \tfrac{1}{2}s^1\omega\right) - \left(\tfrac{1}{2}s^1 - s^2\omega\right)^* \left(\tfrac{1}{2}s^1 - s^2\omega\right)$$

$$= \left(1 - |C_\omega(s)|^2\right)|1 - \tfrac{1}{2}s^1\omega|^2 \qquad \text{for all } s \in G.$$

The left-hand side is strictly positive and $|s^1| < 2$ for $s \in G$. It follows that $|C_\omega(s)| < 1$. $\qquad\qquad\square$

Consider any unitary operator U on a Hilbert space $\mathcal{M}$. Since U is a normal operator having spectrum contained in $\mathbb{T}$, we may apply the spectral theorem, Theorem 1.18, to obtain the spectral decomposition

$$U = \int_{\mathbb{T}} \omega\,dE(\omega), \qquad\qquad (7.21)$$

where E is a $\mathcal{B}(\mathcal{M})$-valued spectral measure on $\mathbb{T}$.

Lemma 7.22. *Let U be a unitary operator with spectral decomposition (7.21). For any $s,t \in G$,*

$$1 - t_U^* s_U = \int_{\mathbb{T}} 1 - \overline{C_\omega(t)}C_\omega(s)\,dE(\omega). \qquad\qquad (7.23)$$

Proof. Consider any $s,t \in G$. By properties of spectral integrals,

$$s_U = \frac{\tfrac{1}{2}s^1 - s^2 U}{1 - \tfrac{1}{2}s^1 U} = \int_{\mathbb{T}} C_\omega(s)\,dE(\omega),$$

and therefore,

$$t_U^* = \int_{\mathbb{T}} \overline{C_\omega(t)}\,dE(\omega).$$

By the multiplicativity of spectral integrals,

$$t_U^* s_U = \int_{\mathbb{T}} \overline{C_\omega(t)}C_\omega(s)\,dE(\omega).$$

Equation (7.23) follows. $\qquad\qquad\square$

We may therefore re-cast G-models as integral representations.

Theorem 7.24. (Fundamental theorem for the symmetrized bidisc— spectral version) *Let $\varphi \in \mathscr{S}(G)$. There exist a Hilbert space $\mathcal{M}$, a $\mathcal{B}(\mathcal{M})$-valued spectral measure E on $\mathbb{T}$, and an analytic map $u\colon G \to \mathcal{M}$ such that*

$$1 - \overline{\varphi(t)}\varphi(s) = \int_{\mathbb{T}} (1 - \overline{C_\omega(t)}C_\omega(s))\,\langle dE(\omega)u_s, u_t\rangle \quad \text{for all } s,t \in G.$$

$$(7.25)$$

176 *Commutative Theory*

Proof. By Theorem 7.18, φ has a G-model $(\mathcal{M}, U, u)$, which is to say that equation (7.17) holds. Let the spectral decomposition of the unitary operator U be given by equation (7.21). By Lemma 7.22, equation (7.23) holds for any $s, t \in G$. On substituting this relation into equation (7.17) we obtain the desired representation (7.25). $\qquad\square$

G-models work equally well for operator-valued functions. Extend Definition 7.16 as follows.

Definition 7.26. *Let $\mathcal{H}, \mathcal{K}$ be Hilbert spaces and let $\varphi \colon G \to \mathcal{B}(\mathcal{H}, \mathcal{K})$ be a map. A G-model for φ is a triple $(\mathcal{M}, U, u)$, where $\mathcal{M}$ is a separable Hilbert space, U is a unitary operator on $\mathcal{M}$ and $u \colon G \to \mathcal{B}(\mathcal{H}, \mathcal{M})$ is a map such that*

$$\mathrm{id}_{\mathcal{H}} - \varphi(t)^* \varphi(s) = u(t)^* (1 - t_U^* s_U) u(s) \qquad \text{for all } s, t \in G.$$

We may then strengthen the fundamental theorem for G (Theorem 7.18) to cover the operator-valued Schur class on G [30, theorem 2.5].

Theorem 7.27. *Let $\mathcal{H}, \mathcal{K}$ be Hilbert spaces. Every function in $\mathscr{S}_{\mathcal{B}(\mathcal{H},\mathcal{K})}(G)$ has a G-model.*

To prove this, invoke the fundamental theorem for $\mathscr{S}_{\mathcal{B}(\mathcal{H},\mathcal{K})}(\mathbb{D}^2)$, Theorem 4.93, and repeat the symmetrization argument in the proof of Lemma 7.3.

7.3 The Network Realization Formula for G

As we showed for the Schur class of the disc, model formulae lead quickly via lurking isometry arguments to realization formulae. The method of Section 3.3 together with Theorem 7.18 produces the following realization.

Theorem 7.28. *Let $\mathcal{H}, \mathcal{K}$ be Hilbert spaces.*

If $\varphi \in \mathscr{S}_{\mathcal{B}(\mathcal{H},\mathcal{K})}(G)$, then there exist a Hilbert space $\mathcal{M}$, a unitary operator U on $\mathcal{M}$, and a unitary operator

$$\begin{bmatrix} A & B \\ C & D \end{bmatrix} \colon \mathcal{H} \oplus \mathcal{M} \to \mathcal{K} \oplus \mathcal{M} \tag{7.29}$$

such that, for all $s \in G$,

$$\varphi(s) = A + B s_U (1 - s_U)^{-1} C.$$

Conversely, if a Hilbert space $\mathcal{M}$, a contraction T on $\mathcal{M}$, and a contraction of the form (7.29) are given, then the function $\varphi \colon G \to \mathcal{B}(\mathcal{H}, \mathcal{K})$ defined by

$$\varphi(s) = A + B s_T (1 - D s_T)^{-1} C$$

belongs to the Schur class $\mathscr{S}_{\mathcal{B}(\mathcal{H},\mathcal{K})}(G)$.

Model Theory on the Symmetrized Bidisc 177

In contrast to the realization formula for $\mathscr{S}(\mathbb{D}^2)$, we require two unitary operators, U and $ABCD$ (or two contractions), to realize a bounded holomorphic function on G. This is due to the fact that the argument makes use of two lurking isometry arguments, reflecting the more subtle geometry of G. Full details are in [29, section 3].

7.4 The Hereditary Functional Calculus on G

In Sections 2.8 and 4.1 we encountered hereditary functions on the disc and bidisc, and their associated functional calculi. More generally, we define a hereditary function on a domain Ω in $\mathbb{C}^d$ to be a complex-valued function h on $\Omega \times \Omega^*$ such that the map

$$(\lambda, \mu) \mapsto h(\lambda, \bar{\mu})$$

is holomorphic on $\Omega \times \Omega$. Here Ω^* denotes $\{\bar{z}: z \in \Omega\}$ and, if $z = (z^1, \ldots, z^d)$, then $\bar{z}$ denotes $(\overline{z^1}, \ldots, \overline{z^d})$.

The set of hereditary functions on Ω is denoted by $\text{Her}(\Omega)$; it is a real Fréchet space under the topology of uniform convergence on compact subsets of $\Omega \times \Omega^*$. Naturally, we desire a functional calculus for $\text{Her}(\Omega)$—we want to make sense of $h(T)$ for any $h \in \text{Her}(\Omega)$ and any commuting tuple T of operators whose joint spectrum is contained in Ω. We outlined one way to do this, for a general Ω, in Remark 4.15. However, the construction depended on a theory that we do not wish to assume, and so we shall present a special argument for the case that $\Omega = G$.

Note that the closure Γ of G in $\mathbb{C}^2$ is $\{(z+w, zw): |z| \le 1, |w| \le 1\}$. For any positive real number r, dilation by r is an action on $\mathbb{C}^2$, which, when factored through π, sends a pair (s^1, s^2) to (rs^1, rrs^2). (It is hard to write this elegantly, since we are using s^2 to denote the second component of s. We multiply s^2 by the square of r.) We shall use the notation

$$r \cdot \Gamma \overset{\text{def}}{=} \{(rs^1, rrs^2): s \in \Gamma\}.$$

Similarly, $r \cdot G$ denotes $\{(rs^1, rrs^2): s \in G\}$.

Lemma 7.30. *The symmetrized bidisc G and its closure Γ are polynomially convex.*

Proof. Consider $\lambda \in \mathbb{C}^2 \setminus \Gamma$. Write $\lambda = (z + w, zw)$ for some $z, w \in \mathbb{C}$. Since $\lambda \notin \Gamma$, at least one of z, w has modulus greater than 1. If both have modulus greater than 1 then the co-ordinate polynomial $f(s^1, s^2) = s^2$ satisfies $|f| \le 1$ on Γ and $|f(\lambda)| > 1$. If $|z| > 1$ but $|w| \le 1$ then consider the polynomial

178 Commutative Theory

$\frac{1}{2}((z^1)^k + (z^2)^k)$ for positive integers k. Since this polynomial is symmetric in z^1 and z^2, it is expressible as $f_k(z^1 + z^2, z^1 z^2)$ for some polynomial f_k in two variables. For any k, $|f_k| \le 1$ on Γ, while for sufficiently large k, $|f_k(\lambda)| > 1$. Hence Γ is polynomially convex.

If K is a compact subset of G then $K \subset r \cdot \Gamma \subset G$ for some $r \in (0,1)$. Hence the polynomial hull of K is contained in G. Thus G is a polynomially convex domain. $\qquad\square$

Clearly $r \cdot \Gamma$ is also polynomially convex for every $r > 0$. The significance of of Lemma 7.30 for us is that it simplifies interpretation of the statement "$\sigma(T) \subset G$" for a commuting tuple T of operators. In principle, the meaning of this statement depends on which notion of joint spectrum is being considered. However, because G is polynomially convex, the statements are all equivalent, whichever spectrum one chooses. We may therefore understand $\sigma(T)$ as the simplest form of spectrum, $\sigma_{\mathrm{alg}}(T)$, as defined in equation (1.5).

An interesting observation is that we *cannot* directly repeat the power series method that we used to define the hereditary functional calculus for the bidisc. It is not the case that every analytic function φ on G can be described by a power series of the form

$$\varphi(s) = \sum_n a_n (s^1, s^2)^n,$$

with n a multi-index (n_1, n_2) ranging over pairs of non-negative integers. For suppose this series converges on G. Then it also converges on iG, and so φ is analytic on iG. However, we can easily produce a function φ, analytic on G, which does not extend to be analytic on iG. Consider, for example, the point $(2r, r^2) \in G$, for some $r \in (0,1)$. One can check that $i(2r, r^2)$ is not in the closure Γ of G for r close to 1. By Lemma 7.30 there is a polynomial f in two variables such that $|f| \le 1$ on Γ and $|f(2ir, ir^2)| > 1$. Let $\varphi(s) = (f(s) - f(2ir, ir^2))^{-1}$ for $s \in \mathbb{C}^2$. Then f is analytic except on a variety and is analytic on G but does not extend analytically to the point $(2ir, ir^2) \in iG$.

Since holomorphic functions on G are not all given by power series, a fortiori not all hereditary functions on G can be expressed by means of a power series in s and $\bar{s}$. Instead one can make use of power series or contour integrals on the bidisc. We shall use the latter. Note that $G^* = G$.

We shall derive an integral formula for h. Recall that $\pi(\lambda) = (\lambda^1 + \lambda^2, \lambda^1 \lambda^2)$ for $\lambda \in \mathbb{C}^2$.

Lemma 7.31. *Let $h \in \mathrm{Her}(G)$ and let $r \in (0,1)$. Let $\gamma(\theta) = re^{i\theta}, 0 \le \theta \le 2\pi$. For every $s, t \in r \cdot G$,*

$$h(s, \bar{t}) = \frac{1}{(2\pi i)^4} \int_\gamma \int_\gamma \int_\gamma \int_\gamma h(\pi(\lambda), \pi(\overline{\mu})) K_{\lambda\mu}(s, \bar{t}) d\lambda^1 \, d\lambda^2 \, d\overline{\mu^1} \, d\overline{\mu^2},$$

$$(7.32)$$

where, for $\lambda, \mu \in \mathbb{D}^2$,

$$K_{\lambda\mu}(s,\bar{t}) = A(\overline{\mu},\bar{t})A(\lambda,s)$$

and

$$A(\lambda,s) = \left(\lambda^1\lambda^2 - \tfrac{1}{2}(\lambda^1+\lambda^2)s^1 + s^2\right)$$
$$\times \left((\lambda^1)^2 - \lambda^1 s^1 + s^2\right)^{-1}\left((\lambda^2)^2 - \lambda^2 s^1 + s^2\right)^{-1}. \qquad (7.33)$$

For brevity we shall in future write

$$\int_\gamma^{(4)} \dots d\lambda\, d\overline{\mu} \quad \text{for} \quad \int_\gamma \int_\gamma \int_\gamma \int_\gamma \dots d\lambda^1\, d\lambda^2\, d\overline{\mu^1}\, d\overline{\mu^2}.$$

Proof. Consider points $s,t \in r \cdot G$ and let $\zeta, \eta \in r\mathbb{D}^2$ be such that $\pi(\zeta) = s$, $\pi(\eta) = t$. Define a function $H \in \text{Her}(\mathbb{D}^2)$ by the formula

$$H(\lambda,\mu) = h(\pi(\lambda),\pi(\mu)) \qquad \text{for all } \lambda, \mu \in \mathbb{D}^2.$$

Since $H(\lambda,\bar{\mu})$ is analytic on $\mathbb{D}^4$, Cauchy's integral formula implies that

$$H(\zeta,\bar{\eta}) = \frac{1}{(2\pi i)^4}\int_\gamma^{(4)} \frac{H(\lambda,\bar{\mu})d\lambda\, d\overline{\mu}}{(\overline{\mu^1} - \overline{\eta^1})(\overline{\mu^2} - \overline{\eta^2})(\lambda^1 - \zeta^1)(\lambda^2 - \zeta^2)}. \qquad (7.34)$$

Recall the notation $\zeta^\sigma = (\zeta^2,\zeta^1)$. Since H is doubly symmetric,

$$H(\zeta,\bar{\eta}) = H(\zeta,\overline{\eta^\sigma}) = H(\zeta^\sigma,\overline{\eta^\sigma}) = H(\zeta^\sigma,\bar{\eta}).$$

We may therefore replace the kernel

$$k_{\lambda\mu}(\zeta,\bar{\eta}) \stackrel{\text{def}}{=} (\overline{\mu^1} - \overline{\eta^1})^{-1}(\overline{\mu^2} - \overline{\eta^2})^{-1}(\lambda^1 - \zeta^1)^{-1}(\lambda^2 - \zeta^2)^{-1}$$

in equation (7.34) by the symmetrized kernel

$$k_{\lambda\mu}^{\text{sym}}(\zeta,\bar{\eta}) \stackrel{\text{def}}{=} \tfrac{1}{4}\left(k_{\lambda\mu}(\zeta,\bar{\eta}) + k_{\lambda\mu}(\zeta,\overline{\eta^\sigma}) + k_{\lambda\mu}(\zeta^\sigma,\overline{\eta^\sigma}) + k_{\lambda\mu}(\zeta^\sigma,\bar{\eta})\right),$$

and the integral formula will continue to hold. We shall express $k_{\lambda\mu}^{\text{sym}}$ in terms of the variables s,t. Let N, D be, respectively, the numerator and denominator of $k_{\lambda\mu}(\zeta,\bar{\eta}) + k_{\lambda,\mu}(\zeta^\sigma,\overline{\eta^\sigma})$. Then

$$D = \overline{P_\mu(\eta)}\,P_\lambda(\zeta),$$

where

$$P_\lambda(\zeta) = (\lambda^1 - \zeta^1)(\lambda^1 - \zeta^2)(\lambda^2 - \zeta^1)(\lambda^2 - \zeta^2)$$
$$= \left((\lambda^1)^2 - s^1\lambda^1 + s^2\right)\left((\lambda^2)^2 - s^1\lambda^2 + s^2\right).$$

180 Commutative Theory

A straightforward calculation shows that

$$N(\zeta,\bar\eta) = 2\overline{\mu^1\mu^2}\lambda^1\lambda^2 - \overline{\mu^1\mu^2}(\lambda^1+\lambda^2)s^1 - (\overline{\mu^1}+\overline{\mu^2})\lambda^1\lambda^2\overline{t^1} + 2\overline{\mu^1\mu^2}s^2$$
$$+ 2\lambda^1\lambda^2\overline{t^2} + (\overline{\mu^1}\lambda^1+\overline{\mu^2}\lambda^2)(\overline{\eta^2}\zeta^1 + \overline{\eta^1}\zeta^2)$$
$$+ (\overline{\mu^1}\lambda^2+\overline{\mu^2}\lambda^1)(\overline{\eta^1}\zeta^2 + \overline{\eta^2}\zeta^1)$$
$$- (\overline{\mu^1}+\overline{\mu^2})\overline{t^1}s^2 - (\lambda^1+\lambda^2)\overline{t^2}s^1 + 2\overline{t^1}s^2.$$

Combine this equation with the fact that

$$k^{\mathrm{sym}}_{\lambda\mu}(\zeta,\bar\eta) = \tfrac{1}{4}\overline{P_\mu(\eta)}^{-1}\left(N(\zeta,\bar\eta) + N(\zeta,\overline{\eta^\sigma})\right)P_\lambda(\zeta)^{-1},$$

and simplify to obtain the relation

$$k^{\mathrm{sym}}_{\lambda\mu}(\zeta,\bar\eta) = A(\overline{\mu},\overline{t})A(\lambda,s).$$

Thus

$$h(s,\overline{t}) = H(\zeta,\bar\eta)$$
$$= \frac{1}{(2\pi i)^4}\int_\gamma^{(4)} H(\lambda,\bar\mu)k^{\mathrm{sym}}_{\lambda\mu}(\zeta,\bar\eta)\,d\lambda\,d\overline{\mu}$$
$$= \frac{1}{(2\pi i)^4}\int_\gamma^{(4)} h(\pi(\lambda),\pi(\bar\mu))A(\overline{\mu},\overline{t})A(\lambda,s)\,d\lambda\,d\overline{\mu}.$$

$\square$

Given Lemma 7.31, there is a natural way to define the hereditary functional calculus on G, analogous to to the definition (4.11) for the bidisc. Consider a commuting pair $T = (T^1, T^2)$ of operators such that $\sigma(T) \subset G$. To define $h(T)$, for any $h \in \mathrm{Her}(G)$, we simply substitute T for s and T^* for $\overline{t}$ in the right-hand side of equation (7.32). To be more precise, choose $r > 0$ such that $\sigma(T) \subset r \cdot G$. For $s, \lambda^1 \in \mathbb{C}$,

$$(\lambda^1)^2 - s^1\lambda^1 + s^2 = 0 \quad \Leftrightarrow \quad \lambda^1 \in \pi^{-1}(\{s\}).$$

Hence $(\lambda^1)^2 - s^1\lambda^1 + s^2 \neq 0$ when $\lambda^1 \in \mathrm{ran}\,\gamma$ and $s \in \sigma(T)$. Apply the spectral mapping theorem to the polynomial function $s \mapsto (\lambda^1)^2 - s^1\lambda^1 + s^2$ and the operator pair T to deduce that

$$0 \notin \sigma((\lambda^1)^2 - T^1\lambda^1 + T^2) \quad \text{for all } \lambda^1 \in \mathrm{ran}\,\gamma.$$

Therefore, by the rational functional calculus, for any $\lambda \in \mathrm{ran}\,\gamma \times \mathrm{ran}\,\gamma$ and for A as in equation (7.33),

$$A(\lambda,T) = \left(\lambda^1\lambda^2 - \tfrac{1}{2}(\lambda^1+\lambda^2)T^1 + T^2\right)$$
$$\times \left((\lambda^1)^2 - \lambda^1 T^1 + T^2\right)^{-1}\left((\lambda^2)^2 - \lambda^2 T^1 + T^2\right)^{-1}$$

Model Theory on the Symmetrized Bidisc

is a well-defined operator, depending continuously on λ. Similarly, $A(\bar{\mu}, T^*)$ is well defined and continuous in μ on $\operatorname{ran} \gamma \times \operatorname{ran} \gamma$.

Definition 7.35. (The hereditary functional calculus on G) *Let h be a hereditary function on G and let T be a commuting pair of operators on $\mathcal{H}$ such that $\sigma(T) \subset G$. The operator $h(T)$ on $\mathcal{H}$ is given by*

$$h(T) \stackrel{\text{def}}{=} \frac{1}{(2\pi i)^4} \int_\gamma^{(4)} h(\pi(\lambda), \pi(\bar{\mu})) A(\bar{\mu}, T^*) A(\lambda, T) d\lambda \, d\bar{\mu}. \tag{7.36}$$

The integral is well defined as a repeated Riemann integral, convergent in the operator norm. In principle $h(T)$ depends on the choice of r in the definition of γ, but in fact, by deformation of contours, it does not depend on r. Nor does it depend on the order of the repeated integrals.

Remark 7.37. As in the case of the disc and bidisc, the hereditary functional calculus on G can easily be enhanced to apply to operator-valued hereditary functions. Once again it suffices to insert the symbol $\otimes$ into the defining equation (7.36).

The standard properties of the hereditary calculus (compare Lemma 4.13) continue to hold in the context of G. Here is a consequence of these simple properties.

Lemma 7.38. *Let $\mathcal{K}, \mathcal{H}$ be Hilbert spaces, let k be a positive semi-definite $\mathcal{B}(\mathcal{K})$-valued hereditary function on G, let T be a commuting pair of operators on $\mathcal{H}$ such that $\sigma(T) \subset G$, and let h be a scalar-valued hereditary function on G. If $h(T) \geq 0$, then $(kh)(T) \geq 0$.*

Here kh is the pointwise product of the functions k, h on $G \times G$, and so is a $\mathcal{B}(\mathcal{K})$-valued hereditary function.

Proof. By Moore's theorem, Theorem 3.6, there is a Hilbert space $\mathcal{M}$ and a map $u: G \to \mathcal{B}(\mathcal{K}, \mathcal{M})$ such that $k(s,t) = u(t)^* u(s)$ for all $s, t \in G$ and, moreover, $\{u(s)x: s \in G, x \in \mathcal{K}\}$ has a dense linear span in $\mathcal{M}$. We claim that u is a holomorphic map. To see this, let us first show that, for any $x \in \mathcal{K}$, the map $s \mapsto u(s)x$ is holomorphic from G to $\mathcal{M}$. Consider any finite subset $t_1, \ldots, t_n$ of points of G and vectors $x_1, \ldots, x_n$ in $\mathcal{K}$. We have, for $s \in G$,

$$\left\langle u(s)x, \sum_j u(t_j)x_j \right\rangle = \sum_j \langle u(t_j)^* u(s)x, x_j \rangle$$

$$= \sum_j \langle k(s, t_j)x, x_j \rangle.$$

182 Commutative Theory

Since k is a hereditary function, the right-hand side of the last equation is holomorphic in s. We have shown that $y^* \circ u(\cdot)x$ is holomorphic for all y^* in a dense subset of $\mathcal{M}^*$. Hence, by Lemma 2.90 and Remark 2.97, $u(\cdot)x$ is holomorphic on G for every $x \in \mathcal{K}$.

A second use of Remark 2.97 shows that u is in fact holomorphic. For non-zero vectors $x \in \mathcal{K}$, $y \in \mathcal{M}$ define a linear functional f_{xy} on $\mathcal{B}(\mathcal{K},\mathcal{M})$ by $f_{xy}(A) = \langle Ax, y \rangle_{\mathcal{Y}}$. We have $\|f_{xy}\| = \|x\|\,\|y\|$, and it is easy to check that the set $E = \{f_{xy}: x \neq 0, y \neq 0\}$ is a norm-generating subset of $\mathcal{B}(\mathcal{K},\mathcal{M})^*$. Since

$$f_{xy} \circ u(s) = \langle u(s)x, y \rangle_{\mathcal{Y}} \qquad \text{for all } s \in G, x \in \mathcal{X} \text{ and } y \in \mathcal{Y}$$

and $u(\cdot)x$ is holomorphic, $f \circ u$ is holomorphic for all $f \in E$. It follows that u is holomorphic.

Now $h1_{\mathcal{M}}$ is a $\mathcal{B}(\mathcal{M})$-valued hereditary function on G, and

$$(kh)(s,t) = u(t)^* \cdot h(s,t)1_{\mathcal{M}} \cdot u(s) \qquad \text{for all } s,t \in G.$$

By the hypothesis that $h(T) \geq 0$ and properties of the hereditary calculus,

$$(kh)(T) = u(T)^* (1_{\mathcal{M}} \otimes h(T)) u(T)$$

$$\geq 0,$$

the final inequality because of Proposition 3.35. $\qquad\qquad\square$

7.5 When Is G a Spectral Domain?

Recall the first and second Holy Grails of spectral set theory (Problems 1.48 and 1.53). The analogs for G are to find concrete necessary and sufficient conditions on a commuting pair T of operators that G be a spectral domain and a complete spectral domain, respectively, for T. The existence of G-models enables us to answer this question.

Theorem 7.39. *Let T be a commuting pair of operators acting on a Hilbert space and let $\sigma(T) \subseteq G$. Then G is a spectral domain for T if and only if*

$$\|C_\omega(T)\| \leq 1 \quad \text{for all } \omega \in \mathbb{T}.$$

Proof. Let $\omega \in \mathbb{T}$. By Lemma 7.20, $|C_\omega| \leq 1$ on G. Since C_ω is clearly analytic on G, $C_\omega(T)$ is well defined, and by the definition of spectral domain, $\|C_\omega(T)\| \leq 1$.

The converse is proven by a positivity argument. Let T act on a Hilbert space $\mathcal{H}$ and assume that $\|C_\omega(T)\| \leq 1$ for all $\omega \in \mathbb{T}$. To show that G is a spectral domain for T, consider a function $\varphi \in \mathscr{S}(G)$. We need to show that

$$\|\varphi(T)\| \leq 1.$$

Model Theory on the Symmetrized Bidisc 183

Step 1. (Invoke the model) By the fundamental theorem for G (Theorem 7.18), φ has a G-model $(\mathcal{M}, U, u)$, so that there exist a Hilbert space $\mathcal{M}$, a unitary U on $\mathcal{M}$, and an analytic map $u\colon G \to \mathcal{M}$ such that equation (7.17) holds. Let the spectral decomposition of U be $\int_{\mathbb{T}} \omega \, dE(\omega)$. By Lemma 7.22

$$1 - t_U^* s_U = \int_{\mathbb{T}} 1 - \overline{C_\omega(t)} C_\omega(s) \, dE(\omega) \quad \text{for all } s,t \in G, \tag{7.40}$$

and by the integral form of the fundamental theorem,

$$1 - \overline{\varphi(t)}\varphi(s) = \int_{\mathbb{T}} (1 - \overline{C_\omega(t)} C_\omega(s)) \, \langle \, dE(\omega) u_s, u_t \rangle_{\mathcal{M}}$$

for all $s,t \in G$.

Step 2. (Approximate the integrand in the integral (7.40)) *Let $\varepsilon > 0$ and let K be a compact subset of G. There exists $\delta > 0$ such that, for every finite partition $\{I_k\}$ of $\mathbb{T}$ by half-open intervals satisfying*

$$\max_k |I_k| < \delta,$$

the inequality

$$\left| 1 - \overline{C_{\omega_1}(t)} C_{\omega_1}(s) - \left(1 - \overline{C_{\omega_2}(t)} C_{\omega_2}(s) \right) \right| < \varepsilon$$

holds for each k, for every $\omega_1, \omega_2 \in I_k$, and every $s,t \in K$.

The assertion follows from the continuity, and hence the uniform continuity, of the map

$$(s,t,\omega) \mapsto 1 - \overline{C_\omega(t)} C_\omega(s)$$

on the compact set $K \times K \times \mathbb{T}$.

Step 3. (Approximate $1 - t_U^* s_U$ locally uniformly by a finite sum) *Let $\varepsilon > 0$ and let K be a compact subset of G. There exists $\delta > 0$ such that, for every finite partition $\{I_k\}$ of $\mathbb{T}$ by half-open intervals satisfying*

$$\max_k |I_k| < \delta \tag{7.41}$$

and every choice of sample points $\{\omega_k^*\}$ with $\omega_k^* \in I_k$, the inequality

$$\left\| 1 - t_U^* s_U - \sum_k \left(1 - \overline{C_{\omega_k^*}(t)} C_{\omega_k^*}(s) \right) E(I_k) \right\| < \varepsilon \tag{7.42}$$

holds for all $s,t \in K$.

To prove this statement, choose δ as in Step 2. Consider a finite partition $\{I_k\}$ of $\mathbb{T}$ by half-open intervals satisfying the inequality (7.41) and choose $\omega_k^* \in I_k$ for each k. By equation (7.23), for all $s,t \in K$,

184 *Commutative Theory*

$$\left\| 1 - t_U^* s_U - \sum_k \left(1 - \overline{C_{\omega_k^*}(t)} C_{\omega_k^*}(s) \right) E(I_k) \right\|$$

$$= \left\| \int_{\mathbb{T}} 1 - \overline{C_\omega(t)} C_\omega(s) \, dE(\omega) - \sum_k \left(1 - \overline{C_{\omega_k^*}(t)} C_{\omega_k^*}(s) \right) E(I_k) \right\|$$

$$= \left\| \sum_k \int_{I_k} \left(1 - \overline{C_\omega(t)} C_\omega(s) - (1 - \overline{C_{\omega_k^*}(t)} C_{\omega_k^*}(s)) \right) dE(\omega) \right\|$$

$$= \left\| \int_{\mathbb{T}} f_{st}(\omega) \, dE(\omega) \right\|,$$

where f_{st} is a measurable scalar function on $\mathbb{T}$ of modulus at most ε everywhere. On approximating f_{st} by simple functions, one readily deduces that the estimate (7.42) holds for all $s, t \in K$.

Step 4. (Approximate $1 - \overline{\varphi(t)}\varphi(s)$ locally uniformly by a finite sum) *Let $\varepsilon > 0$ and let K be a compact subset of G. There exists $\delta > 0$ such that, for every finite partition $\{I_k\}$ of $\mathbb{T}$ by half-open intervals satisfying*

$$\max_k |I_k| < \delta$$

and every choice of sample points $\{\omega_k^*\}$ with $\omega_k^* \in I_k$, the inequality

$$\left| 1 - \overline{\varphi(t)}\varphi(s) - \sum_k \left(1 - \overline{C_{\omega_k^*}(t)} C_{\omega_k^*}(s) \right) \langle E(I_k) u_s, u_t \rangle_{\mathcal{M}} \right| < \varepsilon \qquad (7.43)$$

holds for all $s, t \in K$.

To see this, note that since G-models are automatically holomorphic, u is continuous, and therefore, we may choose M such that $\| u_s \| < M$ for all $s \in K$. Then, by Step 3, we may choose $\delta > 0$ such that, for all finite partitions $\{I_k\}$ with $\max_k |I_k| < \delta$ and every choice of sample points $\{\omega_k^*\}$ with $\omega_k^* \in I_k$, the inequality

$$\left\| 1 - t_U^* s_U - \sum_k \left(1 - \overline{C_{\omega_k^*}(t)} C_{\omega_k^*}(s) \right) E(I_k) \right\| < \varepsilon M^{-2}$$

holds for all $s, t \in K$. It follows that

$$\left| \langle (1 - t_U^* s_U) u_s, u_t \rangle - \sum_k (1 - \overline{C_{\omega_k^*}(t)} C_{\omega_k^*}(s)) \langle E(I_k) u_s, u_t \rangle \right| < \varepsilon M^{-2} \| u_s \| \, \| u_t \|$$

$$< \varepsilon$$

for all $s, t \in K$. On combining this estimate with equation (7.17) we obtain the desired inequality (7.43).

Model Theory on the Symmetrized Bidisc 185

Step 5. (Positivity of certain operators) *If $I \subset \mathbb{T}$ is an interval, $\omega \in \mathbb{T}$ and the hereditary function g on G is given by*

$$g(s,t) = (1 - \overline{C_\omega(t)}C_\omega(s)) \langle E(I)u_s, u_t \rangle,$$

then $g(T) \geq 0$.

This statement is immediate from Lemma 7.38 and the hypothesis that $\|C_\omega(T)\| \leq 1$ for all $\omega \in \mathbb{T}$.

Step 6. (Conclusion) By Step 4 there is a sequence (g_n) of hereditary functions on G such that

(i) $g_n(\lambda, \mu) \to 1 - \overline{\varphi(\mu)}\varphi(\lambda)$ uniformly on compact sets as $n \to \infty$ and
(ii) each g_n is a finite sum of hereditary functions of the type described in Step 5.

By the continuity of the hereditary functional calculus, $g_n(T) \to 1 - \varphi(T)^*\varphi(T)$ in norm as $n \to \infty$. By Step 5, $g_n(T) \geq 0$ for every n. Therefore $1 - \varphi(T)^*\varphi(T) \geq 0$, and so $\|\varphi(T)\| \leq 1$ as required. $\square$

7.6 G Spectral Implies G Complete Spectral

In general, to say that a domain Ω is a *complete* spectral domain for a commuting tuple T of operators is a stronger assertion than to say that Ω is a spectral domain for T. See for example [9, 80]. On the other hand, as we have seen in previous chapters, in the special cases $\Omega = \mathbb{D}$ and $\Omega = \mathbb{D}^2$, the two assertions are equivalent. It is of interest to discover other examples of domains for which equivalence holds. The following strengthening of Theorem 7.39 shows that G is a complete spectral domain for T if and only if it is a spectral domain for T.

Theorem 7.44. *Let T be a commuting pair of operators such that $\sigma(T) \subset G$. Then G is a complete spectral domain for T if and only if $\|C_\omega(T)\| \leq 1$ for all $\omega \in \mathbb{T}$.*

Proof. Necessity is trivial from Theorem 7.39. For sufficiency, consider $\varphi \in \mathscr{S}_{\mathcal{B}(\mathcal{X,Y})}(G)$. By Theorem 7.27, φ has a G-model. The proof of Theorem 7.39 now generalizes easily to show that $\varphi(T)$ is a contraction. Hence G is a complete spectral domain for T. $\square$

7.7 The Spectral Nevanlinna–Pick Problem

An intriguing variant of the classical Nevanlinna–Pick interpolation problem arises in connection with control engineering, specifically, in the design of

186 Commutative Theory

automatic controllers that are robust with respect to various uncertainties in the mathematical model of the device under study. This problem is a particular challenge for analysts: superficially, it is close to well-understood variants of the Nevanlinna–Pick interpolation problem, yet, despite its significance for control engineers, mathematicians have been unable to develop a satisfactory theory of the problem so far.

The "problem of μ synthesis" was formulated by the control engineers John Doyle and Gunter Stein [79, 78] working at the Honeywell Corporation in the early 1980s on control problems in aerospace engineering. Here is an unsolved special case of the problem that is simple to state.

The 2×2 spectral Nevanlinna–Pick problem. *Given n distinct nodes $\lambda_1, \lambda_2, \ldots, \lambda_n \in \mathbb{D}$ and 2×2 target matrices $W_1, W_2, \ldots, W_n$, find necessary and sufficient conditions for the existence of an analytic 2×2 matrix-valued holomorphic function F on $\mathbb{D}$ satisfying*

$$F(\lambda_i) = W_i \qquad \text{for } i = 1, 2, \ldots, n \tag{7.45}$$

and

$$\sigma(F(\lambda)) \subseteq \mathbb{D} \qquad \text{for all } \lambda \in \mathbb{D}. \tag{7.46}$$

Describe all such functions F, when there are some.

Observe that, if A is a 2×2 matrix, then

$$\sigma(A) \subseteq \mathbb{D} \text{ if and only if } (\operatorname{tr} A, \det A) \in G.$$

Therefore, if F is a solution to the spectral interpolation problem (7.45)–(7.46), then the map f defined by $f(\lambda) = (\operatorname{tr} F(\lambda), \det F(\lambda))$ for $\lambda \in \mathbb{D}$ satisfies the following three conditions:

(i) f maps $\mathbb{D}$ to G.
(ii) f is a holomorphic mapping.
(iii) $f(\lambda_i) = (\operatorname{tr} W_i, \det W_i)$ for $i = 1, \ldots, n$.

We are led to study the following variant of the Nevanlinna–Pick problem of Section 2.6.

The Nevanlinna–Pick problem for G. Given n distinct points $\lambda_1, \ldots, \lambda_n \in \mathbb{D}$ and points $s_1, \ldots, s_n \in G$,

(i) determine whether there exists a holomorphic map $f \colon \mathbb{D} \to G$ such that $f(\lambda_i) = s_i$ for $i = 1, \ldots, n$, and
(ii) if so, give a construction of *all* such interpolating functions f.

The discussion shows that a necessary condition for the existence of solutions to the spectral Nevanlinna–Pick problem (7.45)–(7.46) is that the above

Model Theory on the Symmetrized Bidisc 187

Nevanlinna–Pick problem for G have a solution, where $s_i = (\operatorname{tr} W_i, \det W_i)$ for $i = 1, \ldots, n$.

Generically, the converse is also true: provided that none of $W_1, \ldots, W_n$ is a scalar multiple of the identity matrix, then any solution of the Nevanlinna–Pick problem for G can be lifted to a solution of the corresponding *spectral* Nevanlinna–Pick problem. One can therefore hope that the study of holomorphic interpolation into G will be a good way to approach the spectral Nevanlinna–Pick problem. This approach has indeed led to some modest progress on the spectral Nevanlinna–Pick problem [215]. For example, the Nevanlinna–Pick problem for G is solved for the case of two interpolation points ($n = 2$), in the sense that there is an easily-checked criterion for solvability and a recipe for an interpolating function. We shall present the criterion in the next chapter (Corollary 8.34).

7.8 Historical Notes

The fact that a symmetric polynomial in d indeterminates is expressible as a polynomial in the d elementary symmetric functions was first proved by Edward Waring, the Lucasian professor at Cambridge, in 1770 [211]. Theorem 7.1 follows easily from locally uniform approximation on $\mathbb{D}^2$ by polynomials.

The symmetrization argument in Lemma 7.3 was first given in [27], and the existence of G-models in the current form (Theorem 7.18) was given in [30].

Theorem 7.39 was proved by a different method in [27]. That method also depended on a model, but a slightly more complicated one.

The connection between operator theory and robust control has been developed since the 1980s by many mathematicians and engineers. The ideas of Doyle and Stein on the μ-synthesis problem have become an established part of robust control theory [83]. They introduced the notion of the *structured singular value* of a matrix, denoted by μ, which generalizes the operator norm and reflects structural information about uncertainties in the model of a linear system. One instance of μ is the spectral radius of a square matrix, which illustrates the fact that μ is not in general a norm. For this example the μ-synthesis problem becomes the spectral Nevanlinna–Pick problem of Section 7.7. At present, the main drawback of μ as a design tool for engineers is that it remains difficult to solve μ-synthesis problems numerically. For this reason, we believe it is a worthwhile goal for analysts to construct a thoroughgoing theory of such problems. A substantial step in this direction was taken by Bercovici, Foias, and Tannenbaum in a series of papers, including [47], based on the *commutant lifting* approach. There is a Matlab toolbox, developed by engineers, that uses an algorithm close to that proposed by Bercovici, Foias,

188 *Commutative Theory*

and Tannenbaum. However, the algorithm is slow and is not guaranteed to converge. A better theory of μ-synthesis problems would be significant for engineering practice.

The relationship of G with a problem in robust control theory was the original motivation for the study [27] of the geometry, operator theory and function theory of G in 1999. However, in the intervening two decades, there has been an explosion of interest in the theory of the domain G for reasons unconnected with applications. It turns out that G and some similarly defined domains have a rich and often explicit complex geometry, for example [61, 86, 122, 204, 135]. These domains also form the backdrop to some intriguing generalizations to commuting tuples of operators of the Sz.-Nagy–Foias models of contractions [50, 183].

8

Spectral Sets: Three Case Studies

Model theory can be used to discover a variety of interesting applications of von Neumann-like inequalities to complex analysis in a range of settings.

In this chapter we shall give three examples, two in complex geometry and the other in complex algebraic geometry.

8.1 Von Neumann's Inequality and the Pseudo-Hyperbolic Metric on $\mathbb{D}$

If $\dim \mathcal{H} = 2$ and $T \in \mathcal{B}(\mathcal{H})$, we say that T is *diagonalizable* if there exists a pair of eigenvectors for T that span $\mathcal{H}$, that is, there exist linearly independent vectors k_1 and k_2 in $\mathcal{H}$ and scalars z_1 and z_2 in $\mathbb{C}$ such that

$$T k_1 = z_1 k_1 \quad \text{and} \quad T k_2 = z_2 k_2. \tag{8.1}$$

For such an operator, $\sigma(T) = \{z_1, z_2\}$ and when φ is holomorphic on a neighborhood of $\sigma(T)$, $\varphi(T)$ is the unique operator satisfying

$$\varphi(T)k_1 = \varphi(z_1)k_1 \quad \text{and} \quad \varphi(T)k_2 = \varphi(z_2)k_2. \tag{8.2}$$

When is T a contraction? It is necessary that both $|z_1| \leq 1$ and $|z_2| < 1$. If $|z_1| = |z_2| = 1$ then $\|T\| \leq 1$ if and only if either $z_1 = z_2$ or $\langle k_1, k_2 \rangle = 0$. As these trivial cases are of no interest to us, in the following lemma we assume that $|z_1|, |z_2| < 1$.

Lemma 8.3. *If T is defined by equations* (8.1) *and* $|z_1|, |z_2| < 1$*, then* $\|T\| \leq 1$ *if and only if*

$$\frac{\left|\langle k_1, k_2 \rangle\right|^2}{\|k_1\|^2 \|k_2\|^2} \leq \frac{(1 - |z_1|^2)(1 - |z_2|^2)}{\left|1 - z_1 \bar{z}_2\right|^2}. \tag{8.4}$$

189

190   Commutative Theory

Proof. By the "fundamental fact" (1.13), $\|T\| \le 1$ if and only if $1 - T^*T \ge 0$. But if $u = \sum_{i=1}^{2} c_i k_i$ is a general vector in $\mathcal{H}$,

$$\left\langle (1 - T^*T)u, u \right\rangle = \langle u, u \rangle - \langle Tu, Tu \rangle$$

$$= \left\langle \sum_{i=1}^{2} c_i k_i, \sum_{i=1}^{2} c_i k_i \right\rangle - \left\langle \sum_{i=1}^{2} z_i c_i k_i, \sum_{i=1}^{2} z_i c_i k_i \right\rangle$$

$$= \sum_{i,j=1}^{2} (1 - z_j \bar{z}_i) \langle k_j, k_i \rangle c_j \bar{c}_i.$$

Therefore, $1 - T^*T \ge 0$ if and only if the 2×2 matrix

$$A = \begin{bmatrix} (1 - |z_1|^2)\|k_1\|^2 & (1 - z_2\bar{z}_1)\langle k_2, k_1 \rangle \\ (1 - z_1\bar{z}_2)\langle k_1, k_2 \rangle & (1 - |z_2|^2)\|k_2\|^2 \end{bmatrix}$$

is positive semi-definite. Since the diagonal entries are positive, $A \ge 0$ if and only if $\det A \ge 0$, that is,

$$\left| (1 - z_1\bar{z}_2)\langle k_1, k_2 \rangle \right|^2 \le (1 - |z_1|^2)(1 - |z_2|^2)\|k_1\|^2\|k_2\|^2,$$

or equivalently, if and only if the inequality (8.4) holds. $\qquad\square$

Remarkably, the inequality (8.4) can be replaced by a purely geometric condition. For a pair of nonzero vectors $v_1, v_2 \in \mathcal{H}$ we let θ_v denote the angle between v_1 and v_2 when $\mathcal{H}$ is viewed as a real Hilbert space, that is, the quantity,

$$\theta_v = \arccos \left(\mathrm{Re} \left\langle \frac{v_1}{\|v_1\|}, \frac{v_2}{\|v_2\|} \right\rangle \right).$$

For T in equation (8.1), we define the *angle θ_T between the eigenspaces of T* by the formula

$$\theta_T = \min \{\theta_v \colon v_1 \in \ker(T - z_1) \setminus \{0\}, v_2 \in \ker(T - z_2) \setminus \{0\}\}.$$

With these definitions it is easy to see that if $z_1, z_2 \in \mathbb{D}$ and $z_1 \ne z_2$, then

$$\cos^2 \theta_T = \frac{\left| \langle k_1, k_2 \rangle \right|^2}{\|k_1\|^2\|k_2\|^2}, \tag{8.5}$$

the expression on the left-hand side of inequality (8.4). However, if $z_1 = z_2$, then $\theta_T = 0$. Also note that if $z_1 \ne z_2$ and φ is holomorphic on a neighborhood of $\{z_1, z_2\}$, then equations (8.2) imply that

$$\theta_{\varphi(T)} = \begin{cases} \theta_T & \text{if } \varphi(z_1) \ne \varphi(z_2) \\ 0 & \text{if } \varphi(z_1) = \varphi(z_2). \end{cases}$$

Observe that in both cases, $\theta_{\varphi(T)} \le \theta_T$.

Spectral Sets: Three Case Studies 191

Lemma 8.6. *Let $\mathcal{H}$ be a 2-dimensional Hilbert space. Assume that T in $\mathcal{B}(\mathcal{H})$ is diagonalizable and that $\sigma(T) = \{z_1, z_2\} \subseteq \mathbb{D}$. Then*

$$\|T\| \leq 1 \quad \text{if and only if} \quad d(z_1, z_2) \leq \sin\theta_T,$$

where

$$d(z_1, z_2) \overset{\text{def}}{=} \left| \frac{z_1 - z_2}{1 - \bar{z}_2 z_1} \right|. \tag{8.7}$$

Proof. If $z_1 = z_2$, then $\|T\| = |z_1| < 1$, $d(z_1, z_2) = 0$, and $\sin\theta_T = 0$, and the lemma holds trivially. We may therefore assume that $z_1 \neq z_2$, so that equation (8.5) holds. Using Lemma 8.3, we deduce that $\|T\| \leq 1$ if and only if

$$1 - \frac{(1 - |z_1|^2)(1 - |z_2|^2)}{\left|1 - z_1\bar{z}_2\right|^2} \leq 1 - \frac{\left|\langle k_1, k_2 \rangle\right|^2}{\|k_1\|^2 \|k_2\|^2}. \tag{8.8}$$

But equation (8.5) implies that

$$1 - \frac{\left|\langle k_1, k_2 \rangle\right|^2}{\|k_1\|^2 \|k_2\|^2} = 1 - \cos^2\theta_T$$
$$= \sin^2\theta_T$$

and

$$1 - \frac{(1 - |z_1|^2)(1 - |z_2|^2)}{\left|1 - z_1\bar{z}_2\right|^2} = \frac{\left|1 - z_1\bar{z}_2\right|^2 - (1 - |z_1|^2)(1 - |z_2|^2)}{\left|1 - z_1\bar{z}_2\right|^2}$$
$$= \frac{\left|z_1 - z_2\right|^2}{\left|1 - z_1\bar{z}_2\right|^2}$$
$$= d(z_1, z_2)^2.$$

Consequently, inequality (8.8) simplifies to $d(z_1, z_2)^2 \leq \sin^2\theta_T$, which establishes the lemma. $\square$

Note that von Neumann's inequality implies that $\|\varphi(T)\| \leq 1$ whenever $\|T\| \leq 1$ and $\varphi \in \mathscr{S}(\mathbb{D})$. If we reinterpret this assertion using Lemma 8.6 we obtain

$$d(z_1, z_2) \leq \sin\theta_T \implies d(\varphi(z_1), \varphi(z_2)) \leq \sin\theta_{\varphi(T)} \leq \theta_T$$

whenever T is diagonalizable and $\sigma(T) = \{z_1, z_2\}$. Choosing T so that $\theta_T = d(z_1, z_2)$, we obtain

$$d(\varphi(z_1), \varphi(z_2)) \leq d(z_1, z_2),$$

the invariant form of Schwarz's lemma.

192   *Commutative Theory*

Thus von Neumann's inequality implies the Schwarz lemma. The traffic is two way: if one assumes the Schwarz lemma, then Lemma 8.6 implies that von Neumann's inequality holds for diagonalizable T acting on 2-dimensional spaces. In the next section we show how this intimate relationship between von Neumann's inequality and the Schwarz lemma can be generalized to give a spectral domain interpretation of the Carathéodory metric on domains in $\mathbb{C}^d$.

8.2 Spectral Domains and the Carathéodory Metric

Let U be a domain in $\mathbb{C}^d$. The *Carathéodory distance function* (in pseudo-hyperbolic form) is the metric d_U defined on U by the formula

$$d_U(\lambda_1, \lambda_2) = \sup_{\varphi \in \mathscr{S}(U)} d(\varphi(\lambda_1), \varphi(\lambda_2)) \qquad \text{for } \lambda_1, \lambda_2 \in U, \qquad (8.9)$$

where d denotes the *pseudo-hyperbolic metric on* $\mathbb{D}$, that is, the metric d on $\mathbb{D}$ defined in equation (8.7).

For general U, d_U does not necessarily separate points and so is really a "pseudodistance," but for *bounded* domains U, the scaled co-ordinate functions are candidates for φ in the definition (8.9), and so d_U is a true distance. It is an example of an *invariant distance* on U, meaning that it is invariant under automorphisms of U. Here an *automorphism* of U is a bijective holomorphic self-map of U with holomorphic inverse. If α is an automorphism of U, then $d_U(\alpha(\lambda_1), \alpha(\lambda_2)) = d_U(\lambda_1, \lambda_2)$ for all $\lambda_1, \lambda_2 \in U$; this is immediate from the definition, since $\varphi \in \mathscr{S}(U)$ if and only if $\varphi \circ \alpha^{-1} \in \mathscr{S}(U)$. There is an extensive theory of invariant distances [133, 122].[1]

Let us say that a d-tuple T of operators acting on a 2-dimensional Hilbert space $\mathcal{H}$ is *diagonalizable* if there exists a pair of joint eigenvectors for T that span $\mathcal{H}$, that is, there exist linearly independent vectors k_1 and k_2 in $\mathcal{H}$ and a pair of points $\lambda_1, \lambda_2 \in \mathbb{C}^d$ such that the following analog of equations (8.1) holds:

$$T^r k_i = \lambda_i^r k_i \qquad \text{for } i = 1, 2 \text{ and } r = 1, 2, \ldots, d. \qquad (8.10)$$

For such a d-tuple, $\sigma(T) = \{\lambda_1, \lambda_2\}$, and when φ is holomorphic on a neighborhood of $\sigma(T)$, $\varphi(T)$ is the unique operator satisfying

$$\varphi(T)k_1 = \varphi(\lambda_1)k_1 \quad \text{and} \quad \varphi(T)k_2 = \varphi(\lambda_2)k_2.$$

As in the case of a single diagonalizable operator, for a diagonalizable d-tuple T as in equations (8.10) we may define θ_T, the angle between the eigenspaces of T, by the formula

---

[1] In [122] d_U is called the *Möbius distance* and the term *Carathéodory distance* is reserved for arctanh d_U.

Spectral Sets: Three Case Studies

$$\theta_T = \min \{\theta_v \colon v_1 \in \bigcap_{r=1}^{d} \ker(T - \lambda_1^r), v_2 \in \bigcap_{r=1}^{d} \ker(T - \lambda_2^r)\}.$$

As before, if $\lambda_1 \neq \lambda_2$, then equation (8.5) holds, and if $\lambda_1 = \lambda_2$, then $\theta_T = 0$. Consequently, we have

$$d(\varphi(\lambda_1), \varphi(\lambda_2)) \leq \sin \theta_{\varphi(T)} \iff d(\varphi(\lambda_1), \varphi(\lambda_2)) \leq \sin \theta_T \qquad (8.11)$$

whenever φ is holomorphic on a neighborhood of $\{\lambda_1, \lambda_2\}$.

For a domain U in $\mathbb{C}^d$ and a pair of points $\lambda = (\lambda_1, \lambda_2) \in U \times U$, we let

$$\mathcal{S}_\lambda(U) = \{T \colon T \text{ is defined by equations (8.10) and}$$

$$U \text{ is a spectral domain for } T\}. \qquad (8.12)$$

Theorem 8.13. *For any domain U in $\mathbb{C}^d$,*

$$T \in \mathcal{S}_\lambda(U) \text{ if and only if } d_U(\lambda_1, \lambda_2) \leq \sin \theta_T.$$

In particular,

$$d_U(\lambda_1, \lambda_2) = \inf_{T \in \mathcal{S}_\lambda(U)} \sin \theta_T.$$

Proof.

$$T \in \mathcal{S}_\lambda(U)$$

$$\Leftrightarrow \|\varphi(T)\| \leq 1 \text{ for all } \varphi \in \mathscr{S}(U) \qquad \text{by Definition 1.46}$$

$$\Leftrightarrow d(\varphi(\lambda_1), \varphi(\lambda_2)) \leq \sin \theta_{\varphi(T)} \text{ for all } \varphi \in \mathscr{S}(U) \quad \text{by Lemma 8.6}$$

$$\Leftrightarrow d(\varphi(\lambda_1), \varphi(\lambda_2)) \leq \sin \theta_T \text{ for all } \varphi \in \mathscr{S}(U) \qquad \text{by equation (8.11)}$$

$$\Leftrightarrow \sup_{\varphi \in \mathscr{S}(U)} d(\varphi(\lambda_1), \varphi(\lambda_2)) \leq \sin \theta_T$$

$$\Leftrightarrow d_U(\lambda_1, \lambda_2) \leq \sin \theta_T \qquad \text{by equation (8.9).}$$

$\square$

We shall shortly require a natural strengthening of the notion of a spectral domain (compare with Definitions 1.46 and 1.51).

Definition 8.14. *A domain $U \subseteq \mathbb{C}^d$ is a* complete spectral domain *for a commuting d-tuple T of operators if $\sigma(T) \subset U$ and, for every positive integer n and every holomorphic $n \times n$-matrix-valued function $f = [f_{ij}]$ on U,*

$$\|f(T)\| \leq \sup_{\lambda \in U} \|f(\lambda)\|.$$

194 Commutative Theory

8.3 Background Material

We gather together some elements of two well-established theories for application to the theory of the Carathéodory distance.

8.3.1 The Spectrum of a Normal Tuple

In the proof of the main theorem in Section 8.4 we shall make essential use of the spectrum of a commuting tuple, but only for two special cases: first, for operators on a finite-dimensional (indeed, 2-dimensional) space, and second, for a commuting tuple of normal operators. In the finite-dimensional case, as we mentioned in Section 1.2, $\sigma(T)$ contains all the joint eigenvalues of T; in fact, $\sigma(T)$ is *equal* to the set of joint eigenvalues of T [67, page 24].

For normal tuples there is a simple description of the spectrum in terms of a celebrated theory developed primarily by Israel Moiseevich Gelfand. Let $N = (N^1, \ldots, N^d)$ be a d-tuple of commuting normal operators. Fuglede's theorem [91] states that if an operator S commutes with a normal operator T, then S also commutes with T^*. Therefore each N^i commutes with each $(N^j)^*$, and so the C^*-algebra with identity that the operators $N^1, \ldots, N^d$ generate is abelian. Let us call this C^*-algebra $C^*(N)$.

A *character* of a commutative Banach algebra A with identity is a non-zero homomorphism of the algebra onto $\mathbb{C}$. According to the Gerlfand theory, the characters can be identified with the maximal ideals of the algebra, and the space of characters of A is a compact Hausdorff space under a natural topology. The *Gelfand transform* $x \mapsto \hat{x}$ is an algebra-homomorphism from A to the algebra of continuous functions on the character space of A. It is defined by

$$\hat{x}(\varphi) = \varphi(x) \qquad \text{for all } x \in A \text{ and for any character } \varphi \text{ of } A.$$

In the special case that A is a commutative C^*-algebra, the Gelfand transform is an *isomorphism* of A onto the said algebra of continuous functions (with the supremum norm). See [92] for a good treatment of Gelfand theory.

When A is the algebra $C^*(N)$ introduced above, the map $\varphi \mapsto (\varphi(N^1), \ldots, \varphi(N^d))$ identifies the character space of $C^*(N)$ topologically with a compact set in $\mathbb{C}^d$.

Proposition 8.15. *For any d-tuple of commuting normal operators $N = (N^1, \ldots, N^d)$,*

$$\sigma(N) = \sigma_{C^*(N)}(N) = \{(\varphi(N^1), \ldots, \varphi(N^d)) : \varphi \text{ is a character of } C^*(N)\}.$$

See [67, proposition 7.2] for a proof.

Spectral Sets: Three Case Studies

If q is a polynomial in $z^1, \overline{z^1}, \ldots, z^d, \overline{z^d}$, then

$$q(N) \stackrel{\text{def}}{=} q(N^1, (N^1)^*, \ldots, (N^d)^*)$$

is a well-defined element of $C^*(N)$, and

$$\|q(N)\| = \max\{|q(z)|: z \in \sigma(N)\},$$

much as in the case of one variable.

8.3.2 Isometries

The *von Neumann–Wold decomposition* is a canonical way to identify an isometry V on a Hilbert space $\mathcal{H}$ as the direct sum of a unitary operator U on a subspace $\mathcal{H}_0$ of $\mathcal{H}$ and a pure isometry (one with no unitary summand) on $\mathcal{H}_1 = \mathcal{H}_0^{\perp}$. Moreover, $\mathcal{H}_1$ can be identified as

$$\mathcal{H}_1 = \left\{ x \in \mathcal{H}: \lim_{n \to \infty} (V^*)^n x = 0 \right\}.$$

Any pure isometry V is the direct sum of k copies of the unilateral shift S defined in equation (1.31), where k is a cardinal number. Equivalently, one can represent V as the multiplication operator M_z on $\mathrm{H}^2_{\mathcal{L}}$, where $\mathcal{L}$ is a Hilbert space of dimension k. Recall from Section 3.1 that one can think of $\mathrm{H}^2_{\mathcal{L}} = \mathrm{H}^2 \otimes \mathcal{L}$ as

$$\mathrm{H}^2_{\mathcal{L}} = \left\{ \sum_{n=0}^{\infty} a_n z^n : \sum_{n=0}^{\infty} \|a_n\|_{\mathcal{L}}^2 < \infty \right\},$$

the space of square-summable power series with coefficients from $\mathcal{L}$. We define M_z on $\mathrm{H}^2_{\mathcal{L}}$ by

$$(M_z f)(z) = z f(z) \qquad \text{for all } f \in \mathrm{H}^2_{\mathcal{L}} \text{ and } z \in \mathbb{D}.$$

One can show from equation (3.5) that an operator commutes with M_z if and only if it is multiplication by a bounded holomorphic $\mathcal{B}(\mathcal{L})$-valued function on $\mathbb{D}$.

A more detailed account of this material can be found in [194, chapter 1].

8.4 Lempert's Theorem

In this section and the next we shall describe two applications of Theorem 8.13, our operator-theoretic interpretation of the Carathéodory distance. Our first application is to the most striking of all theorems about invariant distances. It establishes a powerful duality between holomorphic maps from U to $\mathbb{D}$ and from $\mathbb{D}$ to U in the case of a bounded convex domain U.

196   Commutative Theory

Let U be a domain in $\mathbb{C}^d$. The *Lempert function* δ_U of U is the map $\delta_U \colon U \times U \to \mathbb{R}^+$ defined by

$$\delta_U(\lambda,\mu) = \inf\{d(z,w)\colon \text{there exists a holomorphic map } \varphi\colon \mathbb{D} \to U$$
$$\text{such that } \varphi(z) = \lambda,\ \varphi(w) = \mu\}. \qquad (8.16)$$

It follows from the connectedness of U and Weierstrass's approximation theorem that δ_U is always finite. It is not necessarily a pseudodistance, however, as it may fail to satisfy the triangle inequality; see [133, 122] for the basic theory of the Lempert function. We form a pseudodistance from δ_U as follows. For any $\lambda,\mu \in U$ the *Kobayashi distance* is the quantity

$$K_U(\lambda,\mu) \overset{\text{def}}{=} \inf \sum_{j=1}^{k-1} \delta_U(\lambda_j,\lambda_{j+1}) \qquad (8.17)$$

over all finite sequences $\lambda_1,\dots,\lambda_k$ in U such that $\lambda_1 = \lambda$ and $\lambda_k = \mu$. It is a simple consequence of the Schwarz-Pick lemma that

$$d_U \ \leq \ K_U \ \leq \ \delta_U \qquad (8.18)$$

for every domain U. Lempert's seminal result [143] is the following.

Theorem 8.19. *If U is a bounded convex domain then*

$$d_U \ = \ K_U \ = \ \delta_U.$$

We give a proof based on the foregoing theory of spectral domains. First we prove a special case. Let $\mathcal{R}(m,n)$ denote the open unit ball of the space of $m \times n$ complex matrices with respect to the operator norm (where an $m \times n$ matrix is identified with an element of $\mathcal{B}(\mathbb{C}^n,\mathbb{C}^m)$).

Proposition 8.20.

$$d_{\mathcal{R}(m,n)} \ = \ \delta_{\mathcal{R}(m,n)}.$$

Proof. Let $U = \mathcal{R}(m,n)$. In the light of inequality (8.18), it suffices to show that $d_U \geq \delta_U$. An easy form of the Schwarz lemma shows that $d_U(0,w) = \|w\|$ for any $w \in U$. If $w \neq 0$ then the holomorphic map $\varphi\colon \mathbb{D} \to U$ given by $\varphi(\lambda) = \lambda w/\|w\|$ maps $0,\|w\|$ to $0,w$, respectively, and hence $\delta_U(0,w) \leq \|w\| = d_U(0,w)$.

Now consider any pair of points $z,w \in U$. Choose an automorphism α of U such that $\alpha(z) = 0$; for example, one could take

$$\alpha_z(x) = (1 - zz^*)^{-\frac{1}{2}}(x - z)(1 - z^*x)^{-1}(1 - z^*z)^{\frac{1}{2}}.$$

Spectral Sets: Three Case Studies

Clearly α_z is holomorphic on U, and an elementary (but entertaining) calculation yields

$$1 - \alpha_z(x)^* \alpha_z(x) = (1 - z^*z)^{\frac{1}{2}} (1 - x^*z)^{-1} (1 - x^*x)(1 - z^*x)^{-1} (1 - z^*z)^{\frac{1}{2}},$$

so that α_z is a self-map of U. Moreover $(\alpha_z)^{-1} = \alpha_{-z}$, and therefore α_z is an automorphism of U. Hence, by the invariance of d_U and δ_U under automorphisms,

$$d_U(z, w) = d_U(0, \alpha_z(w)) = \|\alpha_z(w)\| = \delta_U(0, \alpha_z(w)) = \delta_U(z, w).$$

Thus $d_U = \delta_U$. $\qquad\square$

Proposition 8.21. *Let U be a domain in $\mathbb{C}^d$ and let $\lambda, \mu \in U$. With respect to a fixed basis in $\mathbb{C}^2$, let T^j be the operator on $\mathbb{C}^2$ having matrix* diag (λ^j, μ^j) *for $j = 1, \ldots, d$ and let $T = (T^1, \ldots, T^d)$. If U is a spectral domain for T then U is a complete spectral domain for T.*

Proof. Suppose that U is a spectral domain for T. Consider positive integers m, n and a holomorphic map $h \colon U \to \mathcal{R}(m, n)$. We must show that $h(T)$ is a contraction.

Let θ be the angle between the two fixed basis vectors of $\mathbb{C}^2$. By Theorem 8.13, $d_U(\lambda, \mu) \leq \sin\theta$. Now the Schwarz–Pick lemma implies that any holomorphic map between two domains is a contraction with respect to the Carathéodory distances on its domain and codomain, and therefore

$$d_{\mathcal{R}(m,n)}(h(\lambda), h(\mu)) \leq d_U(\lambda, \mu).$$

Hence, by Proposition 8.20,

$$\delta_{\mathcal{R}(m,n)}(h(\lambda), h(\mu)) = d_{\mathcal{R}(m,n)}(h(\lambda), h(\mu)) \leq d_U(\lambda, \mu).$$

By the definition of the Lempert function, there exist $\zeta, \eta \in \mathbb{D}$, and a holomorphic map $\varphi \colon \mathbb{D} \to \mathcal{R}(m, n)$ such that $\varphi(\zeta) = h(\lambda)$, $\varphi(\eta) = h(\mu)$, and $d(\zeta, \eta) \leq d_U(\lambda, \mu)$. This last inequality implies, by the definition of the Carathéodory distance, that there exists a holomorphic map $g \colon U \to \mathbb{D}$ such that $g(\lambda) = \zeta$ and $g(\mu) = \eta$. We have

$$g(T) = \begin{bmatrix} g(\lambda) & 0 \\ 0 & g(\mu) \end{bmatrix} = \begin{bmatrix} \zeta & 0 \\ 0 & \eta \end{bmatrix}.$$

Since $d(\zeta, \eta) \leq \sin\theta$, it follows from Lemma 8.6 that $g(T)$ is a contraction. Hence, by Theorem 3.42, von Neumann's inequality for operator-valued functions, $\varphi(g(T))$ is also a contraction. Now, we do not claim that $\varphi \circ g = h$, but we do assert that $\varphi \circ g(T) = h(T)$, simply because $\varphi \circ g(\lambda) = h(\lambda)$ and $\varphi \circ g(\mu) = h(\mu)$. Thus $\|h(T)\| \leq 1$. Hence U is a complete spectral domain for T. $\qquad\square$

198   Commutative Theory

Proof of Lempert's theorem. Let U be a bounded convex domain in $\mathbb{C}^d$. We wish to show that $d_U \geq \delta_U$. Consider any pair λ, μ of distinct points in U. Let $\mathcal{H}$ be a 2-dimensional Hilbert space.

Step 1. (Points in U and $\mathbb{D}$) Choose points $\zeta, \eta \in \mathbb{D}$ such that $d(\zeta, \eta) = d_U(\lambda, \mu)$ and a holomorphic map $F: U \to \mathbb{D}$ such that $F(\lambda) = \zeta$, $F(\mu) = \eta$. This is possible, since the supremum in the definition of d_U is attained, by a normal families argument. To show that $d_U(\lambda, \mu) \geq \delta_U(\lambda, \mu)$ we must construct a holomorphic map $\varphi: \mathbb{D} \to U$ such that $\varphi(\zeta) = \lambda$ and $\varphi(\eta) = \mu$.

Step 2. (A $(d+1)$-tuple of operators) Choose a basis k_1, k_2 of unit vectors in $\mathcal{H}$ such that the angle between them is θ, where $\sin\theta = d_U(\lambda, \mu)$, and $\langle k_1, k_2 \rangle_{\mathcal{H}} > 0$. Let T^j in $\mathcal{B}(\mathbb{C}^2)$ have matrix diag (λ^j, μ^j) with respect to k_1, k_2 for $j = 1, \ldots, d$, let $T = (T^1, \ldots, T^d)$, and let Λ have matrix diag (ζ, η). By Lemma 8.6,

$$\|\Lambda\| = 1. \tag{8.22}$$

$(\Lambda, T) \stackrel{\text{def}}{=} (\Lambda, T^1, \ldots, T^d)$ is a commuting $(d+1)$-tuple of operators on $\mathcal{H}$. The spectrum $\sigma(\Lambda, T)$ consists of the two joint eigenvalues of (Λ, T) corresponding to the two joint eigenvectors k_1, k_2, that is,

$$\sigma(\Lambda, T) = \{(\zeta, \lambda), (\eta, \mu)\} \subset \mathbb{D} \times U. \tag{8.23}$$

Observe that

$$F(T) = F(T^1, \ldots, T^d) = \begin{bmatrix} F(\lambda) & 0 \\ 0 & F(\mu) \end{bmatrix} = \begin{bmatrix} \zeta & 0 \\ 0 & \eta \end{bmatrix} = \Lambda.$$

Step 3. $((\mathbb{D} \times U)^-$ is a complete spectral set for $(\Lambda, T))$ By Proposition 8.13, U is a spectral domain for T. Consider any holomorphic map h from $\mathbb{D} \times U$ to $\mathbb{D}^-$. We have

$$h(\Lambda, T) = h(F(T), T) = g(T),$$

where $g: U \to \mathbb{D}^-$ is given by $g(z) = h(F(z), z)$ for $z \in U$. Since U is a spectral domain for T we have $\|g(T)\| \leq 1$, that is, $\|h(\Lambda, T)\| \leq 1$. Hence $\mathbb{D} \times U$ is a spectral domain for (Λ, T). By Proposition 8.21, $\mathbb{D} \times U$ is a *complete* spectral domain for (Λ, T).

To show that $(\mathbb{D} \times U)^-$ is a complete spectral set for (Λ, T) (Definition 1.51), note first that, since U is bounded and convex, $(\mathbb{D} \times U)^-$ is a compact polynomially convex set. Consider a holomorphic matrix-valued map h on a neighborhood of $(\mathbb{D} \times U)^-$ such that $\|h(z)\| \leq 1$ for all $z \in (\mathbb{D} \times U)^-$. According to Definition 8.14, the fact that $\mathbb{D} \times U$ is a complete spectral domain for (Λ, T) implies that $\|h(\Lambda, T)\| \leq 1$. Thus, in accordance with Definition 1.51, $(\mathbb{D} \times U)^-$ is a complete spectral set for (Λ, T).

Spectral Sets: Three Case Studies

Step 4. (Dilation of (Λ, T)) By the inclusion (8.23) and the fact that $(\mathbb{D} \times U)^-$ is a complete spectral set for (Λ, T) we may apply Arveson's dilation theorem (Theorem 1.52) to deduce that (Λ, T) has a normal dilation $(\hat{\Lambda}, \hat{T})$ to the boundary $\partial(\mathbb{D} \times U)$ of $\mathbb{D} \times U$. That is, there exists a Hilbert space $\mathcal{K}$ containing $\mathcal{H}$ and a commuting $(d+1)$-tuple $(\hat{\Lambda}, \hat{T})$ of normal operators on $\mathcal{K}$ such that $\sigma(\hat{\Lambda}, \hat{T}) \subseteq \partial(\mathbb{D} \times U)$ and

$$f(\Lambda, T) = P_{\mathcal{H}} f(\hat{\Lambda}, \hat{T})|\mathcal{H}$$

for all functions f holomorphic on a neighborhood of $\mathbb{D}^- \times U^-$. It follows that $\sigma(\hat{\Lambda}) \subseteq \mathbb{T}$, and therefore that $\hat{\Lambda}$ is a unitary on $\mathcal{K}$, and that $\sigma(\hat{T}) \subseteq \partial U$.

Step 5. (The joint numerical range of $\hat{T}$ is contained in U^-) By definition, the joint numerical range of $\hat{T}$ is the set

$$W(\hat{T}) \overset{\text{def}}{=} \{(\langle \hat{T}^1 x, x \rangle, \ldots, \langle \hat{T}^d x, x \rangle): x \in \mathcal{K}, \|x\| = 1\}.$$

We claim that $W(\hat{T}) \subseteq U^-$ (this is where we use the convexity of U). To prove this statement, let $\mathcal{A}$ be the C^*-subalgebra of $\mathcal{B}(\mathcal{K})$ generated by $\hat{T}^1, \ldots, \hat{T}^d$. Since the $\hat{T}^j$ are commuting normal operators, $\mathcal{A}$ is isometrically isomorphic to the algebra $C(\sigma(\hat{T}))$ of continuous complex-valued functions on $\sigma(\hat{T})$ with supremum norm (see Section 8.3.1).

Consider any unit vector $x \in \mathcal{K}$ and define a linear functional L_x on $\mathcal{A}$ by $L_x(A) = \langle Ax, x \rangle$ for any $A \in \mathcal{A}$. Since the Banach dual space of $C(\sigma(\hat{T}))$ is the space of finite Borel measures on $\sigma(T)$, the functional L_x is given by integration against a finite measure μ on $\sigma(T)$. Since L_x is positive (meaning that it is positive on positive elements of $\mathcal{A}$) and satisfies $L_x(1_{\mathcal{K}}) = 1$, it follows that μ is a probability measure on $\sigma(\hat{T})$. For any polynomial p in d variables, $p(\hat{T})$ in $\mathcal{A}$ corresponds under the Gelfand transform to the polynomial function p on $\sigma(\hat{T})$, and so

$$L_x(p(\hat{T})) = \int_{\sigma(\hat{T})} p \, d\mu.$$

Hence

$$(L_x(\hat{T}^1), \ldots, L_x(\hat{T}^d)) = \int_{\sigma(\hat{T})} (\lambda^1, \ldots, \lambda^d) \, d\mu(\lambda^1, \ldots, \lambda^d).$$

Thus every point of $W(\hat{T})$ lies in the closed convex hull of $\sigma(\hat{T})$. Since $\sigma(\hat{T}) \subseteq \partial U$ and U is convex,

$$W(\hat{T}) \subseteq U^-.$$

Step 6. (Some multiplication operators) In Definition 1.44 the notion of a dilation to the boundary is defined somewhat abstractly. There is a more

200   *Commutative Theory*

concrete description due to Sarason [179]. To say that $(\hat{\Lambda}, \hat{T})$ is a normal dilation of (Λ, T) to $\mathbb{T} \times \partial U$ means that, for some Hilbert space $\mathcal{K}$ containing $\mathcal{H}$,

(i) $\hat{\Lambda}, \hat{T}^j$ are commuting normal operators on $\mathcal{K}$;

(ii) $\sigma(\hat{\Lambda}, \hat{T}) \subseteq \mathbb{T} \times \partial U$; and

(iii) there are subspaces $\mathcal{H}_+, \mathcal{H}_-$ of $\mathcal{K}$ such that $\mathcal{K} = \mathcal{H}_- \oplus \mathcal{H} \oplus \mathcal{H}_+$ and, with respect to this orthogonal decomposition of $\mathcal{K}$,

$$\hat{\Lambda} \sim \begin{bmatrix} * & * & * \\ 0 & \Lambda & * \\ 0 & 0 & * \end{bmatrix}, \quad \hat{T}^j \sim \begin{bmatrix} * & * & * \\ 0 & T^j & * \\ 0 & 0 & * \end{bmatrix}$$

for $j = 1, \ldots, d$, where the $*$ denote unspecified operators.

Observe that $\mathcal{H}_- \oplus \mathcal{H}$ is a common invariant subspace of $\hat{\Lambda}$ and $\hat{T}^j$. Let $\widetilde{\Lambda}, \widetilde{T}^j$ be the restrictions

$$\widetilde{\Lambda} \sim \begin{bmatrix} * & * \\ 0 & \Lambda \end{bmatrix}, \quad \widetilde{T}^j \sim \begin{bmatrix} * & * \\ 0 & T^j \end{bmatrix}$$

of $\hat{\Lambda}, \hat{T}^j$ to $\mathcal{H}_- \oplus \mathcal{H}$. The operators $\widetilde{\Lambda}, \widetilde{T}^j$, $j = 1, \ldots, d$, commute pairwise, and

$$\mathcal{H} \text{ is an invariant subspace for } \widetilde{\Lambda}^*. \tag{8.24}$$

Thus $\widetilde{\Lambda}$ is the restriction of a unitary operator to an invariant subspace, and so it is an isometry. Hence $\widetilde{\Lambda}$ has a von Neumann–Wold decomposition as the orthogonal direct sum of a unitary operator and a shift. That is to say, $\mathcal{H}_- \oplus \mathcal{H}$ can be identified with $\mathcal{H}_0 \oplus \mathrm{H}^2_{\mathcal{L}}$ for some Hilbert spaces $\mathcal{H}_0$ and $\mathcal{L}$, and then $\widetilde{\Lambda}$ is identified with $V \oplus S$ for some unitary V on $\mathcal{H}_0$, where S is the shift operator M_z on $\mathrm{H}^2_{\mathcal{L}}$.

If $x \in \mathcal{H}$ then $(\widetilde{\Lambda}^*)^n x = (\Lambda^*)^n x$, and since

$$\Lambda^* \sim \begin{bmatrix} \bar{\zeta} & 0 \\ 0 & \bar{\eta} \end{bmatrix}$$

with respect to the dual basis k_1^*, k_2^* of k_1, k_2 in $\mathcal{H}$, it is clear that $(\Lambda^*)^n x \to 0$ as $n \to \infty$ for all $x \in \mathcal{H}$. Hence $\mathcal{H} \subset \mathrm{H}^2_{\mathcal{L}}$.

Let X^j be the compression of $\widetilde{T}^j$ to $\mathrm{H}^2_{\mathcal{L}}$ for $j = 1, \ldots, d$:

$$\widetilde{T}^j \sim \begin{bmatrix} * & * \\ * & X^j \end{bmatrix} \quad \text{with respect to the decomposition } \mathcal{H}_0 \oplus \mathrm{H}^2_{\mathcal{L}} \tag{8.25}$$

of $\mathcal{H}_- \oplus \mathcal{H}$. Since $V \oplus S$ commutes with $\widetilde{T}^j$, so does S with X^j. By the remarks in Section 8.3.2, there exists a bounded analytic $\mathcal{B}(\mathcal{L})$-valued function g_j on $\mathbb{D}$ such that $X^j = M_{g_j}$, that is

$$(X^j f)(z) = g_j(z) f(z) \qquad \text{for all } f \in \mathrm{H}^2_{\mathcal{L}}, z \in \mathbb{D}, j = 1, \ldots, d. \tag{8.26}$$

Spectral Sets: Three Case Studies

201

The desired holomorphic map $\varphi\colon \mathbb{D} \to U$ will be created from these functions g_j.

Step 7. (The eigenvectors and eigenvalues of $M_{g_j}^*$) By Step 6, $\mathcal{H} \subset \mathrm{H}_{\mathcal{L}}^2$ and, by statement (8.24), $\mathcal{H}$ is invariant under $\widetilde{\Lambda}^*$. Since $\widetilde{\Lambda}^*|\mathrm{H}_{\mathcal{L}}^2 = S^*$, we have

$$\Lambda^* = \widetilde{\Lambda}^*|\mathcal{H} = S^*|\mathcal{H}. \tag{8.27}$$

Similarly,

$$(T^j)^* = (\widetilde{T}^j)^*|\mathcal{H} = (X^j)^*|\mathcal{H} = M_{g_j}^*|\mathcal{H}. \tag{8.28}$$

Therefore, the eigenvectors and eigenvalues of Λ^* and $(T^j)^*$ are also eigenvectors and eigenvalues of S^* and $M_{g_j}^*$. Now, by definition, Λ has eigenvectors k_1, k_2 and corresponding eigenvalues ζ, η. Hence Λ^* has eigenvectors k_1^*, k_2^*, the dual basis vectors to k_1, k_2 in $\mathcal{H}$, and corresponding eigenvalues $\bar{\zeta}, \bar{\eta}$. The eigenvectors of S^* in $\mathrm{H}_{\mathcal{L}}^2$ with eigenvalue $\bar{\zeta}$ are easily shown to be the vectors as_ζ for some non-zero $a \in \mathcal{L}$, where s_ζ is the Szegő kernel. Hence

$$k_1^* \sim a_1 s_\zeta, \qquad k_2^* \sim a_2 s_\eta \quad \text{in } \mathrm{H}_{\mathcal{L}}^2 \tag{8.29}$$

for some $a_1, a_2 \in \mathcal{L}$.

We claim that a_1, a_2 are linearly dependent. Note first that $\|\Lambda\| = 1$, by equation (8.22), but Λ is not unitary (since its eigenvalues are in $\mathbb{D}$). It follows that the rank of $1 - \Lambda^*\Lambda$ is 1. By reason of equation (8.27),

$$1 - \Lambda^*\Lambda = 1 - (S^*|\mathcal{H})^*S|\mathcal{H} = P_{\mathcal{H}}(1 - SS^*)|\mathcal{H},$$

where $P_{\mathcal{H}}$ is the orthogonal projection operator on $\mathrm{H}_{\mathcal{L}}^2$ with range $\mathcal{H}$. Hence $P_{\mathcal{H}}(1 - SS^*)\mathcal{H}$ has dimension 1, and therefore some vector $f = c_1 a_1 s_\zeta + c_2 a_2 s_\eta$ in $\mathcal{H}$, where c_1, c_2 are not both zero, is in the kernel of $P_{\mathcal{H}}(1 - SS^*)$. The operator $1 - SS^*$ takes f to $(c_1 a_1 + c_2 a_2)s_0$ (where s_0 is the Szegő kernel at 0, which is the constant function 1), and so

$$(c_1 a_1 + c_2 a_2)s_0 \perp \{a_1 s_\zeta, a_2 s_\eta\} \quad \text{in } \mathrm{H}_{\mathcal{L}}^2.$$

It follows that $c_1 a_1 + c_2 a_2$ is orthogonal to both a_1 and a_2 in $\mathcal{L}$, and thus $c_1 a_1 + c_2 a_2 = 0$. Hence a_1, a_2 are linearly dependent, as claimed.

Write

$$a_1 = c_1 a, \qquad a_2 = c_2 a$$

for some unit vector $a \in \mathcal{L}$ and some $c_1, c_2 \in \mathbb{C}$. By the relations (8.29),

$$k_1^*, k_2^* \in \mathcal{H} \text{ correspond to } c_1 a s_\zeta, c_2 a s_\eta \in \mathrm{H}_{\mathcal{L}}^2. \tag{8.30}$$

Since T^j has eigenvectors k_1, k_2 and corresponding eigenvalues λ^j, μ^j, $(T^j)^*$ has eigenvectors k_1^*, k_2^* and corresponding eigenvalues $\overline{\lambda^j}, \overline{\mu^j}$. By virtue of

202   *Commutative Theory*

statements (8.30) and (8.28), the vectors as_ζ, as_η in $H_\mathcal{L}^2$ are eigenvectors of $M_{g_j}^*$ with corresponding eigenvalues $\overline{\lambda^j}, \overline{\mu^j}$. Now, for any $g \in H_{\mathcal{B}(\mathcal{L})}^\infty(\mathbb{D})$, any $u \in \mathcal{L}$ and any $w \in \mathbb{D}$,

$$M_g^*(us_w) = (g(w)^* u)s_w.$$

Thus

$$(g_j(\zeta)^* a)s_\zeta = M_{g_j}^*(as_\zeta) = \overline{\lambda^j} as_\zeta,$$
$$(g_j(\eta)^* a)s_\eta = M_{g_j}^*(as_\eta) = \overline{\mu^j} as_\eta,$$

and therefore

$$g_j(\zeta)^* a = \overline{\lambda^j} a, \qquad g_j(\eta)^* a = \overline{\mu^j} a. \tag{8.31}$$

Step 8. (Construction of a map $\varphi \colon \mathbb{D} \to \mathbb{C}^d$) Define $\varphi \colon \mathbb{D} \to \mathbb{C}^d$ by $\varphi = (\varphi_1, \dots, \varphi_d)$, where

$$\varphi_j(z) = \langle g_j(z)a, a \rangle_\mathcal{L} \qquad \text{for } z \in \mathbb{D}, \ j = 1, \dots, d,$$

and g_j is the $\mathcal{B}(\mathcal{L})$-valued function defined in Step 6, equation (8.26).

Step 9. (Values of φ) Clearly φ is a bounded holomorphic map from $\mathbb{D}$ to $\mathbb{C}^d$. We claim that $\varphi(\zeta) = \lambda$. Indeed,

$$\begin{aligned}
\varphi_j(\zeta) &= \langle g_j(\zeta)a, a \rangle_\mathcal{L} \\
&= \langle a, g_j(\zeta)^* a \rangle_\mathcal{L} \\
&= \langle a, \overline{\lambda^j} a \rangle_\mathcal{L} \qquad \text{by equation (8.31)} \\
&= \lambda^j.
\end{aligned}$$

Similarly $\varphi(\eta) = \mu$.

Step 10. (φ maps $\mathbb{D}$ into U) For any $z \in \mathbb{D}$ and for $j = 1, \dots, d$,

$$\begin{aligned}
\varphi_j(z) &= \langle g_j(z)a, a \rangle_\mathcal{L} \\
&= \langle a, g_j(z)^* a \rangle_\mathcal{L} \\
&= \frac{\langle as_z, (g_j(z)^* a)s_z \rangle_{H_\mathcal{L}^2}}{\|s_z\|^2} \\
&= \frac{\langle as_z, M_{g_j}^*(as_z) \rangle_{H_\mathcal{L}^2}}{\|s_z\|^2} \\
&= \frac{\langle M_{g_j}(as_z), as_z \rangle_{H_\mathcal{L}^2}}{\|s_z\|^2}.
\end{aligned}$$

Spectral Sets: Three Case Studies

Let

$$x_z = \frac{a s_z}{\|s_z\|} \in \mathrm{H}^2_{\mathcal{L}} \quad \text{for all } z \in \mathbb{D}.$$

Then $\|x_z\| = 1$ and

$$\varphi(z) = \left(\langle M_{g_1} x_z, x_z \rangle_{\mathrm{H}^2_{\mathcal{L}}}, \ldots, \langle M_{g_d} x_z, x_z \rangle_{\mathrm{H}^2_{\mathcal{L}}} \right).$$

Thus, for any $z \in \mathbb{D}$,

$$
\begin{aligned}
\varphi(z) &\in W(M_{g_1}, \ldots, M_{g_d}) & \\
&= W(X^1, \ldots, X^d) & \text{by equation (8.26)} \\
&\subseteq W(\widetilde{T}^1, \ldots, \widetilde{T}^d) & \text{by the relation (8.25)} \\
&\subseteq W(\hat{T}^1, \ldots, \hat{T}^d) & \text{since } \widetilde{T}^j \text{ is a restriction of } \hat{T}^j \\
&\subseteq U^- & \text{by Step 5.}
\end{aligned}
$$

We have constructed an analytic map $\varphi \colon \mathbb{D} \to U^-$ such that $\varphi(\zeta) = \lambda$ and $\varphi(\eta) = \mu$. In fact $\varphi(\mathbb{D})$ is contained in U. For suppose that z_0 in $\mathbb{D}$ is such that $\varphi(z_0) \in \partial U$. By the Hahn–Banach theorem (for example, [176, theorem 3.4]) there is a real-linear functional L on $\mathbb{C}^d$ and $\alpha \in \mathbb{R}$ such that $L(\lambda) < \alpha$ for all $\lambda \in U$ and $L(\varphi(z_0)) = \alpha$. The function $L^\sim$ on $\mathbb{C}^d$ given by $L^\sim(\lambda) = L(\lambda) - i L(i\lambda)$ is a complex-linear functional on $\mathbb{C}^d$ such that $\operatorname{Re} L^\sim = L$. Thus $L^\sim \circ \varphi$ is a holomorphic scalar-valued function that maps $\mathbb{D}$ to the set $\{z \in \mathbb{C} \colon \operatorname{Re} z \le \alpha\}$ and maps z_0 to a point with real part α. By the open mapping theorem, $\operatorname{Re} L^\sim \circ \varphi(z) = \alpha$ for all $z \in \mathbb{D}$, and thus $\varphi(\mathbb{D}) \subseteq \partial U$. This contradicts the fact that $\varphi(\zeta) = \lambda$, a point of U. Thus $\varphi(\mathbb{D}) \subseteq U$.

We have completed the construction of the function φ proposed in Step 1, and so we may assert that $d_U(\lambda, \mu) \ge \delta_U(\lambda, \mu)$ for any pair of points λ, μ in U. Hence $d_U = \delta_U$, as required. $\qquad \square$

8.5 The Carathéodory Distance on G

There are few domains for which the Carathéodory distance can be precisely calculated effectively. One for which it *can* is the symmetrized bidisc—thanks largely to the characterization of commuting pairs of operators for which G is a spectral domain, Theorem 7.39.

Theorem 8.32. *For any pair of points $s_1, s_2 \in G$,*

$$d_G(s_1, s_2) = \sup_{\omega \in \mathbb{T}} d(C_\omega(s_1), C_\omega(s_2)), \tag{8.33}$$

where C_ω is defined by equation (7.19).

204 Commutative Theory

Proof. Let us paraphrase Theorem 8.13 and apply it to G. Consider a fixed pair s of distinct points $s_1, s_2 \in G$. For any pair $k = (k_1, k_2)$ of linearly independent vectors in a 2-dimensional Hilbert space, let T_k be the pair of diagonal operators with joint eigenvectors k_1, k_2 and corresponding eigenvalues $s_1 = (s_1^1, s_1^2), s_2 = (s_2^1, s_2^2)$, respectively. By the definition (8.12), $\mathcal{S}_s(G)$ comprises precisely the operator pairs T_k for which G is a spectral domain, and by Theorem 7.39, these are the T_k for which $\|C_\omega(T_k)\| \leq 1$ for all $\omega \in \mathbb{T}$. Now $C_\omega(T_k)$ is the diagonal operator with eigenvectors k_1, k_2 and corresponding eigenvalues $C_\omega(s_1)$, $C_\omega(s_2)$. Hence, by Lemma 8.6, for any $\omega \in \mathbb{T}$,

$$\|C_\omega(T_k)\| \leq 1 \quad \Leftrightarrow \quad d(C_\omega(s_1),\ C_\omega(s_2)) \leq \sin \theta,$$

where θ is the acute angle between k_1 and k_2. Thus

$$T_k \in \mathcal{S}_s(G) \quad \text{if and only if} \quad \sup_{\omega \in T} d(C_\omega(s_1),\ C_\omega(s_2)) \leq \sin \theta.$$

Combining this statement with Theorem 8.13, we deduce that $d_G(s_1, s_2)$ equals the infimum of $\sin \theta$ subject to $\sup_\omega d(C_\omega(s_1),\ C_\omega(s_2)) \leq \sin \theta$. As k varies over all pairs of linearly independent unit vectors, $\sin \theta$ describes $(0, 1]$. Equation (8.33) follows. $\qquad \square$

Remarkably enough, the Lempert function δ_G coincides with the Carathéodory distance d_G [28, corollary 5.7]. This fact cannot be deduced from Lempert's theorem, since it has been shown that G is not biholomorphic to any convex domain [61]. If we expand the right-hand side of equation (8.33), we obtain $d_G(s_1, s_2)$ as the maximum of the modulus of a fractional quadratic function over $\mathbb{T}$. As a consequence we obtain an explicit solvability criterion for the spectral Nevanlinna–Pick problem of Section 7.7 in the case of two interpolation points.

Corollary 8.34. *Let $\lambda_1, \lambda_2 \in \mathbb{D}$ and $s_1, s_2 \in G$. There exists a holomorphic map $f \colon \mathbb{D} \to G$ such that $f(\lambda_1) = s_1$ and $f(\lambda_2) = s_2$ if and only if*

$$\max_{\omega \in \mathbb{T}} \left| \frac{(s_2^1 s_1^2 - s_1^1 s_2^2)\omega^2 + 2(s_2^2 - s_1^2)\omega + s_1^1 - s_2^1}{(s_1^1 - s_2^1 s_1^2)\omega^2 - 2(1 - s_1^2 s_2^2)\omega + s_2^1 - s_1^1 s_2^2} \right| \leq \left| \frac{\lambda_1 - \lambda_2}{1 - \bar{\lambda}_2 \lambda_1} \right|. \tag{8.35}$$

Proof. By the definition of the Lempert function, the sought-for function f exists if and only if

$$\delta_G(s_1, s_2) \quad \leq \quad d(\lambda_1, \lambda_2),$$

that is, in view of the equations $\delta_G = d_G$ and (8.33), if and only if

$$\sup_{\omega \in \mathbb{T}} d(C_\omega(s_1),\ C_\omega(s_2)) \quad \leq \quad d(\lambda_1, \lambda_2),$$

which expands to give the inequality (8.35). $\qquad \square$

Spectral Sets: Three Case Studies 205

8.6 Von Neumann's Inequality on Subvarieties of the Bidisc

In this section we shall use a lurking isometry argument to derive a refinement of Andô's inequality in the special case of a diagonalizable pair of commuting contractions acting on finite-dimensional space. As a corollary we obtain a determinantal representation for algebraic inner varieties in the bidisc.

Definition 8.36. *For any Hilbert space $\mathcal{H}$, we say that $T = (T^1, T^2)$ is a simple pair acting on $\mathcal{H}$ if*

(i) $T^1, T^2 \in \mathcal{B}(\mathcal{H})$;

(ii) $\dim \mathcal{H} = n < \infty$; *and*

(iii) *there exist $\lambda_1, \ldots, \lambda_n \in \mathbb{D}^2$ and $k_1, \ldots, k_n \in \mathcal{H}$ such that*

$$T^r k_i = \lambda_i^r k_i \qquad for \;\; i = 1, \ldots, n \;\; and \;\; r = 1, 2.$$

The vectors $k_1, \ldots, k_n$ are the joint eigenvectors and $\lambda_1, \ldots, \lambda_n$ the joint eigenvalues of T. Recall from Theorem 3.38 that any function φ with a network realization formula having a finite-dimensional model space and an isometric V is a rational inner function. Moreover, the McMillan degree of φ is the minimum of $\dim \mathcal{M}$ over all spaces $\mathcal{M}$ such that there is a network realization of φ with model space $\mathcal{M}$.

Proposition 8.37. *Let T be a contractive simple pair acting on a space $\mathcal{H}$ of dimension n, with joint eigenvectors $k_1, \ldots, k_n$ and corresponding eigenvalues $\lambda_1^r, \ldots, \lambda_n^r$ for $T^r, r = 1, 2$. Assume that*

$$\mathrm{rank}\left[(1 - \bar{\lambda}_i^1 \lambda_j^1)\langle k_j, k_i \rangle\right] = n_1 \quad and \quad \mathrm{rank}\left[(1 - \bar{\lambda}_i^2 \lambda_j^2)\langle k_j, k_i \rangle\right] = n_2. \quad (8.38)$$

There exists a rational $n_1 \times n_1$-matrix-valued inner function $\Phi(z)$ defined on $\mathbb{D}$, of McMillan degree at most n_2, and non-zero vectors $u_1^1, \ldots, u_n^1 \in \mathbb{C}^{n_1}$ such that

$$\Phi(\lambda_j^1) u_j^1 = \lambda_j^2 u_j^1 \tag{8.39}$$

and

$$\lambda_j^2 \in \sigma(\Phi(\lambda_j^1)) \tag{8.40}$$

for $j = 1, \ldots, n$. Moreover, $\|\Phi(z)\| < 1$ for every z in $\mathbb{D}$.

Proof. Since T^1 and T^2 are contractions, by the fundamental fact, statement (1.13), both of the matrices in equations (8.38) are positive semi-definite, and so we may choose n non-zero vectors $u_1^1, \ldots, u_n^1$ in $\mathbb{C}^{n_1}$ such that

$$(1 - \bar{\lambda}_i^1 \lambda_j^1)\langle k_j, k_i \rangle = \langle u_j^1, u_i^1 \rangle. \tag{8.41}$$

206   *Commutative Theory*

It follows from the equation that the vectors u_j^1 are all non-zero. Similarly, there are n non-zero vectors $u_1^2, \ldots, u_n^2$ in $\mathbb{C}^{n_2}$ such that

$$(1 - \bar{\lambda}_i^2 \lambda_j^2)\langle k_j, k_i \rangle = \langle u_j^2, u_i^2 \rangle.$$

Consequently,

$$(1 - \bar{\lambda}_i^2 \lambda_j^2)\langle u_j^1, u_i^1 \rangle = (1 - \bar{\lambda}_i^1 \lambda_j^1)\langle u_j^2, u_i^2 \rangle,$$

which reshuffles to

$$\langle u_j^1, u_i^1 \rangle_{\mathbb{C}^{n_1}} + \langle \lambda_j^1 u_j^2, \lambda_i^1 u_i^2 \rangle_{\mathbb{C}^{n_2}} = \langle u_j^2, u_i^2 \rangle_{\mathbb{C}^{n_2}} + \langle \lambda_j^2 u_j^1, \lambda_i^2 u_i^1 \rangle_{\mathbb{C}^{n_1}}.$$

Therefore,

$$\left\langle \begin{pmatrix} u_j^1 \\ \lambda_j^1 u_j^2 \end{pmatrix}, \begin{pmatrix} u_i^1 \\ \lambda_j^1 u_i^2 \end{pmatrix} \right\rangle_{\mathbb{C}^{n_1} \oplus \mathbb{C}^{n_2}} = \left\langle \begin{pmatrix} \lambda_j^2 u_j^1 \\ u_j^2 \end{pmatrix}, \begin{pmatrix} \lambda_i^2 u_i^1 \\ u_i^2 \end{pmatrix} \right\rangle_{\mathbb{C}^{n_1} \oplus \mathbb{C}^{n_2}}.$$

Let

$$\mathcal{M} = \left[\begin{pmatrix} u_j^1 \\ \lambda_j^1 u_j^2 \end{pmatrix} : j = 1, \ldots, n \right]$$

and

$$\mathcal{N} = \left[\begin{pmatrix} \lambda_j^2 u_j^1 \\ u_j^2 \end{pmatrix} : j = 1, \ldots, n \right].$$

It follows from Lemma 2.18 that there exists an isometry

$$V : \mathcal{M} \to \mathcal{N}$$

satisfying

$$V\left(\begin{pmatrix} u_j^1 \\ \lambda_j^1 u_j^2 \end{pmatrix} \right) = \begin{pmatrix} \lambda_j^2 u_j^1 \\ u_j^2 \end{pmatrix}, \qquad j = 1, \ldots, n.$$

Since V is isometric $\dim \mathcal{M} = \dim \mathcal{N}$. As both $\mathcal{M}$ and $\mathcal{N}$ are subspaces of $\mathbb{C}^{n_1} \oplus \mathbb{C}^{n_2}$, it follows that there exists a unitary W acting on $\mathbb{C}^{n_1} \oplus \mathbb{C}^{n_2}$ such that $V = P_{\mathcal{N}} W | \mathcal{M}$.

If W is represented as a 2×2 block matrix

$$W = \begin{bmatrix} A & B \\ C & D \end{bmatrix} : \mathbb{C}^{n_1} \oplus \mathbb{C}^{n_2} \to \mathbb{C}^{n_1} \oplus \mathbb{C}^{n_2},$$

then

$$\begin{pmatrix} \lambda_j^2 u_j^1 \\ u_j^2 \end{pmatrix} = W \begin{pmatrix} u_j^1 \\ \lambda_j^1 u_j^2 \end{pmatrix} = \begin{bmatrix} A & B \\ C & D \end{bmatrix} \begin{pmatrix} u_j^1 \\ \lambda_j^1 u_j^2 \end{pmatrix}.$$

Spectral Sets: Three Case Studies   207

From the second row of this equation we have

$$u_j^2 = (1 - D\lambda_j^1)^{-1} C u_j^1,$$

and then from the first row,

$$\lambda_j^2 u_j^1 = [A + B\lambda_j^1 (1 - D\lambda_j^1)^{-1} C] u_j^1$$

for $j = 1, \ldots, n$. Consequently, if Φ is defined by the formula

$$\Phi(z) = A + Bz(1 - Dz)^{-1} C \qquad \text{for all } z \in \mathbb{D},$$

then Φ is an $n_1 \times n_1$ matrix-valued inner function defined on $\mathbb{D}$ having McMillan degree at most n_2 and satisfying

$$\Phi(\lambda_j^1) u_j^1 = \lambda_j^2 u_j^1 \qquad \text{for } j = 1, \ldots, n. \tag{8.42}$$

In particular, statement (8.40) holds.

Finally, we must show that $\|\Phi(z)\| < 1$ for all $z \in \mathbb{D}$. Suppose that $\|\Phi(z_0)\| = 1$ for some point $z_0 \in \mathbb{D}$. Then there exist unitary matrices U_1 and U_2 such that $U_1 \Phi(z_0) U_2$ is a diagonal matrix with $(1, 1)$ entry equal to 1 (the singular value decomposition of $\Phi(z_0)$). The $(1, 1)$ entry of $U_1 \Phi(\cdot) U_2$ is holomorphic on $\mathbb{D}$, is bounded by 1 and attains its maximum modulus at z_0. Hence it is constant and equal to 1 on $\mathbb{D}$, and therefore all other entries in the first row and column of $U_1 \Phi(\cdot) U_2$ are identically zero. Thus the rank of

$$\left[(1 - \Phi^*(\lambda_i^1) \Phi(\lambda_j^1) \langle u_j^1, u_i^1 \rangle \right] = \left[(1 - \bar{\lambda}_i^1 \lambda_j^1) \langle k_j, k_i \rangle \right]$$

is at most $n_1 - 1$, a contradiction to equations (8.38). $\qquad \square$

As a consequence of Proposition 8.37 we are able to mimic Sz. Nagy's proof of von Neumann's inequality and derive a refinement of Andô's inequality. Let T be a contractive simple pair and let Φ be the function described in Proposition 8.37. Let $\Phi^\vee$ be defined by the formula $\Phi^\vee(z) = \Phi(\bar{z})^*$ and consider the pair of commuting isometries $(M_z, M_{\Phi^\vee})$ acting on $\mathrm{H}^2_{\mathbb{C}^{n_1}}$. Define n vectors $g_1, \ldots, g_n \in \mathrm{H}^2_{\mathbb{C}^{n_1}}$ by

$$g_j = u_j^1 s_{\bar{\lambda}_j^1} \qquad \text{for } j = 1, \ldots, n, \tag{8.43}$$

where s_λ is the Szegő kernel. Observe that

$$M_z^* g_j = \lambda_j^1 g_j \qquad \text{for } j = 1, \ldots, n \tag{8.44}$$

and

$$M_{\Phi^\vee}^* g_j = M_{\Phi^\vee}^* u_j^1 s_{\bar\lambda_j^1}$$
$$= \left((\Phi^\vee(\bar\lambda_j^1))^* u_j^1\right) s_{\bar\lambda_j^1}$$
$$= \left(\Phi(\lambda_j^1) u_j^1\right) s_{\bar\lambda_j^1}$$
$$= \lambda_j^2 u_j^1 s_{\bar\lambda_j^1} \qquad \text{by equation (8.42)}$$
$$= \lambda_j^2 g_j \qquad \text{for } j = 1,\ldots,n. \qquad (8.45)$$

Next observe that, in view of equation (8.41),

$$\langle g_j, g_i \rangle = \left\langle u_j^1 s_{\bar\lambda_j^1}, u_i^1 s_{\bar\lambda_i^1} \right\rangle_{H_{\mathbb{C}^{n_1}}^2}$$
$$= \left\langle u_j^1, u_i^1 \right\rangle_{\mathbb{C}^{n_1}} \left\langle s_{\bar\lambda_j^1}, s_{\bar\lambda_i^1} \right\rangle_{H^2}$$
$$= \frac{\left\langle u_j^1, u_i^1 \right\rangle}{1 - \bar\lambda_i^1 \lambda_j^1}$$
$$= \langle k_j, k_i \rangle_{\mathcal{H}} \qquad \text{by equation (8.41).}$$

Consequently, by the lurking isometry lemma (Lemma 2.18), there exists an isometry $L: \mathcal{H} \to H_{\mathbb{C}^{n_1}}^2$ such that

$$L k_j = g_j \qquad \text{for } j = 1,\ldots,n.$$

Furthermore, equation (8.44) implies that, for $j = 1,\ldots,n$,

$$M_z^* L\, k_j = M_z^* g_j$$
$$= \lambda_j^1 g_j$$
$$= \lambda_j^1 L k_j$$
$$= L(\lambda_j^1 k_j)$$
$$= L T^1 k_j$$

and equation (8.45) implies that

$$M_{\Phi^\vee}^* L\, k_j = M_{\Phi^\vee}^* g_j$$
$$= \lambda_j^2 g_j$$
$$= \lambda_j^2 L k_j$$
$$= L(\lambda_j^2 k_j)$$
$$= L T^2 k_j.$$

Spectral Sets: Three Case Studies

Let $S = (M_z^*, M_{\Phi^\vee}^*)$ act on $\mathrm{H}^2_{\mathbb{C}^{n_1}}$ and let $\mathcal{M} = L\mathcal{H}$, the span of $\{g_1, \ldots, g_n\}$. Thus $\mathcal{M}$ is invariant for S, and L is a unitary operator from $\mathcal{H}$ to $\mathcal{M}$ that intertwines T and $S|_{\mathcal{M}}$. Since $\mathcal{M}$ is invariant for S, this means that

$$Lp(T) = p(S)L \qquad \text{for all } p \in \mathbb{C}[z_1, z_2].$$

An *extension* of a commuting d-tuple X of operators on a Hilbert space $\mathcal{H}$ is a commuting d-tuple Y acting on a space $\mathcal{K} \supset \mathcal{H}$ with the properties that $\mathcal{H}$ is invariant for Y and $Y|_{\mathcal{H}} = X$. Since we have shown that T is unitarily equivalent to S restricted to an invariant subspace, we can, with a mild abuse of language, consider S to be an extension of T. In turn, S has an extension to the pair of commuting unitary operators

$$U = (L_z^*, L_{\Phi^\vee}^*) \qquad \text{acting on } L^2_{\mathbb{C}^{n_1}}, \tag{8.46}$$

where, for any bounded measurable $n_1 \times n_1$ matrix-valued function G on $\mathbb{T}$, the *Laurent operator* L_G on $L^2_{\mathbb{C}^{n_1}}$ is the operation of pointwise multiplication by G. Note that $(L_G)^* = L_{G^*}$.

Let

$$V = \{\lambda \in \overline{\mathbb{D}}^2 : \det[\lambda^2 - \Phi(\lambda^1)] = 0\}. \tag{8.47}$$

Definition 8.48. *A* distinguished variety *is a set of the form* $\mathcal{A} \cap \overline{\mathbb{D}}^2$, *where* $\mathcal{A}$ *is an algebraic set in* $\mathbb{C}^2$ *with the property that*

$$\mathcal{A} \cap \partial(\mathbb{D}^2) = \mathcal{A} \cap \mathbb{T}^2.$$

The set $\{\lambda : \det[\lambda^2 - \Phi(\lambda^1)] = 0\}$ is algebraic, and V is a distinguished variety, because Φ is inner and $\|\Phi(z)\| < 1$ for $z \in \mathbb{D}$. Indeed, consider $\lambda \in V \cap \partial(\mathbb{D}^2)$. Then λ^2 is an eigenvalue of $\Phi(\lambda^1)$ and either $\lambda^1 \in \mathbb{T}$ or $\lambda^2 \in \mathbb{T}$. In the former case, since Φ is a rational square matrix inner function, $\Phi(\lambda^1)$ is unitary and hence all its eigenvalues lie in $\mathbb{T}$. Thus $\lambda^2 \in \mathbb{T}$ and $\lambda \in V \cap \mathbb{T}^2$. On the other hand, suppose that $|\lambda^2| = 1$. Then $\Phi(\lambda^1)$ has a unimodular eigenvalue, hence has norm at least 1, and so $|\lambda^1| = 1$, and again $\lambda \in \mathbb{T}^2$. Thus $V \cap \partial(\mathbb{D}^2) \subseteq V \cap \mathbb{T}^2$. The reverse inclusion is trivial, and so $V \cap \partial(\mathbb{D}^2) = V \cap \mathbb{T}^2$.

In [13] it is proved that all distinguished varieties arise as V in equation (8.47) for a suitable Φ.

Lemma 8.49. *Let T be a contractive simple pair, let Φ be the function described in Proposition 8.37, and let $p \in \mathbb{C}[z^1, z^2]$ be a polynomial. Then, if U denotes the operator pair defined in equation (8.46),*

$$\|p(U)\| = \sup_{z \in \partial V} |p(z)|, \tag{8.50}$$

and $\sigma(U) = \partial V$.

210   Commutative Theory

Proof. Note that $p(U)^* = p^\vee(U^*) = p^\vee(L_z, L_{\Phi^\vee})$. Let $e_1(\theta), \ldots, e_{n_1}(\theta)$ be an orthonormal basis of $\mathbb{C}^{n_1}$ that is measurable in θ and has the property that each $e_j(\theta)$ is an eigenvector of $\Phi^\vee(e^{i\theta})$; let the corresponding eigenvalue be $\lambda_j(\theta)$ for $0 \le \theta < 2\pi$. (See [128, section II.1]; indeed, one can choose the eigenvectors to be analytic in $e^{i\theta}$, but for a finite number of exceptional points). Thus $\lambda_j(\theta) \in \mathbb{T}$ and

$$\det[\lambda_j(\theta) - \Phi^\vee(e^{i\theta})] = 0$$

for $j = 1, \ldots, n_1$ and $\theta \in [0, 2\pi)$.

If $f \in H^2_{\mathbb{C}^{n_1}}$, then f can be written

$$f(e^{i\theta}) = \sum_{j=1}^{n_1} a_j(\theta) e_j(\theta)$$

for some functions $a_1, \ldots, a_{n_1} \in H^2$. Thus, for any $f \in H^2_{\mathbb{C}^{n_1}}$,

$$\|p(U)^* f\| = p^\vee(L_z, L_{\Phi^\vee}) f(e^{i\theta})$$

$$= \sum_{j=1}^{n_1} a_j(\theta) p^\vee(L_z, L_{\Phi^\vee}) e_j(\theta)$$

$$= \sum_{j=1}^{n_1} a_j(\theta) p^\vee(e^{i\theta}, \lambda_j(\theta)) e_j(\theta),$$

and therefore

$$\|p(U)^* f\|^2 = \sum_j \frac{1}{2\pi} \int_0^{2\pi} |p^\vee(e^{i\theta}, \lambda_j(\theta))\, a_j(\theta)|^2 \, d\theta$$

$$\le \sup\{|p^\vee(e^{i\theta_1}, e^{i\theta_2})|^2 : \det[e^{i\theta_2} - \Phi^\vee(e^{i\theta_1})] = 0\} \, \|f\|^2$$

$$= \sup\{|p(e^{-i\theta_1}, e^{-i\theta_2})|^2 : \det[e^{-i\theta_2} - \Phi(e^{-i\theta_1})] = 0\} \, \|f\|^2$$

$$= \|p\|_{\partial V}^2 \|f\|^2.$$

This proves $\le$ in equation (8.50). For the reverse inequality, choose a point $(e^{i\theta_1}, e^{i\theta_2}) \in \partial V$ at which $|p|$ attains its maximum over ∂V. Let v be an eigenvector of $\Phi^\vee(e^{i\theta_1})$ with eigenvalue $e^{i\theta_2}$. Then evaluate the norm of $p(U)^* f_n \otimes v$, where f_n is a sequence of norm one functions in H^2 whose moduli tend to the point mass at $e^{i\theta_1}$.

Equation (8.50) proves that $\sigma(U) \cap \mathbb{T}^2 = \partial V$. But since U is a unitary pair, its spectrum is contained in $\mathbb{T}^2$, and so we conclude that $\sigma(U) = \partial V$. $\quad\square$

To summarize the above constructions, we have proved the following refinements of Andô's dilation theorem, Theorem 1.54, and Andô's inequality, Theorem 1.55.

Spectral Sets: Three Case Studies   211

Theorem 8.51. *Let T be a contractive simple tuple and let V be given by equation (8.47). Then there exists a Hilbert space $\mathcal{K} \supset \mathcal{H}$ and a pair of commuting unitaries U on $\mathcal{K}$ such that $\sigma(U) = \partial V$ and*

$$p(T) = p(U)|\mathcal{H} \qquad \text{for all } p \in \mathbb{C}[z_1, z_2].$$

By the maximum principle,

$$\sup\{|p(\lambda)|\colon \lambda \in V \cap \mathbb{D}^2\} = \max\{|p(\lambda)|\colon \lambda \in \partial V\},$$

whereby we also obtain a version of Andô's inequality, Theorem 1.55.

Theorem 8.52. *Let T be a contractive simple tuple. Then there is a distinguished variety V such that*

$$\|p(T)\| \leq \sup\{|p(\lambda)|\colon \lambda \in V \cap \mathbb{D}^2\} \qquad \text{for all } p \in \mathbb{C}[z_1, z_2].$$

Remark 8.53. The conclusion of Theorem 8.52 holds for any pair T of commuting contractive matrices, neither of which has an eigenvalue of modulus 1. This follows from an approximation argument and a theorem of J. Holbrook [117], which says that pairs of simultaneously diagonalizable matrices are dense in the set of all pairs of commuting matrices. See [13] for details.

Once one knows Andô's inequality on finite-dimensional spaces, one can deduce it for pairs of commuting contractions on infinite-dimensional spaces by approximating the contractions by commuting matrices. S. Drury gave an explicit construction to do this in [81].

If both T^1 and T^2 are also contractions, then Andô's inequality (Theorem 1.55) states that there is a norm bound whenever f is in $H^\infty(\mathbb{D}^2)$, the set of bounded holomorphic functions on the bidisc, to wit,

$$\text{if } \sigma(T) \subseteq \mathbb{D}^2, \ \|T^1\| \leq 1 \text{ and } \|T^2\| \leq 1 \quad \text{then} \quad \|f(T)\| \leq \|f\|_{H^\infty(\mathbb{D}^2)}. \tag{8.54}$$

If T is a pair of commuting contractions, but we relax the requirement that the spectrum lie in the open bidisc, then inequality (8.54) still holds for f in $A(\mathbb{D}^2)$, the set of functions that are holomorphic on $\mathbb{D}^2$ and continuous on $\overline{\mathbb{D}^2}$, provided that we interpret $f(T)$ as $\lim_{r \uparrow 1} f(rT)$. This assertion follows from the fact that if $f_r(\lambda) \stackrel{\text{def}}{=} f(r\lambda)$, then f_r converges to f in $A(\mathbb{D}^2)$ as r increases to 1.

8.7 Historical Notes

Section 8.3. The elements of the Gelfand theory of commutative C^*-algebras are well described in [74, section I.1.5]. There is also a succinct account in

212 *Commutative Theory*

[144, section 26]. The von Neumann–Wold decomposition was first proved by von Neumann in [209]. The description we give is due to Halmos [98].

Sections 8.1, 8.2, and 8.4. Our approach to the Carathéodory extremal problem and the proof of Lempert's theorem are taken from [8]. Lempert's original proof [143] of his theorem was by a completely different method. Lempert further proved that $\delta_U = d_U$ for "strongly linearly convex domains" U; see the discussions in [122, theorem 11.2.1, remark 11.2.9, and section A.5].

Section 8.5 on the Carathéodory problem on G is taken from [28].

Section 8.6, a refinement of Andô's inequality, is from [13].

9

Calcular Norms

In previous chapters interesting function theory is both discovered and proven with from the viewpoint that holomorphic functions are defined not just on scalars but also, via the functional calculus, on operators as well. In this chapter we pursue the idea further by defining normed spaces of analytic functions through the use of functional calculi. We call norms defined in terms of a functional calculus *calcular norms.*

In the following sections of this chapter we shall first define calcular norms and then explore a number of examples, including the Douglas–Paulsen norm, which arises in the study of K-spectral sets; the B. and F. Delyon norm, which comes up in the theory of the numerical range; the Badea–Beckerman–Crouzeix norm, which is associated with a domain that is the finite intersection of discs; the polydisc norm, which appears when one extends the model theory developed in Chapters 2–4 from $\mathbb{D}$ and $\mathbb{D}^2$ to $\mathbb{D}^d$ for general d; and the Oka norm, which one can use to prove a refinement of the Oka extension theorem containing norm bounds. These examples will illustrate how estimates of calcular norms of functions can have significant operator-theoretic consequences.

9.1 The Taylor Spectrum and Functional Calculus

In this chapter we shall make free use of the Taylor spectrum and functional calculus, which was briefly discussed in Section 1.2. To define it, we shall follow Curto [67]. Let Λ be the exterior algebra on d generators $e_1, \ldots, e_d$, with identity $e_0 = 1$. This is a 2^d-dimensional vector space, which we make into a Hilbert space by declaring the basis vectors e_0 and

$$\{e_{i_1} e_{i_2} \ldots e_{i_k} : 1 \leq i_1 < i_2 < \cdots < i_k \leq d\}$$

to be orthonormal. We define the d *creation operators*

$$E^i \colon \Lambda \to \Lambda \quad \text{by} \quad E^i \xi \overset{\text{def}}{=} e_i \xi.$$

213

214 *Commutative Theory*

These operators anti-commute. If $\mathcal{X}$ is a Banach space, and T is a commuting d-tuple in $\mathcal{B}(\mathcal{X})$, we define $D_T \colon \mathcal{X} \otimes \Lambda \to \mathcal{X} \otimes \Lambda$ by

$$D_T = \sum_{i=1}^{d} T^i \otimes E^i.$$

A calculation shows that

$$D_T(D_T(x \otimes \xi)) = \sum_{i<j} T^i T^j x \otimes (E^i E^j + E^j E^i)\xi = 0,$$

for any $x \in \mathcal{X}$ and $\xi \in \Lambda$, and so the range $\mathrm{ran}(D_T)$ is contained in the kernel $\ker(D_T)$.

Definition 9.1. *Let T be a d-tuple of commuting operators on a Banach space $\mathcal{X}$ as above. The* Taylor spectrum *of T is*

$$\sigma(T) \overset{\mathrm{def}}{=} \{\lambda \in \mathbb{C}^d \colon \mathrm{ran}(D_{T-\lambda}) \neq \ker(D_{T-\lambda})\}.$$

Equivalently, one can construct a cochain complex called the Koszul complex using the sets of k-forms on $\mathcal{X}$, and mapping the k-forms to the $(k+1)$-forms by $D_{T-\lambda}$. The Taylor spectrum is then the set of λ for which the Koszul complex is not exact.

In the case $d = 1$ there is no conflict with established notation, since the Taylor spectrum coincides with the classical spectrum. We shall set out the basic properties of the spectrum and functional calculus that we use in this chapter. For proofs see [198, 67, 89]. In this book we are concerned only with the case that $\mathcal{X}$ is a Hilbert space, though the theory is much more general. Indeed, in [89] the authors develop the functional calculus for $\mathcal{X}$ a Fréchet space.

Property 1. For every non-zero Hilbert space $\mathcal{H}$ and pairwise commuting d-tuple of bounded operators $T = (T^1, \ldots, T^d)$ acting on $\mathcal{H}$ the set $\sigma(T)$ is compact and non-empty.

Property 2. If T is a d-tuple of pairwise commuting operators acting on $\mathcal{H}$ and φ is holomorphic on a neighborhood of $\sigma(T)$, then an operator $\varphi(T) \in \mathcal{B}(\mathcal{H})$ is defined.

Property 3. If Ω is a domain in $\mathbb{C}^d$ and $\sigma(T) \subseteq \Omega$, then the map

$$\varphi \mapsto \varphi(T) \in \mathcal{B}(\mathcal{H}) \quad \text{for all } \varphi \in \mathrm{Hol}(\Omega)$$

is a unital[1] algebra homomorphism.

---

[1] An algebra homomorphism between algebras with multiplicative identities is *unital* if it maps identity to identity

Calcular Norms 215

Property 4. If T is a d-tuple of pairwise commuting operators acting on $\mathcal{H}$ and p is polynomial in d variables (so that $p(T)$ is defined algebraically), and $\varphi \in \text{Hol}(\Omega)$ is defined by the formula

$$\varphi(\lambda) = p(\lambda) \qquad \text{for all } \lambda \in \Omega,$$

then $\varphi(T) = p(T)$. (In other words, the Taylor functional calculus does what you expect on polynomials).

Property 5. If A is an index set, for each $\alpha \in A$, T_α is a d-tuple of pairwise commuting operators acting on $\mathcal{H}_\alpha$, and

$$\sigma\left(\bigoplus_{\alpha \in A} T_\alpha\right) \subseteq \Omega,$$

then[2] $\sigma(T_\alpha) \subseteq \Omega$ for all $\alpha \in A$. Furthermore, if $\varphi \in \text{Hol}(\Omega)$,

$$\varphi\left(\bigoplus_{\alpha \in A} T_\alpha\right) = \bigoplus_{\alpha \in A} \varphi(T_\alpha).$$

Property 6. If T is a d-tuple of pairwise commuting operators acting on $\mathcal{H}$ and $p = (p_1, \ldots, p_m)$ is an m-tuple of polynomials in d variables, then[3]

$$\sigma(p(T)) = p(\sigma(T)).$$

Property 6 is called the *spectral mapping theorem*.

Another significant fact about the Taylor functional calculus is the following striking result of Mihai Putinar [171].

Theorem 9.2. (Uniqueness of the functional calculus) *For each open subset Ω of $\mathbb{C}^d$ that contains the spectrum $\sigma(T)$ of a d-tuple T of commuting operators on a Banach space $\mathcal{X}$, there is exactly one continuous map $f \mapsto f(T) \in \mathcal{B}(\mathcal{X})$, defined for all $f \in \text{Hol}(\Omega)$, such that Properties 3 and 4 hold and such that, for all holomorphic maps F defined on Ω),*

$$\sigma(F(T)) = F(\sigma(T)).$$

9.2 Calcular Norms and Algebras

Definition 9.3. *Let Ω be a domain in $\mathbb{C}^d$. We say that a class[4] of operator tuples $\mathcal{F}$ is subordinate to Ω if*

(i) *the elements of $\mathcal{F}$ are d-tuples of pairwise commuting operators;*
(ii) *if $T \in \mathcal{F}$, then $\sigma(T) \subseteq \Omega$; and*

---

[2] Here, if for $\alpha \in A$, $T_\alpha = (T_\alpha^1, \ldots, T_\alpha^d)$, then we let $\oplus_\alpha T_\alpha = (\oplus_\alpha T_\alpha^1, \ldots, \oplus_\alpha T_\alpha^d)$.
[3] Here, $p(T) = (p_1(T), \ldots, p_m(T))$.
[4] See Remark 9.24 for a discussion of the meaning of the word *class*.

216 *Commutative Theory*

(iii) *for every* $\lambda = (\lambda^1, \ldots, \lambda^d) \in \Omega$ *there exists a Hilbert space* $\mathcal{H}$ *such that*

$$(\lambda^1 \mathrm{id}_{\mathcal{H}}, \ldots, \lambda^d \mathrm{id}_{\mathcal{H}}) \in \mathcal{F}.$$

When $\mathcal{H} = \mathbb{C}$ *we denote this operator tuple by* $([\lambda^1], \ldots, [\lambda^d])$.

If Ω is a domain in $\mathbb{C}^d$ and $\mathcal{F}$ is subordinate to Ω, then as Condition (ii) in Definition 9.3 implies that $\sigma(T) \subset \Omega$ whenever $T \in \mathcal{F}$, by Property 2 of the functional calculus, $\varphi(T)$ is a well-defined operator whenever $\varphi \in \mathrm{Hol}(\Omega)$ and $T \in \mathcal{F}$.

Definition 9.4. *Let* Ω *be a domain in* $\mathbb{C}^d$ *and assume that* $\mathcal{F}$ *is subordinate to* Ω. *For* $\varphi \in \mathrm{Hol}(\Omega)$, *we define* $\|\varphi\|_{\mathcal{F}}$ *by the formula*

$$\|\varphi\|_{\mathcal{F}} = \sup_{T \in \mathcal{F}} \|\varphi(T)\|. \tag{9.5}$$

We refer to the function

$$\varphi \mapsto \|\varphi\|_{\mathcal{F}} \qquad on \ \mathrm{Hol}(\Omega)$$

as the calculAR norm induced by $\mathcal{F}$.

There is no guarantee that equation (9.5) defines a finite quantity. Accordingly, we introduce the following definition.

Definition 9.6. *Let* Ω *be a domain in* $\mathbb{C}^d$ *and assume that* $\mathcal{F}$ *is subordinate to* Ω. *We define* $\mathrm{H}^\infty(\mathcal{F})$ *by*

$$\mathrm{H}^\infty(\mathcal{F}) = \{\varphi \in \mathrm{Hol}(\Omega) : \|\varphi\|_{\mathcal{F}} < \infty\},$$

and we define $\mathscr{S}(\mathcal{F})$ *by*

$$\mathscr{S}(\mathcal{F}) = \{\varphi \in \mathrm{Hol}(\Omega) : \|\varphi\|_{\mathcal{F}} \leq 1\}.$$

Example 9.7. Let Ω be a domain in $\mathbb{C}^d$ and let $\mathcal{F} = \{([\lambda^1], \ldots, [\lambda^d]) : \lambda \in \Omega\}$. Then, for $\varphi \in \mathrm{Hol}(\Omega)$,

$$\begin{aligned}
\|\varphi\|_{\mathcal{F}} &= \sup_{T \in \mathcal{F}} \|\varphi(T)\| \\
&= \sup_{\lambda \in \Omega} \|\varphi\big(([\lambda^1], \ldots, [\lambda^d])\big)\| \\
&= \sup_{\lambda \in \Omega} \|\,[\varphi(\lambda)]\,\| \\
&= \sup_{\lambda \in \Omega} |\varphi(\lambda)| \\
&= \|\varphi\|_{\Omega}.
\end{aligned}$$

Therefore, $\mathrm{H}^\infty(\mathcal{F}) = \mathrm{H}^\infty(\Omega)$ and $\|\varphi\|_{\mathcal{F}} = \|\varphi\|_{\Omega}$ for all $\varphi \in \mathrm{H}^\infty(\Omega)$.

Calcular Norms 217

Example 9.8. Let $\Omega = \mathbb{D}$ and $\mathcal{F}$ be the class of operators T satisfying $\sigma(T) \subseteq \mathbb{D}$ and $\|T\| \le 1$; thus $[\lambda] \in \mathcal{F}$ for all $\lambda \in \mathbb{D}$. On the one hand, if $\varphi \in \mathrm{Hol}(\mathbb{D})$,

$$\|\varphi\|_{\mathcal{F}} = \sup_{T \in \mathcal{F}} \|\varphi(T)\|$$

$$\ge \sup_{\lambda \in \mathbb{D}} \|\varphi([\lambda])\|$$

$$\ge \sup_{\lambda \in \mathbb{D}} \|[\varphi(\lambda)]\|$$

$$= \sup_{\lambda \in \mathbb{D}} |\varphi(\lambda)|$$

$$= \|\varphi\|_{\mathbb{D}}.$$

On the other hand, by the Schur form of von Neumann's inequality (Theorem 1.47),

$$\|\varphi\|_{\mathcal{F}} = \sup_{T \in \mathcal{F}} \|\varphi(T)\| \le \|\varphi\|_{\mathbb{D}}.$$

Therefore, $\mathrm{H}^\infty(\mathcal{F}) = \mathrm{H}^\infty(\mathbb{D})$ as sets, and $\|\varphi\|_{\mathcal{F}} = \|\varphi\|_{\mathbb{D}}$ for all $\varphi \in \mathrm{H}^\infty(\mathbb{D})$.

Example 9.9. Let $\Omega = \mathbb{D}^2$ and $\mathcal{F}$ be the class of commuting pairs $T = (T^1, T^2)$ satisfying $\sigma(T) \subseteq \mathbb{D}^2$ and $\|T^1\|, \|T^2\| \le 1$. Arguing as in the previous example, but using Andô's inequality in place of von Neumann's inequality, one shows that $\mathrm{H}^\infty(\mathcal{F}) = \mathrm{H}^\infty(\mathbb{D}^2)$ as sets and $\|\varphi\|_{\mathcal{F}} = \|\varphi\|_{\mathbb{D}}^2$ for all $\varphi \in \mathrm{H}^\infty(\mathbb{D}^2)$.

Example 9.10. If Ω is a domain in $\mathbb{C}^d$ and $\mathcal{F}$ is the class of operators T such that Ω is a spectral domain for T (cf. Definition 1.46), then, as in the preceding two examples, $\mathrm{H}^\infty(\mathcal{F}) = \mathrm{H}^\infty(\Omega)$ as sets and $\|\varphi\|_{\mathcal{F}} = \|\varphi\|_\Omega$ for all $\varphi \in \mathrm{H}^\infty(\Omega)$.

Example 9.11. Let Ω be a domain in $\mathbb{C}^d$ and let $\mathcal{H}$ be a Hilbert space of holomorphic functions on Ω with reproducing kernel k. Let $\mathrm{Mult}(\mathcal{H})$ denote the space of multipliers of $\mathcal{H}$. In this example, we shall show that $\mathrm{Mult}(\mathcal{H})$ is naturally isometrically isomorphic to $\mathrm{H}^\infty(\mathcal{F})$ for an appropriately chosen $\mathcal{F}$ subordinate to Ω.

If n is a positive integer and $\lambda = (\lambda_1, \ldots, \lambda_n) \in \Omega^n$, we may define a d-tuple of operators

$$X_\lambda = (X_\lambda^1, \ldots, X_\lambda^d)$$

acting on

$$\mathcal{M}_\lambda = [k_{\lambda_i} : i = 1, \ldots, n]$$

by the formulas

$$X_\lambda^r \, k_{\lambda_i} = \bar{\lambda}_i^r k_{\lambda_i} \qquad \text{for } i = 1, \ldots, n, \ r = 1, \ldots, d.$$

218 *Commutative Theory*

Let $\mathcal{F}$ be the class

$$\mathcal{F} = \{X_\lambda^*: n \geq 1 \text{ and } \lambda \in \Omega^n\}.$$

We claim that $\mathcal{F}$ is subordinate to Ω. Since the components of X_λ share the spanning set of simultaneous eigenvectors $k_{\lambda_1}, \ldots, k_{\lambda_n}$, the components of X_λ pairwise commute. Hence the components of X_λ^* commute pairwise, that is, condition (i) in Definition 9.3 holds. Also, as $\sigma(X_\lambda)$ is simply the set of joint eigenvalues of X_λ, $\sigma(X_\lambda) = \{\bar{\lambda}_1, \ldots, \bar{\lambda}_n\}$. Hence $\sigma(X_\lambda^*) = \{\lambda_1, \ldots, \lambda_n\} \subseteq \Omega$, which is condition (ii). Finally, condition (iii) holds since, when $n = 1$, we may take $\mathcal{H} = \mathcal{M}_\lambda$ and then $X_\lambda = (\lambda_1^1 \mathrm{id}_\mathcal{H}, \ldots, \lambda_1^d \mathrm{id}_\mathcal{H})$ for any point λ_1 in Ω.

We claim that, with the above choice of $\mathcal{F}$, $\mathrm{H}^\infty(\mathcal{F})$ is naturally isometrically isomorphic to $\mathrm{Mult}(\mathcal{H})$. This will follow if we can show that $\mathscr{S}(\mathcal{F}) = \mathrm{ball}$ $(\mathrm{Mult}(\mathcal{H}))$. Fix $\varphi \in \mathrm{Hol}(\Omega)$. Then

$$\varphi \in \mathrm{ball} \ (\mathrm{Mult}(\mathcal{H})) \iff \|M_\varphi^*\| \leq 1$$

$$\iff 1 - M_\varphi M_\varphi^* \geq 0$$

$$\iff (1 - \overline{\varphi(\mu)}\varphi(\lambda)) \, \langle k_\mu, k_\lambda \rangle_\mathcal{H} \geq 0$$

$$\iff \forall_{n \geq 1} \, \forall_{\lambda \in \Omega^n} \, \left[(1 - \overline{\varphi(\lambda_i)}\varphi(\lambda_j)) \, \langle k_{\lambda_i}, k_{\lambda_j} \rangle_\mathcal{H} \right] \geq 0$$

$$\iff \forall_{n \geq 1} \, \forall_{\lambda \in \Omega^n} \, 1 - \varphi^\vee(X_\lambda)^* \varphi^\vee(X_\lambda) \geq 0$$

$$\iff \forall_{n \geq 1} \, \forall_{\lambda \in \Omega^n} \, \|\varphi(X_\lambda^*)\| \leq 1$$

$$\iff \forall_{T \in \mathcal{F}} \, \|\varphi(T)\| \leq 1$$

$$\iff \varphi \in \mathscr{S}(\mathcal{F}).$$

Proposition 9.12. *If Ω is a domain in $\mathbb{C}^d$ and $\mathcal{F}$ is subordinate to Ω, then $\mathrm{H}^\infty(\mathcal{F})$ equipped with $\|\cdot\|_\mathcal{F}$ is a Banach algebra. Furthermore,*

$$\mathrm{H}^\infty(\mathcal{F}) \subseteq \mathrm{H}^\infty(\Omega) \tag{9.13}$$

and

$$\|\varphi\|_\Omega \leq \|\varphi\|_\mathcal{F} \qquad \text{for all } \varphi \in \mathrm{H}^\infty(\mathcal{F}). \tag{9.14}$$

Proof. Let $\varphi \in \mathrm{Hol}(\Omega)$. For any $\lambda \in \Omega$, by condition (iii) in Definition 9.5, there is a Hilbert space $\mathcal{H}$ such that $(\lambda^1 \mathrm{id}_\mathcal{H}, \ldots, \lambda^d \mathrm{id}_\mathcal{H}) \in \mathcal{F}$, and therefore

$$\|\varphi\|_\mathcal{F} = \sup_{T \in \mathcal{F}} \|\varphi(T)\|$$

$$\geq \sup_{\lambda \in \Omega} \|\varphi\big((\lambda^1 \mathrm{id}_\mathcal{H}, \ldots, \lambda^d \mathrm{id}_\mathcal{H})\big)\|$$

$$= \sup_{\lambda \in \Omega} \|\varphi(\lambda)\|$$

$$= \|\varphi\|_\Omega.$$

Therefore, the statements (9.13) and (9.14) hold.

Calcular Norms 219

It is immediate from the definition that $H^\infty(\mathcal{F})$ is a normed algebra satisfying $\|1\|_{\mathcal{F}} = 1$ and $\|\varphi\psi\|_{\mathcal{F}} \le \|\varphi\|_{\mathcal{F}}\|\psi\|_{\mathcal{F}}$ for all $\varphi, \psi \in H^\infty(\mathcal{F})$. There remains to show that $H^\infty(\mathcal{F})$ is complete.

Consider a Cauchy sequence $\{\varphi_n\}$ in $H^\infty(\mathcal{F})$. By the inequality (9.14), $\{\varphi_n\}$ is a Cauchy sequence in $H^\infty(\Omega)$. Therefore, as $H^\infty(\Omega)$ is complete, there exists $\varphi \in H^\infty(\Omega)$ such that

$$\sup_{\lambda \in \Omega} |\varphi_n(\lambda) - \varphi(\lambda)| \to 0 \text{ as } n \to \infty. \tag{9.15}$$

We claim that

$$\varphi \in H^\infty(\mathcal{F}) \tag{9.16}$$

and

$$\varphi_n \to \varphi \text{ in } H^\infty(\mathcal{F}). \tag{9.17}$$

To prove statement (9.16), note that for each $T \in \mathcal{F}$, condition (ii) in Definition 9.3 guarantees that Ω is a neighborhood of $\sigma(T)$. Consequently, condition (9.15) and continuity of the functional calculus imply that

$$\varphi_n(T) \to \varphi(T) \qquad \text{for all } T \in \mathcal{F}. \tag{9.18}$$

Also, as $\{\varphi_n\}$ is a Cauchy sequence in $H^\infty(\mathcal{F})$, there exists a constant M such that

$$\|\varphi_n\|_{\mathcal{F}} \le M \qquad \text{for all } n.$$

Therefore, if $T \in \mathcal{F}$,

$$\|\varphi(T)\|_{\mathcal{F}} = \lim_{n \to \infty} \|\varphi_n(T)\| \le \limsup_{n \to \infty} \|\varphi_n\|_{\mathcal{F}} \le M.$$

Hence

$$\|\varphi\|_{\mathcal{F}} = \sup_{T \in \mathcal{F}} \|\varphi(T)\|_{\mathcal{F}} \le M,$$

which proves the membership (9.16).

To prove the limiting relation (9.17), let $\varepsilon > 0$. Choose N such that

$$m, n \ge N \implies \|\varphi_n - \varphi_m\|_{\mathcal{F}} < \varepsilon.$$

Since $\|\varphi_n - \varphi_m\|_{\mathcal{F}} = \sup_{T \in \mathcal{F}} \|\varphi_n(T) - \varphi_m(T)\|$,

$$m, n \ge N \implies \|\varphi_n(T) - \varphi_m(T)\| < \varepsilon \qquad \text{for all } T \in \mathcal{F}.$$

Letting $m \to \infty$ and using statement (9.18), we deduce that

$$n \ge N \implies \|\varphi_n(T) - \varphi(T)\| < \varepsilon \qquad \text{for all } T \in \mathcal{F}.$$

220　　　　　　　　　　　　　　*Commutative Theory*

Hence, since $\|\varphi_n - \varphi\|_{\mathcal{F}} = \sup_{T \in \mathcal{F}} \|\varphi_n(T) - \varphi(T)\|$,

$$n \geq N \implies \|\varphi_n - \varphi\|_{\mathcal{F}} \leq \varepsilon.$$

□

If Ω is a domain in $\mathbb{C}^d$ and $\mathcal{F}$ is subordinate to Ω, then we refer to $H^\infty(\mathcal{F})$ as the *calcular algebra defined by* $\mathcal{F}$. Proposition 9.12 asserts that calcular algebras are Banach algebras. In fact, however, they are much more. For a set A and any positive integer m, we let $\mathcal{M}_m(A)$ denote the set of $m \times m$ matrices with entries in A. We observe that if $\mathcal{H}$ is a Hilbert space and $A = \mathcal{B}(\mathcal{H})$, then $\mathcal{M}_m(A)$ is an algebra. Furthermore, as for each $m \geq 1$, $\mathcal{M}_m(\mathcal{B}(\mathcal{H}))$ and $\mathcal{B}(\mathcal{H}^{(m)})$ are naturally isomorphic and $\mathcal{B}(\mathcal{H}^{(m)})$ is a Banach algebra when equipped with the operator norm, for each $m \geq 1$, $\mathcal{M}_m(A)$ naturally carries the structure of a Banach algebra, its norm being the operator norm of $\mathcal{B}(\mathcal{H}^{(m)})$.

Definition 9.19. *We say that a Banach algebra A with unit together with a sequence, $\{\|\cdot\|_m\}_{m=1}^\infty$, where $\|\cdot\|_m$ is a norm on $\mathcal{M}_m(A)$ for each $m \geq 1$, is an operator algebra if there exists a Hilbert space $\mathcal{H}$ and a unital homomorphism $\rho\colon A \to \mathcal{B}(\mathcal{H})$ such that for each $m \geq 1$,*

$$\|[a_{ij}]\|_m = \|[\rho(a_{ij})]\|_{\mathcal{B}(\mathcal{H}^{(m)})} \tag{9.20}$$

for each $a = [a_{ij}]$ in $\mathcal{M}_m(A)$.

Now, if Ω is a domain in $\mathbb{C}^d$ and $\mathcal{F}$ is subordinate to Ω, then for each positive integer m we may define a norm $\|\cdot\|_{\mathcal{F},m}$ on $\mathcal{M}_m(H^\infty(\mathcal{F}))$ by the formula

$$\|\varphi\|_{\mathcal{F},m} = \sup_{T \in \mathcal{F}} \|[\varphi_{ij}(T)]\| \qquad \text{for } \varphi = [\varphi_{ij}] \in \mathcal{M}_m(H^\infty(\mathcal{F})). \tag{9.21}$$

Thus, as norms have been specified on $\mathcal{M}_m(H^\infty(\mathcal{F}))$ for each m, $H^\infty(\mathcal{F})$ becomes a candidate for being an operator algebra in the sense of Definition 9.19.

Proposition 9.22. *If Ω is a domain in $\mathbb{C}^d$ and $\mathcal{F}$ is subordinate to Ω, then $H^\infty(\mathcal{F})$ equipped with the norms defined in equation (9.21) is an operator algebra.*

Proof. We first prove the statement in the case that $\mathcal{F}$ is a set, then prove the general case after some remarks on set theory.

Suppose that $\mathcal{F}$ is a set. For each $T \in \mathcal{F}$, let $\mathcal{H}_T$ denote the space that T acts on and let

$$\mathcal{H} = \bigoplus_{T \in \mathcal{F}} \mathcal{H}_T. \tag{9.23}$$

Calcular Norms 221

For $h \in H^\infty(\mathcal{F})$ define

$$\rho(h) = \bigoplus_{T \in \mathcal{F}} h(T).$$

Since each summand in the last expression is an operator bounded in norm by $\|h\|_{H^\infty(\mathcal{F})}$, the right-hand side defines a bounded operator. Clearly, ρ is unital, and the properties of the functional calculus imply that ρ is an algebra homomorphism. There remains to show that for each $m \geq 1$ and each $\varphi = [\varphi_{ij}] \in \mathcal{M}_m(H^\infty(\mathcal{F}))$,

$$\|\varphi\|_{\mathcal{F},m} = \|[\rho(\varphi_{ij})]\|.$$

But

$$\|\varphi\|_{\mathcal{F},m} = \sup_{T \in \mathcal{F}} \|[\varphi_{ij}(T)]\|$$

$$= \left\| \bigoplus_{T \in \mathcal{F}} [\varphi_{ij}(T)] \right\|$$

$$= \left\| \left[\bigoplus_{T \in \mathcal{F}} \varphi_{ij}(T) \right] \right\|$$

$$= \|[\rho(\varphi_{ij})]\|.$$

$\square$

Remark 9.24. (Set-theoretic considerations) Ordinarily in analysis one does not need to make the underlying set theory explicit. It is usually evident that reasoning can in principle be formulated within some recognized set theory, such as ZFC, Zermelo–Fraenkel set theory + the axiom of choice. The development in this section, though, can clearly *not* be formulated in ZFC as it stands, since it involves classes $\mathcal{F}$, which need not be sets, as in Examples 9.8–9.10. In the foregoing proof we define the Hilbert space $\mathcal{H}$ by the formula (9.23). This step is legitimate for sets $\mathcal{F}$, but when $\mathcal{F}$ is a class that is not a set, there is no reason to believe that there is such a Hilbert space $\mathcal{H}$. Nevertheless, Proposition 9.22 holds in the stated generality if one works in a suitable version of set theory. We are grateful to George Wilmers and Bob Dumas for guidance on this important topic.

One way to circumvent the difficulties under discussion is simply to restrict consideration to bona fide sets in ZFC. However, this would mean that we could never speak of the classes of, for example, all groups or all Hilbert space contractions. As a result many formulations of definitions and theorems would lose in elegance and simplicity. Mathematicians going back at least to von Neumann, Bernays, and Gödel have felt that this is an unnecessarily drastic solution to the problem and have developed alternative set theories that

222 *Commutative Theory*

better correspond to mathematical intuition. A convenient source for one such theory is Kelley's *General Topology* [129, appendix], a book that is familiar to generations of analysts. *Morse–Kelley set theory* is a strengthening of Gödel–Bernays (GB) set theory obtained by the addition of an axiom schema of comprehension that involves classes; this strengthening entitles us to work freely with *proper classes*, which are not sets, but whose elements are sets.

In this version of set theory Definition 9.3 is perfectly rigorous. The definition of $\|\varphi\|_{\mathcal{F}}$ in equation (9.5) is also well formed. A function defined on a class is itself a class, as is its range. The class of all values $\|\varphi(T)\|$, for $T \in \mathcal{F}$, is a class that is contained in $\mathbb{R}$, hence is a set. It is therefore legitimate to take the supremum of this set.

To conclude the proof of Proposition 9.22, let us fix a separable infinite-dimensional Hilbert space $\mathcal{K}$. Define a class $\tilde{\mathcal{F}}$ of d-tuples of operators by saying that $T \in \tilde{\mathcal{F}}$ if T acts on some closed subspace of $\mathcal{K}$ and T is unitarily equivalent to the restriction of a member of $\mathcal{F}$ to a reducing subspace.

Since

$$\tilde{\mathcal{F}} \subseteq \bigcup \{\mathcal{B}(\mathcal{H})^d : \mathcal{H} \subseteq \mathcal{K}\},$$

$\tilde{\mathcal{F}}$ is a class of operator tuples that is contained in a set, hence is itself a set. We may therefore invoke the special case proved earlier to conclude that $\mathrm{H}^\infty(\tilde{\mathcal{F}})$ is an operator algebra.

We claim that $\mathrm{H}^\infty(\mathcal{F}) = \mathrm{H}^\infty(\tilde{\mathcal{F}})$. Consider $\varphi \in \mathrm{H}^\infty(\mathcal{F})$ and $\tilde{T} \in \tilde{\mathcal{F}}$. Since $\tilde{T}$ is (up to unitary equivalence) the restriction of some element T of $\mathcal{F}$ to a reducing subspace, $\varphi(\tilde{T})$ is a restriction of $\varphi(T)$, and therefore

$$\|\varphi(\tilde{T})\| \quad \leq \quad \|\varphi(T)\| \quad < \quad \infty.$$

Hence $\varphi \in \mathrm{H}^\infty(\tilde{\mathcal{F}})$. Thus $\mathrm{H}^\infty(\mathcal{F}) \subseteq \mathrm{H}^\infty(\tilde{\mathcal{F}})$ and

$$\|\varphi\|_{\mathrm{H}^\infty(\tilde{\mathcal{F}})} \leq \|\varphi\|_{\mathrm{H}^\infty(\mathcal{F})}. \tag{9.25}$$

In the reverse direction, suppose that $\varphi \in \mathrm{H}^\infty(\tilde{\mathcal{F}})$ and consider $T \in \mathcal{F}$, acting on a space $\mathcal{H}_T$. Choose a separable subspace $\mathcal{L}$ of $\mathcal{H}_T$, reducing for T, such that $\|\varphi(T)|\mathcal{L}\| = \|\varphi(T)\|$. Choose a unitary operator U from $\mathcal{L}$ onto a subspace $\tilde{\mathcal{L}}$ of $\mathcal{K}$ and define $\tilde{T}$ to be $U(T|\mathcal{L})U^*$. Then $\tilde{T} \in \tilde{\mathcal{F}}$ and $\varphi(\tilde{T})$ is unitarily equivalent to $\varphi(T|\mathcal{L})$, so that

$$\|\varphi(T)\| \quad = \quad \|\varphi(T|\mathcal{L})\| \quad = \quad \|\varphi(\tilde{T})\| \quad \leq \quad \|\varphi\|_{\mathrm{H}^\infty(\tilde{\mathcal{F}})} \quad < \infty. \tag{9.26}$$

Hence $\varphi \in \mathrm{H}^\infty(\tilde{\mathcal{F}})$, and, by the relations (9.25) and (9.26),

$$\|\varphi\|_{\mathrm{H}^\infty(\tilde{\mathcal{F}})} = \|\varphi\|_{\mathrm{H}^\infty(\mathcal{F})},$$

and so $\mathrm{H}^\infty(\tilde{\mathcal{F}}) = \mathrm{H}^\infty(\mathcal{F})$, as claimed.

Calcular Norms   223

We can slightly strengthen the argument to show that

$$\| \cdot \|_{\tilde{\mathcal{F}},m} = \| \cdot \|_{\mathcal{F},m}$$

as norms on $\mathcal{M}_m(\mathrm{H}^\infty(\mathcal{F}))$ (as defined in equation (9.21)) for every positive integer m.

Hence $\mathrm{H}^\infty(\mathcal{F})$, with the norms $\| \cdot \|_{\mathcal{F},m}$, is indeed an operator algebra in the sense of Definition 9.19.   □

9.3 Halmos's Conjecture and Paulsen's Theorem

A highly successful application of calcular norms to pure operator theory is a theorem of Vern Paulsen that precisely identifies the operators that are similar to contractions in terms of a calcular norm for matrix-valued holomorphic functions.

If $\mathcal{H}$ is a Hilbert space and $C, T \in \mathcal{B}(\mathcal{H})$, let us say that T is *c-similar* to C and write $T \stackrel{c}{\sim} C$, if there exists an invertible $S \in \mathcal{B}(\mathcal{H})$ such that

$$T = STS^{-1} \quad \text{and} \quad \|S\| \|S^{-1}\| \le c.$$

In [99] Paul Halmos observed that if $C, T \in \mathcal{B}(\mathcal{H})$, $T \stackrel{c}{\sim} C$, $\|C\| \le 1$, and p is a polynomial, then von Neumann's inequality (Theorem 1.36) allows the following calculation.

Halmos's observation If p is a polynomial, then

$$
\begin{aligned}
\|p(T)\| &= \|p(SCS^{-1})\| \\
&= \|Sp(C)S^{-1}\| \\
&\le \|S\| \|p(C)\| \|S^{-1}\| \\
&\le c \|p\|_{\mathbb{D}}.
\end{aligned}
\tag{9.27}
$$

Thus if we define an operator T to be *c-polynomially bounded* if

$$\|p(T)\| \le c \|p\|_{\mathbb{D}} \quad \text{for all } p \in \mathbb{C}[z]$$

and to be *polynomially bounded* if there exists a constant c such that T is c-polynomially bounded, then we see that Halmos's calculation (9.27) implies the following observation.

If T is similar to a contraction, then T is polynomially bounded.   (9.28)

This observation led Halmos to ask whether the converse was true, that is, he posed the following question.

224 *Commutative Theory*

Problem 9.29. (problem 6 in [99]) Is every polynomially bounded operator similar to a contraction?

Attempts to answer this famous question led to fruitful activity in the operator theory community over the next decades. The question was eventually answered in the negative in a deep work by Gilles Pisier in 1997 [164].

Presciently, Paulsen observed that Halmos's calculation could be strengthened as follows.

Paulsen's observation If T, C, and S are as in Halmos's observation and $p = [p_{ij}]$ is an $m \times m$ matrix of polynomials, then

$$
\begin{aligned}
\|p(T)\| &= \|[p_{ij}(T)]\| \\
&= \|[p_{ij}(SCS^{-1})]\| \\
&= \|[Sp_{ij}(C)S^{-1}]\| \\
&= \|S^{(m)}[p_{ij}(C)](S^{(m)})^{-1}\| \\
&\leq \|S^{(m)}\| \|[p_{ij}(C)]\| \|(S^{(m)})^{-1}\| \\
&\leq c\|p\|_{\mathbb{D}}.
\end{aligned}
$$

Here $S^{(m)}$ denotes the operator $S \oplus \ldots \oplus S$ acting on $\mathcal{H} \oplus \ldots \oplus \mathcal{H}$.

We are led to define an operator T to be *completely c-polynomially bounded* if

$$\text{for all } m \geq 1 \text{ and for all } p \in \mathcal{M}_m(\mathbb{C}[z]), \quad \|p(T)\| \leq c\|p\|_{\mathbb{D}} \qquad (9.30)$$

and to be *completely polynomially bounded* if there exists a constant c such that T is completely c-polynomially bounded. Halmos's calculation then implies that every operator similar to a contraction is completely polynomially bounded. Paulsen showed in [161] that the converse of this statement holds.

Theorem 9.31. (Paulsen's theorem) *An operator T is similar to a contraction if and only if T is completely polynomially bounded.*

Paulsen deduced Theorem 9.31 from the following much more general result, whose proof we omit.

Theorem 9.32. *Let A be a unital operator algebra, $\mathcal{H}$ a Hilbert space, and let $\rho: A \to \mathcal{B}(\mathcal{H})$ be a unital algebra homomorphism satisfying*

$$\|[\rho(a_{ij})]\| \leq c\|[a_{ij}]\| \quad \text{for all } m \geq 1 \text{ and all } a \in \mathcal{M}_m(A). \qquad (9.33)$$

Then there exists $S \in \mathcal{B}(\mathcal{H})$ such that S is invertible,

$$\|S\| \|S^{-1}\| \leq c,$$

Calcular Norms 225

and

$$\|[S^{-1}\rho(a_{ij})S]\| \leq \|[a_{ij}]\| \qquad \text{for all } m \geq 1 \text{ and all } a \in \mathcal{M}_m(A).$$

To see how Theorem 9.32 implies Theorem 9.31, first recall that Paulsen's calculation implies that if T is similar to a contraction then T is completely polynomially bounded. Conversely, assume that $T \in \mathcal{B}(\mathcal{H})$ is completely polynomially bounded, that is, there exists a constant c such that statement (9.30) holds. Let A denote the *disc algebra*, that is, the Banach algebra of continuous functions on $\mathbb{D}^-$ that are holomorphic on $\mathbb{D}$, equipped with the supremum norm on $\mathbb{D}$. The disc algebra is naturally an operator algebra if, for each $m \geq 1$, we equip $\mathcal{M}_m(A)$ with the norm

$$\|f\|_m = \sup_{\lambda \in \mathbb{D}} \|[f_{ij}(\lambda)]\| \qquad \text{for } f \in \mathcal{M}_m(A).$$

As the polynomials are dense in A, it follows from condition (9.30) that

$$\|f(T)\| \leq c\|f\|_m \qquad \text{for all } m \geq 1 \text{ and all } f \in \mathcal{M}_m(A).$$

Consequently, if we define $\rho \colon A \to \mathcal{B}(\mathcal{H})$ by

$$\rho(f) = f(T) \qquad \text{for all } f \in A,$$

then condition (9.33) holds. Hence by Theorem 9.32, there exists an invertible $S \in \mathcal{B}(\mathcal{H})$ such that

$$\|[S^{-1}f_{ij}(T)S]\| \leq \|[a_{ij}]\| \qquad \text{for all } m \geq 1 \text{ and all } f \in \mathcal{M}_m(A).$$

Letting $m = 1$ and $f(z) = z$, we find that

$$C \overset{\text{def}}{=} S^{-1}TS$$

is a contraction. Consequently, $T = SCS^{-1}$ is similar to a contraction, as was to be proved.

Remark 9.34. If one tracks the quantity $\|S^{-1}\|\|S\|$ in the argument above, one obtains a sharper result than Theorem 9.31: if $T \in \mathcal{B}(\mathcal{H})$ is completely c-bounded, then T is c-similar to a contraction.

The following theorem results if one mimics the proof of Theorem 9.31 while replacing A by $\mathrm{H}^\infty(\mathcal{F})$.

Theorem 9.35. (Paulsen's theorem for calcular algebras) *Let Ω be a domain and assume that $\mathcal{F}$ is a family of operators that is subordinate to Ω. Let $\mathcal{H}$ be a Hilbert space and let $T \in \mathcal{B}(\mathcal{H})$ satisfy $\sigma(T) \subseteq \Omega$. Then*

$$\|\varphi(T)\| \leq c\|\varphi\|_\Omega \qquad \text{for all } m \geq 1 \text{ and all } \varphi \in \mathrm{H}^\infty_{\mathcal{B}(\mathbb{C}^m)}(\Omega)$$

226   *Commutative Theory*

if and only if there exists $X \in \mathcal{B}(\mathcal{H})$ such that $T \overset{c}{\sim} X$ and

$$\|\varphi(X)\| \leq \|\varphi\|_{\Omega} \qquad \text{for all } m \geq 1 \text{ and all } \varphi \in H^{\infty}_{\mathcal{B}(\mathbb{C}^m)}(\Omega).$$

9.4 The Douglas–Paulsen Norm

An interesting choice for $\mathcal{F}$, first studied by Ronald Douglas and Vern Paulsen, arose during their development of a theory of K-spectral sets [77]. For $r \in \mathbb{R}$ satisfying $0 < r < 1$, let $A(r)$ denote the annulus $\{\lambda \in \mathbb{C}: r < |\lambda| < 1\}$. Let $\mathcal{F}_{\mathrm{dp}}(r)$, the *Douglas–Paulsen family with parameter r*, consist of the class of operators T that satisfy

$$\sigma(T) \subseteq A(r) \quad \text{and} \quad r^2 \leq T^*T \leq 1.$$

Noting that $\mathcal{F}_{\mathrm{dp}}(r)$ is subordinate to $A(r)$, we obtain a norm $\| \cdot \|_{\mathcal{F}_{\mathrm{dp}}(r)}$ and a Banach space of analytic functions $H^{\infty}(\mathcal{F}_{\mathrm{dp}}(r))$ as in Definitions 9.4 and 9.6. We will sometimes write $\| \cdot \|_{\mathrm{dp}}$ for $\| \cdot \|_{\mathcal{F}_{\mathrm{dp}}(r)}$ when there is no risk of confusion.

9.4.1 The Douglas–Paulsen Estimate

A significant task is to understand the Douglas–Paulsen norm in function-theoretic terms. In [77] Douglas and Paulsen undertook to develop a Sz.-Nagy–Foias-style dilation model for the elements of $\mathcal{F}_{\mathrm{dp}}(r)$. They accomplished their goal with the aid of the following estimate for $\| \cdot \|_{\mathrm{dp}}$.

Proposition 9.36. (Douglas–Paulsen estimate) *If $T \in \mathcal{F}_{\mathrm{dp}}(r)$, then*

$$\|\varphi(T)\| \leq \left(2 + \frac{1+r}{1-r}\right) \|\varphi\|_{A(r)} \qquad \text{for all } m \geq 1 \text{ and all } \varphi \in H^{\infty}_{\mathcal{B}(\mathbb{C}^m)}(A(r)).$$

The following dilation representation of T is obtained as an immediate corollary of this proposition.

Corollary 9.37. *Let $T \in \mathcal{B}(\mathcal{H})$. If $T \in \mathcal{F}_{\mathrm{dp}}(r)$, then there exists an invertible $S \in \mathcal{H}$ such that*

$$\|S\| \, \|S^{-1}\| \leq 2 + \frac{1+r}{1-r}$$

and STS^{-1} has a dilation[5] to the boundary of $A(r)^-$.

The statement follows from Paulsen's theorem, Theorem 9.35, and Arveson's dilation theorem.

---

[5] This terminology is explained in Definition 1.51.

Corollary 9.37 provides an example of how an estimate of a calcular norm can yield a result in pure operator theory. However, it is not our intention to explore such implications here. We point out that, as the converse to Corollary 9.37 is false, the model yielded by the corollary fails to provide a spatial explanation for why $T \in \mathcal{F}_{\mathrm{dp}}(r)$.

As a corollary of Proposition 9.36 we also obtain the following estimate for $\| \cdot \|_{\mathrm{dp}}$.

Corollary 9.38. *Fix a positive integer m and assume that $\varphi \in \mathrm{Hol}_{\mathcal{B}(\mathbb{C}^m)}(A(r))$.*

$$\|\varphi\|_{\mathrm{dp}} \quad \leq \quad \left(2 + \frac{1+r}{1-r}\right) \|\varphi\|_{A(r)}. \tag{9.39}$$

In particular, $\varphi \in \mathrm{H}^{\infty}_{\mathcal{F}_{\mathrm{dp}}(r)}$ if and only if $\varphi \in \mathrm{H}^{\infty}(A(r))$.

Proof. Fix $\varphi \in \mathrm{Hol}_{\mathcal{B}(\mathbb{C}^m)}(A(r))$. By Proposition 9.36

$$\|\varphi\|_{\mathrm{dp}} = \sup_{T \in \mathcal{F}_{\mathrm{dp}}(r)} \|\varphi(T)\| \leq \left(2 + \frac{1+r}{1-r}\right) \|\varphi\|_{A(r)}.$$

To prove the second statement of the corollary observe that if $\varphi \in \mathrm{Hol}(A(r))$, then inequalities (9.39) and (9.14) imply that $\|\varphi\|_{\mathrm{dp}} < \infty$ if and only if $\|\varphi\|_{A(r)} < \infty$. Thus $\mathrm{H}^{\infty}_{\mathcal{F}_{\mathrm{dp}}(r)} = \mathrm{H}^{\infty}(A(r))$. $\square$

As Corollary 9.38 guarantees that $\mathrm{H}^{\infty}(\mathcal{F}_{\mathrm{dp}}(r)) = \mathrm{H}^{\infty}(A(r))$, it is natural to consider the smallest constant K such that

$$\|\varphi\|_{\mathrm{dp}} \leq K \|\varphi\|_{A(r)} \qquad \text{for all } \varphi \in \mathrm{H}^{\infty}(\mathcal{F}_{A(r)}).$$

The following definition formalizes this idea for a general calcular norm.

Definition 9.40. *Let Ω be a domain and let $\mathcal{F}$ be a class of operators that is subordinate to Ω. If $\mathrm{H}^{\infty}(\mathcal{F}) = \mathrm{H}^{\infty}(\Omega)$ as sets, we define $K(\mathcal{F})$ by*

$$K(\mathcal{F}) = \sup_{\varphi \in \mathscr{S}(\Omega)} \|\varphi\|_{\mathcal{F}}.$$

Otherwise, we let $K(\mathcal{F}) = \infty$. More generally, if m is a positive integer, we define

$$K^m(\mathcal{F}) = \sup_{\varphi \in \mathscr{S}_{\mathcal{B}(\mathbb{C}^m)}(\Omega)} \|\varphi\|_{\mathcal{F}}$$

if $\mathrm{H}^{\infty}(\mathcal{F}) = \mathrm{H}^{\infty}(\Omega)$, and let $K^m(\mathcal{F}) = \infty$ otherwise.

Remarks 9.41.

(i) $K(\mathcal{F})$ is the norm of the identity mapping from

$$(\mathrm{H}^{\infty}(\Omega), \| \cdot \|_{\Omega}) \quad \text{to} \quad (\mathrm{H}^{\infty}(\mathcal{F}), \| \cdot \|_{\mathcal{F}}).$$

228   *Commutative Theory*

Thus $K(\mathcal{F})$ is finite if and only if $H^\infty(\mathcal{F}) = H^\infty(\Omega)$.

(ii) $K^1(\mathcal{F}) = K(\mathcal{F})$.

(iii) $K^1(\mathcal{F}) \le K^2(\mathcal{F}) \le K^3(\mathcal{F}) \le \ldots$

(iv) With the notation of the definition, Lemma 9.38 is equivalent to the assertion that

$$K^m(\mathcal{F}_{\mathrm{dp}}(r)) \quad \le \quad 2 + \frac{1+r}{1-r} \qquad \text{for all } m.$$

Lemma 9.42. *Let Ω be a domain and let $\mathcal{F}$ be a class of operators that is subordinate to Ω. If $K(\mathcal{F}) < \infty$, then $K^m(\mathcal{F}) < \infty$ for all $m \ge 1$.*

Proof. This is because the norm of a matrix of operators is less than or equal to the square root of the sum of the squares of the norms of the entries and greater than or equal to the norm of the largest entry. $\square$

Douglas and Paulsen proved Proposition 9.36 by using a *decomposition constant*. This means they wrote a general function φ in $H^\infty(A(r))$ as the sum of two functions $\psi_1 \in H^\infty(\mathbb{D})$ and $\psi_2 \in H^\infty(\mathbb{C} \setminus \overline{\mathbb{D}(0,r)})$ with

$$\|\psi_1\|_{\mathbb{D}} + \|\psi_2\|_{\mathbb{C} \setminus \overline{\mathbb{D}(0,r)}} \quad \le \quad C \|\varphi\|. \tag{9.43}$$

Then von Neumann's inequality implies that

$$\|\psi_1(T)\| \le \|\psi_1\|_{\mathbb{D}} \quad \text{and} \quad \|\psi_2(T)\| \le \|\psi_2\|_{\mathbb{C} \setminus \overline{\mathbb{D}(0,r)}}.$$

Shields proved in [186] that the best constant in inequality (9.43) was bounded by

$$2 + \left(\frac{1+r}{1-r} \right)^{\frac{1}{2}},$$

and he asked if the optimal constant remains bounded as $r \to 1^-$. This question was answered by Michel Crouzeix in [39], where it was shown that, for all $m \ge 1$,

$$\tfrac{4}{3} < K(\mathcal{F}_{\mathrm{dp}}(r)) \le K^m(\mathcal{F}_{\mathrm{dp}}(r)) \le 2 + \frac{1+\sqrt{r}}{\sqrt{1+\sqrt{r}+r}} \le 2 + \frac{2}{\sqrt{3}}.$$

Crouzeix further improved the above estimates in [65].

The problem of finding the smallest constant C in the inequality (9.43) has a function-theoretic aspect that goes beyond the properties of $\| \cdot \|_{\mathcal{F}_{\mathrm{dp}}(r)}$. Later we shall characterize $\| \cdot \|_{\mathcal{F}_{\mathrm{dp}}(r)}$ in purely function-theoretic terms, as the optimal solution to a problem of extension of a holomorphic function off a subvariety of the bidisc.

Calcular Norms

9.4.2 A Fundamental Theorem for the Douglas–Paulsen Class

Definition 9.44. *For a function $\varphi\colon A(r) \to \mathbb{C}$, we say that $(\mathcal{N},v)$ is a* dp-model *(with parameter r) for φ if $\mathcal{N} = (\mathcal{N}^+,\mathcal{N}^-)$ is an ordered pair of Hilbert spaces, $v = (v^+,v^-)$ is an ordered pair of functions such that $v^+\colon A(r) \to \mathcal{N}^+$ and $v^-\colon A(r) \to \mathcal{N}^-$, and for which the following* dp-model *formula holds for all $\lambda,\mu \in A(r)$:*

$$1 - \overline{\varphi(\mu)}\varphi(\lambda) = (1 - \bar{\mu}\lambda)\langle v_\lambda^+, v_\mu^+\rangle_{\mathcal{N}^+} + (\bar{\mu}\lambda - r^2)\langle v_\lambda^-, v_\mu^-\rangle_{\mathcal{N}^-}. \qquad (9.45)$$

As for previous notions of model, there are low-hanging fruit that can be obtained from this definition via lurking isometry arguments: dp-models are holomorphic (cf. Propositions 2.21 and 4.24) and the converse to the following theorem (cf. Propositions 2.32 and 4.26).

Theorem 9.46. **(Fundamental theorem for $\mathscr{S}(\mathcal{F}_{\mathrm{dp}}(r))$)** *For $r \in (0,1)$, every function in $\mathscr{S}(\mathcal{F}_{\mathrm{dp}}(r))$ has a* dp-*model with parameter r.*

The proof of this theorem makes use of the duality construction, like that of the fundamental theorem for $\mathscr{S}(\mathbb{D}^2)$ (cf. Theorem 4.35). However, it requires the following, often less-than-minor modifications. First, we handle the case when

$$\varphi \text{ is holomorphic on a neighborhood of } A(r)^-. \qquad (9.47)$$

As in the proof of Theorem 4.35 we reason as follows.

1. Fix $r \in (0,1)$ and $\varphi \in \mathscr{S}(\mathcal{F}_{\mathrm{dp}}(r))$.
2. The role of $\mathrm{Hol}(\mathbb{D}^2)$, a space of two-variable functions, is now played by $\mathrm{Hol}(A(r))$, the space of holomorphic functions on the annulus.
3. The ambient space in which the duality argument was executed, $\mathrm{Her}(\mathbb{D}^2)$, is now replaced[6] by $\mathrm{Her}(A(r))$.
4. $\mathcal{R}$ and $\mathcal{P}$ are the same.
5. $\mathcal{W}$, the model cone, is now defined by

$$\mathcal{W} = \{(1 - \bar{\mu}\lambda)A(\lambda,\mu) + (\bar{\mu}\lambda - r^2)B(\lambda,\mu)\colon$$

A and B are positive semi-definite kernels on $A(r)\}$.

6. $\mathcal{W}$ is closed; this depends on the fact that $1 - |\lambda|^2$ and $|\lambda|^2 - r^2$ are jointly bounded away from zero on compact subsets of $A(r)$.

---

[6] The proof makes use of the hereditary functional calculus on $A(r)$; one can easily justify this with the aid of double Laurent series or a suitable integral formula for functions holomorphic on $A(r)^-$.

230   *Commutative Theory*

7. Under the assumptions that $\varphi \in \mathscr{S}(\mathcal{F}_{\mathrm{dp}}(r))$ and $L \in \mathcal{R}^*$, it suffices to show that

$$Lh \geq 0 \text{ for all } h \in \mathcal{W} \tag{9.48}$$

implies

$$L(1 - \overline{\varphi(\mu)}\varphi(\lambda)) \geq 0. \tag{9.49}$$

Accordingly, assume that condition (9.48) holds.

8. H_L^2 and the operator $T = M_\lambda$ on H_L^2 are defined much as before.

9. Since the hereditary functions $\overline{f(\mu)}(1 - \bar{\mu}\lambda)f(\lambda)$ and $\overline{f(\mu)}(\bar{\mu}\lambda - r^2)f(\lambda)$ belong to $\mathcal{W}$ for every $f \in \mathrm{Hol}(A(r))$, L is non-negative on all such functions, so that

$$L\left(\overline{f(\mu)}(1 - \bar{\mu}\lambda)f(\lambda)\right) \geq 0 \quad \text{and} \quad L\left(\overline{f(\mu)}(\bar{\mu}\lambda - r^2)f(\lambda)\right) \geq 0$$

for all $f \in \mathrm{Hol}(A(r))$. In other words,

$$1 - T^*T \geq 0 \quad \text{and} \quad T^*T - r^2 \geq 0. \tag{9.50}$$

10. As $\sigma(T)$ is not necessarily in $A(r)$, statements (9.50) are not enough to guarantee that $T \in \mathcal{F}_{\mathrm{dp}}(r)$. In the proof of Theorem 4.35, we replaced T with rT (with $r < 1$) to guarantee that $\sigma(rT) \subseteq \mathbb{D}^2$. Here, however, such a trick is unavailable as there is no conformal map from $A(r)$ to a smaller approximating annulus. Instead, write the polar decomposition of T as $T = UM$, where $M \geq 0$ and (since T is invertible) U is unitary. For small positive ε, define

$$\rho_\varepsilon(x) = x + \varepsilon\left(\tfrac{2}{1-r}(1-x) - 1\right) \qquad \text{for } x \in [r, 1],$$

so that $\rho_\varepsilon(r) = r + \varepsilon$ and $\rho_\varepsilon(1) = 1 - \varepsilon$. Define T_ε by the formula

$$T_\varepsilon = U\rho_\varepsilon(M),$$

so that $T_\varepsilon^* T_\varepsilon = \rho_\varepsilon(M)^2$. Elementary arguments show that $T_\varepsilon \in \mathcal{F}_{\mathrm{dp}}(r)$ for all sufficiently small $\varepsilon > 0$ and that $T_\varepsilon \to T$ in operator norm as $\varepsilon \to 0$. Therefore, as $\varphi \in \mathscr{S}(\mathcal{F}_{\mathrm{dp}}(r))$, it follows (with the aid of the assumption (9.47)) that

$$\|\varphi(T)\| = \lim_{\varepsilon \to 0+} \|\varphi(T_\varepsilon)\|$$

$$\leq \sup_{X \in \mathcal{F}_{\mathrm{dp}}(r)} \|\varphi(X)\|$$

$$= \|\varphi\|_{\mathcal{F}_{\mathrm{dp}}(r)}$$

$$\leq 1,$$

Calcular Norms 231

and so

$$1 - \varphi(T)^*\varphi(T) \geq 0.$$

11. The inequality (9.49) holds because

$$L(1 - \overline{\varphi(\mu)}\varphi(\lambda)) = \langle (1 - \varphi(T)^*\varphi(T))(1+\mathcal{N}), (1+\mathcal{N})\rangle_L$$

$$\geq 0.$$

These steps prove Theorem 9.46 in the special case when assumption (9.47) holds. To prove the general case, consider $\varphi \in \mathscr{S}(\mathcal{F}_{\mathrm{dp}}(r))$. For $n > 2/(1-r)$ we may consider the annulus $A_n = \{\lambda\colon r + \frac{1}{n} < |\lambda| < 1 - \frac{1}{n}\}$. As φ is holomorphic on a neighborhood of A_n^-, the special case just considered together with a simple scaling argument shows that, for each n, there exist two positive semidefinite kernels on A_n, $P_n^+(\lambda,\mu)$ and $P_n^-(\lambda,\mu)$, such that

$$1 - \overline{\varphi(\mu)}\varphi(\lambda) = ((1-\tfrac{1}{n})^2 - \bar{\mu}\lambda)P_n^+(\lambda,\mu) + (\bar{\mu}\lambda - (r+\tfrac{1}{n})^2)P_n^-(\lambda,\mu) \tag{9.51}$$

for all $\lambda,\mu \in A_n$. These equations are similar to equations (4.30) from the proof of Lemma 4.29, and indeed, the argument employed there demonstrates that there exist a subsequence $\{n_k\}$ and positive semi-definite kernels $P^+(\lambda,\mu)$ and $P^-(\lambda,\mu)$ on $A(r)$ satisfying

$$P_{n_k}^+(\lambda,\mu) \to P^+(\lambda,\mu) \qquad \text{and} \qquad P_{n_k}^-(\lambda,\mu) \to P^-(\lambda,\mu)$$

uniformly on compact subsets of $A(r) \times A(r)$. This implies via equation (9.51) that

$$1 - \overline{\varphi(\mu)}\varphi(\lambda) = (1-\bar{\mu}\lambda)P^+(\lambda,\mu) + (\bar{\mu}\lambda - r^2)P^-(\lambda,\mu) \tag{9.52}$$

for $\lambda,\mu \in A(r)$. If we choose Hilbert spaces $\mathcal{N}^+$ and $\mathcal{N}^-$ and functions $v^+\colon A(r) \to \mathcal{N}^+$ and $v^-\colon A(r) \to \mathcal{N}^-$ such that

$$P^+(\lambda,\mu) = \langle v_\lambda^+, v_\mu^+\rangle_{\mathcal{N}^+} \qquad \text{and} \qquad P^-(\lambda,\mu) = \langle v_\lambda^-, v_\mu^-\rangle_{\mathcal{N}^-},$$

then equation (9.52) becomes equation (9.45). This completes the proof of Theorem 9.46.

9.4.3 Model Theory for $\mathscr{S}(\mathcal{F}_{\mathrm{dp}}(r))$

Armed with a fundamental theorem, one can easily modify the proofs from Chapters 2 and 4 to derive versions of the network realization theorem and the Pick interpolation theorem for holomorphic functions on an annulus in the dp norm.

To achieve a network realization formula the key is to interpret $\lambda \in A(r)$ as an operator on the model space, as was done in the remarks following

232 *Commutative Theory*

Definition 4.18. If $\mathcal{N} = \mathcal{N}^+ \oplus \mathcal{N}^-$ is a decomposed Hilbert space and $\lambda \in A(r)$, we define an operator λ_{dp} on $\mathcal{N}$ by the formula

$$\lambda_{\mathrm{dp}}(v^+ \oplus v^-) = \lambda v^+ \oplus \frac{r}{\lambda} v^- \qquad \text{for } v^+ \in \mathcal{N}^+, v^- \in \mathcal{N}^-. \tag{9.53}$$

In addition, with the setup of Definition 9.44, we define $v \colon A(r) \to \mathcal{N}$ by

$$v_\lambda = v_\lambda^+ \oplus \lambda v_\lambda^- \qquad \text{for all } \lambda \in A(r).$$

With these notations, formula (9.45) becomes

$$1 - \overline{\varphi(\mu)}\varphi(\lambda) = \big\langle (1 - \mu_{\mathrm{dp}}^* \lambda_{\mathrm{dp}}) v_\lambda, v_\mu \big\rangle_{\mathcal{N}},$$

a formula amenable to a lurking isometry argument that enables us to prove the following theorem.

Theorem 9.54. (**A network realization formula for** $\mathscr{S}(\mathcal{F}_{\mathrm{dp}}(r))$) *A scalar function φ on $A(r)$ belongs to $\mathscr{S}(\mathcal{F}_{\mathrm{dp}}(r))$ if and only if there exist a decomposed Hilbert space $\mathcal{N} = \mathcal{N}^+ \oplus \mathcal{N}^-$ and a unitary operator*

$$V = \begin{bmatrix} a & 1 \otimes \beta \\ \gamma \otimes 1 & D \end{bmatrix} \colon \mathbb{C} \oplus \mathcal{N} \to \mathbb{C} \oplus \mathcal{N},$$

such that, for all $\lambda \in A(r)$,

$$\varphi(\lambda) = a + \lambda_{\mathrm{dp}} \big\langle (1 - D\lambda_{\mathrm{dp}})^{-1} \gamma, \beta \big\rangle.$$

In the same way that Theorem 4.44 was used to prove Theorem 4.49, Theorem 9.54 implies the following theorem.

Theorem 9.55. (**A Pick interpolation theorem in the dp norm**) *Let $\lambda_1, \ldots, \lambda_n$ be distinct points in $A(r)$ and let $w_1, \ldots, w_n$ in $\mathbb{C}$. There exists $\varphi \in \mathscr{S}(\mathcal{F}_{\mathrm{dp}}(r))$ such that $\varphi(\lambda_i) = w_i$ for $i = 1, \ldots, n$ if and only if there exist a pair of $n \times n$ positive semi-definite matrices $A^+ = [a_{ij}^+]$ and $A^- = [a_{ij}^-]$ such that*

$$1 - \bar{w}_i w_j = (1 - \bar{\lambda}_i \lambda_j) a_{ij}^+ + (\bar{\lambda}_i \lambda_j - r^2) a_{ij}^-$$

for $i, j = 1, \ldots, n$.

9.4.4 A Function-Theoretic Interpretation of $\| \cdot \|_{\mathcal{F}_{\mathrm{dp}}(r)}$

It is possible, by comparison of Theorems 9.54 and 4.44, to interpret the Douglas–Paulsen norm as the solution to an optimization problem that is formulated in purely function-theoretic terms.

Calcular Norms 233

Theorem 9.56. *Let* $r \in (0,1)$ *and let* φ *be a function on* $A(r)$. *Then* $\varphi \in H^\infty(\mathcal{F}_{dp}(r))$ *if and only if there exists* $\Phi \in H^\infty(\mathbb{D}^2)$ *such that*

$$\varphi(\lambda) = \Phi(\lambda, r/\lambda) \qquad \text{for all } \lambda \in A(r). \tag{9.57}$$

Furthermore, when $\varphi \in H^\infty(\mathcal{F}_{dp}(r))$, Φ *may be chosen so that*

$$\|\Phi\|_{\mathbb{D}^2} = \|\varphi\|_{\mathcal{F}_{dp}(r)}.$$

Finally, if $\Phi \in H^\infty(\mathbb{D}^2)$ *and equation* (9.57) *holds, then* $\varphi \in \mathcal{F}_{dp}(r)$ *and*

$$\|\varphi\|_{\mathcal{F}_{dp}(r)} \leq \|\Phi\|_{\mathbb{D}^2}.$$

Proof. Let us define $g \colon A(r) \to \mathbb{D}^2$ by $g(\lambda) = (\lambda, r/\lambda)$. The theorem is equivalent to the assertion that

$$\varphi \in \mathscr{S}(\mathcal{F}_{dp}(r)) \iff \text{there exists } \Phi \in \mathscr{S}(\mathbb{D}^2) \text{ such that } \varphi = \Phi \circ g,$$

a statement that is easily verified by comparison of the network realization formulas in Theorems 9.54 and 4.44. $\qquad\square$

The theorem can be summarized by the equation

$$\|\varphi\|_{\mathcal{F}_{dp}(r)} = \inf\{\|\Phi\|_{\mathbb{D}^2} \colon \Phi \in H^\infty(\mathbb{D}^2) \text{ and } \Phi(\lambda, \tfrac{r}{\lambda}) = \varphi(\lambda) \text{ for all } \lambda \in A(r)\}. \tag{9.58}$$

In light of this theorem, we can see that estimation of the dp constant by means of a decomposition constant is equivalent to the estimation of the right-hand side of equation (9.58), but with an extension of the form $\Phi(\lambda) = f(\lambda^1) + g(\lambda^2)$.

9.5 The B. and F. Delyon Norm and Crouzeix's Theorem

For $T \in \mathcal{B}(\mathcal{H})$, the *numerical range of* T, $W(T)$, is defined by

$$W(T) = \{\langle Tv, v \rangle_{\mathcal{H}} \colon \|v\| = 1\}.$$

It is well known that $W(T)$ is a bounded convex set in $\mathbb{C}$. For an open bounded convex set $C \subseteq \mathbb{C}$, let $\mathcal{F}_{bfd}(C)$, the *B. and F. Delyon family of* C, denote the class of operators T that satisfy $W(T) \subseteq C$. Since $\sigma(T) \subseteq W(T)$ whenever T is an operator, and $W([\lambda]) = \{\lambda\} \subseteq C$ whenever $\lambda \in C$, it follows immediately that $\mathcal{F}_{bfd}(C)$ is subordinate to C. Thus we may consider the calcular norm $\|\cdot\|_{\mathcal{F}_{bfd}(C)}$ defined by the formula

$$\|\varphi\|_{\mathcal{F}_{bfd}(C)} = \sup_{T \in \mathcal{F}_{bfd}(C)} \|\varphi(T)\| \qquad \text{for } \varphi \in \mathrm{Hol}(C).$$

234   *Commutative Theory*

From time to time we will write simply $\|\varphi\|_{\mathrm{bfd}}$ for $\|\varphi\|_{\mathcal{F}_{\mathrm{bfd}}(C)}$ when there is no danger of confusion.

Theorem 9.59. (B. and F. Delyon theorem) *For C a bounded convex set in $\mathbb{C}$, let*

$$\kappa(C) = 3 + \left(\frac{2\pi \, (\mathrm{diam}(C))^2}{\mathrm{area}(C)} \right)^3 .^7 \qquad (9.60)$$

If $\mathcal{H}$ is a Hilbert space, $T \in \mathcal{B}(\mathcal{H})$, and p is a polynomial, then

$$\|p(T)\| \le \kappa(W(T)) \, \|p\|_{W(T)}.$$

As a consequence of this ground-breaking theorem of the brothers Bernard and François Delyon, the first of its kind in the literature, we obtain the following via a simple approximation argument.

Corollary 9.61. *If C is an open bounded convex set in $\mathbb{C}$ and $\varphi \in \mathrm{Hol}(C)$, then*

$$\|\varphi\|_{\mathcal{F}_{\mathrm{bfd}}(C)} \le \kappa(C)\|\varphi\|_C.$$

In particular, $\mathrm{H}^\infty(\mathcal{F}_{\mathrm{bfd}}(C)) = \mathrm{H}^\infty(C)$ *as sets and*

$$K(\mathcal{F}_{\mathrm{bfd}}(C)) \le \kappa(C),$$

where K is as in Definition 9.40 and κ is as in equation (9.60).

Corollary 9.61 gives rise to a number of questions of great operator-theoretic interest. Three such are as follows: given a fixed convex set, is it possible to compute $K(\mathcal{F}_{\mathrm{bfd}}(C))$? And what of $K^m(\mathcal{F}_{\mathrm{bfd}}(C))$? How does $K^m(\mathcal{F}_{\mathrm{bfd}}(C))$ behave as $m \to \infty$? Another type of problem is to understand how $K(\mathcal{F}_{\mathrm{bfd}}(C))$ behaves as C varies. Here, the famous paper [63] of Crouzeix was a dramatic breakthrough.

Theorem 9.62. (Crouzeix's theorem) *If C is a bounded open convex set in $\mathbb{C}$, then*

$$K(\mathcal{F}_{\mathrm{bfd}}(C)) \le 12.$$

This theorem allows us to define a finite universal constant K_{bfd}, the *Crouzeix constant*, by the formula

$$K_{\mathrm{bfd}} = \sup\{K(\mathcal{F}_{\mathrm{bfd}}(C)) \colon C \text{ is a bounded open convex set in } \mathbb{C}\}.$$

In [63], in addition to showing that $K_{\mathrm{bfd}} \le 12$, Crouzeix made the following fruitful speculation.

---

[7] The 3 here is an exponent, not a footnote marker or superscript.

Calcular Norms 235

Conjecture 9.63. (Crouzeix's conjecture) $K_{\text{bfd}} = 2$.

In [64], Crouzeix improved his original estimate, $K_{\text{bfd}} \leq 12$, to $K_{\text{bfd}} \leq 11.08$, and more recently, he and César Palencia showed that $K_{\text{bfd}} \leq 1 + \sqrt{2}$ [66]. In addition, there has been much numerical work that supports the conjecture [95].

9.6 The Badea–Beckermann–Crouzeix Norm

If $c \in \mathbb{C}$ and $r > 0$, we let

$$D(c,r) = \{z \in \mathbb{C} : |z - c| < r\}$$

and let $\mathcal{F}_{\text{bbc}}(c,r)$ denote the class of operators T such that $\sigma(T) \subseteq D(c,r)$ and $\|T - c\| \leq r$. A *bbc pair* is an ordered pair (c,r) where for some positive integer m, $c = (c_1,\ldots,c_m) \in \mathbb{C}^m$ and $r = (r_1,\ldots,r_m) \in \mathbb{R}^m$ where $r_k > 0$ for each $k = 1,\ldots,m$. If (c,r) is a bbc pair we let

$$D(c,r) = \bigcap_{k=1}^{m} D(c_k,r_k)$$

and

$$\mathcal{F}_{\text{bbc}}(c,r) = \bigcap_{k=1}^{m} \mathcal{F}_{\text{bbc}}(c_k,r_k).$$

Thus $\mathcal{F}_{\text{bbc}}(c,r)$ is the class of operators T with the property that $(T - c_k)/r_k$ is a contraction with spectrum in $\mathbb{D}$ for each k, or equivalently, via von Neumann's inequality, with the property that $D(c_k,r_k)$ is a spectral domain for T for $k = 1,\ldots,m$.

Note that $\mathcal{F}_{\text{bbc}}(c,r)$ is subordinate to $D(c,r)$. We refer to $\mathcal{F}_{\text{bbc}}(c,r)$ as the *Badea–Beckermann–Crouzeix family with pair* (c,r) and refer to the associated calcular norm, $\|\cdot\|_{\mathcal{F}_{\text{bbc}}(c,r)}$, defined by the formula,

$$\|\varphi\|_{\mathcal{F}_{\text{bbc}}(c,r)} = \sup_{T \in \mathcal{F}_{\text{bbc}}(c,r)} \|\varphi(T)\| \qquad \text{for } \varphi \in \text{Hol}(D(c,r)),$$

as the *Badea–Beckermann–Crouzeix norm with pair* (c,r). From time to time, when the underlying bbc pair (c,r) is clear from the context, we shall abuse the notation somewhat and write $\|\varphi\|_{\text{bbc}}$ for $\|\varphi\|_{\mathcal{F}_{\text{bbc}}(c,r)}$.

The following beautiful theorem appears in [39].

236   *Commutative Theory*

Theorem 9.64. (**Badea–Beckermann–Crouzeix theorem**) *Let* (c,r) *be a bbc pair.*[8] *If* $\varphi \in \mathrm{Hol}(D_{c,r})$ *and* $T \in \mathcal{F}_{\mathrm{bbc}}(c,r)$*, then*

$$\|\varphi(T)\| \le \left(m + \frac{m(m+1)}{\sqrt{3}} \right) \|\varphi\|_{D_{c,r}}.$$

As in the previous section, which treated the bfd calcular norm, this theorem implies that $\mathrm{H}^{\infty}(\mathcal{F}_{\mathrm{bbc}}(c,r)) = \mathrm{H}^{\infty}(D(c,r))$. In consequence, all the questions about how the constants $K^n(\mathcal{F}_{\mathrm{bbc}}(c,r))$ behave as n, m, and (c,r) vary are interesting from the point of view of operator theory.

9.6.1 Model Theory for the bbc Class

Model theory for the bbc class can be developed in exactly the same way as for the dp class. The main difference is that the role played in the model theory for the dp class by the ordered pair of functions

$$(\lambda, r/\lambda) \qquad \text{for } \lambda \in A(r)$$

is now played by the m-tuple of functions

$$\big((\lambda - c_1)/r_1, \ldots, (\lambda - c_m)/r_m\big) \qquad \text{for } \lambda \in D(c,r).$$

We first define models.

Definition 9.65. *If* $\varphi \colon D(c,r) \to \mathbb{C}$ *is a function, then we say that a pair* $(\mathcal{M}, u)$ *is a bbc-model for* φ *(with pair* (c,r)*) if* $\mathcal{M} = (\mathcal{M}^1, \ldots, \mathcal{M}^m)$ *is an ordered m-tuple of Hilbert spaces,* $u = (u^1, \ldots, u^m)$ *is an ordered m-tuple of functions with* $u^k \colon D(c,r) \to \mathcal{M}^k$ *for* $k = 1, \ldots, m$*, and for which the following bbc-model formula holds:*

$$1 - \overline{\varphi(\mu)}\varphi(\lambda) = \sum_{k=1}^m (1 - \overline{s_k(\mu)}s_k(\lambda))\big\langle u_{\lambda}^k, u_{\mu}^k \big\rangle_{\mathcal{M}^k} \qquad \text{for all } \lambda, \mu \in D(c,r).$$

$$(9.66)$$

Here, for each k, $s_k \colon D(c,r) \to \mathbb{D}$ *is defined by the formula*

$$s_k(\lambda) = \frac{\lambda - c_k}{r_k} \qquad \text{for } \lambda \in D(c,r).$$

Next one proves the following model theorem for the bbc class.

Theorem 9.67. (**Fundamental theorem for** $\mathscr{S}(\mathcal{F}_{\mathrm{bbc}}(c,r))$) *Let* (c,r) *be a bbc pair. Every function in* $\mathscr{S}(\mathcal{F}_{\mathrm{bbc}}(c,r))$ *has a bbc-model with pair* (c,r)*.*

---

[8] The authors actually treated the more general case in which the discs are taken in the Riemann sphere.

Calcular Norms

As for Theorem 9.46, one copies the argument of the proof of Theorem 4.35, with $\mathrm{Hol}(\mathbb{D}^2)$ replaced by $\mathrm{Hol}(D(c,r))$ and the functions λ^1 and λ^2 replaced by the functions s_k. One runs the cone separation argument, with the pair T becoming the m-tuple of multiplication by $(s_1,\ldots,s_m)$.

As in the remarks following Definition 4.18 it is useful to rewrite the model formula (9.66). Identify the m-tuple of Hilbert spaces $\mathcal{M}$ with the decomposed Hilbert space

$$\mathcal{M} = \mathcal{M}^1 \oplus \cdots \oplus \mathcal{M}^m$$

and, for each $\lambda \in D(c,r)$, let

$$u_\lambda = \bigoplus_{k=1}^m u_\lambda^k,$$

so that $u_\lambda \in \mathcal{M}$. Define an operator $s(\lambda) \in \mathcal{B}(\mathcal{M})$ by the formula

$$s(\lambda) \bigoplus_{k=1}^m x^k = \bigoplus_{k=1}^m s_k(\lambda) x^k. \tag{9.68}$$

With these notations, formula (9.66) is equivalent to

$$1 - \overline{\varphi(\mu)}\varphi(\lambda) = \big\langle (1 - s(\mu)^* s(\lambda)) u_\lambda, u_\mu \big\rangle \quad \text{for all } \lambda, \mu \in D(c,r). \tag{9.69}$$

Lurking isometry arguments applied to the formula (9.69) immediately show that bbc models are automatically holomorphic, a converse to the fundamental theorem above, and when combined with Theorem 9.67, they yield a network realization formula as well.

Theorem 9.70. (A network realization formula for $\mathscr{S}(\mathcal{F}_{\mathrm{bbc}}(c,r))$) *Let (c,r) be a bbc pair. A function $\varphi \colon D(c,r) \to \mathbb{C}$ belongs to $\mathscr{S}(\mathcal{F}_{\mathrm{bbc}}(c,r))$ if and only if there exist a decomposed Hilbert space $\mathcal{M} = \mathcal{M}^1 \oplus \ldots \oplus \mathcal{M}^m$ and a unitary operator*

$$V = \begin{bmatrix} a & 1 \otimes \beta \\ \gamma \otimes 1 & D \end{bmatrix} \colon \mathbb{C} \oplus \mathcal{M} \to \mathbb{C} \oplus \mathcal{M},$$

such that

$$\varphi(\lambda) = a + s(\lambda)\big\langle (1 - Ds(\lambda))^{-1}\gamma, \beta \big\rangle$$

for all $\lambda \in D(c,r)$, where $s(\lambda)$ is interpreted as in equation (9.68).

In the same way that Theorem 4.44 was used to prove Theorem 4.49, Theorem 9.54 implies the following theorem.

Theorem 9.71. (A Pick interpolation theorem in the bbc norm) *Let (c,r) be a bbc pair. Fix n distinct points $z_1,\ldots,z_n \in D(c,r)$ and let $w_1,\ldots,w_n$ be*

238   *Commutative Theory*

in $\mathbb{C}$. There exists $\varphi \in \mathscr{S}(\mathcal{F}_{\mathrm{bbc}}(c,r))$ such that $\varphi(z_i) = w_i$ for $i = 1,\ldots,n$ if and only if there exists an m-tuple $a = (a^1,\ldots,a^m)$ of $n \times n$ positive semi-definite matrices such that

$$1 - \bar{w}_i w_j = \sum_{k=1}^{m} [1 - \overline{s_k(z_i)} s_k(z_j)]\, a_{ij}^k$$

for $i,j = 1,\ldots,n$.

9.6.2 The Crouzeix Constant for the bbc Class

Note that if (c,r) is a bbc pair, then

$$\mathcal{F}_{\mathrm{bbc}}(c,r) \;\subseteq\; \mathcal{F}_{\mathrm{bfd}}(D(c,r)).$$

Hence if $\varphi \in \mathrm{Hol}(D(c,r))$,

$$\|\varphi\|_{\mathcal{F}_{\mathrm{bbc}}(c,r)} \;\leq\; \|\varphi\|_{\mathcal{F}_{\mathrm{bfd}}(D(c,r))} \;\leq\; K_{\mathrm{bfd}} \|\varphi\|_{D(c,r)}.$$

Thus $\mathrm{H}^\infty(\mathcal{F}_{\mathrm{bbc}}(c,r)) = \mathrm{H}^\infty(D(c,r))$ as sets, and the constant $K(\mathcal{F}_{\mathrm{bbc}}(c,r))$, defined in Definition 9.40, satisfies

$$K(\mathcal{F}_{\mathrm{bbc}}(c,r)) \;\leq\; K_{\mathrm{bfd}}$$

for all bbc pairs (c,r). Therefore, as for the bfd class, we may define a Crouzeix constant K_{bbc} for the bbc class by the formula

$$K_{\mathrm{bbc}} = \sup\{K(\mathcal{F}_{\mathrm{bbc}}(c,r)): (c,r) \text{ is a bbc pair}\}. \tag{9.72}$$

Theorem 9.73. $K_{\mathrm{bbc}} = K_{\mathrm{bfd}}$.

The remainder of this section is devoted to the proof of Theorem 9.73.

Remark 9.74. Theorems 9.71 and 9.73 allow for the application of semidefinite programming to the numerical investigation of Crouzeix's conjecture.

Lemma 9.75. *Let $\mathcal{H}$ be a Hilbert space, $T \in \mathcal{B}(\mathcal{H})$, and $\varepsilon > 0$. If $\mathrm{Re}\, T \geq \varepsilon$, then there exists R such that*

$$\|T - r\| \leq r \qquad \text{for all } r \geq R.$$

Proof. Let $R = \|T\|^2/2\varepsilon$. If $r \geq R$, then

$$\begin{aligned}
(T-r)^*(T-r) &= T^*T - 2r\,\mathrm{Re}\,T + r^2 \\
&\leq T^*T - 2r\varepsilon + r^2 \\
&\leq \|T\|^2 - 2r\varepsilon + r^2 \\
&= 2(R-r)\varepsilon + r^2 \\
&\leq r^2.
\end{aligned}$$

$\square$

Calcular Norms

Let us agree to say that an open convex set $C \subseteq \mathbb{C}$ is *regular* if the boundary ∂C is a smooth strictly convex curve in $\mathbb{C}$ or equivalently, there is a differentiable path $\gamma \colon [0,1] \to \partial C$ such that $\operatorname{ran} \gamma = \partial C$, $\gamma'(t) \neq 0$ for all $t \in [0,1]$, γ is oriented counterclockwise, and for which the function $\gamma'(t)/|\gamma'(t)|$ is one-to-one on $[0,1)$. When γ has these properties, we shall say that γ is *regular*. When γ is a regular parametrization of ∂C, $\vec{n}(t)$ defined by

$$\vec{n}(t) = -i \frac{\gamma'(t)}{|\gamma'(t)|}$$

parametrizes the unit outward pointing normal to ∂C at $\gamma(t)$.

If C is regular open convex set in $\mathbb{C}$ and $\varepsilon > 0$, we may consider the set

$$C + \varepsilon \mathbb{D} = \{\lambda + \varepsilon z \colon \lambda \in C \text{ and } z \in \mathbb{D}\}.$$

It is easy to see that $\lambda \in \partial(C + \varepsilon \mathbb{D})$ if and only if $\operatorname{dist}(\lambda, C) = \varepsilon$. Also, if γ is a regular parametrization of ∂C, then

$$\gamma(t) + \varepsilon \vec{n}(t)$$

is a parametrization of $\partial(C + \varepsilon \mathbb{D})$.

Lemma 9.76. *Let C be a regular open convex set in $\mathbb{C}$ and $\varepsilon > 0$. If $\mathcal{H}$ is a Hilbert space, $T \in \mathcal{B}(\mathcal{H})$, and $W(T) \subseteq C$, then there exists a bbc pair (c,r) such that*

$$C \subseteq D(c,r) \subseteq C + \varepsilon \mathbb{D} \tag{9.77}$$

and

$$T \in \mathcal{F}_{\mathrm{bbc}}(c,r). \tag{9.78}$$

Proof. Fix $\varepsilon > 0$ and let C be a regular open convex set in $\mathbb{C}$, regularly parametrized by $\gamma \colon [0,1] \to \mathbb{C}$. Assume that $\mathcal{H}$ is a Hilbert space, $T \in \mathcal{B}(\mathcal{H})$, and $W(T) \subseteq C$.

Fix δ satisfying $0 < \delta < \varepsilon$. For $t \in [0,1]$ define an isometric affine mapping of $\mathbb{C}$ by the formula

$$L_t(z) = -i \frac{|\gamma'(t)|}{\gamma'(t)} (z - \gamma(t) - \delta \vec{n}(t)) \qquad \text{for } z \in \mathbb{C}.$$

By construction, L_t maps the supporting half plane[9] to C at $\gamma(t)$ to $\mathbb{H} + \delta$. Consequently, as $W(T)$ is assumed to lie in C, $W(L_t(T))$ lies in $\mathbb{H} + \delta$. Therefore, $\operatorname{Re} L_t(T) \geq \delta$. By Lemma 9.75, there exists a constant R_t such that

$$\|L_t(T) - r\| \leq r \qquad \text{for all } r \geq R_t. \tag{9.79}$$

As $L_t(C)^-$ is compact and $L_t(C) \subseteq \mathbb{H} + \delta$, by choosing r in inequality (9.79) equal to a sufficiently large value r_t, we may guarantee that

---

[9] That is, the closed half plane that contains C and whose boundary touches C at $\gamma(t)$.

240　　　　　　　　　　　　　*Commutative Theory*

$$\|L_t(T) - r_t\| \le r_t \quad \text{and} \quad L_t(C) \subseteq D(r_t, r_t). \tag{9.80}$$

To summarize, in the previous paragraph we showed that for each $t \in [0, 1]$, there exists a positive constant r_t such that statements (9.80) hold. For each $t \in [0, 1]$ let c_t denote the center of $L_t^{-1}(D(r_t, r_t))$, so that

$$L_t^{-1}(D(r_t, r_t)) = D(c_t, r_t).$$

Then conditions (9.80) imply that

$$\|T - c_t\| \le r_t \quad \text{and} \quad \sigma(T) \subseteq C \subseteq D(c_t, r_t)$$

so that

$$T \in \mathcal{F}_{\mathrm{bbc}}(c_t, r_t) \quad \text{for all } t \in [0, 1] \tag{9.81}$$

and

$$C \subseteq \bigcap_{t \in [0,1]} D(c_t, r_t). \tag{9.82}$$

Moreover,

$$\bigcap_{t \in [0,1]} D(c_t, r_t)^- \subseteq C + \varepsilon \mathbb{D}, \tag{9.83}$$

or equivalently,

$$\bigcup_t (\mathbb{C} \setminus D(c_t, r_t)^-) \quad \supseteq \quad \mathbb{C} \setminus (C + \varepsilon \mathbb{D}).$$

To see this, consider any point $z \in \mathbb{C} \setminus (C + \varepsilon \mathbb{D})$. Choose $t_0 \in [0, 1]$ such that

$$z = \gamma(t_0) + R\vec{n}(t_0);$$

then $R > \varepsilon > \delta$, and hence $z \notin D(c_{t_0}, r_{t_0})$.

Letting $\mathbb{C}^\infty$ denote the extended complex plane, we see from the inclusion (9.83) that

$$\mathbb{C}^\infty \setminus (C + \varepsilon \mathbb{D}) \subseteq \bigcup_{t \in [0,1]} \mathbb{C}^\infty \setminus D(c_t, r_t)^-,$$

that is,

$$\{\mathbb{C}^\infty \setminus D(c_t, r_t)^- : t \in [0, 1]\}$$

is an open cover of $\mathbb{C}^\infty \setminus (C + \varepsilon \mathbb{D})$. As this latter set is compact in $\mathbb{C}^\infty$, it follows that there exist $t_1, \ldots, t_m \in [0, 1]$ such that

$$\mathbb{C}^\infty \setminus (C + \varepsilon \mathbb{D}) \subseteq \bigcup_{k=1}^m \mathbb{C}^\infty \setminus D(c_{t_k}, r_{t_k})^-,$$

Calcular Norms 241

or equivalently,

$$\bigcap_{k=1}^{m} D(c_{t_k}, r_{t_k})^- \subseteq C + \varepsilon \mathbb{D}. \tag{9.84}$$

Define a bbc pair (c, r) by

$$c = (c_{t_1}, \dots, c_{t_m}) \quad \text{and} \quad r = (r_{t_1}, \dots, r_{t_m}).$$

Evidently, as

$$D(c, r) = \bigcap_{k=1}^{m} D(c_{t_k}, r_{t_k}),$$

the inclusion (9.82) implies that $C \subseteq D(c, r)$ and the inclusion (9.83) implies that $D(c, r) \subseteq C + \varepsilon \mathbb{D}$. Thus the relation (9.77) holds. Likewise, as

$$\mathcal{F}_{\text{bbc}}(c, r) = \bigcap_{k=1}^{m} \mathcal{F}_{\text{bbc}}(c_{t_k}, r_{t_k}),$$

the relation (9.81) implies that the membership (9.78) holds. □

Lemma 9.85. *Suppose that C is a regular open convex set in $\mathbb{C}$, $\varepsilon > 0$, and p is a polynomial. There exists a bbc pair (c, r) such that*

$$C \subseteq D(c, r) \subseteq C + \varepsilon \mathbb{D} \tag{9.86}$$

and

$$\|p\|_{\mathcal{F}_{\text{bfd}}(C)} - \varepsilon \leq \|p\|_{\mathcal{F}_{\text{bbc}}(c,r)}. \tag{9.87}$$

Proof. Suppose that C, ε, and p are given. By the definition of $\|\cdot\|_{\mathcal{F}_{\text{bfd}}(C)}$ we may choose an operator T to satisfy $W(T) \subseteq C$ and

$$\|p\|_{\mathcal{F}_{\text{bfd}}(C)} - \varepsilon \leq \|p(T)\|. \tag{9.88}$$

As $W(T) \subseteq C$, Lemma 9.76 implies that there exists a bbc pair (c, r) such that the inclusions (9.86) hold and such that $T \in \mathcal{F}_{\text{bbc}}(c, r)$. It follows from the definition of $\|\cdot\|_{\mathcal{F}_{\text{bbc}}(c,r)}$ that

$$\|p(T)\| \leq \|p\|_{\mathcal{F}_{\text{bbc}}(c,r)}. \tag{9.89}$$

Inequality (9.87) follows by combination of inequalities (9.88) and (9.89). □

Our next lemma removes the hypothesis that C be regular in Lemma 9.85.

Lemma 9.90. *Suppose that C is a bounded open convex set in $\mathbb{C}$, $\varepsilon > 0$, and p is a polynomial. There exists a bbc pair (c, r) such that the relations (9.86) and (9.87) hold.*

242 *Commutative Theory*

Proof. Fix $\varepsilon > 0$ and choose a simple regular closed curve path γ in $(C + \varepsilon\mathbb{D}) \setminus C$ oriented in the counterclockwise direction that bounds a regular open convex set C_1. Since ∂C_1 is a compact subset of $C + \varepsilon\mathbb{D}$, there exists δ such that

$$0 < \delta < \inf\{|\lambda - \mu|: \lambda \in \mathbb{C} \setminus (C + \varepsilon\mathbb{D}), \ \mu \in \partial C_1\}.$$

Application of Lemma 9.85 with C replaced with C_1 and ε replaced with δ yields a bbc pair (c, r) that satisfies

$$C_1 \subseteq D(c, r) \subseteq C_1 + \delta\mathbb{D} \tag{9.91}$$

and

$$\|p\|_{\mathcal{F}_{\mathrm{bfd}}(C_1)} - \delta \leq \|p\|_{\mathcal{F}_{\mathrm{bbc}}(c, r)}. \tag{9.92}$$

From the inclusions (9.91) and the fact that $C \subseteq C_1$, we see that $C \subseteq D(c, r)$. Also, using the definition of δ and the triangle inequality, we see from the inclusions (9.91) that $D(c, r) \subseteq C + \varepsilon\mathbb{D}$. Thus the relation (9.86) holds.

Notice that, by construction, $\delta < \varepsilon$ and, tautologically, as $C \subseteq C_1$, $\|p\|_{\mathcal{F}_{\mathrm{bfd}}(C)} \leq \|p\|_{\mathcal{F}_{\mathrm{bfd}}(C_1)}$. Hence inequality (9.92) implies inequality (9.87). $\qquad\square$

9.6.3 $K_{\mathrm{bbc}} = K_{\mathrm{bfd}}$

We are finally ready to prove Theorem 9.73. By the discussion leading up to Theorem 9.73, $K_{\mathrm{bbc}} \leq K_{\mathrm{bfd}}$. Therefore, it suffices to prove that for each $\varepsilon > 0$,

$$K_{\mathrm{bfd}} - \varepsilon \leq K_{\mathrm{bbc}}. \tag{9.93}$$

Accordingly, fix $\varepsilon > 0$. Use the definition of K_{bfd} to select a bounded open convex set C such that $K_{\mathrm{bfd}} - \varepsilon/2 < K_{\mathrm{bfd}}(C)$. Use the definition of $K_{\mathrm{bfd}}(C)$ to select a polynomial p such that $K_{\mathrm{bfd}}(C) - \varepsilon/2 \leq \|p\|_{\mathcal{F}_{\mathrm{bfd}}(C)}$ and

$$\|p\|_C = 1. \tag{9.94}$$

With these choices, observe that

$$K_{\mathrm{bfd}} - \varepsilon \ \leq \ \|p\|_{\mathcal{F}_{\mathrm{bfd}}(C)}. \tag{9.95}$$

For each positive integer n we now invoke Lemma 9.90 (with $\varepsilon = 1/n$) to obtain a sequence of bbc pairs, $\{(c_n, r_n)\}$, such that

$$C \subseteq D(c_n, r_n) \subseteq C + \tfrac{1}{n}\mathbb{D} \tag{9.96}$$

and

$$\|p\|_{\mathcal{F}_{\mathrm{bfd}}(C)} - \tfrac{1}{n} \leq \|p\|_{\mathcal{F}_{\mathrm{bbc}}(c_n, r_n)}. \tag{9.97}$$

Calcular Norms 243

For each $n \geq 1$ we have

$$
\begin{aligned}
K_{\mathrm{bfd}} - \varepsilon &\leq \|p\|_{\mathcal{F}_{\mathrm{bfd}}(C)} && \text{by inequality (9.95)} \\
&\leq \tfrac{1}{n} + \|p\|_{\mathcal{F}_{\mathrm{bbc}}(c_n, r_n)} && \text{by inequality (9.97)} \\
&\leq \tfrac{1}{n} + K_{\mathrm{bbc}} \|p\|_{D(c_n, r_n)} && \text{by equation (9.72)} \\
&\leq \tfrac{1}{n} + K_{\mathrm{bbc}} \|p\|_{C + \frac{1}{n}\mathbb{D}} && \text{by inclusions (9.96).}
\end{aligned}
$$

Noting that equation (9.94) and the inclusions (9.96) imply that

$$
\|p\|_{C + \frac{1}{n}\mathbb{D}} \to 1 \quad \text{as } n \to \infty,
$$

we deduce that inequality (9.93) holds, as was to be proved. $\qquad\square$

9.7 The Polydisc Norm

Let $\mathcal{F}(d)$ denote the family of pairwise commuting d-tuples, $T = (T^1, \dots, T^d)$, such that

$$
\sigma(T) \subseteq \mathbb{D}^d \qquad \text{and} \qquad \|T^r\| \leq 1 \text{ for } r = 1, \dots, d.
$$

The family $\mathcal{F}(d)$ is subordinate to $\mathbb{D}$; we shall refer to $\| \cdot \|_{\mathcal{F}(d)}$ as the *polydisc norm*. Evidently, when $d = 1$ or 2, $\mathrm{H}^\infty(\mathcal{F}(d)) = \mathrm{H}^\infty(\mathbb{D}^d)$ and $\|\varphi\|_{\mathcal{F}(d)} = \|\varphi\|_{\mathbb{D}^d}$ for all $\varphi \in \mathrm{H}^\infty(\mathbb{D}^d)$ (cf. Examples 9.8 and 9.9). However, when $d > 2$, there is no known function-theoretic description of $\mathrm{H}^\infty(\mathcal{F}(d))$ and in general, $\| \cdot \|_{\mathcal{F}(d)} \neq \| \cdot \|_{\mathbb{D}^d}$. Nevertheless, our experience in Chapter 4 suggests the following definition.

Definition 9.98. *Let d be a positive integer and let $\varphi \colon \mathbb{D}^d \to \mathbb{C}$ be a function. A polymodel for φ is pair $(\mathcal{M}, u)$, where*

$$
\mathcal{M} = \mathcal{M}^1 \oplus \mathcal{M}^2 \oplus \dots \oplus \mathcal{M}^d
$$

is a decomposed Hilbert space,

$$
u = (u^1, u^2, \dots, u^d)
$$

is a d-tuple of functions with $u^r \colon \mathbb{D}^d \to \mathcal{M}^r$ for each r, and such that

$$
1 - \overline{\varphi(\mu)}\varphi(\lambda) = \sum_{r=1}^{d} (1 - \bar{\mu}^r \lambda^r) \langle u_\lambda^r, u_\mu^r \rangle \tag{9.99}
$$

for all $\lambda, \mu \in \mathbb{D}^d$.

244 *Commutative Theory*

Just as in Section 4.3 the model formula (9.99) may be rewritten in the following way. For each $\lambda \in \mathbb{D}^d$, let

$$u_\lambda = \bigoplus_{r=1}^d u_\lambda^r,$$

and interpret λ as an operator on $\mathcal{M}$ via the formula

$$\lambda \bigoplus_{r=1}^d x^r = \bigoplus_{r=1}^d \lambda^r x^r. \tag{9.100}$$

With these notations, formula (9.99) becomes identical to equation (4.22), that is, equation (9.99) is equivalent to

$$1 - \overline{\varphi(\mu)}\varphi(\lambda) = \left\langle (1 - \mu^*\lambda)\, u_\lambda, u_\mu \right\rangle \qquad \text{for all } \lambda, \mu \in \mathbb{D}^d. \tag{9.101}$$

Armed with this formula one uses a lurking isometry argument to show that polymodels are automatically holomorphic (simply follow the proof of Proposition 4.24) and as a consequence, the converse to the fundamental theorem for $\mathscr{S}(\mathcal{F}(d))$ holds as well (follow the proof of Proposition 4.26).

Minor notational changes to the duality argument in Section 4.4 establish the following result.

Theorem 9.102. (Fundamental theorem for $\mathscr{S}(\mathcal{F}(d))$) *Every function* $\varphi \in \mathscr{S}(\mathcal{F}(d))$ *has a polymodel.*

Once Theorem 9.102 is obtained, the proof of Theorem 4.44 can be immediately generalized to obtain a network realization formula for the polydisc class.

Theorem 9.103. (A network realization formula for the polydisc class) *A function* φ *belongs to* $\mathscr{S}(\mathcal{F}(d))$ *if and only if there exist a decomposed Hilbert space* $\mathcal{M}$ *and a unitary operator*

$$V = \begin{bmatrix} a & 1 \otimes \beta \\ \gamma \otimes 1 & D \end{bmatrix} : \mathbb{C} \oplus \mathcal{M} \to \mathbb{C} \oplus \mathcal{M} \tag{9.104}$$

such that

$$\varphi(\lambda) = a + \left\langle \lambda(1 - D\lambda)^{-1}\gamma, \beta \right\rangle \qquad \text{for all } \lambda \in \mathbb{D}^d, \tag{9.105}$$

where λ *is interpreted as in equation* (9.100).

Likewise, once Theorem 9.103 is obtained, the proof of Theorem 4.56 can be immediately generalized to obtain the following solution to the Nevanlinna–Pick problem on a general subset of $\mathbb{D}^d$ in the polydisc norm.

Calcular Norms

Theorem 9.106. (A Pick theorem for the polydisc norm) *Let data* $\Lambda \subseteq \mathbb{D}^d$ *and* $z\colon \Lambda \to \mathbb{C}$ *be given. There exists* $\varphi \in \mathscr{S}(\mathcal{F}(d))$ *such that* $\varphi(\lambda) = z(\lambda)$ *for all* $\lambda \in \Lambda$ *if and only if there exists an ordered d-tuple* $P = (P^1,\dots,P^d)$ *of positive semi-definite kernels on* Λ *such that*

$$1 - \overline{z(\mu)}z(\lambda) = \sum_{r=1}^{d}(1 - \overline{\mu^r}\lambda^r)P^r(\lambda,\mu) \qquad \textit{for all } \lambda,\mu \in \Lambda. \qquad (9.107)$$

Theorem 9.106 has an interesting application to one-variable theory: the following elegant characterization of the bbc class.

Theorem 9.108. *Let* (c,r) *be a bbc pair (where c and r are m-tuples) and let* $s\colon D(c,r) \to \mathbb{D}^m$ *be the function defined by*

$$s(\lambda) = \left(\frac{\lambda - c_1}{r_1},\dots,\frac{\lambda - c_m}{r_m}\right) \qquad \textit{for } \lambda \in D(c,r).$$

If $\varphi\colon D(c,r) \to \mathbb{C}$ *is a function, then* $\varphi \in H^\infty(\mathcal{F}_{\mathrm{bbc}}(c,r))$ *if and only if there exists* $\Phi \in H^\infty(\mathcal{F}(m))$ *such that* $\varphi = \Phi \circ s$. *Furthermore,* Φ *may be chosen so that* $\|\Phi\|_{\mathcal{F}(m)} = \|\varphi\|_{\mathrm{bbc}}$. *Finally, if* $\Phi \in H^\infty(\mathcal{F}(m))$, *then* $\|\Phi \circ s\|_{\mathrm{bbc}} \leq \|\Phi\|_{\mathcal{F}(m)}$.

Proof. The theorem is equivalent to the assertion that

$$\varphi \in \mathscr{S}(\mathcal{F}_{\mathrm{bbc}}(c,r)) \iff \text{ there exists } \Phi \in \mathscr{S}(\mathcal{F}(m)) \text{ such that } \varphi = \Phi \circ s,$$

which is easily verified by comparison of formulas (9.66) and (9.107). $\square$

9.8 The Oka Extension Theorem and Calcular Norms

In this section we shall give a brief exposition of the Oka extension theorem from several complex variables and then show how calcular norms can be used to give a refinement of the theorem with sharp bounds. For more details see [21] and [16].

9.8.1 Oka's Extension Theorem

Definition 9.109. *If* $p = (p_1,\dots,p_m)$ *is an m-tuple of polynomials in d variables we define* $K_p \subseteq \mathbb{C}^d$ *by*

$$K_p = \{\lambda \in \mathbb{C}^d\colon |p_k(\lambda)| \leq 1 \text{ for } k = 1,\dots,m\}.$$

K_p *is referred to as a* p*-polyhedron. We also consider the open* p*-polyhedron*

$$G_p = \{\lambda \in \mathbb{C}^d\colon |p_k(\lambda)| < 1 \text{ for } k = 1,\dots,m\}.$$

246   *Commutative Theory*

If K_p is a p-polyhedron, then the set

$$\Pi = (\mathbb{D}^-)^d \cap K_p$$

is also a p-polyhedron. Indeed, if we define a $d + m$ tuple of polynomials by

$$\delta = (\lambda^1, \ldots, \lambda^d, p_1(\lambda), \ldots, p_m(\lambda)),$$

then $\Pi = K_\delta$.

Theorem 9.110. (**Oka extension theorem**) *Let p be an m-tuple of polynomials in d variables and let*

$$\Pi \overset{\text{def}}{=} (\mathbb{D}^-)^d \cap K_p.$$

If f is holomorphic on a neighborhood of Π, then there exists a function F holomorphic on a neighborhood of $(\mathbb{D}^-)^{d+m}$ such that

$$f(\lambda) = F(\lambda, p(\lambda)) \qquad \text{for all } \lambda \in \Pi. \tag{9.111}$$

9.8.2 Why Is Oka's Theorem Interesting?

Oka's extension theorem has numerous applications in several complex variables. Not only is it a stem theorem in the modern theory of analytic sheaves and the development of the theory of holomorphic functions on domains of holomorphy and, more generally, analytic spaces, it also plays a significant role in the development of functional calculus for commuting tuples in commutative Banach algebras. Perhaps no application of the theorem more dramatically illustrates its power than Oka's original proof of the famous Oka–Weil approximation theorem, the generalization to several complex variables of the classical Runge approximation theorem.

Theorem 9.112. (**Runge's theorem**) *Let K be a compact subset of $\mathbb{C}$ and let E be a set that meets each component of $\mathbb{C}_\infty \setminus K$. If f is holomorphic on a neighborhood of K, then f can be uniformly approximated on K by rational functions whose poles lie in E.*

Here, we wish to restrict ourselves to a special case of Runge's theorem; the case when the complement of K is connected and $E = \{\infty\}$. Noting that

$$\mathbb{C}_\infty \setminus K \text{ is connected} \iff K \text{ is polynomially convex}$$

and that a rational function whose only pole is at infinity is a polynomial, we see that Theorem 9.112 implies the following theorem.

Theorem 9.113. *If K is a polynomially convex compact subset of $\mathbb{C}$, and f is holomorphic on a neighborhood of K, then f can be uniformly approximated on K by polynomials.*

Calcular Norms 247

This theorem has the following elegant generalization to several variables.

Theorem 9.114. (Oka–Weil theorem) *If K is a polynomially convex compact subset of $\mathbb{C}^d$ and f is holomorphic on a neighborhood of K, then f can be uniformly approximated on K by polynomials.*

Proof. **Step 1.** Assume that K is a polynomially convex compact subset of $\mathbb{C}^d$; we may assume without loss that $K \subseteq \mathbb{D}^d$. Let f be holomorphic on a neighborhood U of K.

Step 2. Construct a p-polyhedron K_p such that

$$K \subseteq (\mathbb{D}^-)^d \cap K_p \subseteq U.$$

Step 3. As f is holomorphic on a neighborhood of $(\mathbb{D}^-)^d \cap K_p$, by Oka's extension theorem, there exists F holomorphic on a neighborhood of $(\mathbb{D}^-)^{d+m}$ such that equation (9.111) holds.

Step 4. As F is holomorphic on a neighborhood of $(\mathbb{D}^-)^{d+m}$, F can be represented by a power series

$$F(w,z) = \sum a_{ij} w^i z^j \qquad \text{for all } z, w \in (\mathbb{D}^-)^d, \qquad (9.115)$$

which converges uniformly on $(\mathbb{D}^-)^{d+m}$.

Step 5. It follows that

$$f(\lambda) = \sum a_{ij} \lambda^i p(\lambda)^j$$

represents f as a uniformly convergent series of polynomials. $\square$

9.8.3 The Oka Calcular Norm

If δ is an m-tuple of polynomials in d variables, we let $\mathcal{F}_\delta$ denote the class of pairwise commuting d tuples T that satisfy

$$\sigma(T) \subseteq G_\delta \qquad \text{and} \qquad \|\delta_l(T)\| < 1 \quad \text{for } l = 1, \ldots, m$$

We call $\mathcal{F}_\delta$ the *Oka class* corresponding to δ. Since $\mathcal{F}_\delta$ is subordinate to G_δ, we may introduce the corresponding calcular norm

$$\|\varphi\|_{\mathcal{F}_\delta} = \sup_{T \in \mathcal{F}_\delta} \|\varphi(T)\| \qquad \text{for all } \varphi \in \mathrm{Hol}(G_\delta).$$

For example, if $d = 1$ and

$$\delta = \left(\frac{\lambda - c_1}{r_1}, \ldots, \frac{\lambda - c_m}{r_m} \right),$$

248   *Commutative Theory*

then $\|\varphi\|_{\mathcal{F}_\delta}$ is the bbc norm with pair (c,r), while if

$$\delta = \left(\lambda^1, \ldots, \lambda^d\right),$$

then $\|\varphi\|_{\mathcal{F}_\delta}$ is the polydisc norm.

Definition 9.116. *If δ is an m-tuple of polynomials in d variables and $\varphi\colon G_\delta \to \mathbb{C}$ is a function then a δ-model for φ is pair $(\mathcal{M},u)$ where*

$$\mathcal{M} = \mathcal{M}^1 \oplus \mathcal{M}^2 \oplus \ldots \oplus \mathcal{M}^m$$

is a decomposed Hilbert space, $u\colon G_\delta \to \mathcal{M}$, and, for all $\lambda, \mu \in G_\delta$,

$$1 - \overline{\varphi(\mu)}\varphi(\lambda) = \langle(1 - \delta(\mu)^*\delta(\lambda))u_\lambda, u_\mu\rangle. \tag{9.117}$$

Here, for $\lambda \in G_\delta$, $\delta(\lambda)$ denotes the operator on $\mathcal{M}$ defined by

$$\delta(\lambda)\bigoplus_{l=1}^{m} x^l = \bigoplus_{l=1}^{m} \delta_l(\lambda)x^l. \tag{9.118}$$

Noting the similarity between the formulas (9.117) and (9.101), one can readily duplicate the reasoning in Section 9.7 to obtain the following results. A lurking isometry argument yields the fact that δ-models are automatically holomorphic, and as a consequence, the converse to the fundamental theorem for $\mathcal{S}(\mathcal{F}_\delta)$, Theorem 9.119, holds as well.

Theorem 9.119. (Fundamental theorem for $\mathcal{S}(\mathcal{F}_\delta)$) *For any tuple δ of polynomials in d variables, every function in $\mathcal{S}(\mathcal{F}_\delta)$ has a δ-model.*

Once Theorem 9.119 is obtained, a lurking isometry argument yields a network realization formula for the Oka class $\mathcal{S}(\mathcal{F}_\delta)$.

Theorem 9.120. (Network realization formula for the Oka class) *Let δ be an m-tuple of polynomials in d variables. A function φ on G_δ belongs to $\mathcal{S}(\mathcal{F}_\delta)$ if and only if there exist a decomposed Hilbert space $\mathcal{M} = \mathcal{M}^1 \oplus \ldots \oplus \mathcal{M}^m$ and a unitary operator*

$$V = \begin{bmatrix} a & 1 \otimes \beta \\ \gamma \otimes 1 & D \end{bmatrix}\colon \mathbb{C} \oplus \mathcal{M} \to \mathbb{C} \oplus \mathcal{M} \tag{9.121}$$

such that

$$\varphi(\lambda) = a + \langle\delta(\lambda)(1 - D\delta(\lambda))^{-1}\gamma, \beta\rangle \quad \text{for all } \lambda \in G_\delta, \tag{9.122}$$

where $\delta(\lambda)$ is interpreted as in equation (9.118).

Just as in the case of the polydisc norm, there is a version of Pick's theorem on a general set in the Oka norm $\|\cdot\|_{\mathcal{F}_\delta}$. It is proven using Theorem 9.120 and a lurking isometry argument.

Calcular Norms

Theorem 9.123. **(A Pick theorem for the Oka norm)** *Let data $\Lambda \subseteq G_\delta$ and $z \colon \Lambda \to \mathbb{C}$ be given. There exists $\varphi \in \mathscr{S}(\mathcal{F}_\delta)$ such that $\varphi(\lambda) = z(\lambda)$ for all $\lambda \in \Lambda$ if and only if there exists an ordered m-tuple $P = (P^1, \ldots, P^d)$ of positive semidefinite kernels on Λ such that*

$$1 - \overline{z(\mu)}z(\lambda) = \sum_{l=1}^{m} (1 - \overline{\delta_l(\mu)}\delta_l(\lambda)) P^l(\lambda, \mu) \qquad \text{for all } \lambda, \mu \in \Lambda.$$

9.8.4 A Refinement of Oka's Extension Theorem

When Theorems 9.103 and 9.120 are combined, they have the following interesting corollary. Note that, if δ is an m-tuple of polynomials, then δ maps G_δ to $\mathbb{D}^m$.

Theorem 9.124. *Let δ be an m-tuple of polynomials in d variables. A function f on G_δ belongs to $\mathscr{S}(\mathcal{F}_\delta)$ if and only if there exists a function $F \in \mathscr{S}(\mathcal{F}(d))$ such that*

$$f(\lambda) = F(\delta(\lambda)) \qquad \text{for all } \lambda \in G_\delta. \tag{9.125}$$

Proof. Fix $f \in \mathscr{S}(\mathcal{F}_\delta)$. By Theorem 9.120 there exist a decomposed Hilbert space $\mathcal{M}$ and a unitary operator V acting on $\mathbb{C} \oplus \mathcal{M}$ such that if V is decomposed as in equation (9.121), then equation (9.122) holds. If we define a function $F \colon \mathbb{D}^m \to \mathbb{C}$ by

$$F(z) = a + \langle z(1 - Dz)^{-1}\gamma, \beta \rangle \qquad \text{for all } z \in \mathbb{D}^m, \tag{9.126}$$

then Theorem 9.103 implies that $F \in \mathscr{S}(\mathcal{F}(d))$ and equation (9.122) implies that $f(\lambda) = F(\delta(\lambda))$ for all $\lambda \in G_\delta$.

Conversely, assume that $F \in \mathscr{S}(\mathcal{F}(d))$ and $f = F \circ \delta$. By Theorem 9.103 (with d replaced with m and φ replaced with F) there exist a decomposed Hilbert space $\mathcal{M}$ and a unitary operator V acting on $\mathbb{C} \oplus \mathcal{M}$ such that, if V is decomposed as in equation (9.104), then equation (9.126) holds. Hence equation (9.122) holds and by Theorem 9.120, $f \in \mathscr{S}(\mathcal{F}_\delta)$. $\qquad\square$

A special case of Theorem 9.124 is the following refinement of the classical Oka extension theorem (Theorem 9.110).

Corollary 9.127. *Let $p = (p_1, \ldots, p_m)$ be an m-tuple of polynomials in d variables $\lambda^1, \ldots, \lambda^d$ and let*

$$\delta = (\lambda^1, \ldots, \lambda^d, p_1(\lambda), \ldots, p_m(\lambda)).$$

If $f \in \mathrm{H}^\infty(\mathcal{F}_\delta)$, then there exists $F \in \mathrm{H}^\infty(\mathcal{F}(d+m))$ such that

$$f(\lambda) = F(\lambda, p(\lambda)) \qquad \text{for all } \lambda \in G_\delta. \tag{9.128}$$

250   *Commutative Theory*

Furthermore, F may be chosen so that

$$\|F\|_{\mathcal{F}(d+m)} = \|f\|_{\mathcal{F}_\delta}.$$

Finally, if $F \in H^\infty(\mathcal{F}(d+m))$ and f is defined by equation (9.128), then $f \in H^\infty(\mathcal{F}_\delta)$ and

$$\|f\|_{\mathcal{F}_\delta} \leq \|F\|_{\mathcal{F}(d+m)}.$$

How does this result compare to Theorem 9.110? Since $G_\delta \subseteq \Pi^\circ \subseteq G_{t\delta}$ for all $t < 1$, equations (9.111) and (9.128) are essentially equivalent. In Theorem 9.110 f is assumed to be holomorphic on a neighborhood of $\Pi = K_\delta$, while in Corollary 9.127 f is assumed to lie in $H^\infty(\mathcal{F}_\delta)$, a space of holomorphic functions on G_δ. Similarly, in Theorem 9.110 F is holomorphic on a neighborhood of $(\mathbb{D}^{d+m})^-$, while in Corollary 9.127, $F \in H^\infty(\mathcal{F}(d+m))$, a space of holomorphic functions on $\mathbb{D}^{d+m}$. Finally, the classical Oka theorem lacks the precise bounds expressed in the second and third statements in Corollary 9.127.

Given the observations in the previous paragraph, it seems unlikely that any of the observations of Corollary 9.127 can be deduced in a direct way from the classical Oka theorem. On the other hand it seems feasible that one could execute the reverse, that is, prove Theorem 9.110 using Corollary 9.127. To that end it will be useful to have conditions that ensure the functions holomorphic on a neighborhood of K_δ lie in $H^\infty(\mathcal{F}_\delta)$, a problem we turn to in the next section.

9.8.5 The Boundedness Lemma

Lemma 9.129. *Let δ be an m-tuple of polynomials in d variables. The following statements are equivalent.*

(i) *$f \in H^\infty(\mathcal{F}_\delta)$ whenever f is holomorphic on a neighborhood of K_δ;*

(ii) *$\lambda^r \in H^\infty(\mathcal{F}_\delta)$ for $r = 1, \ldots, d$; and*

(iii) *$\mathcal{F}_\delta$ is bounded.*

Proof. Clearly (i) implies (ii) and (ii) implies (iii). To prove (iii) implies (i) we argue by contradiction. Accordingly, assume that $\mathcal{F}_\delta$ is bounded, f is holomorphic on a neighborhood of K_δ, and $f \notin H^\infty(\mathcal{F}_\delta)$. As $f \notin H^\infty(\mathcal{F}_\delta)$, there exists a sequence $\{T_n\}$ in $\mathcal{F}_\delta$ such that

$$\lim_{n \to \infty} \|f(T_n)\| = \infty. \tag{9.130}$$

Calcular Norms

251

As $\mathcal{F}_\delta$ is bounded, we may form the bounded d-tuple of operators

$$X = \bigoplus_{n=1}^{\infty} T_n.$$

Notice that, for $l = 1, \ldots, m$,

$$\|\delta_l(X)\| = \|\bigoplus_{n=1}^{\infty} \delta_l(T_n)\| = \sup_n \|\delta_l(T_n)\| \leq 1.$$

Consequently, by Property 6 of the functional calculus, it follows that, for $l = 1, \ldots, m$

$$\delta_l(\sigma(X)) = \sigma(\delta_l(X)) \subseteq \mathbb{D}^-.$$

Therefore,

$$\sigma(X) \subseteq \bigcap_{l=1}^{m} \{\lambda \colon |\delta_l(\lambda)| \leq 1\} = K_\delta.$$

As f is holomorphic on a neighborhood of K_δ, we may use Property 2 of the functional calculus to form the well-defined bounded operator $f(X)$. But by Property 5 of the functional calculus,

$$f(X) = f\left(\bigoplus_{n=1}^{\infty} T_n\right) = \bigoplus_{n=1}^{\infty} f(T_n),$$

and the operator on the right-hand side of the above equations is unbounded by equation (9.130). This contradiction proves the lemma. $\qquad \Box$

Example 9.131. If $\delta = (\lambda^1, \ldots, \lambda^d)$, then it is immediate from a comparison of the definitions that $\mathcal{F}_\delta = \mathcal{F}(d)$. Hence, since $\mathcal{F}_\delta$ is obviously bounded, Lemma 9.129 implies that, if φ is holomorphic on a neighborhood of $(\mathbb{D}^d)^-$, then $\varphi \in H^\infty(\mathcal{F}(d))$.

Example 9.132. Let $d = 1$ and let $\delta(\lambda) = \lambda^2$. Then $G_\delta = \mathbb{D}$ and $\mathcal{F}_\delta$ is the class of bounded operators satisfying

$$\sigma(T) \subseteq \mathbb{D} \qquad \text{and} \qquad \|T^2\| \leq 1.$$

As the operator defined on $\mathbb{C}^2$ by the matrix

$$T_c = \begin{bmatrix} \lambda & c \\ 0 & -\lambda \end{bmatrix}$$

satisfies

$$T_c^2 = \begin{bmatrix} \lambda^2 & 0 \\ 0 & \lambda^2 \end{bmatrix},$$

$T_c \in \mathcal{F}_\delta$ if and only if $\lambda \in \mathbb{D}$. In particular, $\mathcal{F}_\delta$ is not bounded.

252   *Commutative Theory*

9.8.6 A Proof of Oka's Theorem

In this section we shall use Corollary 9.127 and Lemma 9.129 to prove Theorem 9.110.

Proof. **Step 1.** Assume that Π is as in the statement of Theorem 9.110 and f is holomorphic on a neighborhood U of Π.

Step 2. Let

$$\delta = (\lambda^1, \ldots, \lambda^d, p_1(\lambda), \ldots, p_m(\lambda))$$

and note that $\Pi = K_\delta$.

Step 3. Choose $r < 1$ such that

$$\Pi \subseteq K_{r\delta} \subseteq U.$$

Step 4. As f is holomorphic on U, f is holomorphic on a neighborhood of $K_{r\delta}$. Hence, as $\mathcal{F}_{r\delta}$ is bounded, Lemma 9.129 implies that $f \in \mathrm{H}^\infty(\mathcal{F}_{r\delta})$.

Step 5. By Corollary 9.127 (with δ replaced by $r\delta$) there exists $F \in \mathrm{H}^\infty(\mathcal{F}_{d+m})$ such that

$$f(\lambda) = F(r\delta(\lambda)) \qquad \text{for all } \lambda \in G_{r\delta}.$$

Step 6. If we define F_r by

$$F_r(w, z) = F(rz, rw),$$

then F_r is holomorphic on a neighborhood of $(\mathbb{D}^-)^{d+m}$ and

$$f(\lambda) = F_r(\delta(\lambda)) \qquad \text{for all } \lambda \in \Pi.$$

We may therefore take $F = F_r$ to obtain the statement in the theorem. $\qquad\square$

9.9 Historical Notes

Section 9.1. The Taylor spectrum and functional calculus were worked out in [198].

Section 9.2, operator algebras. Definition 9.19 is a standard notion in the theory of operator spaces, a popular topic; see the books [162, 165, 87]. If one drops the requirement that A be a Banach algebra and just requires that it be a Banach space and that the norms $\|\cdot\|_m$ on $\mathcal{M}_m l(A)$ satisfy equation (9.20) for some embedding ρ of A in $\mathcal{B}(\mathcal{H})$, then one obtains an *operator space*. In the language of that theory, an operator space that also has the structure of a unital algebra, with respect to which it is a Banach algebra, is defined to be an *operator algebra* if it admits a unital homomorphism ρ onto a subalgebra of $\mathcal{B}(\mathcal{H})$, for some $\mathcal{H}$, which is completely isometric, that is, satisfies the equations (9.20).

Section 9.3. Paulsen's Theorems 9.31 and 9.32 are from [161].

Section 9.4. Proposition 9.36 is from [77].

Section 9.5. Theorem 9.59 is from [70]. Theorem 9.62 is from [63].

Section 9.6. Theorem 9.64 is from [39].

Section 9.7. The polydisc norm was first studied in [7], which contained Theorems 9.102 and 9.103. For this reason the polydisc norm is often called the *Agler norm* and $\mathscr{S}(\mathcal{F}(d))$ is usually called the *Agler–Schur class*. The fact that the polydisc norm differs from the H^∞ norm in dimensions greater than 2 (the fact that $\mathbb{D}^3$ fails to be a spectral domain for some triples of commuting contractions) was shown by N. T. Varopoulos and S. Kaijser [206] and by M. J. Crabb and A. M. Davie [62]. It has been a much-studied phenomenon ever since.

Section 9.8. Corollary 9.127, Lemma 9.129 and their use to prove Oka's theorem in Section 9.8.6 are from [21] and [16].

10

Operator Monotone Functions

In 1934, Karl Löwner characterized functions that preserve matrix ordering [145]. In this chapter we shall prove Löwner's theorems and generalize them to several variables.

10.1 Löwner's Theorems

We shall let $SA\mathbb{M}_n$ denote the set of $n \times n$ self-adjoint matrices.

Definition 10.1. *Let E be an open set in $\mathbb{R}$. A function $f \colon E \to \mathbb{R}$ is called n-matrix monotone if, for all matrices S, T in $SA\mathbb{M}_n$ such that $\sigma(S) \cup \sigma(T) \subset E$,*

$$S \leq T \;\Rightarrow\; f(S) \leq f(T). \tag{10.2}$$

The function f is called operator monotone *if, whenever S, T are self-adjoint operators with spectra contained in E, the implication (10.2) holds.*

When $n = 1$, being 1-monotone is just being an increasing function. As n grows, the set of n-matrix monotone functions shrinks (indeed, it can be shown that for each n there is a function that is n-matrix monotone but not $(n + 1)$-matrix monotone). It is easy to show that x^3 is not 2-monotone.

It can be shown that if f is 2-matrix monotone then it is continuously differentiable [76, theorem VII.III]; we shall avoid some technicalities by making the blanket assumption that all functions that we consider are continuously differentiable everywhere.[1]

---

[1] The C^1 hypothesis can be removed by a mollification argument; this is straightforward for functions of one variable but more subtle for several variables [159].

Operator Monotone Functions

Definition 10.3. *Let E be an open set in $\mathbb{R}$, and let $n \geq 2$. The Löwner class $\mathcal{L}_n(E)$ is the set of C^1 functions $f \colon E \to \mathbb{R}$ with the property that, for every n distinct points $x_1, \ldots, x_n$ in E, the divided difference matrix A defined by*

$$A_{ij} = \begin{cases} \dfrac{f(x_j) - f(x_i)}{x_j - x_i} & \text{if} \quad i \neq j \\[2mm] f'(x_i) & \text{if} \quad i = j \end{cases} \tag{10.4}$$

is positive semi-definite.

Löwner proved:

Theorem 10.5. **(Löwner)** *Let E be an open interval in $\mathbb{R}$, and $f \colon E \to \mathbb{R}$. Let $n \geq 2$. Then f is n-matrix monotone on E if and only if f is in $\mathcal{L}_n(E)$.*

Since E is connected, we can move from S to T along the path $(1 - t)S + tT$.

Definition 10.6. *Let E be an open interval in $\mathbb{R}$, and $f \colon E \to \mathbb{R}$. We say that f is* locally n-matrix monotone on E *if, whenever $t \mapsto S(t)$ is a C^1 curve in $SA\mathbb{M}_n$ and $S'(t_0) \geq 0$, then*

$$\frac{d}{dt} f(S(t))|_{t_0} \geq 0. \tag{10.7}$$

If f is locally n-matrix monotone for every n, we say that f is locally operator monotone.

Technically we should distinguish between f as a function from E to $\mathbb{R}$ and the induced function, $f^\sharp$ say, that maps self-adjoint matrices with spectra in E into self-adjoint matrices. It is not immediately obvious that if f is C^1, then $f^\sharp$ is, but it is true—see Lemma 10.15. The left-hand side of inequality (10.7) is really the derivative of $f^\sharp(S(t))$, and by the chain rule we have

$$\frac{d}{dt} f^\sharp(S(t))|_{t_0} = Df^\sharp(S(t_0))[S'(t_0)],$$

where the right-hand side means the derivative of $f^\sharp$ in the direction $S'(t_0)$. For legibility, we will just write f for both the function on E and the function on matrices unless it is important to distinguish them.

Löwner also characterized operator monotone functions. Recall that $\mathbb{H}$ denotes the upper half plane, and $\mathscr{P}(\mathbb{H})$ denotes the set of holomorphic functions that map $\mathbb{H}$ to the closed upper half plane.

Definition 10.8. *Let E be an open set in $\mathbb{R}$. The* Pick class of E, *denoted by $\mathcal{P}(E)$, is the set of functions $f \colon E \to \mathbb{R}$ for which there exists an analytic function $F \colon \mathbb{H} \to \mathbb{H}$ that satisfies*

256 Commutative Theory

$$\lim_{y \downarrow 0} F(x + iy) = f(x) \quad \text{for all } x \in E.$$

Löwner's second main theorem says that a function is operator monotone if and only if it extends to the Pick class of the upper half plane.

Theorem 10.9. (**Löwner**) *Let E be an open interval in $\mathbb{R}$, and $f \colon E \to \mathbb{R}$ be differentiable. The following statements are equivalent.*

(i) *The function f is locally operator monotone on E.*
(ii) *The function f is operator monotone on E.*
(iii) *The function f is in $\mathcal{L}_n(E)$ for every $n \in \mathbb{N}$.*
(iv) *The function f is in $\mathcal{P}(E)$.*

Remark 10.10. Being n-matrix monotone for a fixed n is a real analysis condition—it can be thought of as a strong form of concavity. But requiring it for all n makes a function analytically continuable to the upper half plane—complex analysis enters the picture.

Remark 10.11. Löwner defined a function to be *operator monotone* if it is n-matrix monotone for every n. It is easy to see that this condition implies that the function is operator monotone in our sense. Indeed, suppose S, T are self-adjoint operators with spectra in the open interval E, and satisfying $S \leq T$. Let (P_n) be a sequence of projections converging to the identity in the strong operator topology, where P_n has rank n. By n-monotonicity, we have

$$f(P_n S P_n) \leq f(P_n T P_n) \tag{10.12}$$

for all n, since the spectrum of the compression $P_n S_n P$ is contained in the convex hull of the spectrum of S. But approximating f uniformly by polynomials on a closed interval $[a,b] \subset E$ with $\sigma(S) \cup \sigma(T) \subset (a,b)$, one sees that in the strong operator topology the left-hand side of equation (10.12) tends to $f(S)$ and the right-hand side to $f(T)$.

To show that Definition 10.6 makes sense, let us start with the following lemma.

Lemma 10.13. *Suppose S is in SAM_n and has eigenvalues $x_1 \leq x_2 \cdots \leq x_n$. Let $\rho = \min\{|x_i - x_j| \colon x_i \neq x_j\}$. Let $0 < \varepsilon < \rho/4$. If H is in SAM_n with $\|H\| < \varepsilon$, then the eigenvalues $x_1' \leq x_2' \leq \cdots \leq x_n'$ of the operator $S' = S + H$ satisfy*

$$|x_j - x_j'| \leq \varepsilon \quad \text{for } j = 1, \ldots, n.$$

Proof. For any number z in $\mathbb{C} \setminus \{\sigma(S') \cup \sigma(S)\}$, we have

$$\|(z - S')^{-1}\| = \|(z - S)^{-1}[(z - S') + (S' - S)](z - S')^{-1}\|$$
$$< \|(z - S)^{-1}\| \big[1 + \varepsilon \|(z - S')^{-1}\|\big].$$

Operator Monotone Functions 257

Therefore

$$\|(z-S')^{-1}\| < \frac{\|(z-S)^{-1}\|}{1-\varepsilon\|(z-S)^{-1}\|}. \tag{10.14}$$

Observe that

$$\text{dist}(z,\sigma(S)) = \|(z-S)^{-1}\|^{-1}.$$

So if $r = \text{dist}(z,\sigma(S))$, then inequality (10.14) gives

$$\text{dist}(z,\sigma(S')) > r - \varepsilon.$$

By continuity, this means that $\text{dist}(x'_j,\sigma(S)) < \varepsilon$.

If the eigenvalues of S are all distinct, we are done. If there is some eigenvalue x_j of multiplicity m say, we must show that there are m eigenvalues of S' within ε of x_j. To prove this, let Γ be a circle of radius $R = \rho/2$ centered at x_j. Let P be projection onto the span of the eigenvectors of S with eigenvalue x_j, and let P' be the projection onto the span of the eigenvectors of S' with eigenvalues in $(x_j - \varepsilon, x_j + \varepsilon)$. Then

$$\|P - P'\| = \left\|\frac{1}{2\pi i} \int_\Gamma (z-S)^{-1} - (z-S')^{-1}dz\right\|$$

$$\leq R \sup_{z\in\Gamma} \|(z-S)^{-1}\|\,\|S-S'\|\,\|(z-S')^{-1}\|$$

$$< R\,\frac{1}{R}\,\varepsilon\,\frac{1}{R-\varepsilon}$$

$$< 1.$$

Therefore, the ranks of P and P' are the same. $\qquad\square$

Lemma 10.15. *Let E be an open interval in $\mathbb{R}$, suppose $f\colon E \to \mathbb{R}$ is C^1. Let $S \in SAM_n$ have $\sigma(S) \subset E$. Then $f^\sharp$ is differentiable at S.*

Proof. Suppose that the eigenvalues of S are $x_1 \leq \cdots \leq x_n$. Let

$$\rho = \min\{\|x_i - x_j\|\colon x_i \neq x_j\},$$

let $H \in SAM_n$, and choose t small enough that $\|tH\| < \frac{\rho}{4}$. Suppose first that f and f' both vanish at all the x_j's. Let $S+tH$ have eigenvalues $x'_1 \leq \cdots \leq x'_n$. The norm of $f(S+tH)$ is the maximum of $|f(x'_i)|$; assume this occurs at x'_j. Since f is C^1, by the mean value theorem we have

$$|f(x'_j)| = |f(x'_j) - f(x_j)|$$

$$= |x_j - x'_j| \cdot |f'(x''_j)|,$$

258 *Commutative Theory*

where x''_j is some point between x_j and x'_j. By Lemma 10.13, we have $|x'_j - x_j| \le t$. Since f' is continuous and $f'(x_j)$ is assumed to be 0, we obtain

$$f(S+tH) - f(S) = o(t),$$

and so $f^\sharp$ is differentiable at S, with derivative 0.

For a general f, choose a polynomial p that agrees with f to first order at each x_j. As the derivative of $(f - p) \circ S(t)$ has been shown to be zero, we can finish the proof of the lemma by assuming that $f = p$. By linearity, we can take p to be a monomial, $p(x) = x^m$, say. Expanding $(S+tH)^m = S^m + tR + O(t^2)$ we see that the derivative of $S(t)^m$ is the sum as k goes from 0 to $m - 1$ of the product of k copies of S, one copy of H, and then another $m - k - 1$ copies of S. The proof is now complete. $\square$

We shall now stop writing $f^\sharp$, and just use the notation $f(S(t))$. We want a formula for the derivative.

Lemma 10.16. *Suppose $S(t)$ is a C^1 path in $S\mathbb{A}\mathbb{M}_n$. Let $S(t_0)$ be diagonal, with diagonal entries $x_1, \ldots, x_n$. Let f be a C^1 function in an open interval containing $\{x_1, \ldots, x_n\}$. Let A be the divided difference matrix (10.4) for f and the points $x_1, \ldots, x_n$, where $A_{ij} = f'(x_j)$ if $x_i = x_j$. Then*

$$\frac{d}{dt} f(S(t))|_{t_0} = A \cdot S'(t_0) \qquad (10.17)$$

and $f(S(t))$ is C^1.

The right-hand side of equation (10.17) denotes the *Schur product* of the matrices A and $S'(t_0)$, that is, the matrix with (i, j) entry equal to the product of the (i, j) entries of A and $S'(t_0)$. An important property of the Schur product is that the Schur product of two positive semi-definite matrices is positive semi-definite. This can be proved in several ways; here is one.

Theorem 10.18. *The Schur product of two positive semi-definite matrices is positive semi-definite.*

Proof. Let A and B be positive semi-definite $n \times n$ matrices. By Moore's theorem, Theorem 2.15, there are vectors $\{u_j\} \subseteq \mathbb{C}^r$ and $\{v_j\} \subseteq \mathbb{C}^s$ such that

$$A_{ij} = \langle u_j, u_i \rangle_{\mathbb{C}^r}$$
$$B_{ij} = \langle v_j, v_i \rangle_{\mathbb{C}^s}.$$

(We can choose r and s to be the ranks of A and B, but we don't need to use that fact for this proof). Now define new vectors x_j in $\mathbb{C}^r \otimes \mathbb{C}^s$ by $x_j = u_j \otimes v_j$. The Schur product $A \cdot B$ is then the Gramian of the vectors $\{x_j\}$, and hence is positive semi-definite. $\square$

Operator Monotone Functions

Proof of Lemma 10.16. We shall show that the curve $S(t)$ can be explicitly diagonalized, to first order, in a neighborhood of t_0.

We can suppose that $t_0 = 0$. Let $S_0 = S(0)$, $\Delta = S'(0)$ and let Λ denote the block diagonal matrix with $\Lambda_{ij} = \Delta_{ij}$ if $x_i = x_j$, and 0 otherwise. By construction, Λ commutes with S_0, so $Df(S_0)[\Lambda]$, the derivative of f at S_0 in the direction of Λ, is just the product $f'(S_0)\Lambda$. (One can see this by choosing a basis that simultaneously diagonalizes both S_0 and Λ.)

Let Y be the skew-adjoint matrix given by

$$Y_{ij} = \frac{\Delta_{ij}}{x_j - x_i} \qquad \text{for } x_i \neq x_j$$

and $Y_{ij} = 0$ if $x_i = x_j$.

Then e^{tY} is unitary for real t. We have

$$
\begin{aligned}
f\left(e^{tY}(S_0 + t\Lambda)e^{-tY}\right) &= e^{tY} f(S_0 + t\Lambda)e^{-tY} \\
&= e^{tY}\left(f(S_0) + tf'(S_0)\Lambda + o(t)\right)e^{-tY} \\
&= (1 + tY + o(t))\left(f(S_0) + tf'(S_0)\Lambda + o(t)\right)(1 - tY + o(t)) \\
&= f(S_0) + t\left(Yf(S_0) + f'(S_0)\Lambda - f(S_0)Y\right) + o(t).
\end{aligned}
$$

Since

$$f(S_0) = \text{diag}\,\{f(x_1), \ldots, f(x_n)\}, \quad f'(S_0) = \text{diag}\,\{f'(x_1), \ldots, f'(x_n)\},$$

it can easily be verified that

$$Yf(S_0) + f'(S_0)\Lambda - f(S_0)Y = A \cdot \Delta.$$

Thus

$$f\left(e^{tY}(S_0 + t\Lambda)e^{-tY}\right) = f(S_0) + tA \cdot \Delta + o(t). \qquad (10.19)$$

In particular, this relation holds in the case that f is the identity function $f(x) = x$, in which case the corresponding divided difference matrix "A" is the matrix all of whose entries are 1. Thus, in this case, equation (10.19) becomes

$$e^{tY}(S_0 + t\Lambda)e^{-tY} = S_0 + t\Delta + o(t).$$

Since also $S(t) = S_0 + t\Delta + o(t)$, it follows that

$$S(t) = e^{tY}(S_0 + t\Lambda)e^{-tY} + o(t), \qquad (10.20)$$

which is the promised diagonalization of $S(t)$ to first order. In light of Lemma 10.15, we have

$$
\begin{aligned}
f(S(t)) &= f\left(e^{tY}(S_0 + t\Lambda)e^{-tY}\right) + o(t) \qquad && \text{by equation (10.20)} \\
&= f(S_0) + tA \cdot \Delta + o(t) && \text{by equation (10.19)}.
\end{aligned}
$$

260   *Commutative Theory*

Therefore

$$\frac{d}{dt} f(S(t))|_{t=0} = A \cdot \Delta.$$

Finally, we observe that if $S(t)$ is C^1, then the spectrum of $S(t)$ changes continuously by Lemma 10.13, so the divided difference matrix A changes continuously with t_0. As both functions on the right-hand side of equation (10.17) are continuous, so is the left-hand side and therefore $f(S(t))$ is C^1. □

This allows us to prove that locally monotone and monotone are the same.

Lemma 10.21. *Suppose E is an open interval in $\mathbb{R}$ and $f \colon E \to \mathbb{R}$ is C^1. Then f is locally n-matrix monotone on E if and only if it is n-matrix monotone.*

Proof. Suppose f is locally n-monotone, and $S \leq T$ are in SAM_n. Define $S(t) = (1 - t)S + tT$. Notice that all the eigenvalues of $S(t)$ lie between the smallest eigenvalue of S and the largest eigenvalue of T, so will remain in the interval E. We have

$$S'(t_0) = T - S \geq 0$$

for every t_0. By the fundamental theorem of calculus (which we can use since the integrand is continuous)

$$f(T) - f(S) = \int_0^1 \frac{d}{dt} f(S(t)) dt,$$

and the integral of positive matrices is positive, so f is n-matrix monotone. The converse is obvious.   □

Proof of Theorem 10.5. Suppose f is in $\mathcal{L}_n(E)$. By Lemma 10.21, it is enough to show that f is locally n-matrix monotone. But by Lemma 10.16, if $S'(t_0) = \Delta$ is positive, then the derivative of $f(S(t))$ is the Schur product of the divided difference matrix, which is positive since f is in $\mathcal{L}_n(E)$, with Δ, and hence is positive.

Conversely, suppose f is locally n-matrix monotone. Let $x_1, \ldots, x_n$ be distinct points in E, let Δ be the matrix of all 1's, let $S(0)$ be the diagonal matrix with entries $x_1, \ldots, x_n$, and let $S(t) = S(0) + t\Delta$. By Lemma 10.16, the derivative of $f(S(t))$ at 0 is the divided difference matrix of f, which is therefore positive semi-definite.   □

To prove Löwner's second theorem, Theorem 10.9, in view of the preceding remarks and his first theorem, we need to show that (iii) and (iv) are equivalent: all the difference matrices in the definition (10.4) are positive if and only if f extends to a holomorphic function from $\mathbb{H}$ to $\mathbb{H}$.

Operator Monotone Functions 261

Let $\alpha\colon \overline{\mathbb{D}} \to \overline{\mathbb{H}} \cup \{\infty\}$ and $\beta\colon \overline{\mathbb{H}} \to \overline{\mathbb{D}}$ be the Cayley transform and its inverse.

$$\alpha(\lambda) = i\frac{1+\lambda}{1-\lambda},$$

$$\beta(z) = \frac{z-i}{z+i}. \tag{10.22}$$

Proof of Theorem 10.9. Suppose f is in $\mathcal{L}_n(E)$ for every $n \in \mathbb{N}$. Let $\psi = \beta \circ f \circ \alpha$. Let $J = \beta(E) \subset \mathbb{T}$. On $E \times E$, let

$$A(x,y) = \begin{cases} \dfrac{f(x)-f(y)}{x-y} & \text{if } x \neq y \\[2mm] f'(x) & \text{if } x = y, \end{cases}$$

and on $J \times J$ let

$$P(\tau,\sigma) = \begin{cases} \dfrac{1-\psi(\tau)\overline{\psi(\sigma)}}{1-\tau\overline{\sigma}} & \text{if } \tau \neq \sigma \\[2mm] |\psi'(\tau)| & \text{if } \tau = \sigma. \end{cases}$$

We may call A and P, the *Löwner* and *Pick kernels* corresponding to f and ψ on E and J, respectively. By direct calculation we find that

$$P(\tau,\sigma) = \frac{\alpha(\tau)+i}{f(\alpha(\tau))+i} \frac{\overline{\alpha(\sigma)-i}}{\overline{f(\alpha(\sigma))-i}} A(\alpha(\tau),\alpha(\sigma)).$$

Thus a Löwner kernel on E is positive semi-definite if and only if the corresponding Pick kernel on J is.

Suppose (iii). Then the Löwner kernel A is positive semi-definite and so therefore is the Pick kernel P on J. By Theorem 5.33, there is a function φ in $\mathscr{S}(\mathbb{D})$ that has ψ as its non-tangential limit function on J. So if we define $F = \alpha \circ \varphi \circ \beta$, then $F\colon \mathbb{H} \to \mathbb{H}$ and

$$\lim_{y\downarrow 0} F(x+iy) = f(x) \quad \text{for all } x \in E.$$

Thus $f \in \mathcal{P}(E)$, which is to say that (IV) holds.

Conversely, suppose $f \in \mathcal{P}(E)$. Then f has an extension to $F\colon \mathbb{H} \to \mathbb{H}$. Let $\varphi = \beta \circ F \circ \alpha$. Since φ is in $\mathscr{S}(\mathbb{D})$ it has a model $(\mathcal{M},u)$:

$$1 - \varphi(\lambda)\overline{\varphi(\mu)} = (1 - \lambda\bar{\mu})\langle u(\lambda),u(\mu)\rangle \quad \text{for all } \lambda,\mu \in \mathbb{D}. \tag{10.23}$$

Let $z = \alpha(\lambda)$, $w = \alpha(\mu)$. Define

$$v(z) = \frac{F(z)+i}{z+i} u(\beta(z)).$$

262 *Commutative Theory*

Then changing variables in equation (10.23) and rearranging, we get a model for F:

$$F(z) - \overline{F(w)} = (z - \bar{w})\langle v(z), v(w)\rangle_{\mathcal{M}}. \tag{10.24}$$

By the Schwarz reflection principle, F extends to be analytic on the union of the upper half plane, the lower half plane, and E. So in particular, every point of E is a B-point.[2] Therefore for any points x_j in E, the vectors v_{x_j+iy} have a weak limit v_{x_j} as $y \to 0^+$, and we find that the Löwner matrix A from equation (10.4) is just the Gramian of the vectors v_{x_j}. $\qquad\square$

Remark 10.25. Another way to prove that functions in $\mathcal{P}(E)$ are operator monotone is to use the Nevanlinna representation from Theorem 6.3, prove that μ puts no mass on E [76, lemma II.2] and then show that when you substitute operators S and T into the integral in the Nevanlinna representation theorem, the difference is an integral of $(t - S)^{-1} - (t - T)^{-1}$. Since $-1/z$ is easily seen to be monotone, this means that the difference is an integral of positive operators, hence positive.

10.2 An Interlude on Linear Programming

In the proof of Theorem 10.72 we need a result from linear programming, Richard Duffin's strong duality theorem, originally proved in 1956 [82]. It is an important result, and the proof is based on a cone separation argument, so we shall give the proof in the case that we need, in Theorem 10.40. For a more detailed account, see [94, chapter 4].

10.2.1 Strong Duality in Linear Programming

We shall prove the strong duality theorem in $\mathbb{R}^n$, originally due to von Neumann. In this section all our variables will be considered real, and for $x \in \mathbb{R}^n$, inequalities like $x > 0$ are interpreted componentwise: $x^r > 0$, $1 \le r \le n$. The basic linear programming problem is the following.

Given an $m \times n$ matrix A with real entries, and vectors $b \in \mathbb{R}^m$, $c \in \mathbb{R}^n$, find

$$\sup \{c^T x \colon Ax \le b, \ x \ge 0\}. \tag{10.26}$$

A vector $x \ge 0$ such that $Ax \le b$ is called *feasible*; if there are no feasible vectors, the problem is called *infeasible*. The supremum in the expression

---

[2] The definition of a B-point at x for a function in $\mathcal{P}(\mathbb{H})$ is the obvious modification of the definition in Section 5.2 for functions in $\mathcal{S}(\mathbb{D})$; equivalently, it is the same as saying φ has a B-point at $\beta(x)$.

Operator Monotone Functions 263

(10.26) is called the *value* of the program. It is possible that the supremum is infinite, if the feasible region is unbounded in the direction of c.

The inequality $Ax \le b$ can be replaced by equalities by the introduction of *slack variables*. This means rewriting

$$\sum_{j=1}^{n} A_{ij}x_j \le b_i, \quad x \ge 0 \qquad \text{for } i = 1,\ldots,m$$

as

$$\sum_{j} A_{ij}x_j = b_i - z_i, \quad x \ge 0, z \ge 0 \qquad \text{for } i = 1,\ldots,m.$$

So by adjoining z to x to give a vector in $\mathbb{R}^{n+m}$, adjoining $-I_m$ to the right of A to make it an $m \times (n+m)$ matrix, and adding zeroes to c, the programming problem (10.26) can also be written in equational form as

$$\sup \{c^T x : Ax = b, \ x \ge 0\}. \tag{10.27}$$

The problem (10.26) is called the *primal*. The dual problem is to find

$$\inf \{b^T y : A^T y \ge c\}. \tag{10.28}$$

The weak duality theorem says that the value (10.26) is always less than or equal to the value (10.28), assuming both programs are feasible.

Theorem 10.29. (Weak duality) *Let x be feasible for the primal problem, and y for the dual problem. Then*

$$c^T x \ \le \ b^T y.$$

Proof. For such x, y, we have

$$c^T x \ \le \ y^T Ax \ \le \ y^T b.$$

$\square$

The strong duality theorem, Theorem 10.32, says that if the supremum (10.26) is attained, then it equals the infimum (10.28). To prove this, we start with Farkas's lemma. Note that $x \ge 0$ is the same as $x^T \ge 0$.

Lemma 10.30. *Let A be an $m \times n$ matrix, and let $b \in \mathbb{R}^m$. Exactly one of the following two alternatives holds:*

(i) *The system $Ax \le b$, $x \ge 0$ has a solution $x \in \mathbb{R}^n$.*
(ii) *The system $A^T y \ge 0$, $b^T y < 0$ has a solution $y \in \mathbb{R}^m$.*

264 *Commutative Theory*

Proof. Both cannot occur simultaneously, because in that case

$$0 \le y^T A x$$
$$= y^T b$$
$$< 0.$$

Now assume (i) fails. The set $\{Ax: x \ge 0\}$ is a closed wedge, and by assumption it does not contain b. So, by the Hahn–Banach theorem, there is some $y \in \mathbb{R}^m$ such that

$$y^T A x \ge 0 \qquad \text{for all } x \ge 0 \tag{10.31}$$

and $y^T b < 0$. But the assertion (10.31) is the same as $A^T y \ge 0$, so (ii) holds. $\qquad \Box$

Theorem 10.32. (**Strong duality**) *Suppose the supremum* (10.26) *is attained at some vector x^* and equals μ. Then the infimum* (10.28) *is also attained and equals μ.*

Proof. Let $\varepsilon > 0$. Let A' be A with the row $-c^T$ adjoined at the bottom and b' be b with the number $-\mu - \varepsilon$ adjoined at the bottom. Since μ is the supremum, the system $A'x \le b'$, $x \ge 0$ has no solution. By Lemma 10.30, there exists $y' = (y, \alpha)$ such that

$$A^T y \ge \alpha c \qquad \text{and} \qquad b^T y < \alpha(\mu + \varepsilon). \tag{10.33}$$

When $\varepsilon = 0$, since (i) holds in Farkas's lemma, for any (y, α),

$$A^T y \ge \alpha c \qquad \Rightarrow \qquad b^T y \ge \alpha \mu.$$

Therefore $\alpha \varepsilon$ in statement (10.33) is positive. Now replace y by $(1/\alpha)y$, to get a solution to the dual problem with value less than or equal to $\mu + \varepsilon$.

Finally observe that the feasible set for the dual $\{y: A^T y \ge c\}$ is closed, convex and non-empty, and its boundary is given by a finite number of hyperplanes. The set may be unbounded, but since we know $b^T y$ is bounded below by μ, we know that the infimum (10.28) is attained. $\qquad \Box$

10.2.2 Semi-Definite Programming

A semi-definite program is similar to a linear program, but we optimize over positive semi-definite matrices. The basic problem is the following.

Given symmetric $n \times n$ matrices $C, A_1, \ldots, A_m$ and real numbers $b_1, \ldots, b_m$, find

$$\sup\{\text{tr}(CX): X \ge 0 \text{ and } \text{tr}(A_j X) = b_j \text{ for } j = 1, \ldots, m\}. \tag{10.34}$$

Notice that, formally, this is similar to the expression (10.27). However, the supremum (10.34) is not necessarily attained even for feasible bounded problems with finite values, because the boundary of the region is no longer given by finitely many linear equations, and may be curved.

Example 10.35. When $n = 2$, the value of

$$\sup\{-x_{11}: x_{12} = 1,\ X \geq 0\}$$

is zero, but it is never attained.

A greater subtlety is that that the limit value may not equal the value. A *limit sequence* for the program (10.34) is a sequence X_k of positive semi-definite matrices such that

$$\lim_{k \to \infty} \mathrm{tr}(A_j X_k) = b_j, \qquad 1 \leq j \leq m.$$

The program is called *limit feasible* if a limit sequence exists. The *limit value* of the program (10.34) is

$$\sup \{\limsup_{k \to \infty} \mathrm{tr}(C X_k) : X_k \text{ is a limit sequence}\}.$$

Example 10.36. Suppose $n = 2, m = 1$,

$$C = \begin{pmatrix} 0 & 1 \\ 1 & 0 \end{pmatrix}, \quad A = \begin{pmatrix} 1 & 0 \\ 0 & 0 \end{pmatrix}$$

and $b = 0$. The only feasible X's are

$$\begin{pmatrix} 0 & 0 \\ 0 & * \end{pmatrix},$$

so the value of the program is 0. But if

$$X_k = \begin{pmatrix} \frac{1}{k} & k \\ k & k^3 \end{pmatrix},$$

then X_k is a limit sequence, and the limit value is $+\infty$.

The dual of the program (10.34) is

$$\inf \left\{ \sum_j b_j y_j : \sum_j y_j A_j \geq C \right\}. \tag{10.37}$$

Exercise 10.38. What are the duals of Examples 10.35 and 10.36?

266 *Commutative Theory*

The weak duality theorem is:

Theorem 10.39. *If the dual program* (10.37) *is feasible and if the primal program* (10.34) *is limit feasible, then the limit value of the primal is less than or equal to the value of the dual.*

Proof. The proof is immediate. Suppose y is feasible for the dual and X_k is a limit sequence for the primal. Then

$$\limsup_{k \to \infty} \operatorname{tr}(C X_k) \le \limsup_{k \to \infty} \operatorname{tr}\left(\sum y_j A_j X_k\right)$$
$$= \sum y_j b_j.$$

Finally, observe that the possible limit values for the primal are bounded above and the possible values of the dual are bounded below, so the supremum of the former and the infimum of the latter are both finite. □

The strong duality theorem for semi-definite programming is the following.

Theorem 10.40. (Duffin) *Suppose the primal program* (10.34)

$$\sup\{\operatorname{tr}(CX): \operatorname{tr}(A_j X) = b_j,\ 1 \le j \le m,\ X \ge 0\}$$

is feasible and has a finite value μ. Assume also that there is a feasible matrix X that is positive definite. Then the dual program (10.37)

$$\inf\left\{\sum_j b_j y_j \colon \sum_j y_j A_j \ge C\right\}$$

is feasible and also has value μ.

Example 10.36 shows that the theorem is false if we omit the hypothesis that there is a strictly positive definite feasible matrix.

We shall prove the theorem in a similar way to the proof of Theorem 10.32. First let us establish some notation. The set $SA\mathbb{M}_n$ is a real vector space, with inner product

$$\langle X, Y \rangle \overset{\text{def}}{=} \operatorname{tr}(XY).$$

Let Pos_n denote the positive semi-definite cone in $SA\mathbb{M}_n$. This cone is self-dual:

Lemma 10.41. *Let $Y \in SA\mathbb{M}_n$. Then $\langle X, Y \rangle \ge 0$ for all $X \in \operatorname{Pos}_n$ if and only if $Y \in \operatorname{Pos}_n$.*

We leave the proof of Lemma 10.41 as an exercise. Let $A_1, \ldots, A_m$ be given matrices in $SA\mathbb{M}_n$. Define $L_A \colon SA\mathbb{M}_n \to \mathbb{R}^m$ by

Operator Monotone Functions 267

$$(L_A(X))_j = \langle A_j, X \rangle \qquad \text{for } j = 1, \ldots, m. \tag{10.42}$$

It is straightforward to verify that the adjoint L_A^* is given by

$$L_A^*(y) = \sum_{j=1}^m y_j A_j.$$

Here is the version of Farkas's lemma that we need.

Lemma 10.43. *Let L_A be given by equation* (10.42) *and let $b \in \mathbb{R}^m$. Exactly one of the following two alternatives holds:*

(i) *The program $L_A(X) = b, X \geq 0$ is limit feasible.*
(ii) *The program $L_A^*(y) \geq 0$, $\langle b, y \rangle < 0$ is feasible.*

Proof. If (i) holds, choose $X_k \geq 0$ so that $L_A(X_k) \to b$. Then for all $y \in \mathbb{R}^m$ with $L_A^* y \geq 0$, we have

$$\langle b, y \rangle = \lim \langle L_A(X_k), y \rangle$$
$$= \lim \langle (X_k), L_A^* y \rangle$$
$$\geq 0.$$

So (ii) fails.

Conversely, if (i) fails, then b is not in the closure of the cone $\{L_A(X): X \in \mathrm{Pos}_n\}$. So, by the Hahn–Banach theorem, there exists $y \in \mathbb{R}^m$ such that $\langle b, y \rangle < 0$ and

$$\langle L_A(X), y \rangle = \langle X, L_A^* y \rangle \geq 0 \quad \text{for all } X \geq 0.$$

By Lemma 10.41 this means that (ii) is feasible. $\qquad\square$

We use the hypothesis that there is a strictly positive feasible X to show that the limit value equals the value.

Lemma 10.44. *Suppose the program* (10.34) *has a positive definite feasible matrix $\tilde{X}$. Then the limit value and the value coincide.*

Proof. Let γ be the limit value, which must be greater than or equal to the value. Assume first that $\gamma < \infty$. Let $\varepsilon > 0$. Choose a limit sequence X_k so that $\lim_{k \to \infty} \mathrm{tr}(CX_k) > \gamma - \varepsilon$. Let $\mathcal{M}$ be the subspace of $\mathbb{R}^m$ that is the range of L_A; let the dimension of $\mathcal{M}$ be ℓ. Let $\mathcal{N}$ be an ℓ dimensional subspace of $SA\mathbb{M}_n$ chosen so that

$$L_A: \mathcal{N} \to \mathcal{M}$$

is an isomorphism. Let $L_A^{-1}: \mathcal{M} \to \mathcal{N}$ be the inverse, which is bounded since the spaces are finite dimensional.

268 *Commutative Theory*

Choose k large enough that

$$\|L_A^{-1}(b - L_A(X_k))\| < \varepsilon, \tag{10.45}$$

and

$$\mathrm{tr}(C X_k) \geq \gamma - \varepsilon.$$

Let

$$\delta = \|\tilde{X}^{-1}\|\varepsilon.$$

Now define

$$W_k = (1 - \delta)[X_k + L_A^{-1}(b - L_A(X_k))] + \delta \tilde{X}.$$

Note first that $L_A(W_k) = b$. Second, $(1 - \delta)X_k + \delta \tilde{X}$ is positive definite, so that W_k is also positive definite provided the norm of the perturbation

$$(1 - \delta)L_A^{-1}(b - L_A(X_k))$$

is smaller than the smallest eigenvalue of $\delta \tilde{X}$. The last statement does hold, by inequality (10.45) and the definition of δ. Therefore W_k is feasible, and

$$\mathrm{tr}(C W_k) \geq (1 - \delta)(\gamma - \varepsilon) - \|C\|\varepsilon + \delta \mathrm{tr}(C W_k)$$
$$= \gamma + O(\varepsilon).$$

Therefore we have feasible solutions whose values are arbitrarily close to γ, and so the value equals the limit value.

If $\gamma = \infty$, a similar argument shows that the value of the program is also infinite. $\qquad \square$

Now we can prove Theorem 10.40.

Proof. By Lemma 10.44, the limit value of the primal is also μ. First let us prove that the dual program is feasible. Otherwise, for every $y \in \mathbb{R}^m$, we have

$$L_A^*(y) - \alpha C \geq 0 \quad \Rightarrow \quad \alpha \leq 0. \tag{10.46}$$

Define $L^* \colon \mathbb{R}^{m+1} \to S\mathbb{M}_n$ by $L^*(y, \alpha) = L_A^*(y) - \alpha C$, where $y \in \mathbb{R}^m$ and $\alpha \in \mathbb{R}$. Then the implication (10.46) is the same as saying that the program

$$L^* \begin{pmatrix} y \\ \alpha \end{pmatrix} \geq 0, \quad \left\langle \begin{pmatrix} 0 \\ -1 \end{pmatrix}, \begin{pmatrix} y \\ \alpha \end{pmatrix} \right\rangle < 0$$

is not feasible. By Lemma 10.43, this means that

$$L(X) = \begin{pmatrix} 0 \\ -1 \end{pmatrix}, \quad X \geq 0$$

Operator Monotone Functions   269

is limit feasible, which means there is a sequence $X_k \geq 0$ such that $L_A(X_k) \to 0$ and $\langle C, X_k \rangle \to 1$. But if W_k is any limit feasible sequence for the original program (10.34) and $\lim \langle C, W_k \rangle = \gamma$, then $W_k + X_k$ is also limit feasible, and $\lim \langle C, W_k \rangle = \gamma + 1$. This implies that the limit value of the primal is infinite, a contradiction.

So the dual is feasible; suppose its value is γ, which we know is greater than or equal to μ by the weak duality (10.39). Since the value is γ, we have

$$L_A^*(y) \geq C \Rightarrow \langle b, y \rangle \geq \gamma.$$

This implies that if we introduce another real variable α, then

$$L_A^* y - \alpha C \geq 0, \ \alpha \geq 0 \Rightarrow \langle b, y \rangle \geq \alpha \gamma. \tag{10.47}$$

(When $\alpha > 0$ this is just scaling; when $\alpha = 0$ you argue that otherwise you could reduce the value of the dual below γ). As in the proof of Theorem 10.32, we rewrite this with an augmented matrix as

$$\begin{pmatrix} L_A^* & -C \\ 0 & 1 \end{pmatrix} \begin{pmatrix} y \\ \alpha \end{pmatrix} \in \begin{pmatrix} \mathrm{Pos}_n \\ \mathbb{R}_+ \end{pmatrix} \ \Rightarrow \ \left\langle \begin{pmatrix} b \\ -\gamma \end{pmatrix}, \begin{pmatrix} y \\ \alpha \end{pmatrix} \right\rangle \geq 0.$$

So by Farkas's lemma, Lemma 10.43, we conclude that the program

$$\begin{pmatrix} L_A & 0 \\ -L_C & 1 \end{pmatrix} \begin{pmatrix} X \\ z \end{pmatrix} = \begin{pmatrix} b \\ -\gamma \end{pmatrix}$$

is limit feasible. This means there are sequences $X_k \geq 0$ and $z_k \geq 0$ such that $L_A X_k \to b$ and

$$\langle C, X_k \rangle - z_k \to \gamma.$$

This means $\limsup \langle C, X_k \rangle \geq \gamma$, so that $\mu \geq \gamma$, and therefore $\mu = \gamma$. $\square$

For a more thorough discussion, see [94, chapter 4]. The treatment there is more general: instead of just looking at maps from SAM_n to $\mathbb{R}$, the authors consider maps between general ordered vector spaces.

10.3 Locally Matrix Monotone Functions in d Variables

Let $CSAM_n^d$ denote the set of commuting d-tuples of self-adjoint $n \times n$ matrices. If S is in $CSAM_n^d$, we can choose an orthonormal basis of eigenvectors that diagonalize all the S^r's simultaneously, so that, up to unitary conjugation,

$$S^r = \begin{pmatrix} x_1^r & & \\ & \ddots & \\ & & x_n^r \end{pmatrix} \qquad \text{for } 1 \leq r \leq d. \tag{10.48}$$

270   Commutative Theory

The n points $x_j = (x_j^1, \ldots, x_j^d)$, $j = 1, \ldots, n$, comprise the spectrum of S, which is contained in $\mathbb{R}^d$ since the matrices are all self-adjoint. If $S(t)$ is a C^1 curve of commuting self-adjoints, then $S(0) + t \frac{d}{dt} f(S(t))|_{t=0}$ commutes to first order.

For any $X \in \mathbb{M}_n$ we define diag X to be the diagonal matrix in $\mathbb{M}_n$ with diagonal entries X_{ii}, and for any $\Delta \in SA\mathbb{M}_n^d$ we define diag Δ to be $(\text{diag } \Delta^1, \ldots, \text{diag } \Delta^d)$.

Definition 10.49. *A d-tuple S in $CSA\mathbb{M}_n^d$ is generic if its spectrum consists of n distinct points.*

Definition 10.50. *Let E be an open set in $\mathbb{R}^d$ and f be a real-valued C^1 function on E. Then f is locally $\mathbb{M}_n$-monotone on E if, whenever S is in $CSA\mathbb{M}_n^d$ with $\sigma(S) = \{x_1, \ldots, x_n\}$ consisting of n distinct points in E, and $S(t)$ is a C^1 curve in $CSA\mathbb{M}_n^d$ such that $S(0) = S$ and $\frac{d}{dt} S(t)|_{t=0} \geq 0$, then $\frac{d}{dt} f(S(t))|_{t=0}$ exists and is positive semi-definite.*

The advantage of working with generic tuples is that it is much easier to understand paths and derivatives when eigenvalues do not cross. The non-generic version of the following lemma was given by Kelly Bickel [51]; see Theorem 10.99.

Lemma 10.51. *Let S be in $CSA\mathbb{M}_n^d$ and Δ be in $SA\mathbb{M}_n$, with S generic. Then there exists a C^1 curve $S(t)$ of commuting self-adjoints with $S(0) = S$ and $S'(0) = \Delta$ if and only if*

$$[S^r, \Delta^s] = [S^s, \Delta^r] \qquad \text{for } 1 \leq r \neq s \leq d. \tag{10.52}$$

Proof. $(\Rightarrow)$: If $S(t) = S + t\Delta + o(t)$ is commutative, calculate

$$[S^r(t), S^s(t)] = t \left([S^r, \Delta^s] - [S^s, \Delta^r]\right) + o(t).$$

The coefficient of t must vanish, giving equation (10.52).

$(\Leftarrow)$: Suppose S is as in equation (10.48), and equation (10.52) holds. This means that

$$\Delta_{ij}^s(x_j^r - x_i^r) = \Delta_{ij}^r(x_j^s - x_i^s) \qquad \text{for } r \neq s,$$

and so

$$\Delta_{ij}^r \frac{1}{x_j^r - x_i^r} = \Delta_{ij}^s \frac{1}{x_j^s - x_i^s} \qquad \text{if } x_j^r - x_i^r \neq 0 \neq x_j^s - x_i^s. \tag{10.53}$$

Define a skew-symmetric matrix Y by

$$Y_{ij} = \Delta_{ij}^r \frac{1}{x_j^r - x_i^r} \qquad \text{for any } r \text{ such that } x_j^r - x_i^r \neq 0. \tag{10.54}$$

Operator Monotone Functions   271

For any $i \neq j$, there is some r with $x_j^r - x_i^r \neq 0$, so equation (10.54) defines Y_{ij}; and equation (10.53) says it doesn't matter which r we choose. Let all the diagonal terms of Y be 0.

Define

$$S^r(t) = e^{tY}(S^r + t \operatorname{diag} \Delta^r)e^{-tY}. \tag{10.55}$$

Since e^{tY} is a unitary matrix and $S^r + t \operatorname{diag} \Delta^r$ is diagonal, $S(t) \in CSAM_n^d$ and

$$\frac{d}{dt}S^r(t)|_{t=0} = [Y, S^r] + \operatorname{diag} \Delta^r = \Delta^r.$$

$\square$

If S and Δ satisfy equation (10.52) and S is generic, then for any function f that is C^1 on a neighborhood of $\sigma(S)$, we define the directional derivative of f at S in direction Δ by

$$Df(S)[\Delta] = \frac{d}{dt}f(S(t))|_{t=0}, \tag{10.56}$$

where $S(t)$ is the curve given by equations (10.55) and (10.54). We shall show in Proposition 10.64 that the right-hand side of equation (10.56) is unchanged if $S(t)$ is replaced by any other curve that agrees with it to first order. (Bickel proved that this statement is true without the genericity assumption; see Theorem 10.100.)

First, let us show that the right-hand side of equation (10.56) exists. Indeed,

$$f(S(t)) = e^{tY}f(S + t \operatorname{diag} \Delta)e^{-tY}. \tag{10.57}$$

Since $S + t \operatorname{diag} \Delta$ is diagonal, $f(S + t \operatorname{diag} \Delta)$ is diagonal, with ith entry

$$f(x_i + t\Delta_{ii}) = f(x_i) + t\sum_{r=1}^{d}\Delta_{ii}^r \frac{\partial f}{\partial x^r}(x_i) + o(t).$$

In other words,

$$f(S + t \operatorname{diag} \Delta) = f(S) + t\sum_{r=1}^{d}(\operatorname{diag} \Delta^r)\frac{\partial f}{\partial x^r}(S) + o(t).$$

Hence on differentiating equation (10.57) at 0 we obtain

$$\frac{d}{dt}f(S(t))|_{t=0} = [Y, f(S)] + \sum_{r=1}^{d}(\operatorname{diag} \Delta^r)\frac{\partial f}{\partial x^r}(S).$$

We have shown the following.

272   *Commutative Theory*

Proposition 10.58. *Let S be a generic d-tuple of commuting self-adjoint matrices in $\mathbb{M}_n$. Fix an orthonormal basis of eigenvectors such that every S^r is diagonal:*

$$S^r = \begin{pmatrix} x_1^r & & \\ & \ddots & \\ & & x_n^r \end{pmatrix}.$$

Let Δ be a d-tuple of self-adjoints satisfying equation (10.52). Let f be C^1 on a neighborhood of $\{x_1, \dots, x_n\}$ in $\mathbb{R}^d$, where x_j is the d-tuple $(x_j^1, \dots, x_j^d)$. Then

$$\left[Df(S)[\Delta]\right]_{ij} = \begin{cases} \Delta_{ij}^r \frac{f(x_j) - f(x_i)}{x_j^r - x_i^r} & \text{if } i \neq j, \text{ where } x_j^r \neq x_i^r \\ \sum_{r=1}^d \Delta_{ii}^r \frac{\partial f}{\partial x^r}|_{x_i} & \text{if } i = j. \end{cases} \tag{10.59}$$

Corollary 10.60. *For S, Δ as in Proposition 10.58, if f, g are C^1 functions that agree to first order on $\sigma(S)$, then $Df(S)[\Delta] = Dg(S)[\Delta]$.*

The next lemma asserts the unsurprising fact that the spectrum of an element in $CSA\mathbb{M}_n^d$ is continuous.

Lemma 10.61. *Let R and S be in $CSA\mathbb{M}_n^d$. For every point μ in the joint spectrum of R there is an x_p in the joint spectrum of S with*

$$\|\mu - x_p\| \leq \sqrt{dn} \|R - S\|.$$

Proof. Choose an orthonormal basis that diagonalizes S, so that S is as in equation (10.48). Let $\Delta = R - S$. Let μ be a joint eigenvalue of R with corresponding eigenvector $\xi = (\xi_1, \dots, \xi_n)^t$. Choose p so that $|\xi_p| \geq |\xi_j|$ for $1 \leq j \leq n$.

Then, for $1 \leq r \leq d$, we have

$$R^r \xi = \mu^r \xi = (S^r + \Delta^r)\xi.$$

In particular,

$$\sum_{j=1}^n R_{pj}^r \xi_j = \mu^r \xi_p.$$

Therefore

$$(\mu^r - x_p^r)\xi_p = \sum_{j=1}^n \Delta_{pj}^r \xi_j.$$

Operator Monotone Functions   273

So

$$|\mu^r - x_p^r| \le \sum_{j=1}^{n} |\Delta_{pj}^r|$$

$$\le \sqrt{n} \sqrt{\sum_j |\Delta_{pj}^r|^2}$$

$$\le \sqrt{n} \|\Delta^r\|,$$

and hence

$$\sum_{r=1}^{d} |\mu^r - x_p^r|^2 \le dn \|\Delta\|^2.$$

$\square$

Lemma 10.62. *Let $R(t)$ be a Lipschitz path in $CSA\mathbb{M}_n^d$ for $0 \le t < 1$. Assume that $R(0) = S$ is generic. Then there exists $\varepsilon > 0$ and Lipschitz maps $X_1, \ldots, X_n \colon [0, \varepsilon) \to \mathbb{R}^d$ such that $\sigma(R(t)) = \{X_j(t) \colon j = 1, \ldots, n\}$.*

Proof. Choose an orthonormal basis that diagonalizes S, so that S is as in equation (10.48). The joint eigenvalues of S are the points $x_i = (x_i^1, \ldots, x_i^d)$, and genericity means $\|x_i - x_j\| > 0$ if $i \ne j$. Choose ε so that for all t such that $0 \le \varepsilon$,

$$\sqrt{dn} \|R(t) - S\| \le \frac{1}{3} \min_{i \ne j} \|x_i - x_j\|. \tag{10.63}$$

By Lemma 10.61, for every joint eigenvalue x of S there is a joint eigenvalue μ of $R(t)$ within $\sqrt{dn} \|R(t) - S\|$ of it. By inequality (10.63), this means that $R(t)$ is also generic, and it makes sense to talk of the joint eigenvalue of $R(t)$ that is closest to x_j. Let us call these joint eigenvalues $X_j(t)$. We have proved that

$$\|X_j(t) - x_j\| \le \sqrt{dn} \|R(t) - S\| \qquad \text{for } 0 \le t \le \varepsilon.$$

Repeating the argument with $R(t_1)$ in place of S, we find

$$\|X_j(t_2) - X_j(t_1)\| \le \sqrt{dn} \|R(t_2) - R(t_1)\| \qquad \text{for } 0 \le t_1, t_2 \le \varepsilon.$$

As R is assumed to be Lipschitz, each X_j is Lipschitz also. $\square$

Proposition 10.64. *If S is generic in $CSA\mathbb{M}_n^d$, Δ is in $SA\mathbb{M}_d^n$, and they satisfy the commutation relations (10.52), then for any C^1 path $R(t) \in CSA\mathbb{M}_n^d$ such that $R(0) = S$, $R'(0) = \Delta$ and any $f \in C^1$,*

$$\frac{d}{dt} f(R(t))|_{t=0} = Df(S)[\Delta]. \tag{10.65}$$

274 *Commutative Theory*

Proof. If g is a monomial then a simple calculation shows that

$$\frac{d}{dt}g(R(t))|_{t=0}$$

exists and depends only on g, S, and Δ. It follows that, for any polynomial g,

$$\frac{d}{dt}g(R(t))|_{t=0} = \frac{d}{dt}g(S(t))|_{t=0} = Dg(S)[\Delta]. \qquad (10.66)$$

Consider any $f \in C^1$ and pick a polynomial g that agrees with f to first order on $\sigma(S)$. By Corollary 10.60,

$$Df(S)[\Delta] = Dg(S)[\Delta]. \qquad (10.67)$$

We claim that

$$\frac{d}{dt}g(R(t))|_{t=0} = \frac{d}{dt}f(R(t))|_{t=0}.$$

For by Lemma 10.62, there exist Lipschitz functions $X_1, \ldots, X_n \colon [0, \varepsilon) \to \mathbb{R}^d$ such that $\sigma(R(t)) = \{X_1(t), \ldots, X_n(t)\}$ for all t. Then $f(S) = g(S)$ and

$$\|(f-g)(R(t))\| = \max_i |(f-g)(X_i(t))|$$

$$= o(\max_i \|X_i(t) - X_i(0)\|)$$

$$= o(t).$$

Hence

$$\left\| \frac{f(R(t)) - f(S)}{t} - \frac{g(R(t)) - g(S)}{t} \right\| \to 0 \text{ as } t \to 0.$$

In view of equation (10.66),

$$\frac{f(R(t)) - f(S)}{t} \to \frac{d}{dt}g(R(t))|_{t=0} = Dg(S)[\Delta] \text{ as } t \to 0.$$

On combining this relation with equation (10.67) we obtain equation (10.65). $\square$

Corollary 10.68. *A real-valued C^1 function f on an open set $E \subseteq \mathbb{R}^d$ is locally $\mathbb{M}_n$-monotone if and only if*

$$Df(S)[\Delta] \geq 0$$

for every generic S in $CSAM_n^d$ with spectrum in E and every Δ in SAM_n^d such that $\Delta \geq 0$ and

$$[S^r, \Delta^s] = [S^s, \Delta^r] \qquad \text{for } 1 \leq r \neq s \leq d.$$

The statement follows immediately from the Definition 10.50 and Proposition 10.64.

We define the Löwner classes in d variables, $\mathcal{L}_n^d(E)$.

Operator Monotone Functions 275

Definition 10.69. *Let E be an open subset of $\mathbb{R}^d$. The set $\mathcal{L}_n^d(E)$ consists of all real-valued C^1-functions on E that have the following property: whenever $\{x_1,\ldots,x_n\}$ are n distinct points in E, there exist positive semi-definite $n \times n$ matrices $A^1,\ldots,A^d$ such that*

$$f(x_j) - f(x_i) = \sum_{r=1}^{d}(x_j^r - x_i^r)A^r(i,j) \qquad \text{for } 1 \le i,j \le n \qquad (10.70)$$

$$\text{and} \quad A^r(i,i) = \frac{\partial f}{\partial x^r}\bigg|_{x_i} \qquad \text{for } 1 \le i \le n,\ 1 \le r \le d. \qquad (10.71)$$

Theorem 10.72. *Let E be an open set in $\mathbb{R}^d$ and f a real-valued C^1 function on E. Then f is locally $\mathbb{M}_n$-monotone if and only if f is in $\mathcal{L}_n^d(E)$.*

Proof. ($\Leftarrow$) We must show: if S is generic with $\sigma(S) \subset E$, if Δ is a positive d-tuple and $[S^r, \Delta^s] = [S^s, \Delta^r]$ for all r,s, then $Df(S)[\Delta] \ge 0$.

Let $\sigma(S) = \{x_1,\ldots,x_n\}$. Choose A^r as in Definition 10.69. For $i \neq j$, assume without loss of generality that $x_j^1 \neq x_i^1$. Then

$$[Df(S)[\Delta]]_{ij} = \Delta_{ij}^1 \frac{f(x_j) - f(x_i)}{x_j^1 - x_i^1}$$

$$= \frac{\Delta_{ij}^1}{x_j^1 - x_i^1}\left(\sum_{r=1}^{d}(x_j^r - x_i^r)A^r(i,j)\right)$$

$$= \sum_{r=1}^{d} \Delta_{ij}^r A^r(i,j).$$

(We get the last line by using equation (10.53)). By equation (10.59) the same formula holds for $[Df(S)[\Delta]]_{ij}$ when $i = j$, so $Df(S)[\Delta]$ is the sum of the Schur products of Δ^r with A^r, so is positive.

($\Rightarrow$) Let f be locally $\mathbb{M}_n$-monotone, and fix distinct points $\{x_1,\ldots,x_n\}$ in E. Let S be given by equation (10.48). We wish to find positive matrices A^r such that equations (10.70) and (10.71) hold.

Let $\mathcal{G}$ be the set of all skew-symmetric real $n \times n$ matrices Γ with the property that there exists a d-tuple A of real positive semi-definite matrices satisfying

$$A^r(i,i) = \frac{\partial f}{\partial x^r}\bigg|_{x_i} \qquad \text{for } 1 \le i \le n,\ 1 \le r \le d, \qquad (10.74)$$

$$\sum_{r=1}^{d}(x_j^r - x_i^r)A^r(i,j) = \Gamma_{ij} \qquad \text{for } 1 \le i \neq j \le n. \qquad (10.76)$$

Let Λ be the matrix $\Lambda_{ij} = f(x_j) - f(x_i)$. We wish to show Λ is in $\mathcal{G}$.

276   *Commutative Theory*

Notice that $\mathcal{G}$ is a closed convex set. Moreover, it is non-empty, because $\frac{\partial f}{\partial x^r}\big|_{x_i}$ is always greater than or equal to 0, so that $\mathcal{G}$ contains 0. Without loss of generality, we shall assume that

$$f_{r,i} \stackrel{\text{def}}{=} \frac{\partial f}{\partial x^r}\bigg|_{x_i} > 0, \qquad \text{for } 1 \le r \le d,\, 1 \le i \le n.$$

(This can be done by the addition of $\varepsilon(x^1 + \cdots + x^d)$ to f, together with the choice of $\varepsilon \to 0$ at the end of the argument.)

If Λ is not in $\mathcal{G}$, by the Hahn–Banach theorem there is a real linear functional $L: \mathbb{M}_n \to \mathbb{R}$ and some $\delta \ge 0$ such that $L(\Gamma) \ge -\delta$ for all Γ in $\mathcal{G}$, and $L(\Lambda) < -\delta$. So we can assume that there is a real skew-symmetric matrix K such that $\operatorname{tr}(\Gamma K) \ge -\delta$ for all Γ in $\mathcal{G}$, and $\operatorname{tr}(\Lambda K) < -\delta$.

Define Δ by

$$\Delta_{ij}^r = (x_j^r - x_i^r)K_{ji} \quad \text{for } i \ne j,$$

and with the diagonal entries Δ_{ii}^r chosen such that each $\Delta^r \ge 0$ and such that

$$\mu^r \stackrel{\text{def}}{=} \sum_{i=1}^{n} f_{r,i}\Delta_{ii}^r \tag{10.77}$$

is minimal over all choices of $\Delta_{11}^r, \ldots, \Delta_{nn}^r$ subject to $\Delta^r \ge 0$ for each r. A minimal choice exists, since all the $f_{r,i}$ are strictly positive by assumption.

We can rewrite the choice of the Δ_{ii}^r in the language of Theorem 10.40 by saying we want to maximize $-\mu^r = \sum_{i=1}^n f_{r,i}\Delta_{ii}^r$. We have feasible positive definite candidates, since we can choose Δ^r with arbitrarily large diagonal entries. The value of the program is finite, since it is bounded by 0, so we can use Theorem 10.40 to conclude that

$$-\mu^r = \min \sum_{i \ne j} \Delta_{ij} A^r(i,j), \tag{10.78}$$

where A^r range over the set of real positive matrices such that the diagonal entries of A^r are $f_{r,1}, \ldots, f_{r,n}$.

With Δ chosen this way, Δ is in $S A \mathbb{M}_n$, and

$$[\Delta^s, S^r]_{ij} = (x_j^s - x_i^s)K_{ji}(x_j^r - x_i^r) = [\Delta^r, S^s]_{ij},$$

so that Δ satisfies equation (10.52). As f is locally $\mathbb{M}_n$-monotone, $Df(S)[\Delta] \ge 0$ by Lemma 10.51.

Operator Monotone Functions 277

By assumption,

$$
\begin{aligned}
-\delta > \mathrm{tr}(\Lambda K) &= \sum_{1 \le i \ne j \le n} [f(x_j) - f(x_i)]K_{ji} \\
&= \sum_{1 \le i \ne j \le n} \frac{f(x_j) - f(x_i)}{x_j^r - x_i^r} \Delta_{ij}^r \qquad (10.79) \\
&= \sum_{1 \le i,j \le n} [Df(S)[\Delta]]_{ij} - \sum_{r=1}^{d}\sum_{i=1}^{n} \Delta_{ii}^r f_{r,i},
\end{aligned}
$$

where in equations (10.79), for each (i,j) we choose a (perhaps different) r such that the denominator is non-zero. Therefore

$$
\sum_{r=1}^{d} \mu^r - \delta > \sum_{1 \le i,j \le n} [Df(S)[\Delta]]_{ij} \ge 0. \qquad (10.81)
$$

Thus, if we can prove that

$$
\sum_{r=1}^{d} \mu^r \le \delta, \qquad (10.82)
$$

we shall derive a contradiction.

Let $A = (A^1, \dots, A^d)$ have the diagonal entries of each A^r equal to $f_{r,1}, \dots, f_{r,n}$ for each r. Let Γ be the corresponding element of $\mathcal{G}$: $\Gamma_{ii} = 0$ and

$$
\Gamma_{ij} = \sum_{r=1}^{d} (x_j^r - x_i^r) A^r(i,j) \quad \text{for } i \ne j.
$$

We have

$$
\begin{aligned}
-\delta \le \mathrm{tr}\, \Gamma K \\
&= \sum_{i \ne j}\sum_{r=1}^{d} (x_j^r - x_i^r) A^r(i,j) K_{ji} \\
&= \sum_{r=1}^{d}\sum_{i \ne j} \Delta_{ij}^r A^r(i,j) \\
&\le \sum_{r=1}^{d} (-\mu^r),
\end{aligned}
$$

where the last inequality is from equation (10.78). Therefore $\sum_{r=1}^{d} \mu^r \le \delta$. This contradicts inequality (10.81), and so it follows that $\Lambda \in \mathcal{G}$. Necessity is proved. $\qquad \square$

278 *Commutative Theory*

10.4 The Löwner Class in d Variables

Throughout this section, E will be an open set in $\mathbb{R}^d$. In Definition 10.69 we defined the Löwner class $\mathcal{L}_n^d(E)$ for each n. We would like to relate $\mathcal{L}_n^d(E)$ to some class of functions that maps $\mathbb{H}^d$ to $\mathbb{H}$, as happens in the $d = 1$ case of Theorem 10.9.

Let $\alpha \colon \mathbb{D} \to \mathbb{H}$ and $\beta \colon \mathbb{H} \to \mathbb{D}$ be the Cayley transform and its inverse, defined by equations (10.22). Let us define the Löwner class to be the set of Cayley transforms of the class $\mathcal{F}(d)$ defined in Section 9.7,

$$\mathcal{L}^d = \{\alpha \circ \varphi \circ \beta \colon \varphi \in \mathcal{F}(d)\}.$$

Equivalently, after a change of variables in equation (9.99), $\mathcal{L}^d$ is the set of functions f defined on $\mathbb{H}^d$ that have a model of the form

$$f(z) - \overline{f(w)} = \sum_{r=1}^{d} \langle (z^r - \bar{w}^r)\, v_z^r, v_w^r \rangle_{\mathcal{M}^r} \tag{10.83}$$

where $v^r \colon \mathbb{H}^d \to \mathcal{M}^r$ is a map into a Hilbert space. Since $\mathscr{S}(\mathcal{F}(d)) = \mathscr{S}(\mathbb{D}^d))$ for $d = 1$ and 2, and is a proper subset when $d \geq 3$, the set $\mathcal{L}^d$ coincides with the set $\mathcal{P}(\mathbb{H}^d)$ defined in Section 6.1 for $d = 1$ and 2, but for $d \geq 3$, we have $\mathcal{L}^d \subsetneq \mathcal{P}(\mathbb{H}^d)$.

Definition 10.84. *The Löwner class of E, denoted $\mathcal{L}(E)$, is the set of functions $f \colon E \to \mathbb{R}$ which have the property that there is a function $F \in \mathcal{L}^d$ which extends analytically across E and such that*

$$\lim_{y \downarrow 0} F(x^1 + iy, \dots, x^d + iy) = f(x^1, \dots, x^d) \quad \text{for all } x \in E.$$

Another possible definition for the Löwner class associated to E is:

Definition 10.85. *The class $\mathcal{L}_\partial(E)$ is the set of C^1 functions $f \colon E \to \mathbb{R}$ such that there are positive semi-definite functions $A^1, \dots, A^d \colon E \times E \to \mathbb{R}$ that satisfy*

$$A^r(z,z) = \left. \frac{\partial f}{\partial x^r} \right|_z$$

and

$$f(z) - f(w) = \sum_{r=1}^{d} (z^r - w^r) A^r(z,w) \quad \text{for all } z, w \in E.$$

Fortunately, all these classes coincide.

Theorem 10.86. *Let E be an open non-empty set in $\mathbb{R}^d$. The following classes all coincide.*

Operator Monotone Functions 279

(i) $\bigcap_{n=1}^{\infty} \mathcal{L}_n^d(E)$.

(ii) $\mathcal{L}_\partial(E)$.

(iii) $\mathcal{L}(E)$.

(iv) *The set of functions that are locally $\mathbb{M}_n$-monotone on E for every n.*

(v) *The set of functions that are locally operator monotone on E.*

We have already shown that (i) and (iv) are equal. To show that (i) is the same as (ii), one must show that if there are suitable matrices A^r on every finite subset of E, then there is a single choice of the A^r that works on all of E. This can be done by some form of transfinite induction.

To prove (ii) $\subseteq$ (iii), use a lurking isometry argument to get a model, and then use the model to get v_z (or equivalently, the kernel $A^r(z,w)$ on $\mathbb{H}^d$). Then one argues that the function just constructed in $\mathbb{H}^d$ does indeed analytically continue to E.

Clearly (v) $\subseteq$ (iv), so it remains to show that (iii) $\subseteq$ (v). Do this by writing down a model for the function, substituting in a path $S(t)$ of commuting d-tuples, differentiating, and showing that the result is of the form $v^* S'(0)v$, so manifestly positive if $S'(0)$ is.

For the details of the proof of Theorem 10.86, see [24].

10.5 Globally Monotone Rational Functions in Two Variables

It is an open question whether locally monotone functions in d variables are the same as globally monotone ones. However, the answer is yes for rational functions in two variables.

Theorem 10.87. *Let F be a rational function of two variables, and let Γ be the zero set of the denominator of F. Assume that F is real-valued on $\mathbb{R}^2 \setminus \Gamma$. Let E be an open rectangle in $\mathbb{R}^2 \setminus \Gamma$. Then F is globally operator monotone on E if and only if f is in $\mathcal{L}(E)$.*

Proof. Suppose F is of degree $n = (n_1, n_2)$. Let $\varphi = \beta \circ F \circ \alpha$, a rational inner function. This has a network realization, by Theorem 4.44. More holds: since we are in two variables, by a theorem of A. Kummert [138] (see also [44, 132]), one can find a realization as in Theorem 4.44 with model space $\mathcal{M} = \mathcal{M}^1 \oplus \mathcal{M}^2$, where $\mathcal{M}^r$ has dimension equal to n_r, and a unitary operator V on $\mathbb{C} \oplus \mathcal{M}$ satisfying equation (4.48) for all $\lambda \in E$:

$$V \begin{pmatrix} 1 \\ \lambda u_\lambda \end{pmatrix} = \begin{pmatrix} \varphi(\lambda) \\ u_\lambda \end{pmatrix}.$$

280   *Commutative Theory*

Let us now make two assumptions:

$$\text{Assumption:} \quad 1 \notin \sigma(V) \quad \text{and} \quad \varphi(1,1) \neq 1. \tag{10.88}$$

Under these assumptions, define $A = -i(V-1)^{-1}(V+1)$. This operator is self-adjoint, and some algebra shows that

$$A\begin{pmatrix} 1 \\ v_z \end{pmatrix} = \begin{pmatrix} F(z) \\ -zv_z \end{pmatrix}, \tag{10.89}$$

where

$$v_z = \frac{1 - \beta(z)}{\beta \circ F(z) - 1} \, u_{\beta(z)}.$$

Let $X = -P_{\mathcal{M}} A P_{\mathcal{M}}$ be the compression of A to $\mathcal{M}$. Then equation (10.89) can be unwound to yield the fact that there is some $c \in \mathbb{R}$ and a vector $\gamma \in \mathcal{M}$ such that F has a representation of the form

$$F(z) = c + \langle (X-z)^{-1}\gamma, \gamma \rangle. \tag{10.90}$$

We are using z of course to represent $z^1 P_{\mathcal{M}^1} + z^2 P_{\mathcal{M}^2}$. In the language of Chapter 6, we are saying that F has a representation of type 2, and even in the subclass of type 2 in which Y_1, Y_2 are projections. If $S = (S^1, S^2)$ is a pair of commuting self-adjoints on some Hilbert space $\mathcal{H}$, then one can use equation (10.90) to define a functional calculus:

$$\langle F(S)\xi, \eta \rangle_{\mathcal{H}} = c \, \langle \xi, \eta \rangle + \left\langle \left(\begin{smallmatrix} X \\ \otimes \\ \mathrm{id}_{\mathcal{H}} \end{smallmatrix} - \begin{smallmatrix} P_{\mathcal{M}^1} \\ \otimes \\ S^1 \end{smallmatrix} - \begin{smallmatrix} P_{\mathcal{M}^2} \\ \otimes \\ S^2 \end{smallmatrix} \right)^{-1} \begin{smallmatrix} \gamma \\ \otimes \\ \xi \end{smallmatrix}, \begin{smallmatrix} \gamma \\ \otimes \\ \eta \end{smallmatrix} \right\rangle_{\mathcal{M} \otimes \mathcal{H}},$$
$$\tag{10.91}$$

provided S is such that the expression in parentheses on the right is invertible, and where we write our tensors vertically to enhance legibility. More compactly, if we write

$$S_P := \begin{smallmatrix} P_{\mathcal{M}^1} \\ \otimes \\ S^1 \end{smallmatrix} + \begin{smallmatrix} P_{\mathcal{M}^2} \\ \otimes \\ S^2 \end{smallmatrix}$$

and

$$R_\gamma \colon \mathcal{H} \to \mathcal{M} \otimes \mathcal{H}$$
$$\xi \mapsto \begin{smallmatrix} \gamma \\ \otimes \\ \xi \end{smallmatrix},$$

then equation (10.91) becomes

$$F(S) = c \, \mathrm{id}_{\mathcal{H}} + R_\gamma^* \left(\begin{smallmatrix} X \\ \otimes \\ \mathrm{id}_{\mathcal{H}} \end{smallmatrix} - S_P \right)^{-1} R_\gamma. \tag{10.92}$$

Operator Monotone Functions 281

From equation (10.90), we see that the poles of F all lie in the set

$$\{z: \det(X - z) = 0\}. \tag{10.93}$$

Because of the dimension of the spaces $\mathcal{M}^r$, the latter set is algebraic of degree at most (n_1, n_2). Since F has degree n, this means that the polar set of F is algebraic of degree n, so in fact equation (10.93) equals the polar set of F and hence is disjoint from the rectangle E. This allows us to go from local to global.

By rescaling, we can assume that E is the square $(-1, 1) \times (-1, 1)$. Since $(X - z)$ is invertible for all z of the form (s, s), where $-1 < s < 1$, we can conclude that $\|Xm\| \geq \|m\|$ for every $m \in \mathcal{M}$.

Assume that S and T are pairs of commuting self-adjoint matrices with joint spectrum in E and satisfy $S \leq T$. Define $S(t) = (1-t)S + tT$. Then, for $0 \leq t \leq 1$, the pair $S(t) = (S^1(t), S^2(t))$ is a pair of self-adjoint matrices *each of which has norm less than* 1. They may commute only at the end points, but it *is* true that $\underset{\mathrm{id}_{\mathcal{H}}}{\overset{X}{\otimes}} - S_P(t)$ is invertible for $0 \leq t \leq 1$. Therefore

$$F(T) - F(S) = R_\gamma^* \left(\underset{\mathrm{id}_{\mathcal{H}}}{\overset{X}{\otimes}} - T_P \right)^{-1} R_\gamma - R_\gamma^* \left(\underset{\mathrm{id}_{\mathcal{H}}}{\overset{X}{\otimes}} - S_P \right)^{-1} R_\gamma$$

$$= \int_0^1 \frac{d}{dt} R_\gamma^* \left(\underset{\mathrm{id}_{\mathcal{H}}}{\overset{X}{\otimes}} - S(t)_P \right)^{-1} R_\gamma \, dt. \tag{10.94}$$

Since $S(t)$ is increasing, so is $\left(\underset{\mathrm{id}_{\mathcal{H}}}{\overset{X}{\otimes}} - S(t)_P \right)^{-1}$, and so the integrand in equation (10.94) is positive, and hence $F(T) \geq F(S)$. $\square$

Remark 10.95. We skipped a few details in the proof. If Assumption 10.88 does not hold, let

$$\rho_t(z) = \frac{z+t}{1-tz}.$$

Replace F by $F_t = \rho_t \circ F \circ \rho_t$. Then for all but finitely many values of t, Assumption 10.88 does hold for F_t. Prove the theorem for these F_t and take a limit.

We also asserted that equation (10.91) follows from equation (10.90). For a proof of this, and explanation of why the argument using F_t works, see [24].

Remark 10.96. The key point in the proof is that the equation (10.92) makes sense only for a pair of *commuting* self-adjoints, since otherwise the left-hand side is not defined. However the right-hand side makes sense even when S^1 and S^2 don't commute, so can be used to *define* F on such pairs.

Remark 10.97. It is unknown if Theorem 10.87 holds in three or more variables. The difficulty with the preceding proof is that one can no longer

282   *Commutative Theory*

necessarily get a model on a space $\mathcal{M}$ whose dimension exactly matches the degree of F [132], so we don't know if $\left(\underset{\mathrm{id}_{\mathcal{H}}}{\overset{X}{\otimes}} - S_P(t) \right)$ must be invertible along the whole arc.

10.6 Historical Notes

Theorems 10.5 and 10.9 were proved by K. Löwner in [145]. See the books [76, 173]. Lemma 10.13 can be strengthened to the following theorem of Rellich [128, theorem 6.8].

Theorem 10.98. (*Rellich*) *Assume $S(t)$ is a C^1 path in $SA\mathbb{M}_n$ for t in an open interval $E \subseteq \mathbb{R}$. Then there are n C^1 functions $\lambda_j(\cdot)$ on E such that, for every $t \in E$, the eigenvalues of $S(t)$ are $\lambda_1(t),\ldots,\lambda_n(t)$.*

Another proof of Lemma 10.16 (the first part of which is our proof of Lemma 10.15) is in [76, theorem VIII.I] .

Theorem 10.40 is a special case of Duffin's strong duality theorem, which applies to ordered topological vector spaces [82]. The treatment in Section 10.2 is based on [94].

The non-generic version of differentiating along curves was analyzed by K. Bickel [51]. Among other results, she proved the following. Let S be in $CSA\mathbb{M}_n^d$. Let U be a unitary matrix such that $U^*SU = D$ is a diagonal tuple, with diagonal entries $x_j = (x_j^1,\ldots,x_j^d)$, for $1 \le j \le n$.

Let Δ be in $SA\mathbb{M}_n$. Let $\Gamma^r = U^*\Delta^r U$. Define $\tilde{\Gamma}_{ij}^r$ to equal Γ_{ij}^r if $x_i = x_j$, and to be zero otherwise.

Theorem 10.99. *There exists a C^1 curve $S(t)$ in $CSA\mathbb{M}_n^d$ with $S(0) = S$ and $S'(0) = \Delta$ if and only if*

$$[D^r,\Gamma^s] = [D^s,\Gamma^r] \quad \text{and} \quad [\tilde{\Gamma}^r,\tilde{\Gamma}^s] = 0 \qquad \text{for } 1 \le r,s \le d.$$

Theorem 10.100. *Let $S(t)$ be a C^1 curve in $CSA\mathbb{M}_n^d$ defined on an open interval I, and let E be an open set in $\mathbb{R}^d$ containing $\sigma(S(t))$ for $t \in I$. If $f \in C^1(E,\mathbb{R})$, then*

(i) $\frac{d}{dt}f(S(t))|_{t_0}$ *exists for every t_0 in I and is continuous in t_0.*
(ii) *If $T(t)$, $t \in I$ is another C^1 curve in $CSA\mathbb{M}_n^d$, satisfying $T(t_0) = S(t_0)$ and $T'(t_0) = S'(t_0)$, then*

$$\frac{d}{dt}f(T(t))|_{t_0} = \frac{d}{dt}f(S(t))|_{t_0}.$$

PART TWO

Non-Commutative Theory

11

Motivation for Non-Commutative Functions

The philosophy of Part I is that, to study functions, it pays to think of their actions on operators. This helps us both to understand functions better and to understand operators better. An important, but less explicit, moral of Part I is that the way in which a function is presented can determine how easy it is to understand what happens when the function is evaluated on operators.

If f is a function of a single variable and T is a single operator and if there is more than one way to make sense of $f(T)$, then normally all definitions will agree. In several variables, a new difficulty emerges, even for polynomials. Suppose $p \in \mathbb{C}[z, w]$ is given by

$$p(z, w) = zw = wz.$$

If now S and T are a pair of operators on some space $\mathcal{H}$, what should $p(S, T)$ mean? It could reasonably be ST, TS, or $\frac{1}{2}(ST + TS)$.[1] One solution of course is to restrict yourself to operators that commute. But there are many reasons why you might want to work with functions of operators that don't commute. Let us start with the simplest functions that we can apply to non-commuting operators, the non-commutative (nc) polynomials.

11.1 Non-Commutative Polynomials

A non-commutative polynomial, also called a free polynomial, is an expression like

$$p(z, w) = 2 - z + 2w + 2wz - 3zw + zzw + zwz + 5wzz.$$

Formally, a non-commutative polynomial in d variables is an element of the semigroup algebra over the free monoid with d generators; by default, our

---

[1] You could make the case for $\sqrt{2}(ST) + (1 - \sqrt{2})TS$, but you are unlikely to have many supporters.

285

286   *Non-Commutative Theory*

coefficients will be complex numbers. This means you choose d generators, $z^1, \ldots, z^d$, and then form words of length $\ell \in \mathbb{N} \cup \{0\}$ in these generators by writing a string of ℓ generators; these form the monomials (the choice $\ell = 0$ corresponding to the constant monomial 1). The semi-group algebra then comprises all finite linear combinations of monomials. You add and multiply in the obvious way (the product of two monomials is the concatenation of their strings), and this gives an algebra over $\mathbb{C}$ (or whatever ground field you chose for your coefficients). We shall let $\mathcal{P}_d$ denote the algebra of all non-commutative polynomials in d variables with complex coefficients.

Given a non-commutative polynomial,[2] one can evaluate it on any d-tuple of elements from any algebra. We shall primarily be concerned with matrices and operators, but we want to allow the matrices to have arbitrary dimension.

Definition 11.1. *Let* $\mathbb{M}_n$ *denote the set of* $n \times n$ *matrices with complex coefficients, equipped with the operator norm as operators on* $\mathbb{C}^n$.

Then $\mathbb{M}_n^d$ consists of d-tuples of $n \times n$ matrices, and we let

$$\mathbb{M}^d \stackrel{\text{def}}{=} \bigcup_{n=1}^{\infty} \mathbb{M}_n^d.$$

The set $\mathbb{M}^d$ will play the rôle that $\mathbb{C}^d$ does for holomorphic functions. For $x \in \mathbb{M}^d$, we define

$$\|x\| \stackrel{\text{def}}{=} \max_{1 \leq j \leq d} \|x^j\|.$$

We shall let SAM^d denote the set of self-adjoint d-tuples:

$$\mathrm{SAM}^d = \{z \in \mathbb{M}^d \colon z^j = (z^j)^* \text{ for } 1 \leq j \leq d\},$$

and CM^d the commuting ones:

$$\mathrm{CM}^d = \{z \in \mathbb{M}^d \colon z^i z^j = z^j z^i \text{ for } 1 \leq i, j \leq d\}.$$

We shall say that the *level* of $x \in \mathbb{M}^d$ is n if $x \in \mathbb{M}_n^d$, and say that two points $x, y \in \mathbb{M}^d$ are on the same level if they are both in some $\mathbb{M}_n^d$. There are a couple of natural operations on $\mathbb{M}^d$ that we shall use repeatedly. If $x \in \mathbb{M}_n^d$ and $y \in \mathbb{M}_m^d$, then $x \oplus y$ is the element of $\mathbb{M}_{n+m}^d$ obtained by taking the direct sums of the matrices:

$$x \oplus y = \left(\begin{pmatrix} x^1 & 0 \\ 0 & y^1 \end{pmatrix}, \ldots, \begin{pmatrix} x^d & 0 \\ 0 & y^d \end{pmatrix} \right).$$

If $x \in \mathbb{M}_n^d$ and s is in $\mathbb{M}_n$, then we let sx denote the d-tuple $(sx^1, \ldots, sx^d)$, and similarly xs means the d-tuple where each entry is multiplied on the right by s.

---

[2] We shall always assume d is the number of variables.

Motivation for Non-Commutative Functions

11.2 Sums of Squares

Hilbert's 17th problem concerns real polynomials p in d variables that are non-negative on $\mathbb{R}^d$. A *certificate* that $p \geq 0$ is a collection of functions $q_1, \ldots, q_n$ such that

$$p(x) = \sum_{j=1}^{n} q_j(x)^2. \tag{11.2}$$

Hilbert knew that for $d \geq 2$, one cannot always find polynomials q_j that satisfy an identity of the form (11.2); he asked whether there exist *rational* functions q_j with this property. Artin proved in 1927 that there do.

Now, suppose we are working with non-commutative polynomials in $2d$ variables, which we label $x^1, \ldots, x^d, x^{1*}, \ldots, x^{d*}$. We can define a $*$-operation on monomials in these variables by reversing the order of the terms and replacing each x^j with x^{j*} and each x^{j*} with x^j; we can then extend this $*$-operation to polynomials by defining $(am_1 + bm_2)^* = \bar{a}m_1^* + \bar{b}m_2^*$ for $a, b \in \mathbb{C}$ and monomials m_1, m_2. Given such a polynomial p and a d-tuple λ in $\mathbb{M}^d$, we define

$$p(\lambda) \overset{\text{def}}{=} p(\lambda^1, \ldots, \lambda^d, (\lambda^1)^*, \ldots, (\lambda^d)^*).$$

When does it happen that $p(\lambda) \geq 0$ for every $\lambda \in \mathbb{M}^d$? It will if p has a sum-of-squares representation, that is, if there exist non-commutative polynomials q_j in the $2d$ variables such that

$$p(\lambda) = \sum_{j=1}^{n} q_j(\lambda)^* q_j(\lambda). \tag{11.3}$$

In [102], Helton proved the remarkable converse statement: $p(\lambda) \geq 0$ for every $\lambda \in \mathbb{M}^d$ if and only if p has a representation of the form (11.3). In contrast to the scalar case, one does not need rational functions. Helton's paper addressed the case that all the matrices had real entries and the coefficients were real. The same theorem holds in the complex case, as was proved by McCullough and Putinar [150].

Helton's work has inspired many developments. See [111] for a survey up to 2007 and [105] for a more recent overview.

11.3 Nullstellensatz

The classical Nullstellensatz says that if one starts with an ideal $I \lhd \mathbb{C}[z_1, \ldots, z_d]$, passes to the variety $V(I)$ consisting of the set of points in $\mathbb{C}^d$ at which all the functions in I vanish and then considers those polynomials

288 *Non-Commutative Theory*

that vanish on $V(I)$, one recovers the radical of I. To recover I exactly, one needs to look at the set of commuting matrix d-tuples on which all functions in I vanish and then consider the polynomials that annihilate those matrices. This gives a complete correspondence between varieties and ideals [88].

There are various non-commutative versions of the Nullstellensatz. See, for example, [170, 110, 59, 178].

11.4 Linear Matrix Inequalities

A linear matrix inequality, or LMI, is a problem of the following form. Given $\lambda \in \mathrm{SAM}^d$, find $x \in \mathbb{R}^{d-1}$ such that

$$\sum_{j=1}^{d-1} x^j \lambda^j + \lambda^d \geq 0, \tag{11.4}$$

or show no such x exists. Many engineering problems can be reduced to LMIs, and there are efficient algorithms for solving them [54].

Sometimes, the stability of a system can be reduced to the question of whether a non-linear matrix-valued function F can be made positive for some choice of argument. A question that has attracted a lot of attention is to determine when such a problem can be reduced to an LMI—see, for example, [112, 109, 106, 107, 108].

As a simple example, consider the Riccati inequality. Given $(a,b,c) \in \mathrm{M}_n^3$, does there exist $x \in \mathrm{SAM}_n$ such that

$$F(x) \stackrel{\text{def}}{=} ax + xa^* - xbb^*x + c^*c > 0? \tag{11.5}$$

The function F is not linear in x, but inequality (11.5) is equivalent to the LMI

$$\begin{pmatrix} ax + xa^* + c^*c & xb \\ b^*x & 1 \end{pmatrix} > 0. \tag{11.6}$$

This is an LMI (good!) but if we treat it as being of the form (11.4), it is in $\binom{d}{2}$ variables (bad!) If we naively used Sylvester's criterion to check the inequality (11.6), we would have $2d$ polynomial inequalities, of degree up to $2d$, in these variables. This is clearly not a good approach, as it does not take full advantage of the matrix structure of the problem. One motivation for non-commutative functions is to study objects like the relation (11.5) in terms of functions of matrices, in a dimension-independent way.

Motivation for Non-Commutative Functions 289

11.5 The Implicit Function Theorem

Consider the following equation for $(x,y) \in \mathbb{M}^2$.

$$p(x,y) \stackrel{\text{def}}{=} x^3 + 2xy - 4yx + 5x - 6y + 7 = 0. \tag{11.7}$$

For every $x \in \mathbb{M}^1$, the equation can be solved; we shall see in Section 15.5 that, for most choices of x, all the solutions y must commute with x. We will provide details later, but the outline of the proof is as follows. Suppose one wants to apply an implicit function theorem and write y locally as a function of x. The condition one needs is that the derivative of p in the direction y have full rank. At a particular point (x,y),

$$\frac{\partial p}{\partial y}[h] = Dp(x,y)[(0,h)]$$

$$= \lim_{t \to 0} \frac{p(x,y+th) - p(x,y)}{t}$$

$$= 2xh - 4hx - 6h. \tag{11.8}$$

It is easy to check when the right-hand side of equation (11.8) has full rank. By a theorem of Sylvester [192], the map

$$\mathbb{M}^1 \ni h \mapsto ah - hb$$

has full rank if and only if a and b have no common eigenvalues. So the right-hand side of equation (11.8) has full rank when $2x$ and $4x+6$ have no common eigenvalues, which is generically true. So we conclude that if (x,y) solves equation (11.7) and $\sigma(x) \cap \sigma(2x+3) = \emptyset$, then $xy = yx$, and therefore

$$y = (x^3 + 5x + 7)(2x+6)^{-1}.$$

This is an algebraic result, but the proof uses analysis, and moreover uses functions that are potentially more general than polynomials or rational functions.

11.6 Matrix Monotone Functions

We saw in Chapter 10 that there is a problem in replacing "local" with "global" in Theorem 10.86. If one wants to imitate Lemma 10.21, one must evaluate the function f on d-tuples of matrices that do not commute (except at the two end points, $S(0)$ and $S(1)$). Theorem 10.87 did this in a special case. To do it in general, we need a better understanding of functions of non-commuting matrix variables.

290   *Non-Commutative Theory*

11.7 The Functional Calculus

Much of the first part of the book is devoted to the idea that, for any space of holomorphic functions on a domain U in $\mathbb{C}^d$, it is fruitful to substitute d-tuples of commuting operators in as arguments of functions in the space. Non-commutative functions (which we will define in Chapter 12) are an analog to holomorphic functions when the variables do not commute. A priori we define them on d-tuples of matrices, but it makes sense to try to evaluate them on d-tuples of operators or, indeed, d-tuples of elements from a Banach algebra.

In Chapter 14 we shall develop a model formula and a network realization formula for non-commutative functions that are bounded on free polynomial polyhedra—we call these free holomorphic functions. In Chapter 16 we shall extend this network realization formula to d-tuples of operators on infinite-dimensional spaces and describe which functions arise this way. Finally, in Section 16.4, we sketch how we can use free functions to develop a functional calculus for d-tuples in an arbitrary Banach algebra.

11.8 Historical Notes

The original idea of a non-commutative function, or a function of noncommuting variables, is due to Joseph Taylor [199, 200]. The ideas lay fallow for the next 30 years, but then several authors independently revisited them. Here is a (far from exhaustive) sampling: Voiculescu [207, 208], in the context of free probability; Popescu [166–169], in the context of extending classical function theory to d-tuples of bounded operators; Ball, Groenewald, and Malakorn [42], in the context of extending realization formulas from functions of commuting operators to functions of non-commuting operators; Alpay and Kalyuzhnyi-Verbovetzkii [31] in the context of realization formulas for rational functions that are J-unitary on the boundary of the domain; Helton [102] in proving positive matrix-valued functions are sums of squares; Helton and Vinnikov [112], Helton, Klep, and McCullough [103, 104] and Helton and McCullough [107] in the context of LMIs; and Kaliuzhnyi-Verbovetskyi and Vinnikov in the context of factoring in non-commutative settings [126].

The monograph [127] by Kaliuzhnyi-Verbovetskyi and Vinnikov was the first book on non-commutative functions and laid a common groundwork for the theory.

Strictly speaking, there is a distinction between a free polynomial and the polynomial function of non-commuting variables that it defines, but in the context of this book the distinction is not significant.

12

Basic Properties of Non-Commutative Functions

12.1 Definition of an nc-Function

We shall start by listing some properties of free polynomials, as defined in Section 11.1. Suppose $\varphi \in \mathcal{P}_d$. It is a function from $\mathbb{M}^d$ to $\mathbb{M}^1$, and the following statements hold.

(i) The function is *graded*, which means that, if $x \in \mathbb{M}_n^d$, then $\varphi(x) \in \mathbb{M}_n$.

(ii) The function preserves direct sums: $\varphi(x \oplus y) = \varphi(x) \oplus \varphi(y)$.

(iii) The function preserves joint similarities: if $x, y \in \mathbb{M}_n^d$ and s is invertible in $\mathbb{M}_n$, then $\varphi(s^{-1}xs) = s^{-1}\varphi(x)s$.

(iv) The function preserves intertwining: if $x \in \mathbb{M}_n^d$, $y \in \mathbb{M}_m^d$, and $Lx = yL$ for some $m \times n$ matrix L, then $L\varphi(x) = \varphi(y)L$.

(v) For every $x \in \mathbb{M}^d$, $\varphi(x)$ is in the unital algebra generated by the components of x.

We shall define an nc-function to be one that shares some of these properties. In general, the functions will be defined only on a subset E of $\mathbb{M}^d$, but to take advantage of the algebraic properties, and to do analysis, we will want E to have some structure.

Let $E \subseteq \mathbb{M}^d$. For each $n \in \mathbb{N}$, we shall let

$$E_n \stackrel{\text{def}}{=} E \cap \mathbb{M}_n^d,$$

and informally call this set "E at level n." When choosing a pair of elements from E, normally we will want them to be at the same level, so we shall let $E^{[2]} = E \boxtimes E$ denote the set of pairs of elements from the same level. More generally,

$$E \boxtimes F \stackrel{\text{def}}{=} \bigcup_{n=1}^{\infty} E_n \times F_n. \tag{12.1}$$

291

292 Non-Commutative Theory

Definition 12.2. *Let* $E \subseteq \mathbb{M}^d$. *We say that* E *is* closed with respect to direct sums *if*

$$x \oplus y \in E \qquad \text{for all } x, y \in E.$$

We say E *is* closed with respect to unitary conjugation *if*

$$u^* x u \in E \qquad \text{for all } n \in \mathbb{N}, x \in E_n \text{ and for every unitary } u \in \mathbb{M}_n.$$

We say that E *is* closed with respect to similarities *if*

$$s^{-1} x s \in E \quad \text{for all } n \in \mathbb{N}, x \in E_n \text{ and for every non-singular matrix } s \in \mathbb{M}_n.$$

Definition 12.3. *The* finite topology, *also called the* finitely open topology *or the* disjoint union topology, *on* $\mathbb{M}^d$ *is the topology consisting of the sets* Ω *such that*

$$\Omega \cap \mathbb{M}_n^d \text{ is open in the Euclidean topology on } \mathbb{M}_n^d \text{ for all } n \in \mathbb{N}.$$

Definition 12.4. *A set* $E \subseteq \mathbb{M}^d$ *is called an* nc-set *if it is closed with respect to direct sums. We say that* Ω *is an* nc-domain *if it is an nc-set that is open in the finite topology.*

On nc-sets, Properties (ii) and (iii) are equivalent to (iv). If φ is defined only on a subset E of $\mathbb{M}_n^d$, then to say that φ preserves similarities means that if both x and $s^{-1} x s$ are in E, for some non-singular matrix s, then $\varphi(s^{-1} x s) = s^{-1} \varphi(x) s$. Likewise, φ is said to preserve direct sums if, whenever $x, y, x \oplus y$ are all in E, then $\varphi(x \oplus y) = \varphi(x) \oplus \varphi(y)$, and φ is said to preserve intertwinings if, whenever $x, y \in E$ and $Lx = yL$ for some matrix L, then $L\varphi(x) = \varphi(y)L$.

Lemma 12.5. *Suppose* $E \subset \mathbb{M}^d$ *is closed with respect to direct sums, and* $\varphi \colon E \to \mathbb{M}^1$ *is a graded function. Then* φ *preserves intertwinings if and only if it preserves direct sums and similarities.*

Proof. If φ preserves intertwinings, then a fortiori it preserves similarities. If $x \in \mathbb{M}_m^d$, $y \in \mathbb{M}_n^d$, and $z = x \oplus y$, let P be projection from $\mathbb{C}^{m+n}$ onto $\mathbb{C}^m$. Then $Pz = xP$ and $P^\perp z = yP^\perp$, so that $\varphi(z) = \varphi(x) \oplus \varphi(y)$.

Conversely, if φ preserves direct sums and similarities, and $Lx = yL$, then

$$\begin{pmatrix} I & L \\ 0 & I \end{pmatrix} \begin{pmatrix} y & 0 \\ 0 & x \end{pmatrix} \begin{pmatrix} I & -L \\ 0 & I \end{pmatrix} = \begin{pmatrix} y & 0 \\ 0 & x \end{pmatrix}.$$

Apply φ to both sides, to get

$$\begin{pmatrix} \varphi(y) & L\varphi(x) - \varphi(y)L \\ 0 & \varphi(x) \end{pmatrix} = \begin{pmatrix} \varphi(y) & 0 \\ 0 & \varphi(x) \end{pmatrix},$$

and so φ preserves intertwinings. $\square$

Basic Properties of Non-Commutative Functions 293

We shall define an nc-function on an nc-set $E \subseteq \mathbb{M}^d$ to be a function that is graded and intertwining-preserving, that is, has Properties (i) and (iv). By Lemma 12.5, an nc-function is one that is graded and preserves similarities and direct sums. We shall not require (v). This property will turn out to be satisfied automatically for certain nc-functions, called free functions (see Theorem 14.4), and will clearly be a necessary hypothesis for any theorem that states that a certain nc-function can be approximated by free polynomials, even pointwise. However, as we shall see in Section 15.4, if we want to use the implicit function theorem, we cannot have Property (v).

Definition 12.6. *Let E be an* nc-set. *An* nc-function φ *on E is a graded function $\varphi\colon E \to \mathbb{M}^1$ that is intertwining-preserving.*

One sometimes wants to define a function ψ on some set F that is not nc—a finite set, for example—and then extend it to be nc on a set that is nc. What properties must ψ have to be the restriction to F of an nc-function φ on an nc-set E containing F? Lemma 12.5 fails on sets that are not nc; it turns out that preservation of intertwinings is the correct notion.

Lemma 12.7. *Let $F \subseteq \mathbb{M}^d$ and $\psi\colon F \to \mathbb{M}^d$. There exists an* nc-set $E \supseteq F$ *and an* nc-function $\varphi\colon E \to \mathbb{M}^1$ *that extends ψ if and only if ψ is graded and preserves intertwinings.*

Proof. Necessity is obvious. To see sufficiency, first note that the smallest nc-set containing F is

$$\{x_1 \oplus \cdots \oplus x_m \colon m \in \mathbb{N}, x_1, \ldots, x_m \in F\}. \tag{12.8}$$

So it suffices to take E to be the set (12.8). Then define

$$\varphi(x_1 \oplus \cdots \oplus x_m) \stackrel{\text{def}}{=} \psi(x_1) \oplus \cdots \oplus \psi(x_m),$$

and observe that this extension is intertwining preserving on E. Indeed, suppose

$$L(x_1 \oplus \cdots \oplus x_m) = (y_1 \oplus \cdots \oplus y_n)L,$$

where each x_i and y_j is in F. Then L can be thought of as an $n \times m$ block matrix (L_{ij}), where

$$L_{ij}x_j = y_i L_{ij}$$

and each L_{ij} is a rectangular matrix that maps the space on which x_j operates to the one on which y_i operates. Since ψ is intertwining-preserving,

$$L_{ij}\psi(x_j) = \psi(y_i)L_{ij}.$$

294 *Non-Commutative Theory*

Therefore

$$L \, \varphi(x_1 \oplus \cdots \oplus x_m) = \varphi(y_1 \oplus \cdots \oplus y_n) \, L.$$

$\square$

With Lemma 12.7 in hand, we extend Definition 12.6 to an arbitrary set:

Definition 12.9. *Let* $F \subseteq \mathbb{M}^d$. *An* nc-function *is a graded function* $\varphi\colon F \to \mathbb{M}^1$ *that preserves intertwinings.*

Given an nc-set E, its *unitary envelope* is the set

$$\{u^* x u \colon x \in E, \, u \text{ is unitary}\}. \tag{12.10}$$

More formally, we should write the set (12.10) as

$$\bigcup_{n=1}^{\infty} \{u^* x u \colon x \in E \cap \mathbb{M}_n^d, \, u \text{ is a unitary in } \mathbb{M}_n\}.$$

However this is cumbersome, so when we write an expression like (12.10), we assume that only algebraically sensible expressions are included. Note that the set (12.10) is an nc-set.

Just as we extended an nc-function from an arbitrary set to its nc-hull in Lemma 12.7, we can extend any nc-function φ on E to its unitary envelope, simply by defining

$$\varphi(u^* x u) = u^* \varphi(x) u.$$

This will be an nc-function on a larger domain, and it is often convenient to assume therefore that E is closed with respect to unitary conjugation. (Indeed, this is sometimes included in the definition of an nc-set). Likewise the *similarity envelope* of a set $E \subseteq \mathbb{M}^d$ is defined to be the set

$$\{s^{-1} x s \colon x \in E, \, s \text{ is invertible}\},$$

and one can similarly extend φ to the similarity envelope of E. We do not, however, want to restrict ourselves to similarity-invariant sets, because even in the case $d = 1$, the similarity orbit of any non-scalar operator will be unbounded, and it will not be reasonable to assume that φ is bounded on this set. Boundedness will play a key role in our analysis (see Chapter 14).

We shall sometimes want to consider vector-valued or matrix-valued nc-functions. If $\mathcal{K}, \mathcal{L}$ are Hilbert spaces, an element of the algebraic tensor product $\mathcal{B}(\mathcal{K}, \mathcal{L}) \otimes \mathbb{M}_n$ can naturally be interpreted as an $n \times n$ matrix of operators, and we will make the identification

$$\mathcal{B}(\mathcal{K}, \mathcal{L}) \otimes \mathbb{M}_n = \mathcal{B}(\mathcal{K} \otimes \mathbb{C}^n, \mathcal{L} \otimes \mathbb{C}^n). \tag{12.11}$$

Basic Properties of Non-Commutative Functions 295

The right-hand side of equation (12.11) comes with a norm, so we can assign a norm to any element in $\mathcal{B}(\mathcal{K},\mathcal{L}) \otimes \mathbb{M}^1$. We now define a *graded* $\mathcal{B}(\mathcal{K},\mathcal{L})$-*valued function* on a set $E \subseteq \mathbb{M}^d$ to be a map φ such that if $x \in E \cap \mathbb{M}_n^d$, then $\varphi(x) \in \mathcal{B}(\mathcal{K},\mathcal{L}) \otimes \mathbb{M}_n$.

In the special case that $\mathcal{K}$ is 1-dimensional, we identify $\mathcal{B}(\mathbb{C},\mathcal{L})$ with the Hilbert space $\mathcal{L}$, and call the graded $\mathcal{B}(\mathbb{C},\mathcal{L})$-valued functions *graded vector-valued* or $\mathcal{L}$-*valued functions*. Thus, for such a function φ, if $x \in E \cap \mathbb{M}_n^d$, then we can regard $\varphi(x)$ as an element of either $\mathcal{B}(\mathbb{C},\mathcal{L}) \otimes \mathbb{M}_n$ or $\mathcal{B}(\mathbb{C}^n,\mathcal{L} \otimes \mathbb{C}^n)$.

When dealing with elements of $\mathcal{B}(\mathcal{K},\mathcal{L}) \otimes \mathbb{M}_n$ we shall have a lot of tensor symbols. To enhance legibility we will write them vertically:

$$\begin{matrix} T \\ \otimes \\ s \end{matrix} \overset{\text{def}}{=} T \otimes s \qquad \text{for } T \in \mathcal{B}(\mathcal{K},\mathcal{L}) \quad \text{and } s \in \mathbb{M}_n.$$

Thus

$$\begin{matrix} T \\ \otimes \\ s \end{matrix} \in \mathcal{B}\left(\begin{matrix} \mathcal{K} \\ \otimes \\ \mathbb{C}^n \end{matrix} , \begin{matrix} \mathcal{L} \\ \otimes \\ \mathbb{C}^n \end{matrix} \right).$$

Definition 12.12. *Let* $\mathcal{K},\mathcal{L}$ *be Hilbert spaces. A* $\mathcal{B}(\mathcal{K},\mathcal{L})$-*valued nc-function on a set* $E \subseteq \mathbb{M}^d$ *is a map* φ *taking values in* $\mathcal{B}(\mathcal{K},\mathcal{L}) \otimes \mathbb{M}^1$ *that satisfies*

(i) φ *is a graded* $\mathcal{B}(\mathcal{K},\mathcal{L})$-*valued function; and*

(ii) *if* $x \in E_m, y \in E_n$, *and* $s \in \mathcal{B}(\mathbb{C}^m,\mathbb{C}^n)$ *satisfies* $sx = ys$, *then*

$$\begin{matrix} \text{id}_{\mathcal{L}} \\ \otimes \\ s \end{matrix} \varphi(x) = \varphi(y) \begin{matrix} \text{id}_{\mathcal{K}} \\ \otimes \\ s \end{matrix}. \tag{12.13}$$

Normally we shall assume that E is an nc-set, in which case Condition (ii) is equivalent to the following two conditions:

(iia) The map φ preserves direct sums.

(iib) If x and $s^{-1}xs$ are in E, then

$$\varphi(s^{-1}xs) = \begin{matrix} \text{id}_{\mathcal{L}} \\ \otimes \\ s^{-1} \end{matrix} \varphi(x) \begin{matrix} \text{id}_{\mathcal{K}} \\ \otimes \\ s \end{matrix}.$$

12.2 Locally Bounded nc-Functions Are Holomorphic

Let Ω be an nc-domain. We shall say that an nc-function is *finitely locally bounded on* Ω if, at every point $x \in \Omega$, there is a finite neighborhood of x on which the function is bounded. (This implies that if x is in Ω_n, then there is an open neighborhood U of x in Ω_n on which the function is bounded). The main result in this section is that finitely locally bounded nc-functions are automatically continuous and holomorphic, by which we mean that, for all $n \in \mathbb{N}$ and $x \in \Omega \cap \mathbb{M}_n^d$ and for all $h \in \mathbb{M}_n^d$,

296 Non-Commutative Theory

$$\lim_{t \to 0} \frac{\varphi(x + th) - \varphi(x)}{t} \text{ exists.} \qquad (12.14)$$

In particular, this means that $\varphi|_{\Omega \cap \mathbb{M}_n^d}$ is a holomorphic function in the dn^2 entries of the matrices.

We shall repeatedly use the following lemma. Note that in equation (12.16), by $\varphi(y)L - L\varphi(x)$ we mean

$$\varphi(y) \overset{\mathrm{id}_{\mathcal{K}}}{\underset{L}{\otimes}} - \overset{\mathrm{id}_{\mathcal{L}}}{\underset{L}{\otimes}} \varphi(x).$$

Lemma 12.15. *Let Ω be an nc-domain in $\mathbb{M}^d$, let $\mathcal{K}$ and $\mathcal{L}$ be Hilbert spaces, and let φ be a $\mathcal{B}(\mathcal{K}, \mathcal{L})$-valued nc-function on Ω. Fix $n \geq 1$ and $L \in \mathbb{M}_n$. If $x, y \in \Omega \cap \mathbb{M}_n^d$ and*

$$\begin{bmatrix} y & yL - Lx \\ 0 & x \end{bmatrix} \in \Omega \cap \mathbb{M}_{2n}^d,$$

then

$$\varphi\left(\begin{bmatrix} y & yL - Lx \\ 0 & x \end{bmatrix}\right) = \begin{bmatrix} \varphi(y) & \varphi(y)L - L\varphi(x) \\ 0 & \varphi(x) \end{bmatrix}. \qquad (12.16)$$

Proof. Let

$$s = \begin{bmatrix} \mathrm{id}_{\mathbb{C}^n} & L \\ 0 & \mathrm{id}_{\mathbb{C}^n} \end{bmatrix}$$

so that

$$\begin{bmatrix} y & yL - Lx \\ 0 & x \end{bmatrix} = s^{-1} \begin{bmatrix} y & 0 \\ 0 & x \end{bmatrix} s.$$

Using Properties (ii) and (iiii), we have

$$\varphi\left(\begin{bmatrix} y & yL - Lx \\ 0 & x \end{bmatrix}\right) = \varphi(s^{-1}(y \oplus x)s)$$

$$= \begin{bmatrix} \mathrm{id}_{\mathcal{L}} \underset{\mathrm{id}_{\mathbb{C}^n}}{\otimes} & \underset{L}{\overset{-\mathrm{id}_{\mathcal{L}}}{\otimes}} \\ 0 & \mathrm{id}_{\mathcal{L}} \underset{\mathrm{id}_{\mathbb{C}^n}}{\otimes} \end{bmatrix} \begin{bmatrix} \varphi(y) & 0 \\ 0 & \varphi(x) \end{bmatrix} \begin{bmatrix} \mathrm{id}_{\mathcal{K}} \underset{\mathrm{id}_{\mathbb{C}^n}}{\otimes} & \underset{L}{\overset{\mathrm{id}_{\mathcal{K}}}{\otimes}} \\ 0 & \mathrm{id}_{\mathcal{K}} \underset{\mathrm{id}_{\mathbb{C}^n}}{\otimes} \end{bmatrix}$$

$$= \begin{bmatrix} \varphi(y) & \varphi(y)L - L\varphi(x) \\ 0 & \varphi(x) \end{bmatrix}.$$

$\square$

Theorem 12.17. *Let Ω be an nc-domain in $\mathbb{M}^d$, let $\mathcal{K}$ and $\mathcal{L}$ be Hilbert spaces, and let φ be a $\mathcal{B}(\mathcal{K}, \mathcal{L})$-valued nc-function on Ω. If φ is finitely locally bounded on Ω, then φ is holomorphic on Ω.*

Basic Properties of Non-Commutative Functions   297

Proof. The proof will proceed in two steps. We first show that if φ is finitely locally bounded, then φ is finitely continuous. Then we show that φ is holomorphic.

Step 1. Fix $x \in \Omega \cap \mathbb{M}_n^d$ and let $\varepsilon > 0$. As φ is finitely locally bounded, there exist $r, M > 0$ such that

$$\text{ball}\left(\begin{bmatrix} x & 0 \\ 0 & x \end{bmatrix}, r\right) \subseteq \Omega \cap \mathbb{M}_{2n}^d$$

and

$$y \in \text{ball}\left(\begin{bmatrix} x & 0 \\ 0 & x \end{bmatrix}, r\right) \implies \|\varphi(y)\| < M. \tag{12.18}$$

Choose δ sufficiently small that $0 < \delta < \min\{r\varepsilon/2M, r/2\}$ and ball$(x, \delta) \subseteq \Omega$. Let $y \in \mathbb{M}_n^d$ have $\|y - x\| < \delta$. Then

$$\left\| \begin{bmatrix} y & (M/\varepsilon)(y-x) \\ 0 & x \end{bmatrix} - \begin{bmatrix} x & 0 \\ 0 & x \end{bmatrix} \right\| < r,$$

since both $\|y - x\|$ and $\|(M/\varepsilon)(y - x)\|$ are less than $r/2$.

By Lemma 12.15,

$$\varphi\left(\begin{bmatrix} y & (M/\varepsilon)(y-x) \\ 0 & x \end{bmatrix}\right) = \begin{bmatrix} \varphi(y) & (M/\varepsilon)(\varphi(y)-\varphi(x)) \\ 0 & \varphi(x) \end{bmatrix},$$

and by the implication (12.18), the last matrix has norm less than M, so that $\|\varphi(y) - \varphi(x)\| < \varepsilon$. As ε was arbitrary, this proves that φ is continuous.

Step 2. To see that φ is holomorphic, fix $x \in \Omega \cap \mathbb{M}_n^d$. If $h \in \mathbb{M}_n^d$ is selected sufficiently small, then

$$\begin{bmatrix} x+th & h \\ 0 & x \end{bmatrix} \in \Omega \cap \mathbb{M}_{2n}^d$$

for all sufficiently small $t \in \mathbb{C}$. But

$$\begin{bmatrix} x+th & h \\ 0 & x \end{bmatrix} = \begin{bmatrix} x+th & (1/t)\big((x+th)-x\big) \\ 0 & x \end{bmatrix}.$$

Hence Lemma 12.15 implies that

$$\varphi\left(\begin{bmatrix} x+th & h \\ 0 & x \end{bmatrix}\right) = \begin{bmatrix} \varphi(x+th) & (1/t)\big(\varphi(x+th)-\varphi(x)\big) \\ 0 & \varphi(x) \end{bmatrix}. \tag{12.19}$$

As the left-hand side of equation (12.19) is continuous at $t = 0$, it follows that the 12-entry of the right-hand side of equation (12.19) must converge. As h is arbitrary, this implies that φ is holomorphic. $\qquad\square$

298   *Non-Commutative Theory*

If we take the limit as $t \to 0$ in equation (12.19), we get a formula that we shall use later. We write $D\varphi(x)[h]$ to denote the derivative of φ at x in the direction h.

Proposition 12.20. *With the hypotheses of Theorem 12.17,*

$$\varphi\left(\begin{bmatrix} x & h \\ 0 & x \end{bmatrix}\right) = \begin{bmatrix} \varphi(x) & D\varphi(x)[h] \\ 0 & \varphi(x) \end{bmatrix}.$$

Without local boundedness, the algebraic properties of being nc do not guarantee that a function be well-behaved.

Example 12.21. Let $d = 1$, and define a function φ on Jordan blocks by sending a Jordan block with 0 eigenvalues to the zero matrix of the same size, and a Jordan block with non-zero eigenvalues to the identity matrix of that size. Extend φ by direct sums to any matrix in Jordan canonical form and then by similarity to any matrix. The function φ is then an nc-function, since it preserves direct sums and similarities. However, it is neither continuous nor locally bounded. Indeed, let $\varepsilon > 0$, and apply Lemma 12.15 with $y = \varepsilon, x = 0, L = 1/\sqrt{\varepsilon}$. This gives

$$\varphi\left(\begin{bmatrix} \varepsilon & \sqrt{\varepsilon} \\ 0 & 0 \end{bmatrix}\right) = \begin{bmatrix} 1 & \frac{1}{\sqrt{\varepsilon}} \\ 0 & 0 \end{bmatrix}. \tag{12.22}$$

As ε tends to 0, the right-hand side of equation (12.22) diverges unboundedly, whereas

$$\varphi\left(\begin{bmatrix} 0 & 0 \\ 0 & 0 \end{bmatrix}\right) = \begin{bmatrix} 0 & 0 \\ 0 & 0 \end{bmatrix}.$$

12.3 Nc Topologies

In light of Theorem 12.17, we want to restrict our attention to finitely locally bounded nc-functions.

Definition 12.23. *An nc topology on $\mathbb{M}^d$ is a topology that has a basis of nc-domains. If τ is an nc topology and Ω is in τ, a function φ with domain Ω is called τ-holomorphic if it is an nc-function that is τ-locally bounded, that is, for all $x \in \Omega$ there exists $U \in \tau$ such that $x \in U$ and $B \in \mathbb{R}$ such that $\|\varphi(y)\| \leq B$ for all $y \in U$.*

By Theorem 12.17, if τ is an nc topology, every τ-holomorphic function is finitely-finitely continuous, and holomorphic in the sense of statement (12.14).

Basic Properties of Non-Commutative Functions 299

Example 12.24. The *fine* topology is the topology that has all nc-domains as a basis. This is the largest nc topology, so has the largest supply of holomorphic functions. We shall see in Theorems 15.1 and 15.11 that the inverse function theorem and implicit function theorem hold for finely holomorphic maps.

Definition 12.25. *If* $x \in \mathbb{M}_n^d$, *the k-ampliation of x, which we denote by* $x^{(k)}$, *is the element of* $\mathbb{M}_{nk}^d$ *obtained by taking the direct sum of k copies of x.*

Example 12.26. The *fat topology*, also called the *uniformly open topology*, is the topology generated by balls centered at a point and their ampliations. For $a \in \mathbb{M}_n^d$, let

$$F(a,r) = \bigcup_{k=1}^{\infty} \{y \in \mathbb{M}_{nk}^d : \|y - a^{(k)}\| < r\}. \tag{12.27}$$

Lemma 12.28. *Sets of the form* (12.27) *form a basis for a topology.*

Proof. We need to show that if $x \in F(a,r) \cap F(b,s)$, then for some $\varepsilon > 0$, we have $F(x,\varepsilon) \subseteq F(a,r) \cap F(b,s)$. We will leave the verification of this as an exercise. $\square$

The fat topology is the topology with the basis $\{F(a,r): a \in \mathbb{M}^d, r > 0\}$. We shall see in Theorems 15.9 and 15.13 that the inverse function theorem and implicit function theorem also hold for fat holomorphic maps. An advantage of the fat category is that the implicit function theorem holds provided the derivative has full rank at a point; in the fine category, it is required to have full rank in a neighborhood of the point.

Example 12.29. Let δ be an $I \times J$ matrix of free polynomials in d variables. Let

$$B_\delta \stackrel{\text{def}}{=} \{x \in \mathbb{M}^d : \|\delta(x)\| < 1\}. \tag{12.30}$$

The norm on the right-hand side is in $\mathcal{B}(\mathbb{C}^J \otimes \mathbb{C}^n, \mathbb{C}^I \otimes \mathbb{C}^n)$, for $x \in \mathbb{M}_n^d$. A set of the form (12.30) is called a *polynomial polyhedron*. The *free topology* is the topology generated by all polynomial polyhedra. They clearly form the basis for a topology, as the intersection $B_\delta \cap B_\gamma$ is $B_{\delta \oplus \gamma}$. (They are the non-commutative analog of the polynomial polyhedra G_δ we encountered in Section 9.8.)

If $a \in \mathbb{C}^d$ and we let $\delta(x)$ be the $d \times 1$ column with entries $\frac{1}{r}(x^j - a^j)$, then, at level 1, B_δ will be the Euclidean ball centered at a with radius r. If a is 0 and r is 1, we get the set of all strict column contractions, the set of x such that

$$\sum_{j=1}^{d} x^{j*} x^j < 1.$$

300 Non-Commutative Theory

If $\delta(x)$ is the diagonal $d \times d$ matrix with entries x^j, then B_δ is the set of all d-tuples of strict contractions. If $\delta(x_1, x_2)$ is the 1×1 matrix

$$(x_1 x_2 - x_2 x_1)(x_1 x_2 - x_2 x_1) - 1,$$

then B_δ is empty at level 1 and non-empty at level 2 and above.

We shall study free holomorphic functions in Chapter 14 and show that there is a realization formula for a free holomorphic function that is bounded on a polynomial polyhedron (Theorem 14.33). As a consequence, we shall prove in Theorem 14.39 that free holomorphic functions can be locally approximated by polynomials and, conversely, that this characterizes these functions. The reason we work with different nc topologies, rather than just fixing one, is that there is *no* nc topology for which both the implicit function theorem and polynomial approximation hold, Theorem 15.18. The former is useful for non-commutative algebraic geometry, the latter for non-commutative complex analysis and functional calculus. So we must be willing to choose our topology, depending on the problem at hand. This is why we omitted Property (v) of non-commutative polynomials when we defined nc-functions.

Notice that no nc topology is Hausdorff, or even T_1: any open set containing x must also contain $x \oplus x$.

12.4 Historical Notes

Lemma 12.5 is proved in [200]. See also [127, proposition 2.1]. Lemma 12.15 is in [104, 127]. Theorem 12.17, in the continuous case, is proved in [104], and under just the local boundedness hypothesis, [127, theorem 7.2]. Example 12.21 is from [148].

In [127, theorem 4.1], Kaliuzhnyi-Verbovetskyi and Vinnikov prove what they call the Taylor–Taylor formula, which is a foundation for much of the book (e.g., they use it to prove Theorem 12.17). To explain the Taylor–Taylor formula, named after Brook and Joseph, one must first develop the nc-analog of higher derivatives. Consider a locally bounded nc-function f on a domain in $\mathbb{M}^d$. The analog of the classical first derivative is the "nc difference–differential operator" $\Delta_R f(x, y)(z)$, defined for any pair x, y in $\mathbb{M}^d$ and any d-tuple z of (rectangular) matrices of suitable size by the equation

$$f\left(\begin{bmatrix} x & z \\ 0 & y \end{bmatrix}\right) = \begin{bmatrix} f(x) & \Delta_R f(x, y)(z) \\ 0 & f(y) \end{bmatrix}.$$

More precisely, $\Delta_R f(x, y)(z)$ is the analog of a divided difference operator when $x \neq y$ and of a derivative at x when $x = y$ (as in Proposition 12.20).

Basic Properties of Non-Commutative Functions 301

Higher-order nc difference–differential operators are defined similarly, with larger block upper triangular matrices. For example,

$$f\left(\begin{bmatrix} x_0 & z_1 & 0 \\ 0 & x_1 & z_2 \\ 0 & 0 & x_2 \end{bmatrix}\right) = \begin{bmatrix} f(x_0) & \Delta_R f(x_0,x_1)(z_1) & \Delta_R^2 f(x_0,x_1,x_2)(z_1,z_2) \\ 0 & f(x_1) & \Delta_R f(x_1,x_2)(z_2) \\ 0 & 0 & f(x_2) \end{bmatrix}.$$

Here, for $x_0,x_1,x_2 \in \mathbb{M}^d$, the second-order nc difference–differential operator $\Delta_R^2 f(x_0,x_1,x_2)$ is bilinear on suitable pairs (z_1,z_2) of d-tuples of matrices and has certain grading and intertwining properties. The finite Taylor–Taylor expansion of a locally bounded nc-function f about a center $y \in \mathbb{M}^d$ is

$$f(x) = f(y) + \Delta_R f(y,y)(x-y) + \cdots + \Delta_R^N f(y,\ldots,y)(x-y,\ldots,x-y)$$
$$+ \Delta_R^{N+1} f(y,\ldots,y,x)(x-y,\ldots,x-y).$$

Under appropriate assumptions, one can use the finite Taylor–Taylor series to construct an infinite power series expansion that converges in an appropriate sense. Note, however, that such expansions are not in general valid in all dimensions: if y is in $\mathbb{M}_s^d$, then the power series expansion (finite or infinite) holds only in dimensions that are multiples of s. For x in $\mathbb{M}_{ms}^d$ one has an expansion (Theorem 7.8)

$$f(x) = \sum_{\ell=0}^{\infty} \left(x - \bigoplus_{\alpha=1}^{m} y\right)^{\odot_s \ell} \Delta_R^\ell f(y,\ldots,y),$$

where the symbol $\odot_s \ell$ denotes the ℓth power in the tensor algebra over $\mathbb{C}^s$. Conditions for the convergence of the series are delicate; they are analyzed in chapters 7 and 8 of [127].

13
Montel Theorems

13.1 Overview

Let Ω be an open set in $\mathbb{C}^d$ and assume that $\{u^k\}$ is a sequence in $\mathrm{Hol}(\Omega)$, the algebra of of holomorphic functions on Ω equipped with the topology of uniform convergence on compact subsets. The classical Montel theorem asserts that if $\{u^k\}$ is locally uniformly bounded on Ω, then there exists a subsequence $\{u^{k_l}\}$ that converges in $\mathrm{Hol}(\Omega)$.

It is well known that if $\mathcal{H}$ is an infinite-dimensional Hilbert space, then Montel's theorem breaks down for $\mathrm{Hol}(\Omega, \mathcal{H})$, the space of $\mathcal{H}$-valued holomorphic functions. For example, if $\mathcal{H} = \ell^2$ and $\{f^k\}$ is a locally uniformly bounded sequence of holomorphic functions on Ω, then the sequence

$$\begin{pmatrix} f^1(\lambda) \\ 0 \\ 0 \\ \vdots \end{pmatrix}, \begin{pmatrix} 0 \\ f^2(\lambda) \\ 0 \\ \vdots \end{pmatrix}, \begin{pmatrix} 0 \\ 0 \\ f^3(\lambda) \\ \vdots \end{pmatrix}, \cdots$$

is a locally uniformly bounded sequence that has a convergent subsequence only if there exists a subsequence $\{f^{k_l}\}$ that converges uniformly to 0 on Ω.

One way to get around this difficulty is to assume that the ranges of u^k stay inside a compact set, as for example in this theorem of Arendt and Nikol'skiĭ [34] (see Section 13.2 for a proof).

Theorem 13.1. *Assume that Ω is an open set in $\mathbb{C}^d$, $\{\lambda_i\}$ is a dense sequence in Ω, and $\mathcal{H}$ is a Hilbert space. If $\{u^k\}$ is a sequence in $\mathrm{Hol}(\Omega, \mathcal{H})$ that is locally uniformly bounded on Ω, and for each i, $\{u^k(\lambda_i)\}$ is a convergent sequence in $\mathcal{H}$, then $\{u^k\}$ is a convergent sequence in $\mathrm{Hol}(\Omega, \mathcal{H})$.*

The main idea of this chapter is that, without the assumption on $\{u^k(\lambda_i)\}$, one can still deduce a modified convergence result. There is a sequence of

302

Montel Theorems 303

unitaries that "pulls the ranges of the u^k back" to produce a convergent sequence. In the commutative case, this yields:

Theorem 13.2. *If Ω is an open set in $\mathbb{C}^d$, $\mathcal{H}$ is a Hilbert space, and $\{u^k\}$ is a locally uniformly bounded sequence in $\mathrm{Hol}(\Omega, \mathcal{H})$, then there exists a sequence $\{U^k\}$ of unitary operators on $\mathcal{H}$ such that $\{U^k u^k\}$ has a subsequence that converges in $\mathrm{Hol}(\Omega, \mathcal{H})$.*

13.2 Wandering Isometries

If Ω is an open set in $\mathbb{C}^d$, and $\mathcal{H}$ is a Hilbert space, let $\mathrm{Hol}(\Omega, \mathcal{H})$ denote the space of holomorphic $\mathcal{H}$-valued functions on Ω. If $u \in \mathrm{Hol}(\Omega, \mathcal{H})$ and $E \subseteq \Omega$, let

$$\|u\|_E = \sup_{\lambda \in E} \|u(\lambda)\|_{\mathcal{H}}.$$

If $\{u^k\}$ is sequence in $\mathrm{Hol}(\Omega, \mathcal{H})$, we say that $\{u^k\}$ is *uniformly bounded on Ω* if

$$\sup_k \|u^k\|_\Omega < \infty,$$

and we say that $\{u^k\}$ is *locally uniformly bounded on Ω* if for each $\lambda \in \Omega$, there exists a neighborhood G of λ such that $\{u^k\}$ is uniformly bounded on G. Recall that if such a neighborhood exists, then a Cauchy estimate implies that $\{u^k\}$ is equicontinuous at λ, that is, for each $\varepsilon > 0$ there exists a ball G_0 such that $\lambda \in G_0 \subseteq G$ and

$$\|u^k(\mu) - u^k(\lambda)\| < \varepsilon \qquad \text{for all } \mu \in G_0 \text{ and all } k.$$

We equip $\mathrm{Hol}(\Omega, \mathcal{H})$ with the usual topology of uniform convergence on compacta. Thus a sequence $\{u^k\}$ in $\mathrm{Hol}(\Omega, \mathcal{H})$ is convergent precisely when there is a function $u \in \mathrm{Hol}(\Omega, \mathcal{H})$ such that

$$\lim_{k \to \infty} \|u^k - u\|_E = 0$$

for every compact $E \subseteq \Omega$. We say that a sequence $\{u^k\}$ in $\mathrm{Hol}(\Omega, \mathcal{H})$ is a *Cauchy sequence* if, for each compact $E \subseteq \Omega$, $\{u^k\}$ is uniformly Cauchy on E, that is, for each $\varepsilon > 0$, there exists N such that

$$k, l \geq N \implies \|u^k - u^l\|_E < \varepsilon.$$

It is well known that $\mathrm{Hol}(\Omega, \mathcal{H})$ is complete.

304 Non-Commutative Theory

Proof of Theorem 13.1. Fix a compact set $E \subseteq \Omega$ and let $\varepsilon > 0$. Note that, as $\{u^k\}$ is assumed to be locally uniformly bounded on Ω, $\{u^k\}$ is equicontinuous at each point of Ω. Hence as E is compact, we may construct a finite collection $\{B_r : r = 1, \ldots, m\}$ of open balls in $\mathbb{C}^d$ such that

$$E \subseteq \bigcup_{r=1}^{m} B_r \subseteq \Omega \tag{13.3}$$

and, for all r, all $\mu_1, \mu_2 \in B_r$ and all k,

$$\|u^k(\mu_1) - u^k(\mu_2)\| < \varepsilon/3. \tag{13.4}$$

As $\{\lambda_i\}$ is assumed dense in Ω, for every r there is an index i_r such that

$$\lambda_{i_r} \in B_r. \tag{13.5}$$

Consequently, as we assume that, for each i, $\{u^k(\lambda_i)\}$ is a convergent (and hence Cauchy) sequence in $\mathcal{H}$, there exists $N \in \mathbb{N}$ such that, for each r,

$$k, j \geq N \implies \|u^k(\lambda_{i_r}) - u^j(\lambda_{i_r})\| < \varepsilon/3. \tag{13.6}$$

Consider any $\lambda \in E$. Use the inclusion (13.3) to choose r such that $\lambda \in B_r$. Use the relation (13.5) to choose i_r such that $\lambda_{i_r} \in B_r$. As λ and λ_{i_r} are both in B_r, by the inequality (13.4),

$$\|u^k(\lambda) - u^k(\lambda_{i_r})\| < \varepsilon/3 \qquad \text{for all } k.$$

Hence, by the implication (13.6), if $k, j \geq N$, then

$$\|u^k(\lambda) - u^j(\lambda)\| \leq \|u^k(\lambda) - u^k(\lambda_{i_r})\| + \|u^k(\lambda_{i_r}) - u^j(\lambda_{i_r})\| + \|u^j(\lambda_{i_r}) - u^j(\lambda)\|$$
$$< \varepsilon.$$

Since this inequality holds for an arbitrary point $\lambda \in E$, $\{u^k\}$ is uniformly Cauchy on E. Since E is an arbitrary compact subset of Ω, $\{u^k\}$ is a Cauchy sequence in $\mathrm{Hol}(\Omega, \mathcal{H})$. Therefore, $\{u^k\}$ converges in $\mathrm{Hol}(\Omega, \mathcal{H})$. $\square$

The key lemma needed to prove Theorem 13.2 is the following one, which we call the *wandering isometries lemma*.

Lemma 13.7. *Assume that Ω is an open set in $\mathbb{C}^d$, $\{\lambda_i\}$ is a sequence in Ω, and $\mathcal{H}$ is a Hilbert space. If $\{u^k\}$ is sequence in $\mathrm{Hol}(\Omega, \mathcal{H})$ that is locally uniformly bounded on Ω, then there exists a subsequence $\{u^{k_l}\}$ of $\{u^k\}$ and a sequence $\{V^l\}$ of unitary operators on $\mathcal{H}$ such that, for each i, $\{V^l u^{k_l}(\lambda_i)\}$ is a convergent sequence in $\mathcal{H}$.*

Montel Theorems 305

Proof. If $\mathcal{H}$ is finite-dimensional, one can let each unitary be the identity, and the result is the familiar Montel theorem. So we shall assume that $\mathcal{H}$ is infinite-dimensional. Let $\{e_i\}$ be an orthonormal basis for $\mathcal{H}$. For each k let

$$\mathcal{H}^k = \mathrm{span}\,\{e_1, e_2, \ldots, e_k\},$$
$$\mathcal{M}_i^k = \mathrm{span}\,\{u^k(\lambda_1), u^k(\lambda_2), \ldots, u^k(\lambda_i)\} \qquad \text{for } i = 1, \ldots, k.$$

For each k choose a unitary $U^k \in \mathcal{B}(\mathcal{H})$ satisfying

$$U^k \mathcal{M}_i^k \subseteq \mathcal{H}^i \qquad \text{for } i = 1, \ldots, k.$$

Observe that with this construction, for each i,

$$\{U^k u^k(\lambda_i)\}_{k=i}^{\infty}$$

is a bounded sequence in $\mathcal{H}^i$, a finite-dimensional Hilbert space. Therefore, there exist $v_i \in \mathcal{H}$ and an increasing sequence of indices $\{k_l\}$ such that

$$U^{k_l} u^{k_l}(\lambda_i) \to v_i \quad \text{in } \mathcal{H} \quad \text{as } l \to \infty.$$

Applying this fact successively with $i = 1$, $i = 2$, and so on, at each stage taking a subsequence of the previously selected subsequence, we arrive at a sequence $\{v_i\}$ in $\mathcal{H}$ and an increasing sequence of indices $\{k_l\}$ such that

$$U^{k_l} u^{k_l}(\lambda_i) \to v_i \quad \text{in } \mathcal{H} \quad \text{as } l \to \infty$$

for all i. The lemma then follows if we let $V^l = U^{k_l}$. $\qquad\square$

Proof of Theorem 13.2. Assume that Ω is an open set in $\mathbb{C}^d$, $\mathcal{H}$ is a Hilbert space and $\{u^k\}$ is a locally uniformly bounded sequence in $\mathrm{Hol}(\Omega, \mathcal{H})$. The theorem follows from the classical Montel theorem (with $U^k = \mathrm{id}_{\mathcal{H}}$ for all k) if $\dim \mathcal{H} < \infty$. Therefore, we may assume that $\dim \mathcal{H} = \infty$.

Fix a dense sequence $\{\lambda_i\}$ in Ω. By Lemma 13.7, there exists a subsequence $\{u^{k_l}\}$ and a sequence $\{V^l\}$ of unitary operators on $\mathcal{H}$ such that, for each i, $\{V^l u^{k_l}(\lambda_i)\}$ is a convergent sequence in $\mathcal{H}$. Furthermore, as $\{u^k\}$ is locally uniformly bounded, $\{V^l u^{k_l}\}$ is also locally uniformly bounded. Therefore, Theorem 13.1 implies that $\{V^l u^{k_l}\}$ is a convergent sequence in $\mathrm{Hol}(\Omega, \mathcal{H})$. The theorem then follows by choice of $\{U^k\}$ to be any sequence of unitaries in $\mathcal{B}(\mathcal{H})$ such that $U^{k_l} = V^l$ for all l. $\qquad\square$

13.3 A Graded Montel Theorem

Now we shall study graded functions on domains in $\mathbb{M}^d$. Let Ω be a finitely open set in $\mathbb{M}^d$. Recall that, by Definition 12.12 and the preceding remarks, an $\mathcal{H}$-valued nc-function on an nc-set $E \subseteq \mathbb{M}^d$ is a function φ on E such

306   *Non-Commutative Theory*

that $x \in E_n$ gets mapped to an element of $\mathcal{H} \otimes \mathbb{M}_n = \mathcal{B}(\mathbb{C}^n, \mathcal{H} \otimes \mathbb{C}^n)$, and if $sx = ys$, then

$$\overset{\mathrm{id}_\mathcal{H}}{\underset{s}{\otimes}} \varphi(x) = \varphi(y)s.$$

Let $\mathrm{Holg}(\Omega)$ denote the collection of graded functions on Ω that are holomorphic in the sense of Section 12.2 and let $\mathrm{Holg}_\mathcal{H}(\Omega)$ denote the collection of graded $\mathcal{H}$-valued functions on Ω that are holomorphic.

We view $\mathrm{Holg}(\Omega)$ and $\mathrm{Holg}_\mathcal{H}(\Omega)$ as complete metric spaces endowed with the topology of uniform convergence on finitely compact subsets of Ω. Note that the functions in $\mathrm{Holg}(\Omega)$ and $\mathrm{Holg}_\mathcal{H}(\Omega)$ are *not* assumed to be nc. We shall let $\mathrm{Holg}_\mathcal{H}^{\mathrm{nc}}(\Omega)$ denote the set of functions in $\mathrm{Holg}_\mathcal{H}(\Omega)$ that are nc.

If τ is an nc-topology, and Ω is τ-open, we denote by $\mathrm{Holg}_\mathcal{H}^{\tau}(\Omega)$ the space of $\mathcal{H}$-valued τ-holomorphic functions in the sense of Definition 12.23, that is, the $\mathcal{B}(\mathbb{C}, \mathcal{H})$-valued nc-functions on Ω that are τ-locally bounded. Thus

$$\mathrm{Holg}_\mathcal{H}^{\tau}(\Omega) \subseteq \mathrm{Holg}_\mathcal{H}^{\mathrm{nc}}(\Omega) \subseteq \mathrm{Holg}_\mathcal{H}(\Omega).$$

If Ω is finitely open in $\mathbb{M}^d$, we may construct a *finitely compact-open exhaustion of* Ω, that is, an increasing sequence of compact sets $\{K_i\}$ that satisfies

$$K_1 \subset K_2^\circ \subset K_2 \subset K_3^\circ \subset \cdots$$

and $\Omega = \bigcup_i K_i$, where K° denotes the interior of a set K. For a set $E \subseteq \Omega$ and $f \in \mathrm{Holg}(\Omega)$ we let

$$\|f\|_E = \sup_{\lambda \in E} \|f(\lambda)\|,$$

and then in the usual way define a metric d on $\mathrm{Holg}(\Omega)$ with the formula

$$d(f, g) = \sum_{n=1}^{\infty} \frac{1}{2^n} \frac{\|f - g\|_{K_n}}{1 + \|f - g\|_{K_n}}.$$

It then follows that $f_k \to f$ in the metric space $(\mathrm{Holg}(\Omega), d)$ if and only if, for each finitely compact set K in Ω, $\{f_k\}$ converges uniformly to f on K, that is,

$$\lim_{k \to \infty} \|f - f_k\|_K = 0.$$

Furthermore, $\mathrm{Holg}(\Omega)$ is a complete metric space when endowed with the topology of uniform convergence on finitely compact subsets of Ω. Moreover, the corresponding definitions and statements apply to $\mathrm{Holg}_\mathcal{H}(\Omega)$.

It is a straightforward exercise to extend Montel's theorem to the space $\mathrm{Holg}_\mathcal{H}(\Omega)$ when $\dim \mathcal{H}$ is finite.

Montel Theorems

Proposition 13.8. *If Ω is a finitely open set in $\mathbb{M}^d$, $\mathcal{H}$ is a Hilbert space with $\dim \mathcal{H} < \infty$, and $\{u^k\}$ is a finitely locally uniformly bounded sequence in $\mathrm{Holg}_{\mathcal{H}}(\Omega)$, then $\{u^k\}$ has a convergent subsequence.*

Also, with the setup we have just described, mere notational changes to the proof of Theorem 13.1 yield a proof of the following proposition.

Proposition 13.9. *Assume that Ω is a finitely open set in $\mathbb{M}^d$, $\{\lambda_i\}$ is a dense sequence in Ω with $\lambda_i \in \mathbb{M}_{n_i}^d$ for each i, and $\mathcal{H}$ is a Hilbert space. If $\{u^k\}$ is sequence in $\mathrm{Holg}_{\mathcal{H}}(\Omega)$ that is finitely locally uniformly bounded on Ω, and for each fixed i, $\{u^k(\lambda_i)\}$ is a convergent sequence in $\mathcal{B}(\mathbb{C}^{n_i}, \mathcal{H} \otimes \mathbb{C}^{n_i})$, then $\{u^k\}$ is a convergent sequence in $\mathrm{Holg}_{\mathcal{H}}(\Omega)$.*

We now turn to an analog of Theorem 13.2 in the graded setting.

Lemma 13.10. (Wandering isometries lemma—non-commutative case) *Assume that Ω is a finitely open set in $\mathbb{M}^d$ and $\{\lambda_i\}$ is a sequence in Ω (where, for each i, $\lambda_i \in \mathbb{M}_{n_i}^d$). If $\mathcal{H}$ is an infinite-dimensional Hilbert space and $\{u^k\}$ is sequence in $\mathrm{Holg}_{\mathcal{H}}(\Omega)$ with the property that $\{u^k(\lambda_i)\}$ is bounded for each i, then there exists a subsequence $\{u^{k_l}\}$ and a sequence $\{V^l\}$ of unitary operators on $\mathcal{H}$ such that, for each i, $\{(V^l \otimes \mathrm{id}_{n_i}) u^{k_l}(\lambda_i)\}$ is a convergent sequence in $\mathcal{B}(\mathbb{C}^{n_i}, \mathcal{H} \otimes \mathbb{C}^{n_i})$.*

Proof. Choose an increasing sequence $\{\mathcal{H}_i\}$ of subspaces of $\mathcal{H}$ with the property that

$$\dim \mathcal{H}_1 = n_1^2 \quad \text{and, for } i \geq 1, \quad \dim(\mathcal{H}_{i+1} \ominus \mathcal{H}_i) = n_{i+1}^2.$$

For each n, let $\{e_1, \ldots, e_n\}$ denote the standard basis of $\mathbb{C}^n$.

Fix k. For $i = 1, \ldots, k$, as $u^k(\lambda_i): \mathbb{C}^{n_i} \to \mathcal{H} \otimes \mathbb{C}^{n_i}$, there exist n_i^2 vectors $x_{r,s}^{k,i} \in \mathcal{H}$, $r, s = 1, \ldots, n_i$, such that

$$u^k(\lambda_i) e_r = \sum_{s=1}^{n_i} x_{r,s}^{k,i} \otimes e_s \qquad \text{for } r = 1, \ldots, n_i. \tag{13.11}$$

For each $i = 1, \ldots k$, define $\mathcal{M}_i^k$ by

$$\mathcal{M}_i^k = \mathrm{span}\{x_{r,s}^{k,i}: r, s = 1, \ldots, n_i\},$$

and then define a sequence of spaces $\{\mathcal{N}_i^k\}$ by setting $\mathcal{N}_1^k = \mathcal{M}_1^k$ and

$$\mathcal{N}_i^k = \left(\mathcal{M}_1^k + \mathcal{M}_2^k + \cdots + \mathcal{M}_i^k\right) \ominus \left(\mathcal{M}_1^k + \mathcal{M}_2^k + \cdots + \mathcal{M}_{i-1}^k\right),$$

for $i = 2, \ldots, k$. For $i = 1, \ldots, k$, as $\dim \mathcal{M}_i^k \leq n_i^2$, so also $\dim \mathcal{N}_i^k \leq n_i^2$. Consequently, as the spaces $\{\mathcal{N}_i^k\}$ are also pairwise orthogonal, it follows that there exists a unitary $U^k \in \mathcal{B}(\mathcal{H})$ such that

308   *Non-Commutative Theory*

$U^k(\mathcal{N}_1^k) \subseteq \mathcal{H}_1$ and $U^k(\mathcal{N}_i^k) \subseteq \mathcal{H}_i \ominus \mathcal{H}_{i-1}$ for $i = 2,\ldots,k$.

With this construction it follows by use of equation (13.11) that

$$\left(\begin{matrix} U^k \\ \otimes \\ \mathrm{id}_{n_i} \end{matrix} \right) u^k(\lambda_i)(\mathbb{C}^{n_i}) \subseteq \begin{matrix} \mathcal{H}_i \\ \otimes \\ \mathbb{C}^{n_i} \end{matrix} \qquad \text{for } i = 1,\ldots,k. \qquad (13.12)$$

Now observe that, as the inclusion (13.12) holds for each k, for each i,

$$\left(\begin{matrix} U^k \\ \otimes \\ \mathrm{id}_{n_i} \end{matrix} \right) u^k(\lambda_i)(\mathbb{C}^{n_i}) \subseteq \begin{matrix} \mathcal{H}_i \\ \otimes \\ \mathbb{C}^{n_i} \end{matrix} \qquad \text{for } k = i, i+1, \ldots,$$

that is,

$$\left\{ \begin{matrix} U^k \\ \otimes \\ \mathrm{id}_{n_i} \end{matrix} u^k(\lambda_i) \right\}_{k=i}^{\infty}$$

is a bounded sequence in $\mathcal{B}(\mathbb{C}^{n_i}, \mathcal{H}_i \otimes \mathbb{C}^{n_i})$, a finite-dimensional Hilbert space. Therefore, for each i, there exist $L \in \mathcal{H}$ and an increasing sequence of indices $\{k_l\}$ such that

$$U^{k_l} u^{k_l}(\lambda_i) \to L \quad \text{in } \mathcal{B}\left(\mathbb{C}^{n_i}, \begin{matrix} \mathcal{H}_i \\ \otimes \\ \mathbb{C}^{n_i} \end{matrix} \right) \quad \text{as } l \to \infty.$$

Applying this fact successively with $i = 1$, $i = 2$, and so on, at each stage taking a subsequence of the previously selected subsequence, we construct a sequence $\{L_i\}$ with $L_i \in \mathcal{B}(\mathbb{C}^{n_i}, \mathcal{H}_i \otimes \mathbb{C}^{n_i})$ for each i and an increasing sequence of indices $\{k_l\}$ such that, for all i,

$$U^{k_l} u^{k_l}(\lambda_i) \to L_i \quad \text{in } \mathcal{B}\left(\mathbb{C}^{n_i}, \begin{matrix} \mathcal{H}_i \\ \otimes \\ \mathbb{C}^{n_i} \end{matrix} \right) \quad \text{as } l \to \infty.$$

The lemma then follows if we let $V^l = U^{k_l}$. $\qquad\square$

Lemma 13.10 suggests the following notation. Let Ω be a finitely open set in $\mathbb{M}^d$ and let $\mathcal{H}$ be a Hilbert space. If U is a unitary acting on $\mathcal{H}$, and $f \in \mathrm{Holg}_{\mathcal{H}}(\Omega)$ then we may define $U * f \in \mathrm{Holg}_{\mathcal{H}}(\Omega)$ by the formula

$$(U * f)|\Omega_n = \left(\begin{matrix} U \\ \otimes \\ \mathrm{id}_n \end{matrix} \right) f|\Omega_n \qquad \text{for all } n.$$

With this notation we may formulate a graded analog of Theorem 13.2.

Theorem 13.13. *If Ω is a finitely open set in $\mathbb{M}^d$, $\mathcal{H}$ is a Hilbert space, and $\{u^k\}$ is a finitely locally uniformly bounded sequence in $\mathrm{Holg}_{\mathcal{H}}(\Omega)$, then there exists a sequence $\{U^k\}$ of unitary operators on $\mathcal{H}$ such that $\{U^k * u^k\}$ has a convergent subsequence.*

Proof. Assume that Ω is an open set in $\mathbb{M}^d$, $\mathcal{H}$ is a Hilbert space and $\{u^k\}$ is a finitely locally uniformly bounded sequence in $\mathrm{Holg}_{\mathcal{H}}(\Omega)$. If $\dim \mathcal{H} < \infty$, then the theorem follows from Proposition 13.8 if we choose $U^k = \mathrm{id}_{\mathcal{H}}$ for all k. Therefore, we assume that $\dim \mathcal{H} = \infty$.

Montel Theorems 309

Fix a dense sequence $\{\lambda_i\}$ in Ω. By Lemma 13.10, there exists a subsequence $\{u^{k_l}\}$ and a sequence $\{V^l\}$ of unitary operators on $\mathcal{H}$ such that for each fixed i, $\{V^l u^{k_l}(\lambda_i)\}$ is a convergent sequence in $\mathcal{B}(\mathbb{C}^{n_i}, \mathcal{H} \otimes \mathbb{C}^{n_i})$. Furthermore, as $\{u^k\}$ is locally uniformly bounded, so also is $\{V^l u^{k_l}\}$. Therefore, Proposition 13.9 implies that $\{V^l u^{k_l}\}$ is a convergent sequence in $\mathrm{Holg}_{\mathcal{H}}(\Omega)$. The theorem then follows by choice of $\{U^k\}$ to be any sequence of unitaries in $\mathcal{B}(\mathcal{H})$ such that $U^{k_l} = V^l$ for all l. $\square$

13.4 An nc Montel Theorem

In this section, we are going to prove Theorem 13.13 in the nc and τ-holomorphic setting. Let us fix throughout this section a Hilbert space $\mathcal{H}$, a finitely open nc-set Ω, and an nc-topology τ. We shall prove:

Theorem 13.14. *Suppose $\{u^k\}$ is a finitely locally uniformly bounded sequence in $\mathrm{Holg}_{\mathcal{H}}^{\mathrm{nc}}(\Omega)$. Then there exist $u \in \mathrm{Holg}_{\mathcal{H}}^{\mathrm{nc}}(\Omega)$, a sequence $\{U^k\}$ of unitary operators on $\mathcal{H}$, and an increasing sequence of indices $\{k_l\}$ such that $U^{k_l} * u^{k_l} \to u$ in $\mathrm{Holg}(\Omega)$.*

If in addition Ω is τ-open and $\{u^k\}$ is a τ-locally uniformly bounded sequence in $\mathrm{Holg}_{\mathcal{H}}^{\tau}(\Omega)$, then the subsequence can be chosen so that u is in $\mathrm{Holg}_{\mathcal{H}}^{\tau}(\Omega)$.

Let us emphasize that, in the second case, u is in $\mathrm{Holg}_{\mathcal{H}}^{\tau}(\Omega)$, but the convergence is in $\mathrm{Holg}(\Omega)$, that is, it is uniform convergence on finitely compact subsets of Ω, not on τ-compact sets.

Definition 13.15. *Assume that τ is an nc-topology and $\Omega \in \tau$. If $\{u^k\}$ is a sequence in $\mathrm{Holg}_{\mathcal{H}}^{\tau}(\Omega)$, we say that $\{u^k\}$ is τ-locally uniformly bounded on Ω if for each $\lambda \in \Omega$, there exists a τ-open set $G \subseteq \Omega$ such that $\lambda \in G$ and*

$$\sup_k \|u^k\|_G < \infty.$$

Lemma 13.16. *Let $u \in \mathrm{Holg}_{\mathcal{H}}(\Omega)$ and $\{u^k\}$ be a sequence in $\mathrm{Holg}_{\mathcal{H}}^{\mathrm{nc}}(\Omega)$. If $\{u^k\}$ is finitely locally uniformly bounded on Ω and $u^k \to u$ in $\mathrm{Holg}_{\mathcal{H}}(\Omega)$, then $u \in \mathrm{Holg}_{\mathcal{H}}^{\mathrm{nc}}(\Omega)$. If in addition $\Omega \in \tau$ and $\{u^k\}$ is finitely locally uniformly bounded, then $u \in \mathrm{Holg}_{\mathcal{H}}^{\tau}(\Omega)$.*

Proof. To prove the first assertion, note that as $u \in \mathrm{Holg}_{\mathcal{H}}(\Omega)$, it is graded. To verify that it preserves direct sums, assume that $\lambda, \mu, \lambda \oplus \mu \in \Omega$. Then, as $u^k \to u$ in $\mathrm{Holg}_{\mathcal{H}}(\Omega)$ and $u^k \in \mathrm{Holg}_{\mathcal{H}}^{\mathrm{nc}}(\Omega)$ for all k,

$$u(\lambda \oplus \mu) = \lim_{k \to \infty} u^k(\lambda \oplus \mu)$$
$$= \lim_{k \to \infty} \left(u^k(\lambda) \oplus u^k(\mu) \right)$$

310 · Non-Commutative Theory

$$= \lim_{k \to \infty} u^k(\lambda) \oplus \lim_{k \to \infty} u^k(\mu)$$

$$= u(\lambda) \oplus u(\mu).$$

Finally, note that if $n \geq 1$, $s \in \mathbb{M}_n$ is invertible, and both λ and $s\lambda s^{-1}$ are in Ω_n, then

$$u(s\lambda s^{-1}) = \lim_{k \to \infty} u^k(s\lambda s^{-1})$$

$$= \lim_{k \to \infty} \overset{\mathrm{id}_{\mathcal{H}}}{\underset{s}{\otimes}} u^k(\lambda) s^{-1}$$

$$= \overset{\mathrm{id}_{\mathcal{H}}}{\underset{s}{\otimes}} u(\lambda) s^{-1},$$

which proves that u preserves similarities, and hence u is nc.

To prove the second assertion, fix $\lambda \in \Omega$. As $\{u^k\}$ is τ-locally uniformly bounded on Ω, there exist $G \subseteq \Omega$ and a constant ρ such that $\lambda \in G \in \tau$ and

$$\sup_k \|u^k\|_G \leq \rho.$$

Consider $\mu \in G$. As we assume that $u^k \to u$ in $\mathrm{Holg}_{\mathcal{H}}(\Omega)$, it follows that

$$\|u(\mu)\| = \lim_{k \to \infty} \|u^k(\mu)\| \leq \rho.$$

But then,

$$\|u\|_G \leq \rho.$$

As $G \in \tau$, this proves that u is τ-locally bounded on Ω. $\qquad\square$

Definition 13.15 and Lemma 13.16, allow one to deduce Theorem 13.14 as a corollary of Theorem 13.13.

Proof of Theorem 13.14. As $\{u^k\}$ is finitely locally uniformly bounded in $\mathrm{Holg}_{\mathcal{H}}(\Omega)$, Theorem 13.13 implies that there exists a sequence $\{U^k\}$ of unitary operators on $\mathcal{H}$ such that $\{U^k * u^k\}$ has a subsequence that converges in $\mathrm{Holg}_{\mathcal{H}}(\Omega)$. Consequently, we may choose $u \in \mathrm{Holg}_{\mathcal{H}}(\Omega)$ and an increasing sequence of indices $\{k_l\}$ such that $U^{k_l} * u^{k_l} \to u$ in $\mathrm{Holg}(\Omega)$. The proof is completed by the observation that Lemma 13.16 implies that u can be taken to be in $\mathrm{Holg}_{\mathcal{H}}^{\mathrm{nc}}(\Omega)$ or $\mathrm{Holg}_{\mathcal{H}}^{\tau}(\Omega)$ if the u^k are. $\qquad\square$

13.5 Closed Cones

In Chapter 14, we shall want to know that certain cones are closed, so that we can employ the duality construction. In this section, Ω will be a finitely open set. We shall let $\mathrm{Holg}(\Omega)^*$ be the space of functions $\lambda \mapsto f(\lambda)^*$ for $f \in \mathrm{Holg}(\Omega)$. The algebraic tensor product $\mathrm{Holg}(\Omega)^* \otimes \mathrm{Holg}(\Omega)$

Montel Theorems

can be concretely realized as the set of functions A defined on $\Omega \boxtimes \Omega$ such that there exist a finite-dimensional Hilbert space $\mathcal{H}$ and $u, v \in \mathrm{Holg}_{\mathcal{H}}(\Omega)$ such that

$$A(\lambda, \mu) = v(\mu)^* u(\lambda) \qquad \text{for } (\lambda, \mu) \in \Omega \boxtimes \Omega.$$

As the functions in $\mathrm{Holg}(\Omega)^* \otimes \mathrm{Holg}(\Omega)$ are holomorphic in λ for each fixed μ and anti-holomorphic in μ for each fixed λ, we may complete $\mathrm{Holg}(\Omega)^* \otimes \mathrm{Holg}(\Omega)$ in the topology of uniform convergence on finitely compact subsets of $\Omega \boxtimes \Omega$ to obtain the space of *graded hereditary functions on* Ω, $\mathrm{Herg}(\Omega)$. Inside $\mathrm{Herg}(\Omega)$, we may define a cone $\mathcal{P}$ by

$$\mathcal{P} = \{u(\mu)^* u(\lambda): \ u \in \mathrm{Holg}_{\mathcal{H}}(\Omega) \text{ for some Hilbert space } \mathcal{H}\}. \tag{13.17}$$

Theorem 13.18. $\mathcal{P}$ *is closed in* $\mathrm{Herg}(\Omega)$.

Proof. Assume that $\{v^k\}$ is a sequence with $v^k \in \mathrm{Hol}_{\mathcal{H}_k}(\Omega)$ for each k and with $v^k(\mu)^* v^k(\lambda) \to A$ in $\mathrm{Herg}(\Omega)$. We may assume that $\mathcal{H}_k$ is separable for each k. Fix a separable infinite dimensional Hilbert space $\mathcal{H}$ and for each each k choose an isometry $V^k: \mathcal{H}_k \to \mathcal{H}$. If for each k we let $u^k = V^k * v^k$, then $\{u^k\}$ is a sequence in $\mathrm{Holg}_{\mathcal{H}}(\Omega)$ and $u^k(\mu)^* u^k(\lambda) \to A$ in $\mathrm{Herg}(\Omega)$.

Now, as $u^k(\mu)^* u^k(\lambda) \to A$ in $\mathrm{Herg}(\Omega)$ it follows that $\{u^k\}$ is a finitely locally uniformly bounded sequence in $\mathrm{Holg}_{\mathcal{H}}(\Omega)$. Hence, by Theorem 13.13, there exists a sequence U^k of unitary operators on $\mathcal{H}$ such that $\{U^k * u^k\}$ has a convergent subsequence, that is, there exists $u \in \mathrm{Holg}_{\mathcal{H}}(\Omega)$ and an increasing sequence of indices $\{k_l\}$ such that $U^{k_l} * u^{k_l} \to u$. But then, for each $(\lambda, \mu) \in \Omega \boxtimes \Omega$,

$$\begin{aligned}
A(\lambda, \mu) &= \lim_{k \to \infty} u^k(\mu)^* u^k(\lambda) \\
&= \lim_{l \to \infty} u^{k_l}(\mu)^* u^{k_l}(\lambda) \\
&= \lim_{l \to \infty} (U^{k_l} * u^{k_l})(\mu)^* (U^{k_l} * u^{k_l})(\lambda) \\
&= u(\mu)^* u(\lambda),
\end{aligned}$$

that is, $A \in \mathcal{P}$. $\qquad\square$

We also may use wandering Montel theorems to study sums of τ-holomorphic dyads. We let $\mathrm{Herg}^\tau(\Omega)$ denote the closure of

$$\{v(\mu)^* u(\lambda): \ u, v \in \mathrm{Holg}_{\mathcal{H}}^\tau(\Omega) \text{ for some finite-dimensional Hilbert space } \mathcal{H}\}$$

inside $\mathrm{Herg}(\Omega)$ and define $\mathcal{P}^\tau$ in $\mathrm{Herg}^\tau(\Omega)$ by

$$\mathcal{P}^\tau = \{u(\mu)^* u(\lambda): \ u \in \mathrm{Holg}_{\mathcal{H}}^\tau(\Omega) \text{ for some Hilbert space } \mathcal{H}\}.$$

Theorem 13.19. *Let* τ *be an nc-topology and let* $\Omega \in \tau$. *Then* $\mathcal{P}^\tau$ *is closed in* $\mathrm{Herg}^\tau(\Omega)$.

312   *Non-Commutative Theory*

Proof. Assume that $u^k(\mu)^* u^k(\lambda) \to A$ in Herg(Ω), where, as in the proof of Theorem 13.18, we may assume that $u^k \in \text{Holg}^\tau_{\mathcal{H}}(\Omega)$ for each k. By Theorem 13.13, there exist $u \in \text{Holg}^\tau_{\mathcal{H}}(\Omega)$, a sequence U^k of unitary operators on $\mathcal{H}$, and and an increasing sequence of indices $\{k_l\}$ such that $U^{k_l} * u^{k_l} \to u$. But then as in the proof of Theorem 13.18, $A(\lambda, \mu) = u(\mu)^* u(\lambda)$ for all $(\lambda, \mu) \in \Omega \boxtimes \Omega$, that is, $A \in \mathcal{P}^\tau$. $\qquad\square$

Finally, we shall prove that the model cone is closed; we shall need this result in Sections 14.3 and 14.4. Let δ be a $I \times J$ matrix whose entries are free polynomials in d variables, and let B_δ be the polynomial polyhedron defined in equation (12.30). The model cone $\mathcal{C}$ is the set of graded hereditary functions on B_δ of the form

$$\mathcal{C} \stackrel{\text{def}}{=} \left\{ \begin{array}{c} \text{id}_{\mathbb{C}^I} \\ \otimes \\ u(\mu)^* \end{array} \left(\text{id} - \begin{array}{c} \delta(\mu)^* \delta(\lambda) \\ \otimes \\ \text{id}_{\mathcal{H}} \end{array} \right) \begin{array}{c} \text{id}_{\mathbb{C}^I} \\ \otimes \\ u(\lambda) \end{array} : u \in \text{Holg}_{\mathcal{H}}(B_\delta), \ u \text{ is nc,} \right.$$

$$\left. \mathcal{H} \text{ is a Hilbert space} \right\}. \tag{13.21}$$

Theorem 13.22. *The model cone $\mathcal{C}$, defined in equation* (13.21), *is closed in* Her(B_δ).

Proof. Suppose u^k is a sequence of nc functions in $\text{Holg}_{\mathcal{H}}(B_\delta)$ (we may assume the space $\mathcal{H}$ is the same for each u^k, as in the proof of Theorem 13.18), so that

$$\begin{array}{c} \text{id}_{\mathbb{C}^I} \\ \otimes \\ u^k(\mu)^* \end{array} \left(\text{id} - \begin{array}{c} \delta(\mu)^* \delta(\lambda) \\ \otimes \\ \text{id}_{\mathcal{H}} \end{array} \right) \begin{array}{c} \text{id}_{\mathbb{C}^I} \\ \otimes \\ u^k(\lambda) \end{array} \tag{13.23}$$

converges in Herg(B_δ) to $A(\lambda, \mu)$. On any finitely compact set, $\|\delta(x)\|$ will be bounded by a constant that is strictly less than one. Since the sequence (13.23) converges uniformly on finitely compact subsets of $B_\delta \boxtimes B_\delta$, this means that u^k is a finitely locally uniformly bounded sequence. Therefore by Theorem 13.13, there exist unitaries U^k such that $U^k * u^k$ has a convergent subsequence, which converges to some nc function $u \in \text{Holg}_{\mathcal{H}}(B_\delta)$. Then

$$A(\lambda, \mu) = \begin{array}{c} \text{id}_{\mathbb{C}^I} \\ \otimes \\ u(\mu)^* \end{array} \left(\text{id} - \begin{array}{c} \delta(\mu)^* \delta(\lambda) \\ \otimes \\ \text{id}_{\mathcal{H}} \end{array} \right) \begin{array}{c} \text{id}_{\mathbb{C}^I} \\ \otimes \\ u(\lambda) \end{array},$$

as desired. $\qquad\square$

13.6 Historical Notes

Theorem 13.1 is from [34]. Most of the material in this chapter on wandering isometries comes from [19].

14

Free Holomorphic Functions

Throughout this chapter, δ will be a fixed $I \times J$ matrix of free polynomials in $\mathcal{P}_d$ and B_δ the polynomial polyhedron $\{x \in \mathbb{M}^d \colon \|\delta(x)\| < 1\}$. One can add rows or columns of 0's to δ to make it a square matrix, but it is easier to keep track of the algebra if we allow I and J to be distinct.

A free[1] holomorphic function is one that is, locally, a function that is bounded on some B_δ, so we shall study $H^\infty(B_\delta)$, the bounded nc-functions on B_δ. We shall let $H_1^\infty(B_\delta)$ denote the closed unit ball, the space of functions that are bounded by 1, and we shall develop a model formula and representation theorem for them analogous to the network realization formulas in earlier chapters.

We shall use these results to show that holomorphic functions that are bounded on the commutative d-tuples in B_δ extend to bounded free holomorphic functions on all of B_δ (Theorem 14.41), and that graded functions on free open sets are freely holomorphic if and only if they are locally approximable (in the free topology) by free polynomials (Theorem 14.39).

14.1 The Range of a Free Holomorphic Function

We shall show that free holomorphic functions φ have Property (v) in Section 12.1: for every $x \in \mathbb{M}^d$, the image $\varphi(x)$ is in the algebra generated by x for every x in the domain $\mathrm{dom}(\varphi)$ of φ. For an element x of $\mathbb{M}_n^d$, let $\mathrm{Alg}(x)$ denote the unital subalgebra of $\mathbb{M}_n$ generated by x, that is, $\{p(x) \colon p \in \mathcal{P}_d\}$.

Let $e_1, \ldots, e_n$ be the standard basis for $\mathbb{C}^n$. Recall that $x^{(k)}$ is the direct sum of k copies of x.

---

[1] Strictly speaking, the terminology *freely holomorphic* is more logical, meaning "holomorphic with respect to the free topology," but the present terminology has become established.

313

314 Non-Commutative Theory

Lemma 14.1. *Let* $z \in \mathbb{M}_n^d$, *with* $\|z\| < 1$. *If* $w \in \mathbb{M}_n$ *and* $w \notin \text{Alg}(z)$ *then there is an invertible* $s \in \mathbb{M}_{n^2}$ *such that* $\|s^{-1}z^{(n)}s\| < 1$ *and* $\|s^{-1}w^{(n)}s\| > 1$.

Proof. Let $\mathcal{A} = \text{Alg}(z)$. Since $w \notin \mathcal{A}$, and $\mathcal{A}$ is finite-dimensional and therefore closed, the Hahn–Banach theorem implies that there is a matrix $K \in \mathbb{M}_n$ such that $\text{tr}(aK) = 0$ for all $a \in \mathcal{A}$ and $\text{tr}(wK) \neq 0$. Let $u \in \mathbb{C}^n \otimes \mathbb{C}^n$ be the direct sum of the columns of K, and $v = e_1 \oplus e_2 \oplus \ldots e_n$. Then, for any $b \in \mathbb{M}_n$, we have

$$\text{tr}(bK) = \langle b^{(n)}u, v \rangle.$$

We have $\langle a^{(n)}u, v \rangle = 0$ for all $a \in \mathcal{A}$, but $\langle w^{(n)}u, v \rangle \neq 0$.

Let $\mathcal{A} \otimes \text{id}$ denote the subalgebra $\{a^{(n)} : a \in \mathcal{A}\}$ of $\mathcal{M}_{n^2}$ and let $\mathcal{M} = (\mathcal{A} \otimes \text{id})u$. This is an $\mathcal{A} \otimes \text{id}$-invariant subspace of $\mathbb{C}^{n^2}$, but it is not $w^{(n)}$-invariant (since $v \perp \mathcal{M}$, but v is not perpendicular to $w^{(n)}u$). With respect to the decomposition $\mathcal{M} \oplus \mathcal{M}^{\perp}$ of $\mathbb{C}^n \otimes \mathbb{C}^n$, every matrix in $\mathcal{A} \otimes \text{id}$ has 0 in the $(2,1)$ entry, but $w^{(n)}$ does not.

Let $s = x\text{id}_{\mathcal{M}} + y\text{id}_{\mathcal{M}^{\perp}}$, with $x \gg y > 0$. Then

$$s^{-1}\begin{bmatrix} A & B \\ C & D \end{bmatrix} s = \begin{bmatrix} A & \frac{y}{x}B \\ \frac{x}{y}C & D \end{bmatrix}.$$

If the ratio x/y is large enough, then for each of the d matrices z^r, the corresponding $s^{-1}(z^r \otimes \text{id})s$ has strict contractions in the $(1,1)$ and $(2,2)$ slots, and each $(1,2)$ entry will be so small that $s^{-1}(z^r \otimes \text{id})s$ is a contraction.

For w, however, as C is non-zero, the norm of $s^{-1}w^{(n)}s$ can be made arbitrarily large. $\square$

Lemma 14.2. *Let* $z \in B_\delta \cap \mathbb{M}_n^d$ *and let* $w \in \mathbb{M}_n$ *not be in* $\text{Alg}(z)$. *Then there is an invertible* $s \in \mathbb{M}_{n^2}$ *such that* $s^{-1}z^{(n)}s \in B_\delta$ *and* $\|s^{-1}w^{(n)}s\| > 1$.

Proof. As in the proof of Lemma 14.1, we can find an invariant subspace $\mathcal{M}$ for $\mathcal{A} \otimes \text{id}$ that is not w-invariant. Decompose $\delta(z^{(n)})$ as a map from $(\mathcal{M} \otimes \mathbb{C}^J) \oplus (\mathcal{M}^{\perp} \otimes \mathbb{C}^J)$ into $(\mathcal{M} \otimes \mathbb{C}^I) \oplus (\mathcal{M}^{\perp} \otimes \mathbb{C}^I)$. With s as in Lemma 14.1, and $x \gg y > 0$, and P the projection from $\mathbb{C}^n \otimes \mathbb{C}^n$ onto $\mathcal{M}$, we get

$$\delta(s^{-1}z^{(n)}s) = \begin{bmatrix} P\delta(z^{(n)})P & \frac{y}{x}P\delta(z^{(n)})P^{\perp} \\ 0 & P^{\perp}\delta(z^{(n)})P^{\perp} \end{bmatrix}. \qquad (14.3)$$

For x/y large enough, every matrix of the form (14.3) is a contraction, so $s^{-1}z^{(n)}s \in B_\delta$. But $s^{-1}w^{(n)}s$ will contain a non-zero element multiplied by $\frac{x}{y}$, so we achieve the claim. $\square$

Theorem 14.4. *If* φ *is a free holomorphic function then* $\varphi(z) \in \text{Alg}(z)$ *for all* $z \in \text{dom}(\varphi)$.

Free Holomorphic Functions 315

Proof. We can assume that $z \in B_\delta$ and that $\|\varphi\| \leq 1$ on B_δ. Let $w = \varphi(z)$. If $w \notin \text{Alg}(z)$, then by Lemma 14.2, there is an s such that $s^{-1}z^{(n)}s \in B_\delta$ and $\|\varphi(s^{-1}z^{(n)}s)\| = \|s^{-1}w^{(n)}s\| > 1$, a contradiction. $\quad\square$

In Section 15.4, we shall show that this result holds for fine holomorphic functions when $d = 1$, but fails when $d \geq 2$.

14.2 Nc-Models and Free Realizations

Let $\mathcal{K}$ and $\mathcal{L}$ be separable Hilbert spaces, and let φ be a $\mathcal{B}(\mathcal{K}, \mathcal{L})$-valued graded function on B_δ.

Definition 14.5. *An nc-model* for φ consists of an auxiliary Hilbert space $\mathcal{M}$ and a $\mathcal{B}(\mathcal{K}, \mathcal{M} \otimes \mathbb{C}^J)$-valued nc-function u on B_δ such that, for all pairs $x, y \in B_\delta$ that are on the same level,

$$1 - \varphi(y)^*\varphi(x) = u(y)^* \left[\underset{1-\delta(y)^*\delta(x)}{\overset{1}{\otimes}} \right] u(x). \tag{14.6}$$

We shall say that the model is free *if, in addition, u is a free holomorphic $\mathcal{B}(\mathcal{K}, \mathcal{M} \otimes \mathbb{C}^J)$-valued function.*

The 1's in equation (14.6) have to be interpreted appropriately. If x, y are at level n, then equation (14.6) means

$$\text{id}_{\mathcal{K} \otimes \mathbb{C}^n} - \varphi(y)^*\varphi(x) = u(y)^* \left[\underset{\text{id}_{\mathbb{C}^J \otimes \mathbb{C}^n} - \delta(y)^*\delta(x)}{\overset{\text{id}_\mathcal{M}}{\otimes}} \right] u(x).$$

Definition 14.7. *A free realization* of φ consists of an auxiliary Hilbert space $\mathcal{M}$ and an isometry

$$\begin{array}{cc} & \mathcal{K} \quad \mathcal{M} \otimes \mathbb{C}^I \\ \begin{array}{c} \mathcal{L} \\ \mathcal{M} \otimes \mathbb{C}^J \end{array} & \begin{pmatrix} A & B \\ C & D \end{pmatrix} \end{array} \tag{14.8}$$

such that, for all $x \in B_\delta$,

$$\varphi(x) = \underset{1}{\overset{A}{\otimes}} + \underset{1}{\overset{B}{\otimes}} \underset{\delta(x)}{\overset{1}{\otimes}} \left[1 - \underset{1}{\overset{D}{\otimes}} \underset{\delta(x)}{\overset{1}{\otimes}} \right]^{-1} \underset{1}{\overset{C}{\otimes}}. \tag{14.9}$$

Again, the 1's need to be interpreted appropriately. If $x \in \mathbb{M}_n^d$, then $\varphi(x)$ is an operator from $\underset{\mathbb{C}^n}{\overset{\mathcal{K}}{\otimes}}$ to $\underset{\mathbb{C}^n}{\overset{\mathcal{L}}{\otimes}}$, and equation (14.9) means

$$\varphi(x) = \underset{\text{id}_{\mathbb{C}^n}}{\overset{A}{\otimes}} + \underset{\text{id}_{\mathbb{C}^n}}{\overset{B}{\otimes}} \underset{\delta(x)}{\overset{\text{id}_\mathcal{M}}{\otimes}} \left[\begin{array}{c} \text{id}_\mathcal{M} \\ \otimes \\ \text{id}_{\mathbb{C}^J} \\ \otimes \\ \text{id}_{\mathbb{C}^n} \end{array} - \underset{\text{id}_{\mathbb{C}^n}}{\overset{D}{\otimes}} \underset{\delta(x)}{\overset{\text{id}_\mathcal{M}}{\otimes}} \right]^{-1} \underset{\text{id}_{\mathbb{C}^n}}{\overset{C}{\otimes}}.$$

316   *Non-Commutative Theory*

We shall prove in Theorem 14.33 that every function in $H_1^\infty(B_\delta)$ has an nc-model and a free realization. First, we show that these latter two are equivalent and that any such function lies in $H_1^\infty(B_\delta)$. Indeed, we shall do more. A partially defined free realization or nc-model can be used to give a formula that extends the function to all of B_δ. If $E \subseteq B_\delta$, and φ is a $\mathcal{B}(\mathcal{K},\mathcal{L})$-valued graded function on E, let us say that φ has a *free realization on E* if there is an isometry as in formula (14.8) such that equation (14.9) holds for all $x \in E$. We say that φ has an *nc-model on E* if there is a $\mathcal{B}(\mathcal{K}, \mathcal{M} \otimes \mathbb{C}^J)$-valued nc-function u on E such that equation (14.6) holds for all $(x, y) \in E^{[2]}$.

Theorem 14.10. *Let $E \subseteq B_\delta$, and let φ be a $\mathcal{B}(\mathcal{K},\mathcal{L})$-valued graded function on E. If φ has a free realization on E, then φ extends to a function in $H_1^\infty(B_\delta)$ that has an nc-model.*

Proof. Extend φ to all of B_δ by equation (14.9). For all $x \in B_\delta$, define

$$u(x) = \left[1 - \underset{1}{\overset{D}{\otimes}}\ \underset{\delta(x)}{\overset{1}{\otimes}} \right]^{-1} \underset{1}{\overset{C}{\otimes}}.$$

Use the fact that

$$\begin{bmatrix} A^* & C^* \\ B^* & D^* \end{bmatrix} \begin{bmatrix} A & B \\ C & D \end{bmatrix} = \begin{bmatrix} 1 & 0 \\ 0 & 1 \end{bmatrix}$$

to write $A^*A = 1 - C^*C$, $A^*B = -C^*D$, $B^*A = -D^*C$, $B^*B = 1 - D^*D$. Then, after some cancellations,

$$1 - \varphi(y)^*\varphi(x) = \underset{1}{\overset{C^*}{\otimes}} \left[1 - \underset{\delta(y)^*}{\overset{1}{\otimes}}\ \underset{1}{\overset{D^*}{\otimes}} \right]^{-1} \left[1 - \underset{\delta(y)^*\delta(x)}{\overset{1}{\otimes}} \right] \left[1 - \underset{1}{\overset{D}{\otimes}}\ \underset{\delta(x)}{\overset{1}{\otimes}} \right]^{-1} \underset{1}{\overset{C}{\otimes}}$$

$$= u(y)^* \left[1 - \underset{\delta(y)^*\delta(x)}{\overset{1}{\otimes}} \right] u(x).$$

This proves that φ has an nc-model. Moreover, letting $y = x$, we see that $\|\varphi(x)\| \leq 1$. Since any function of the form (14.9) is clearly nc, this proves that these functions all lie in $H_1^\infty(B_\delta)$. $\qquad\square$

To go from an nc-model to a free realization, we need a little structure on E. We do not, however, need it to be open. First, let us prove a preliminary lemma.

Lemma 14.11. *Let $\mathcal{H},\mathcal{K},\mathcal{L}$ be Hilbert spaces. Suppose $T \in \mathcal{B}(\mathcal{K} \otimes \mathcal{H}, \mathcal{L} \otimes \mathcal{H})$ satisfies*

$$(\mathrm{id}_\mathcal{L} \otimes w^*)T(\mathrm{id}_\mathcal{K} \otimes w) = T \qquad \text{for every unitary } w \text{ on } \mathcal{H}.$$

Then T decomposes as $T' \otimes \mathrm{id}_\mathcal{H}$ for some $T' \in \mathcal{B}(\mathcal{K},\mathcal{L})$.

Proof. Since the unitaries span $\mathcal{B}(\mathcal{H})$, we have

$$T(\mathrm{id}_\mathcal{K} \otimes x) = (\mathrm{id}_L \otimes x)T \qquad \text{for all } x \in \mathcal{B}(\mathcal{H}).$$

Free Holomorphic Functions 317

The result follows from letting x range over the matrix units and the permutations. $\square$

Theorem 14.12. *Let $E \subseteq B_\delta$ be an nc-set that is closed with respect to unitary conjugation. Let φ be a $\mathcal{B}(\mathcal{K}, \mathcal{L})$-valued graded function on E that preserves direct sums and conjugation by unitaries. Suppose φ has an nc-model u on E. Then φ has an extension to a function in $H_1^\infty(B_\delta)$ that has a free realization.*

Proof. Assume that equation (14.6) holds on $E \boxtimes E$. The lurking isometry argument shows that, at each level n, there is an isometry V_n such that

$$
V_n = \begin{bmatrix} A_n & B_n \\ C_n & D_n \end{bmatrix} : \begin{pmatrix} \mathrm{id}_{\mathcal{K}} \\ \otimes \\ \mathrm{id}_{\mathbb{C}^n} \\ \\ \mathrm{id}_{\mathcal{M}} \\ \otimes \\ \delta(x) \end{pmatrix} \begin{matrix} k \\ \otimes \\ h \end{matrix} \mapsto \begin{pmatrix} \varphi(x) \\ u(x) \end{pmatrix} \begin{matrix} k \\ \otimes \\ h \end{matrix} \tag{14.13}
$$

for all $x \in E \cap \mathbb{M}_n^d$, all $k \in \mathcal{K}$ and all $h \in \mathbb{C}^n$. We would like to find a single matrix V of the form (14.8) such that $V_n = V \otimes \mathrm{id}_{\mathbb{C}^n}$. If we can, then by solving equation (14.13) for φ we derive a free realization.

At each level n, we will average V_n over the unitary group $\mathcal{U}_n$. Let m_n denote Haar measure on the compact group $\mathcal{U}_n$ of unitary $n \times n$ matrices. Define

$$
\tilde{V}_n = \begin{bmatrix} \tilde{A}_n & \tilde{B}_n \\ \tilde{C}_n & \tilde{D}_n \end{bmatrix} = \int_{\mathcal{U}_n} \begin{bmatrix} \mathrm{id}_{\mathcal{L}} \otimes z^* & 0 \\ 0 & \mathrm{id}_{\mathcal{M}} \otimes \mathrm{id}_{\mathbb{C}^J} \otimes z^* \end{bmatrix} \begin{bmatrix} A_n & B_n \\ C_n & D_n \end{bmatrix} \begin{bmatrix} \mathrm{id}_{\mathcal{K}} \otimes z & 0 \\ 0 & \mathrm{id}_{\mathcal{M}} \otimes \mathrm{id}_{\mathbb{C}^I} \otimes z \end{bmatrix} dm_n(z).
$$

If $y = z^* x z$, then the formula (14.13) gives

$$
\begin{bmatrix} \frac{1}{z^*} \otimes & 0 \\ 0 & \underset{z^*}{\otimes} \frac{1}{} \end{bmatrix} \begin{bmatrix} A_n & B_n \\ C_n & D_n \end{bmatrix} \begin{bmatrix} \frac{1}{z} \otimes & 0 \\ 0 & \underset{z}{\otimes} \frac{1}{} \end{bmatrix} : \begin{pmatrix} 1 \\ \underset{\delta(y)}{\otimes} u(y) \end{pmatrix} \begin{matrix} k \\ \otimes \\ h \end{matrix} \mapsto \begin{pmatrix} \varphi(y) \\ u(y) \end{pmatrix} \begin{matrix} k \\ \otimes \\ h \end{matrix}.
$$

Since this holds for all $y \in E$, we find that, just as in equation (14.13),

$$
\tilde{V}_n : \begin{pmatrix} 1 \\ \underset{\delta(x)}{\otimes} u(x) \end{pmatrix} \begin{matrix} k \\ \otimes \\ h \end{matrix} \mapsto \begin{pmatrix} \varphi(x) \\ u(x) \end{pmatrix} \begin{matrix} k \\ \otimes \\ h \end{matrix} \tag{14.14}
$$

for all $x \in E \cap \mathbb{M}_n^d$, $k \in \mathcal{K}, h \in \mathbb{C}^n$.

Since Haar measure is translation-invariant, for every $w \in \mathcal{U}_n$, we have

$$
\begin{bmatrix} \tilde{A}_n & \tilde{B}_n \\ \tilde{C}_n & \tilde{D}_n \end{bmatrix} = \begin{bmatrix} \frac{1}{w^*} \otimes & 0 \\ 0 & \underset{w^*}{\otimes} \frac{1}{} \end{bmatrix} \begin{bmatrix} \tilde{A}_n & \tilde{B}_n \\ \tilde{C}_n & \tilde{D}_n \end{bmatrix} \begin{bmatrix} \frac{1}{w} \otimes & 0 \\ 0 & \underset{w}{\otimes} \frac{1}{} \end{bmatrix}.
$$

318 *Non-Commutative Theory*

By Lemma 14.11, this means that there is a decomposition

$$
\begin{bmatrix} \tilde{A}_n & \tilde{B}_n \\ \tilde{C}_n & \tilde{D}_n \end{bmatrix} = \begin{bmatrix} A_n' \underset{\mathrm{id}_{\mathbb{C}^n}}{\otimes} & B_n' \underset{\mathrm{id}_{\mathbb{C}^n}}{\otimes} \\ C_n' \underset{\mathrm{id}_{\mathbb{C}^n}}{\otimes} & D_n' \underset{\mathrm{id}_{\mathbb{C}^n}}{\otimes} \end{bmatrix} = V_n'.
\tag{14.15}
$$

$\tilde{V}_n$ is an isometry on the span of the left-hand side of the display (14.14), so we can extend V_n' to be an isometry from $\mathcal{K} \oplus (\mathcal{M} \otimes \mathbb{C}^I)$ to $\mathcal{L} \oplus (\mathcal{M} \otimes \mathbb{C}^J)$. (As usual, we add an infinite-dimensional summand to $\mathcal{M}$, if necessary, so that we can do this).

Finally, consider the sequence V_n' in the closed unit ball of $\mathcal{B}(\mathcal{K} \oplus (\mathcal{M} \otimes \mathbb{C}^I), \mathcal{L} \oplus (\mathcal{M} \otimes \mathbb{C}^J))$, which is compact in the weak-* and weak operator topologies. Let n_j be a sequence of natural numbers tending to ∞ such that, for every $m \in \mathbb{N}$, every n_j is divisible by m for j large enough (e.g., $n_j = j!$). Let V be a weak-* cluster point of V_{n_j}'.

Let $x \in B_\delta \cap \mathbb{M}_m^d$. For j large enough, the direct sum of n_j/m copies of x will be in $B_\delta \cap \mathbb{M}_{n_j}^d$. Since the formula (14.14) applies for every n_j we have the map

$$
\underset{\mathrm{id}_{\mathbb{C}^m}}{\overset{V}{\otimes}} : \begin{pmatrix} \overset{1}{\underset{\delta(x)}{\overset{1}{\otimes}} u(x)} \end{pmatrix} \overset{k}{\underset{h}{\otimes}} \mapsto \begin{pmatrix} \varphi(x) \\ u(x) \end{pmatrix} \overset{k}{\underset{h}{\otimes}}.
\tag{14.16}
$$

Writing

$$
V = \begin{array}{c} \\ \mathcal{L} \\ \mathcal{M} \otimes \mathbb{C}^J \end{array} \overset{\mathcal{K} \quad \mathcal{M} \otimes \mathbb{C}^I}{\begin{pmatrix} A & B \\ C & D \end{pmatrix}}
\tag{14.17}
$$

we get the desired free realization. $\qquad\qquad\square$

Notice that we end up with a formula for u. From the formulae (14.16) and (14.17) we obtain the equation

$$
\underset{\mathrm{id}_{\mathbb{C}^n}}{\overset{C}{\otimes}} + \underset{\mathrm{id}_{\mathbb{C}^n}}{\overset{D}{\otimes}} \underset{\delta(x)}{\overset{\mathrm{id}_{\mathcal{M}}}{\otimes}} u(x) = u(x) \qquad \text{for all } x \in E_n,
$$

so that

$$
u(x) = \left(1 - \underset{\mathrm{id}_{\mathbb{C}^n}}{\overset{D}{\otimes}} \underset{\delta(x)}{\overset{\mathrm{id}_{\mathcal{M}}}{\otimes}}\right)^{-1} \underset{\mathrm{id}_{\mathbb{C}^n}}{\overset{C}{\otimes}} \qquad \text{for all } x \in E_n.
$$

Therefore u is a free holomorphic function automatically.

Corollary 14.18. *An nc-model on B_δ is a free model.*

Free Holomorphic Functions   319

Combining Theorems 14.10 and 14.12, we get:

Theorem 14.19. *A $\mathcal{B}(\mathcal{K},\mathcal{L})$-valued graded function on B_δ has an nc-model if and only if it has a free realization.*

14.3 Free Pick Interpolation

The free Pick interpolation problem on B_δ is to determine, given N points $\lambda_1,\dots,\lambda_N$ in B_δ and N matrices w_i of corresponding dimension, whether there exists $\varphi \in H_1^\infty(B_\delta)$ such that $\varphi(\lambda_i) = w_i$, $1 \le i \le N$.

Note first that by letting $\lambda = \oplus_{i=1}^N \lambda_i$ and $w = \oplus_{i=1}^N w_i$, we convert the original N point problem to the one-point Pick problem of mapping λ to w. Second, in contrast to the scalar case, one cannot always solve the Pick problem if one drops the norm constraint. For example, by Theorem 14.4, no free function maps

$$\begin{pmatrix} 0 & 1 \\ 0 & 0 \end{pmatrix} \quad \text{to} \quad \begin{pmatrix} 1 & 0 \\ 0 & 0 \end{pmatrix}.$$

To state the theorem, let us make the following definitions for λ in $\mathbb{M}^d$. Define

$$I_\lambda = \{p \in \mathcal{P}_d : p(\lambda) = 0\},$$

and define an nc version of the Zariski closure of λ by

$$V_\lambda = \{x \in \mathbb{M}^d : p(x) = 0 \text{ whenever } p \in I_\lambda\}.$$

Theorem 14.20. *Let $\lambda \in B_\delta \cap \mathbb{M}_n^d$ and let $w \in \mathbb{M}_n$. There exists a function φ in the closed unit ball of $H^\infty(B_\delta)$ such that $\varphi(\lambda) = w$ if and only if*

(i) $w \in \mathrm{Alg}(\lambda)$, *so that there exists $p_0 \in \mathcal{P}_d$ such that $p_0(\lambda) = w$; and*
(ii) $\sup\{\|p_0(x)\|: x \in V_\lambda \cap B_\delta\} \le 1$.

Moreover, if these two conditions are satisfied, φ can be chosen to have a free realization.

Let $E = V_\lambda \cap B_\delta$, and observe that E is closed under direct sums and unitary conjugation.

Lemma 14.21. *Let $\lambda, x \in \mathbb{M}^d$. The following conditions are equivalent.*

(i) $x \in V_\lambda$;
(ii) *there is a homomorphism $\alpha : \mathrm{Alg}(\lambda) \to \mathrm{Alg}(x)$ such that $\alpha(\lambda^r) = x^r$ for $r = 1,\dots,d$;*

320 Non-Commutative Theory

(iii) *the map $p(\lambda) \mapsto p(x)$ is a well-defined map from* $\mathrm{Alg}(\lambda)$ *to* $\mathrm{Alg}(x)$*; and*

(iv) *the map $p(\lambda) \mapsto p(x)$ is a completely bounded homomorphism.*

Proof. The equivalence of (i)–(iii) is by definition. That (iii) is equivalent to (iv) is because of R. Smith's result that every bounded homomorphism defined on a finite-dimensional algebra is automatically completely bounded [162, exercise 3.11]. □

We let $\mathcal{V}$ denote the vector space of free polynomials on E, where we identify polynomials that agree on E, and we let $\mathcal{V}_{\mathcal{L}(\mathcal{H},\mathcal{M})}$ denote the vector space of $\mathcal{L}(\mathcal{H},\mathcal{M})$-valued free polynomials on E. As any such polynomial on E is uniquely determined by its values on λ, the space of such functions is finite-dimensional whenever $\mathcal{H}$ and $\mathcal{M}$ are finite-dimensional.

Consider the following sets of functions on $E^{[2]}$, where all sums are over a finite set of indices:

$$H(E) = \left\{ h(y,x) = \sum g_i(y)^* f_i(x): f_i, g_i \in \mathcal{P}_d \right\}$$

$$R(E) = \{ h \in H(E): h(x,y) = h(y,x)^* \}$$

$$C(E) = \left\{ h(y,x) = \sum u_i(y)^*[\mathrm{id} - \delta(y)^*\delta(x)]u_i(x): \right.$$
$$\left. u_i \text{ is an } \mathcal{L}(\mathbb{C}, \mathbb{C}^J) - \text{ valued nc polynomial} \right\}$$

$$P(E) = \left\{ h(y,x) = \sum f_i(y)^* f_i(x): f_i \in \mathcal{P}_d \right\}.$$

We topologize $H(E)$ with the norm

$$\|h(y,x)\| = \|h(\lambda,\lambda)\|.$$

We want to prove that $C(E)$ is closed, Lemma 14.24. This is essentially Theorem 13.22, except now we want the theorem on the finite set E, instead of the open set Ω. Once we show $C(E)$ is closed, a cone separation argument (Lemma 14.27) shows that p_0 has a model, and this gives the desired extension to φ.

We shall first bound the total number of terms in certain sums of squares.

Lemma 14.22. *Let $\mathcal{H}, \mathcal{M}$ be finite-dimensional Hilbert spaces, and let $F(y,x)$ be an arbitrary graded $\mathcal{L}(\mathcal{M})$-valued function on $E^{[2]}$. Let $N_0 = \dim \mathcal{V}_{\mathcal{L}(\mathcal{H},\mathcal{M})}$. If G can be represented in the form*

$$G(y,x) = \sum_{i=1}^{m} g_i(y)^* F(y,x) g_i(x) \qquad \text{for } (x,y) \in E^{[2]},$$

Free Holomorphic Functions 321

where $m \in \mathbb{N}$ and each $g_i \in \mathcal{V}_{\mathcal{L}(\mathcal{H},\mathcal{M})}$, then G can be represented in the form

$$G(y,x) = \sum_{i=1}^{N_0} f_i(y)^* F(y,x) f_i(x) \qquad \text{for } (x,y) \in E^{[2]}, \tag{14.23}$$

where $f_i \in \mathcal{V}_{\mathcal{L}(\mathcal{H},\mathcal{M})}$ for $i = 1,\ldots,N_0$.

Proof. Let $e_1(\cdot),\ldots,e_{N_0}(\cdot)$ be a basis of $\mathcal{V}_{\mathcal{L}(\mathcal{H},\mathcal{M})}$. For $i = 1,\ldots,m$, let

$$g_i(x) = \sum_{l=1}^{N_0} c_{il}\, e_l(x).$$

Form the $m \times N_0$ matrix $C = [c_{il}]$. As C^*C is an $N_0 \times N_0$ positive semi-definite matrix, there exists an $N_0 \times N_0$ matrix $A = [a_{kl}]$ such that $C^*C = A^*A$. This leads to the formula,

$$\sum_{i=1}^{m} \overline{c}_{il_1} c_{il_2} = \sum_{k=1}^{N_0} \overline{a}_{kl_1} a_{kl_2},$$

valid for all $l_1, l_2 = 1,\ldots,N_0$. If $(x,y) \in E^{[2]}$, then

$$G(y,x) = \sum_{i=1}^{m} g_i(y)^* F(y,x) g_i(x)$$

$$= \sum_{i=1}^{m} \left(\sum_{l=1}^{N_0} c_{il}\, e_l(y)\right)^* F(y,x) \left(\sum_{l=1}^{N_0} c_{il}\, e_l(x)\right)$$

$$= \sum_{l_1,l_2=1}^{N_0} \left(\sum_{i=1}^{m} \overline{c}_{il_1} c_{il_2}\right) e_{l_1}(y)^* F(y,x) e_{l_2}(x)$$

$$= \sum_{l_1,l_2=1}^{N_0} \left(\sum_{k=1}^{N_0} \overline{a}_{kl_1} a_{kl_2}\right) e_{l_1}(y)^* F(y,x) e_{l_2}(x)$$

$$= \sum_{k=1}^{N_0} \left(\sum_{l=1}^{N_0} a_{kl}\, e_l(y)\right)^* F(y,x) \left(\sum_{l=1}^{N_0} a_{kl}\, e_l(x)\right).$$

This proves that equation (14.23) holds with $f_i = \sum_{l=1}^{N_0} a_{il}\, e_l$. $\qquad\square$

We have shown that every element of $C(E)$ can be modeled as

$$u(y)^* \left(1 - \frac{1_{\mathcal{M}}}{\delta(y)^*} \otimes \frac{1_{\mathcal{M}}}{\delta(x)}\right) u(x),$$

where u is an $\mathcal{M} \otimes \mathbb{C}^J$-valued free polynomial and $\dim \mathcal{M} \le N_0$.

322   Non-Commutative Theory

Lemma 14.24. $C(E)$ *is closed.*

Proof. By Lemma 14.22, every element in $C(E)$ can be represented in the form

$$\sum_{i=1}^{N_0} u_i(y)^*[\mathrm{id} - \delta(y)^*\delta(x)]u_i(x), \tag{14.25}$$

where $N_0 = \dim \mathcal{V}_{\mathcal{L}(\mathbb{C},\mathbb{C}^J)}$. Suppose a sequence of elements of the form (14.25) approaches some h in $H(E)$ at the point (λ, λ):

$$\sum_{i=1}^{N_0} u_i^{(k)}(\lambda)^*[\mathrm{id} - \delta(\lambda)^*\delta(\lambda)]u_i^{(k)}(\lambda) \to h(\lambda, \lambda) \text{ as } k \to \infty.$$

Since $\lambda \in B_\delta$, there is a constant M such that, for each i and k,

$$\|u_i^{(k)}(\lambda)\| \leq M.$$

Passing to a subsequence, one can assume that each $u_i^{(k)}(\lambda)$ converges to some $u_i(\lambda)$ (since $u_i^{(k)}$ is a graded $\mathcal{L}(\mathbb{C}, \mathbb{C}^J)$ valued function and $J < \infty$). By Lemma 14.21, we have

$$u_i^{(k)}(x) \to u_i(x) \qquad \text{for all } x \in E.$$

Therefore, for all $(x, y) \in E^{[2]}$, we have

$$\sum_{i=1}^{N_0} u_i^{(k)}(y)^*[\mathrm{id} - \delta(y)^*\delta(x)]u_i^{(k)}(x) \to \sum_{i=1}^{N_0} u_i(y)^*[\mathrm{id} - \delta(y)^*\delta(x)]u_i(x)$$

$$= h(y, x).$$

$\square$

Lemma 14.26. $P(E) \subseteq C(E)$.

Proof. We have

$$f(y)^*f(x) - \sum_{k=0}^{m-1} f(y)^*\delta(y)^{k*}[\mathrm{id} - \delta(y)\delta(x)]\delta(x)^k f(x)$$

$$= f(y)^*\delta(y)^m \delta(x)^m f(x).$$

As $m \to \infty$, the right-hand side goes to zero for every $(x, y) \in E^{[2]}$. Since $C(E)$ is closed, by Lemma 14.24, this proves that $f(y)^*f(x) \in C(E)$, and hence so are finite sums of this form. $\square$

Lemma 14.27. *Suppose* $\sup\{\|p_0(x)\|: x \in E\} \leq 1$. *Then the function*

$$h(y, x) = \mathrm{id} - p_0(y)^*p_0(x)$$

is in $C(E)$.

Free Holomorphic Functions 323

Proof. Since $C(E)$ is closed, this will follow from the Hahn–Banach theorem [176, theorem 3.3.4] if we can can show that $L(h(y,x)) \geq 0$ whenever

$$L \in R(E)^* \qquad \text{and} \qquad L(g) \geq 0 \text{ for all } g \in C(E). \tag{14.28}$$

Assume statements (14.28) hold, and define $L^\sharp \in H(E)^*$ by the formula

$$L^\sharp(h(y,x)) = L\left(\frac{h(y,x) + h(x,y)^*}{2}\right) + iL\left(\frac{h(y,x) - h(x,y)^*}{2i}\right),$$

and then define sesquilinear forms on $\mathcal{V}$ and $\mathcal{V}_{\mathcal{L}(\mathbb{C},\mathbb{C}^J)}$ by the formulas

$$\langle f, g \rangle_{L_1} = L^\sharp(g(y)^* f(x)) \qquad \text{for all } f, g \in \mathcal{V}$$

$$\langle F, G \rangle_{L_2} = L^\sharp(G(y)^* F(x)) \qquad \text{for all } F, G \in \mathcal{V}_{\mathcal{L}(\mathbb{C},\mathbb{C}^J)}.$$

Observe that Lemma 14.26 implies that $f(y)^* f(x) \in C(E)$ whenever $f \in \mathcal{V}$ or $\mathcal{V}_{\mathcal{L}(\mathbb{C},\mathbb{C}^J)}$. Hence statements (14.28) imply that $\langle f, f \rangle_{L_1} \geq 0$ for all $f \in \mathcal{V}$, and $\langle F, F \rangle_{L_2} \geq 0$ for all $F \in \mathcal{V}_{\mathcal{L}(\mathbb{C},\mathbb{C}^J)}$, that is, $\langle \cdot, \cdot \rangle_{L_1}$ and $\langle \cdot, \cdot \rangle_{L_2}$ are semi-inner products on $\mathcal{V}$ and $\mathcal{V}_{\mathcal{L}(\mathbb{C},\mathbb{C}^J)}$, respectively.

To make them into inner products, choose $\varepsilon > 0$ and define

$$\langle f, g \rangle_1 = L^\sharp(g(y)^* f(x)) + \varepsilon \operatorname{tr}(g(\lambda)^* f(\lambda)) \text{ for } f, g \in \mathcal{V} \tag{14.29}$$

$$\langle F, G \rangle_2 = L^\sharp(G(y)^* F(x)) + \varepsilon \operatorname{tr}(G(\lambda)^* F(\lambda))$$

$$\text{for } F, G \in \mathcal{V}_{\mathcal{L}(\mathbb{C},\mathbb{C}^J)}. \tag{14.30}$$

We let $H_{L_1}^2$ and $H_{L_2}^2$ denote the Hilbert spaces that are the completions of $\mathcal{V}$ and $\mathcal{V}_{\mathcal{L}(\mathbb{C},\mathbb{C}^J)}$ equipped with the inner products (14.29) and (14.30).

The fact that L is non-negative on $C(E)$ means that

$$L^\sharp(G(y)^* F(x)) \geq L^\sharp(G(y)^* \delta(y)^* \delta(x) F(x)) \tag{14.31}$$

for all F in $\mathcal{V}_{\mathcal{L}(\mathbb{C},\mathbb{C}^J)}$. Since $\|\delta(\lambda)\| < 1$, we also have

$$\operatorname{tr}(F(\lambda)^* F(\lambda)) > \operatorname{tr}(F(\lambda)^* \delta(\lambda)^* \delta(\lambda) F(\lambda)), \tag{14.32}$$

if $F \neq 0$. Combining inequalities (14.31) and (14.32), we find that multiplication by δ is a strict contraction on $H_{L_2}^2$.

Let M denote the d-tuple of multiplication by the co-ordinate functions x^r on $H_{L_1}^2$. We have just shown that $\|\delta(M)\| < 1$, so M is in B_δ. As M is also in V_λ, it follows that M is in E. Thus $\|p_0(M)\| \leq 1$, by hypothesis. Therefore

$$\operatorname{id} - p_0(M)^* p_0(M) \geq 0,$$

and so for all f in $\mathcal{V}$ we have

$$L^\sharp(f(y)^* f(x)) + \varepsilon \operatorname{tr}(f(\lambda)^* f(\lambda))$$

$$\geq L^\sharp(f(y)^* p_0(y)^* p_0(x) f(x)) + \varepsilon \operatorname{tr}(f(\lambda)^* p_0(\lambda)^* p_0(\lambda) f(\lambda)).$$

324   *Non-Commutative Theory*

Letting f be the function 1 and letting $\varepsilon \to 0$, we deduce that

$$L(\mathrm{id} - p_0(y)^* p_0(x)) \geq 0,$$

as desired. $\hfill\square$

We can now prove Theorem 14.20.

Proof. (Necessity). Condition (i) follows from Theorem 14.4.

To see condition (ii), note that for all $x \in V_\lambda \cap B_\delta$. we have $p_0(x) = \varphi(x)$. Indeed, by Theorem 14.4, there is a polynomial q such that $q(\lambda \oplus x) = \varphi(\lambda \oplus x)$. Therefore $q(\lambda) = p_0(\lambda)$, and so, since $x \in V_\lambda$, we also have $q(x) = p_0(x)$, and hence $p_0(x) = \varphi(x)$. But φ is in the unit ball of $H_1^\infty(B_\delta)$, and so $\|\varphi(x)\| \leq 1$ for every x in B_δ.

(Sufficiency). Suppose (i) and (ii) hold. By Lemma 14.27, the function

$$h(y,x) = \mathrm{id} - p_0(y)^* p_0(x)$$

is in $C(E)$. This means that p_0 has an nc-model on E, and so, by Theorem 14.12, it can be extended to a function in $H_1^\infty(B_\delta)$ that has a free realization. $\hfill\square$

14.4 Free Realizations Redux

By combining Theorem 14.20 and Theorem 13.22 we can prove that every function in $H_1^\infty(B_\delta)$ has a free realization.

Theorem 14.33. *A graded function φ on B_δ is in $H_1^\infty(B_\delta)$ if and only if it has a free realization as in Definition 14.7.*

Proof. Sufficiency is proved in Theorem 14.10. For necessity, assume that φ is in $H_1^\infty(B_\delta)$. Let (λ_i) be a sequence in B_δ that is dense in the finite topology. For each $k \in \mathbb{N}$, let $\Lambda_k = \oplus_{i=1}^k \lambda_i$, and let $W_k = \oplus_{i=1}^k \varphi(\lambda_i)$.

By Theorem 14.20, for each k, there is a function φ_k that agrees with φ on Λ_k and has an nc-model, so that

$$1 - \varphi_k(y)^* \varphi_k(x) = u^k(y)^* \left(1 - \underset{\delta(y)^*}{\overset{1}{\otimes}} \ \underset{\delta(x)}{\overset{1}{\otimes}} \right) u^k(x). \tag{14.34}$$

By Theorem 13.22, we can conclude that there is a separable Hilbert space $\mathcal{M}$ and an $\mathcal{M} \otimes \mathbb{C}^J$-valued nc-function v such that

$$v(y)^* \left(1 - \underset{\delta(y)^*}{\overset{1_\mathcal{M}}{\otimes}} \ \underset{\delta(x)}{\overset{1_\mathcal{M}}{\otimes}} \right) v(x) = 1 - \varphi(y)^* \varphi(x) \qquad \text{for all } (x,y) \in B_\delta \boxtimes B_\delta.$$

Thus φ has an nc-model, and hence, by Theorem 14.12, it has a free realization. $\hfill\square$

Free Holomorphic Functions

Suppose that φ has a free realization as in equation (14.9), and expand

$$\left[1 - \mathop{\otimes}_{1}^{D} \mathop{\otimes}_{\delta(x)}^{1} \right]^{-1} = \sum_{k=0}^{\infty} \left(\mathop{\otimes}_{1}^{D} \mathop{\otimes}_{\delta(x)}^{1} \right)^{k}$$

as a Neumann series. This gives a series representation for φ:

$$\varphi(x) = \mathop{\otimes}_{1}^{A} + \sum_{k=0}^{\infty} \mathop{\otimes}_{1}^{B} \mathop{\otimes}_{\delta(x)}^{1} \left(\mathop{\otimes}_{1}^{D} \mathop{\otimes}_{\delta(x)}^{1} \right)^{k} \mathop{\otimes}_{1}^{C}. \tag{14.35}$$

The series converges absolutely and uniformly on $\{x \colon \|\delta(x)\| \leq r\}$ for each $r < 1$. Moreover, each partial sum is a free polynomial. We have proved that free functions are locally uniform limits of free polynomials. Let us state this as a corollary.

Definition 14.36.

$$K_\delta = \{x \in \mathbb{M}^d \colon \|\delta(x)\| \leq 1\}.$$

Corollary 14.37. *Let φ be in $H^\infty(B_\delta)$. For every $r < 1$ there is a sequence of free polynomials that converges absolutely and uniformly to φ on $K_{\delta/r}$.*

To get convergence on all of $K_{\delta/r}$, we need to group the terms as in equation (14.35). For example, suppose δ is the diagonal $d \times d$ matrix with x^j on the diagonal, for $1 \leq j \leq d$. A power series on B_δ is a series of the form

$$f(x) = \sum_{\ell=0}^{\infty} \sum_{|\alpha|=\ell} c_\alpha x^\alpha. \tag{14.38}$$

The number of words x^α with length ℓ is d^ℓ; so if all you know is that the coefficients c_α are bounded, you can conclude only that the series (14.38) converges if $\|\delta(x)\| < 1/\ell$. However the sequence of partial sums

$$\sum_{\ell=0}^{N} \sum_{|\alpha|=\ell} c_\alpha x^\alpha$$

converges, as $N \to \infty$, on $\{\|\delta(x)\| \leq r\}$ for every $r < 1$.

In Example 12.29 there is an example of a polynomial polyhedron that is empty at level 1, so free holomorphic functions cannot be represented by power series on this set, although they have a series representation as in equation (14.35).

The following result gives a characterization of free holomorphic functions. If φ is a graded function defined on a free open set Ω, we say that φ *is locally*

326 *Non-Commutative Theory*

approximable by polynomials if for each $x \in \Omega$ and $\varepsilon > 0$, there exists a free neighborhood U of x contained in Ω and a free polynomial p such that

$$\sup_{x \in U} \|\varphi(x) - p(x)\| < \varepsilon.$$

Theorem 14.39. *Let Ω be a free open set and let φ be a graded function defined on Ω. Then φ is a free holomorphic function if and only if φ is locally approximable by polynomials.*

Proof. Sufficiency follows because the uniform limit of free polynomials is nc and bounded. For necessity, let x be in Ω. Then since Ω is open, there exists a matrix δ of free polynomials, and $r < 1$, such that $x \in B_{\delta/r} \subseteq B_{\delta}$. Now apply Corollary 14.37. $\square$

14.5 Extending off Varieties

Suppose $\mathfrak{A}$ is a subset of $\mathcal{P}_d$, and let $\mathfrak{V} = \mathrm{Var}(\mathfrak{A})$ be given by

$$\mathfrak{V} = \{x \in \mathbb{M}^d \colon p(x) = 0 \text{ for all } p \in \mathfrak{A}\}. \tag{14.40}$$

If λ is in $\mathfrak{V}$, then $\mathfrak{A} \subseteq I_\lambda$, and $V_\lambda \subseteq \mathfrak{V}$. Let Ω be a freely open set in $\mathbb{M}^d$; we shall say that a function f defined on $\mathfrak{V} \cap \Omega$ is free holomorphic if, for every point x in $\mathfrak{V} \cap \Omega$ there is a basic free open set $B_\delta \subseteq \Omega$ containing x and a free holomorphic function ψ defined on B_δ such that $\psi|_{\mathfrak{V} \cap B_\delta} = f|_{\mathfrak{V} \cap B_\delta}$.

In the scalar case, every holomorphic function defined on an analytic variety inside a domain of holomorphy extends to a holomorphic function on the whole domain, by a celebrated theorem of H. Cartan [58]. The geometric conditions that guarantee that all bounded holomorphic functions extend to be bounded on the whole domain have been investigated by Henkin and Polyakov [113] and Knese [131]; however, even when bounded extensions exist, the extension is almost never isometric [12]. But in the matrix case, any bounded free holomorphic function on $\mathfrak{V} \cap B_\delta$ does extend to a free holomorphic function on B_δ with the same norm.

Theorem 14.41. *Let $\mathfrak{V}$ be as in equation* (14.40) *and let δ be a matrix of free polynomials such that $\mathfrak{V} \cap B_\delta$ is non-empty. Let f be a bounded free holomorphic function defined on $B_\delta \cap \mathfrak{V}$. Then there is a free holomorphic function φ on B_δ that extends f and such that*

$$\|\varphi\|_{H^\infty(B_\delta)} = \sup_{x \in \mathfrak{V} \cap B_\delta} \|f(x)\|. \tag{14.42}$$

The proof is a repeat of the prooof of Theorem 14.33, where we choose (λ_i) to be dense in $\mathfrak{V} \cap B_\delta$.

Free Holomorphic Functions 327

A special case of Theorem 14.41 is the statement that, if δ is a matrix of commuting polynomials, any bounded holomorphic function defined on the set of commuting d-tuples λ satisfying $\|\delta(\lambda)\| < 1$ can be extended to a bounded nc-function of the same norm on all of B_δ.

14.6 The nc Oka–Weil Theorem

Let us define polynomial convexity in the free topology. Recall that K_δ is defined in Definition 14.36.

Definition 14.43. *Let* $E \subset \mathbb{M}^d$. *The* polynomial hull *of E is defined to be*

$$\hat{E} \stackrel{\text{def}}{=} \bigcap \{K_\delta : E \subseteq K_\delta\}.$$

(If E is not contained in any K_δ, we declare $\hat{E}$ to be $\mathbb{M}^d$.) We say that a compact set is polynomially convex *if it equals its polynomial hull.*

Note that $\hat{E}$ is always an nc-set.

Definition 14.44. *Let* $y \in \mathbb{M}_n^d$. *The* polynomial dilation hull *of y is the set of $x \in \mathbb{M}^d$ with the property that, if $x \in \mathbb{M}_m^d$, then there exists $k \in \mathbb{N}$ and an isometry* $V \colon \mathbb{C}^m \to \mathbb{C}^{nk}$ *such that*

$$p(x) = V^* p(y^{(k)}) V \qquad \textit{for all } p \in \mathcal{P}_d. \tag{14.45}$$

(Recall that $y^{(k)}$ is the k-ampliation of y). Augat, Balasubramanian, and McCullough proved in [38] that freely compact nc-sets are very small.

Theorem 14.46. (i) *The polynomial dilation hull of y is exactly the polynomial hull of $\{y\}$.*

(ii) *An nc-set E in $\mathbb{M}^d$ is freely compact if and only if it is contained in the polynomial dilation hull of some y in E.*

Given Theorem 14.46, one can prove a version of the Oka–Weil theorem, with the assumption of polynomial convexity hidden in the assumption of being nc.

Theorem 14.47. *Let* $E \subseteq \mathbb{M}^d$ *be a compact set in the free topology and assume that E is nc. Let Ω be a free open set containing E, and let φ be a free holomorphic function defined on Ω. Then φ can be uniformly approximated on E by free polynomials.*

Proof. Assume that E is in the polynomial hull of $\{y\}$. There is some B_δ containing y such that φ is bounded on B_δ and has a realization formula. By

328   *Non-Commutative Theory*

equation (14.45), we have $E \subseteq B_\delta$. By Theorem 14.39 (or Theorem 14.4), for each n there is a free polynomial p_n such that $\|p_n(y) - \varphi(y)\| \leq \frac{1}{n}$.

To see that p_n converges uniformly on E, let $x \in E$ be arbitrary. As $x \oplus y$ is in B_δ, there is a free polynomial q_n such that

$$\|q_n(x \oplus y) - \varphi(x \oplus y)\| \leq \frac{1}{n}.$$

Therefore $\|q_n(y) - p_n(y)\| \leq \frac{2}{n}$, and since x is in the polynomial hull of $\{y\}$, we have

$$\|p_n(x) - \varphi(x)\| \leq \|p_n(x) - q_n(x)\| + \|q_n(x) - \varphi(x)\| \leq \frac{3}{n}.$$

$\square$

14.7 Additional Results

There is a strengthened version of Theorem 14.33 that, in the commutative case, is a version of Leech's theorem, Theorem 3.56. We shall not give the proof here. See [14] or [43].

Theorem 14.48. *Let $\mathcal{H}, \mathcal{K}, \mathcal{L}$ be Hilbert spaces, let f be a $\mathcal{B}(\mathcal{L}, \mathcal{H})$-valued free holomorphic function that is bounded on B_δ, and let g be a $\mathcal{B}(\mathcal{K}, \mathcal{H})$-valued free holomorphic function that is bounded on B_δ. Then there exists φ in the closed unit ball of the $\mathcal{B}(\mathcal{K}, \mathcal{L})$-valued free holomorphic functions on B_δ such that $f\varphi = g$ if and only if*

$$f(x)^* f(x) - g(x)^* g(x) \geq 0 \qquad \text{for all } x \in B_\delta.$$

Moreover, if this inequality holds, then φ has a free realization.

An *nc-rational function* is a function that can be built up from free polynomials in a finite number of arithmetic operations. Such functions are freely holomorphic away from their poles (suitably understood). We leave the proof of the following as an exercise (or see [14, theorem 10.1]).

Theorem 14.49. *Let φ and ψ be free holomorphic functions on a free open set Ω. Then $\psi\varphi^{-1}$ and $\varphi^{-1}\psi$ are free holomorphic off the zero set of φ.*

Free holomorphic functions can be handled locally by free realizations. But when these are pieced together, the function can be unbounded on basic free open sets, as the following example of James Pascoe [to appear in *J. Op. Thy.*] shows. The *row ball* in $\mathbb{M}^d$ is the basic free open set B_δ corresponding to the free polynomial $\delta(x) = [x^1 \ x^2 \dots x^d]$.

Free Holomorphic Functions 329

Theorem 14.50. *(Pascoe) Let $d \geq 2$. There is a free holomorphic function defined on all of $\mathbb{M}^d$ that is unbounded on the row ball.*

14.8 Historical Notes

Theorem 14.19 was first proved in [14], under the assumption that $\mathcal{K}$ and $\mathcal{L}$ are finite-dimensional and that E is all of B_δ. The extension to arbitrary dimensions, and the proof we give here by averaging over the unitaries, is due to S. Balasubramanian [40]. It is also proved in arbitrary dimensions in [43].

Theorem 14.20 is from [17]. Lemma 14.22 is proved in [150, theorem 2.1].

Theorem 14.33, and 14.48 in the case that all the Hilbert spaces are finite-dimensional was proved in [14]; it was extended to infinite dimensional spaces in [43]. Theorem 14.46 is from [38].

15

The Implicit Function Theorem

15.1 The Fine Inverse Function Theorem

Recall from Section 12.3 that the fine topology is the topology generated by all nc-domains, and a τ-holomorphic function is an nc-function that is τ-locally bounded. For $\Omega \subseteq \mathbb{M}^d$, we shall say that a map $\Phi \colon \Omega \to \mathbb{M}^{d'}$ is an nc-*map* if each component of Φ is an nc-function.

The following inverse function theorem is due to Helton, Klep, and McCullough [104] and Pascoe [160].

Theorem 15.1. *Let $\Omega \subseteq \mathbb{M}^d$ be an nc-domain. Let Φ be a fine holomorphic map from Ω to $\mathbb{M}^d$. The following conditions are equivalent.*

(i) Φ *is injective on* Ω;

(ii) $D\Phi(x)$ *is non-singular for every* $x \in \Omega$; *and*

(iii) *the function* Φ^{-1} *exists and is a fine holomorphic map.*

Proof. (i) $\Rightarrow$ (ii) Suppose $D\Phi(x)h = 0$ for some $x \in \Omega_n$ and some non-zero $h \in \mathbb{M}_n^d$. By Proposition 12.20, this means that

$$\Phi\left(\begin{bmatrix} x & h \\ 0 & x \end{bmatrix}\right) = \begin{bmatrix} \Phi(x) & D\Phi(x)[h] \\ 0 & \Phi(x) \end{bmatrix}$$

$$= \Phi\left(\begin{bmatrix} x & 0 \\ 0 & x \end{bmatrix}\right),$$

and so Φ is not injective.

(ii) $\Rightarrow$ (i) Let

$$s = \begin{bmatrix} 1 & 0 & 0 & 1 \\ 0 & 1 & 0 & 0 \\ 0 & 0 & 1 & 0 \\ 0 & 0 & 0 & 1 \end{bmatrix},$$

The Implicit Function Theorem

and observe that

$$s^{-1}\begin{bmatrix} a & 0 & 0 & 0 \\ 0 & b & 0 & 0 \\ 0 & 0 & c & 0 \\ 0 & 0 & 0 & d \end{bmatrix}s = \begin{bmatrix} a & 0 & 0 & a-d \\ 0 & b & 0 & 0 \\ 0 & 0 & c & 0 \\ 0 & 0 & 0 & d \end{bmatrix}.$$

Therefore

$$\Phi\left(\begin{bmatrix} x_1 & 0 & 0 & x_1-x_2 \\ 0 & x_2 & 0 & 0 \\ 0 & 0 & x_1 & 0 \\ 0 & 0 & 0 & x_2 \end{bmatrix}\right)$$

$$= \begin{bmatrix} \Phi(x_1) & 0 & 0 & \Phi(x_1)-\Phi(x_2) \\ 0 & \Phi(x_2) & 0 & 0 \\ 0 & 0 & \Phi(x_1) & 0 \\ 0 & 0 & 0 & \Phi(x_2) \end{bmatrix}$$

$$= \begin{bmatrix} \Phi\left(\begin{bmatrix} x_1 & 0 \\ 0 & x_2 \end{bmatrix}\right) & D\Phi\left(\begin{bmatrix} x_1 & 0 \\ 0 & x_2 \end{bmatrix}\right)\begin{bmatrix} 0 & x_1-x_2 \\ 0 & 0 \end{bmatrix} \\ \begin{bmatrix} 0 & 0 \\ 0 & 0 \end{bmatrix} & \Phi\left(\begin{bmatrix} x_1 & 0 \\ 0 & x_2 \end{bmatrix}\right) \end{bmatrix}.$$

So if $\Phi(x_1) = \Phi(x_2)$, then

$$D\Phi\left(\begin{bmatrix} x_1 & 0 \\ 0 & x_2 \end{bmatrix}\right)\begin{bmatrix} 0 & x_1-x_2 \\ 0 & 0 \end{bmatrix} = 0.$$

(i) $\Rightarrow$ (iii) Let $G = \Phi(\Omega)$. By Theorem 12.17 and the invariance of domain theorem, G is open in the finite topology, and $\Phi\colon \Omega \to G$ is a homeomorphism in that topology. Since Φ preserves direct sums and intertwinings, G is an nc-set, and Φ^{-1} preserves direct sums. To see that Φ^{-1} preserves intertwinings, assume that $x,y \in \Omega$, that $\lambda = \Phi(x)$, $\mu = \Phi(y)$, and that $\mu L = L\lambda$ for some matrix L. Since Ω is an nc-domain, for ε small enough, we have

$$\begin{bmatrix} y & \varepsilon(yL-Lx) \\ 0 & x \end{bmatrix}$$

is in Ω. By Lemma 12.15, for each component Φ^r of Φ, we have

$$\Phi^r\left(\begin{bmatrix} y & \varepsilon(yL-Lx) \\ 0 & x \end{bmatrix}\right) = \begin{bmatrix} \Phi^r(y) & \varepsilon(\Phi^r(y)L-L\Phi^r(x)) \\ 0 & \Phi^r(x) \end{bmatrix}$$

$$= \begin{bmatrix} \mu^r & 0 \\ 0 & \lambda^r \end{bmatrix}$$

$$= \Phi^r\left(\begin{bmatrix} y & 0 \\ 0 & x \end{bmatrix}\right).$$

As Φ is injective, this means that $yL = Lx$, and so $\Phi^{-1}(\mu)L = L\Phi^{-1}(\lambda)$.

332 Non-Commutative Theory

All that remains is to show that Φ^{-1} is finely locally bounded. Let $\mu = \Phi(y)$ be a point in G. Let $\Omega_1 = \Omega \cap \{x: \|x\| < \|y\| + 1\}$, and let $G_1 = \Phi(\Omega_1)$. Then G_1 is a fine open set that contains μ, and $\Phi^{-1}|_{G_1}$ is bounded. $\qquad\square$

15.2 The Fat Inverse Function Theorem

Recall from Example 12.26 that the fat topology is generated by the sets $F(a,r)$, which are the unions of all balls centered at ampliations of a that have radius r. Our next objective is to prove the inverse function theorem in the fat category. The difficulty is showing that $\Phi(\Omega)$ is fat open.

The reason for the effort is that if $D\Phi(a)$ is non-singular for some fat holomorphic function Φ, then it is non-singular in a fat neighborhood of a (Theorem 15.8); so the fat version of the implicit function theorem requires only the checking of the non-singularity of the derivative at a point. In the fine case, we need to know the derivative is non-singular in a neighborhood.

If τ is a topology on $\mathcal{M}^d$ and σ is a topology on $\mathbb{M}^b$, then we let $\tau \boxtimes \sigma$ be the topology on $\mathbb{M}^{d+b}$ that has a basis

$$\tau \boxtimes \sigma = \bigcup \{A \boxtimes B: A \in \tau, B \in \sigma\}.$$

(Recall from equation (12.1) that $A \boxtimes B$ means the union of products of couples from the same level). If τ and σ are admissible, then $\tau \boxtimes \sigma$ is admissible. The following lemma is easily verified.

Lemma 15.2. *Let τ be an admissible topology on $\mathbb{M}^d$ and assume that $\varphi: \Omega \to \mathbb{M}^1$ is a τ-holomorphic function. If σ is any admissible topology, then g defined on $\Omega \boxtimes \mathbb{M}^d$ by the formula*

$$g(x,h) = D\varphi(x)[h] \qquad for\ all\ (x,h) \in \Omega \boxtimes \mathbb{M}^d$$

is a $\tau \boxtimes \sigma$-holomorphic function. Furthermore, for each $n \in \mathbb{N}$ and $x \in \Omega \cap \mathbb{M}_n^d$, the function $h \mapsto g(x,h)$ is a bounded linear map from $\mathbb{M}_n^d$ to $\mathbb{M}_n^1$.

Let φ be an nc-function defined on a nc-domain $\Omega \subseteq \mathbb{M}^d$, and let $a \in \Omega$. We define the *Hessian of φ at a* to be the bilinear form $H\varphi(a)$ defined on $\mathbb{M}^d \boxtimes \mathbb{M}^d$ by the formula

$$H\varphi(a)[h,k] = \lim_{t \to 0} \frac{D\varphi(a+tk)[h] - D\varphi(a)[h]}{t} \qquad for\ all\ h,k \in \mathbb{M}^d.$$

Lemma 15.3. *Let $\Omega \subseteq \mathbb{M}^d$ be a fine open set, $\varphi: \Omega \to \mathbb{M}^1$ a fine holomorphic function, and $a \in \Omega$. If h and k are sufficiently small, then*

$$\varphi\left(\begin{bmatrix} a & k & h & 0 \\ 0 & a & 0 & h \\ 0 & 0 & a & k \\ 0 & 0 & 0 & a \end{bmatrix}\right) = \begin{bmatrix} \varphi(a) & D\varphi(a)[k] & D\varphi(a)[h] & H\varphi(a)[h,k] \\ 0 & \varphi(a) & 0 & D\varphi(a)[h] \\ 0 & 0 & \varphi(a) & D\varphi(a)[k] \\ 0 & 0 & 0 & \varphi(a) \end{bmatrix}.$$

The Implicit Function Theorem

Proof. Let

$$X = \begin{bmatrix} a & k \\ 0 & a \end{bmatrix}$$

and

$$H = \begin{bmatrix} h & 0 \\ 0 & h \end{bmatrix}.$$

Define a function $g(x,h)$ by

$$g(x,h) = D\varphi(x)[h] \quad \text{for all } (x,h) \in \Omega \boxtimes \mathbb{M}^d.$$

By Lemma 15.2 g is a fine holomorphic function of $2d$ variables. Hence, by Proposition 12.20,

$$g(X,H) = g\left(\begin{bmatrix} (a,h) & (k,0) \\ 0 & (a,h) \end{bmatrix}\right)$$

$$= \begin{bmatrix} g(a,h) & Dg(a,h)[k,0] \\ 0 & g(a,h) \end{bmatrix}.$$

But

$$Dg(a,h)[k,0] = \lim_{t \to 0} \frac{g(a+tk,h) - g(a,h)}{t}$$

$$= \lim_{t \to 0} \frac{D\varphi(a+tk)[h] - D\varphi(a)[h]}{t}$$

$$= H\varphi(a)[h,k].$$

Therefore,

$$g(X,H) = \begin{bmatrix} D\varphi(a)[h] & H\varphi(a)[h,k] \\ 0 & D\varphi(a)[h] \end{bmatrix}.$$

Using this last formula and Proposition 12.20 several times, we have

$$\varphi\left(\begin{bmatrix} a & k & h & 0 \\ 0 & a & 0 & h \\ 0 & 0 & a & k \\ 0 & 0 & 0 & a \end{bmatrix}\right) = \varphi\left(\begin{bmatrix} X & H \\ 0 & X \end{bmatrix}\right)$$

$$= \begin{bmatrix} \varphi(X) & g(X,H) \\ 0 & \varphi(X) \end{bmatrix}$$

$$= \begin{bmatrix} \varphi(a) & D\varphi(a)[k] & D\varphi(a)[h] & H\varphi(a)[h,k] \\ 0 & \varphi(a) & 0 & D\varphi(a)[h] \\ 0 & 0 & \varphi(a) & D\varphi(a)[k] \\ 0 & 0 & 0 & \varphi(a) \end{bmatrix}.$$

$\square$

334 *Non-Commutative Theory*

The Hessian is a second derivative; if we can control it, we can ensure that the first derivative does not change rapidly.

Lemma 15.4. *Suppose that $\varphi \colon \Omega \to \mathbb{M}^1$ is a fat holomorphic function. For each $a \in \Omega$, there exists $r \in \mathbb{R}^+$ such that $H\varphi$ is a uniformly bounded bilinear form on $F(a,r)$.*

Proof. Fix $a \in \Omega$. Since φ is a fat holomorphic function, there exist $s, \rho \in \mathbb{R}^+$ such that $F(a,s) \subseteq \Omega$, φ is a fine holomorphic function on $F(a,s)$, and

$$\sup_{x \in F(a,s)} \|\varphi(x)\| \leq \rho.$$

Let $r = s/2$. If $x \in F(a,r)$, then by the triangle inequality, if $\|h\|, \|k\| < r/2$, then

$$\begin{bmatrix} x & k & h & 0 \\ 0 & x & 0 & h \\ 0 & 0 & x & k \\ 0 & 0 & 0 & x \end{bmatrix} \in F(a,s).$$

Hence, by Lemma 15.3,

$$\|H\varphi(x)[h,k]\| \leq \rho$$

whenever $x \in F(a,r)$ and $\|h\|, \|k\| < r/2$. It follows that if $x \in F(a,r)$, then

$$\|H\varphi(x)[h,k]\| \leq \frac{4\rho}{r^2} \|h\| \|k\|$$

for all h and k. $\qquad\square$

Let $\Omega \subseteq \mathbb{M}^d$ be a fine domain, $\varphi \colon \Omega \to \mathbb{M}^1$ a fine holomorphic function, and $a \in \Omega \cap \mathbb{M}_n^d$. We set $L = D\varphi(a)$. If L is non-singular (i.e., surjective), then for each $k \in \mathbb{N}$, $\mathrm{id}_k \otimes L$ is non-singular as well. Thus, if we set $L_k = \mathrm{id}_k \otimes L$, then for each k, L_k has a right inverse, that is, there is a bounded transformation $R \colon \mathbb{M}_{kn}^1 \to \mathbb{M}_{kn}^d$ such that $L_k R = 1$.

Definition 15.5. *We say that L is* completely non-singular *if*

$$\sup_k \{\inf\{\|R\| \colon R \text{ is a right inverse of } L_k\}\} < \infty.$$

If L is completely non-singular, we define $c(L)$ by

$$c(L) = \left(\sup_k \{\inf\{\|R\| \colon R \text{ is a right inverse of } L_k\}\} \right)^{-1}.$$

Lemma 15.6. *If $L \colon \mathbb{M}_n^d \to \mathbb{M}_n^1$ is linear and has a right inverse R, then L is completely non-singular and $c(L) \geq 1/(n\|R\|)$.*

The Implicit Function Theorem 335

Proof. Note that $\mathrm{id}_k \otimes R$ is a right inverse of $\mathrm{id}_k \otimes L$. Therefore $c(L)$ is at least the reciprocal of

$$\|R\|_{cb} \overset{\mathrm{def}}{=} \sup_k \|\mathrm{id}_k \otimes R\|.$$

By a result of R. Smith [190], [162, proposition 8.11], any linear operator T defined on an operator space and with range $\mathbb{M}_n$ has $\|T\|_{cb} = \|\mathrm{id}_n \otimes T\| \leq n\|T\|$. But R is just a d-tuple of linear operators from $\mathbb{M}_n$ to $\mathbb{M}_n$, and so $\|R\|_{cb} \leq n\|R\|$. $\qquad\square$

Lemma 15.7. *If L is completely non-singular, $k \in \mathbb{N}$, $E\colon \mathbb{M}_{kn}^d \to \mathbb{M}_{kn}$ is linear, and $\|E\| < c(L)$, then $L_k + E$ is non-singular.*

Proof. Assume that L is completely non-singular, $k \in \mathbb{N}$, $E\colon \mathbb{M}_{kn}^d \to \mathbb{M}_{kn}$, and $\|E\| < c(L)$. Choose $R\colon \mathbb{M}_{kn}^1 \to \mathbb{M}_{kn}^d$ to satisfy $L_k R = 1$ and $\|R\| \leq c(L)^{-1}$.

If $\|E\| < c(L)$, then $\|ER\| < 1$ and, as a consequence, $1 + ER$ is invertible. But

$$\begin{aligned}(L_k + E)R(1 + ER)^{-1} &= (L_k R + ER)(1 + ER)^{-1} \\ &= (1 + ER)(1 + ER)^{-1} \\ &= 1.\end{aligned}$$

Hence, if $\|E\| < c(L)$, then $L_k + E$ is surjective. $\qquad\square$

Theorem 15.8. *Let $U \subseteq \mathbb{M}^d$ be a fat nc-domain and assume that $\Phi\colon U \to \mathbb{M}^\ell$ is a fat holomorphic function. Let $a \in U \cap \mathbb{M}_n^d$.*

(i) *If $D\Phi(a)$ has full rank, then there exists a fat domain Ω such that $a \in \Omega \subseteq U$ and $D\Phi(x)$ has full rank for all $x \in \Omega$.*

(ii) *If $\ell \leq d$ and $D\Phi(a)$ is an isomorphism from*

$$0^{d-\ell} \times \mathbb{M}_n^\ell \overset{\mathrm{def}}{=} \{(0,\dots,0,h^{d-\ell+1},\dots,h^d)\colon h^r \in \mathbb{M}_n, d-\ell+1 \leq r \leq d\}$$

onto $\mathbb{M}_n^\ell$, then there is a fat domain Ω such that $a \subset \Omega \subseteq U$ and such that, for all $p \in \mathbb{N}$ and for all $x \in \Omega_p$, $D\Phi(x)$ is an isomorphism from $0^{d-\ell} \times \mathbb{M}_p^\ell$ onto $\mathbb{M}_p^\ell$.

Proof. Let $\Phi = (\Phi^1,\dots,\Phi^\ell)^t$. By Lemma 15.4, there exist $s, M \in \mathbb{R}^+$ such that, for each $1 \leq j \leq \ell$,

$$\|H\Phi^j(x)[h,k]\| \leq M\|h\|\|k\|$$

for all $x \in F(a,s)$ and all $h, k \in \mathbb{M}^d$ that are on the same level as x. Choose $r \in \mathbb{R}^+$ satisfying

$$r < \min\left\{s, \frac{c(Df(a))}{M\sqrt{\ell}}\right\}.$$

336 Non-Commutative Theory

Let $m \in \mathbb{N}$ and $x \in F(a,r) \cap \mathbb{M}_{mn}^d$ (so that $\|x - a^{(m)}\| < r$). We have, for each j,

$$\|D\Phi^j(x)[h] - D\Phi^j(a^{(m)})[h]\| = \left\| \int_0^1 \frac{d}{dt} D\Phi^j\left(a^{(m)} + t(x - a^{(m)})\right)[h]\,dt \right\|$$

$$= \left\| \int_0^1 H\Phi^j\left(a^{(m)} + t(x - a^{(m)})\right)[h, x - a^{(m)}]\,dt \right\|$$

$$\leq M \|h\| \|x - a^{(m)}\|$$

$$< \frac{c(D\Phi(a))}{\sqrt{\ell}} \|h\|.$$

Thus

$$\|D\Phi(x) - D\Phi(a^{(m)})\| < c(D\Phi(a)).$$

Hence, by Lemma 15.7, $D\Phi(x)$ is non-singular, which proves (i).

Part (ii) follows in the same way, by consideration of $D\Phi(x)|_{0^{d-\ell} \times \mathbb{M}_m^\ell}$. By hypothesis, this map has a right inverse at a, and so, by Lemma 15.6, is completely non-singular. Therefore there is a fat neighborhood of a (perhaps smaller than in case (i)) on which $D\Phi(x)|_{0^{d-\ell} \times \mathbb{M}_m^\ell}$ is non-singular. □

We can now prove a fat version of the inverse function theorem, Theorem 15.1. For any $a \in \mathbb{M}_n^d$ and $r > 0$, we shall let

$$\mathrm{Ball}(a,r) = \{x \in \mathbb{M}_n^d : \max_{1 \leq j \leq d} \|a^j - x^j\| < r\}.$$

Theorem 15.9. *Let $\Omega \subseteq \mathbb{M}^d$ be a fat nc-domain. Let Φ be a fat holomorphic map from Ω to $\mathbb{M}^d$. The following statements are equivalent:*

(i) *Φ is injective on Ω;*
(ii) *$D\Phi(a)$ is non-singular for every $a \in \Omega$;*
(iii) *the function Φ^{-1} exists and is a fat holomorphic map.*

Proof. In light of Theorem 15.1, all that remains to prove is that assumption (ii) implies that Φ^{-1} is fatly holomorphic. Let $U = \Phi(\Omega)$ and let $b = \Phi(a) \in U \cap \mathbb{M}_n^d$. We must find a fat neighborhood of b on which Φ^{-1} is bounded. This in turn will follow if we can find $r, s > 0$ such that

$$\Phi(F(a,r)) \supseteq F(b,s). \tag{15.10}$$

By Lemma 15.4, there exists $r_1 > 0, M$ such that the Hessian of f is bounded by M on $F(a,r_1)$. Choose $0 < r < r_1$ so that

$$Mr < \frac{1}{2} c(D\Phi(a)),$$

The Implicit Function Theorem

and choose $s > 0$ so that

$$s \; < \; \frac{r}{2} c(D\Phi(a)).$$

We claim that, with these choices, the inclusion (15.10) holds.

Indeed, choose $k \in N$, and let $x \in \mathbb{M}_{kn}$ satisfy $\|x - a^{(k)}\| < r$. Then

$$
\begin{aligned}
\|\Phi(x) - \Phi(a^{(k)})\| &= \left\| \int_0^1 \frac{d}{dt} \Phi(a^{(k)} + t(x - a^{(k)}))dt \right\| \\
&= \left\| \int_0^1 D\Phi(a^{(k)} + t(x - a^{(k)}))[x - a^{(k)}]dt \right\| \\
&= \left\| D\Phi(a^{(k)})[x - a^{(k)}] + \int_0^1 D\Phi(a^{(k)} \right. \\
&\qquad \left. + t(x - a^{(k)}))[x - a^{(k)}] - D\Phi(a^{(k)})[x - a^{(k)}]dt \right\| \\
&\geq \|D\Phi(a^{(k)})[x - a^{(k)}]\| - M\|x - a^{(k)}\|^2 \\
&\geq \left(c(D\Phi(a)) - M\|x - a^{(k)}\| \right) \|x - a^{(k)}\| \\
&\geq \frac{1}{2} c(D\Phi(a))\|x - a^{(k)}\|.
\end{aligned}
$$

Since $D\Phi$ is non-singular, $\Phi(\mathrm{Ball}(a^{(k)}, r))$ is an open connected set, and by the last inequality it contains $\mathrm{Ball}(b^{(k)}, s)$. $\qquad \square$

15.3 The Implicit Function Theorem

Let $f = (f_1, \ldots, f_k)$ be a $\mathcal{B}(\mathbb{C}, \mathbb{C}^k)$-valued nc-function. Let Z_f denote the zero set $\cap_{i=1}^k Z_{f_i}$ of f. If $a \in \mathbb{M}_n^d$ then the derivative $Df(a)$ of f at a, if it exists, is a linear map from $\mathbb{M}_n^d$ to $\mathbb{M}_n^k$. We shall say that $Df(a)$ is of *full rank* if the rank of this linear map is kn^2.

For convenience in the following theorem, we shall write h in $\mathbb{M}_n^k$ as $h = (h^{d-k+1}, \ldots, h^d)$.

Theorem 15.11. *Let Ω be an nc-domain and let f be a $\mathcal{B}(\mathbb{C}, \mathbb{C}^k)$-valued fine holomorphic function on Ω, for some k such that $1 \leq k \leq d - 1$. Suppose that, for all $n \in \mathbb{N}$, all $a \in \Omega \cap \mathbb{M}_n^d$ and all $h \in \mathbb{M}_n^k \setminus \{0\}$,*

$$Df(a)[(0, \ldots, 0, h^{d-k+1}, \ldots, h^d)] \neq 0. \tag{15.12}$$

338 *Non-Commutative Theory*

Let W be the projection onto the first $d - k$ co-ordinates of $Z_f \cap \Omega$. Then there is a $\mathcal{B}(\mathbb{C}, \mathbb{C}^k)$-valued fine holomorphic function g on W such that

$$Z_f = \{(y, g(y)): y \in W\}.$$

Proof. Let $\Phi(x) = (x^1, \dots, x^{d-k}, f(x))^t$ be the fine holomorphic map defined on Ω by prepending the first $d - k$ co-ordinate functions. By the relation (15.12), $D\Phi$ is non-singular on Ω, and so, by Theorem 15.1, it is a fine holomorphic map from Ω onto $U = \Phi(\Omega)$, with inverse Ψ.

Let us write points x in $\mathbb{M}_n^d$ as (y, z), where $y \in \mathbb{M}_n^{d-k}$ and $z \in \mathbb{M}_n^k$. Then y is in W if and only if there is some z such that $(y, z) \in \Omega$ and $f(y, z) = 0$.

Let $\Psi = \psi_1 \oplus \psi_2$, where ψ_1 is Ψ followed by projection onto the first $d - k$ co-ordinates, and ψ_2 is Ψ followed by projection onto the last k co-ordinates. Define $g(y) = \psi_2(y^1, \dots, y^{d-k}, 0, \dots, 0)$.

If $(y, z) \in Z_f \cap \Omega$, then $\Phi(y, z) = (y, 0)$ and

$$\Psi \circ \Phi(y, z) = (y, z) = (\psi_1(y, 0), g(y)),$$

so $z = g(y)$.

Conversely, if $y \in W$ and $z = g(y)$, then $\Psi(y, 0) = (\psi_1(y, 0), g(y))$, so

$$\Phi \circ \Psi(y, 0) = (y, 0) = (\psi_1(y, 0), f(\psi_1(y, 0), g(y)).$$

Therefore $f(y, g(y)) = 0$. $\qquad\square$

In the fat category, we need only to know $Df(a)$ is full rank at one point. In the following statement, P_{d-k} is the map from $\mathbb{M}^d$ to $\mathbb{M}^{d-k}$ that projects onto the first $d - k$ components.

Theorem 15.13. *Let Ω be an nc-domain in $\mathbb{M}^d$ and let f be a $\mathcal{B}(\mathbb{C}, \mathbb{C}^k)$-valued fat holomorphic function on Ω for some k with $1 \leq k \leq d - 1$. Let $a \in \Omega_n$. Suppose that, for all $h \in \mathbb{M}_n^k \setminus \{0\}$,*

$$Df(a)[(0, \dots, 0, h^{d-k+1}, \dots, h^d)] \neq 0. \tag{15.14}$$

There exist a fat open neighborhood U of a in Ω and a $\mathcal{B}(\mathbb{C}, \mathbb{C}^k)$-valued fat holomorphic function g on $P_{d-k}(U)$ such that

$$Z_f \cap U = \{(y, g(y)): y \in P_{d-k}(U)\}.$$

Proof. By Theorem 15.8, we can find a fat open neighborhood U of a on which Df is full rank. We now follow the proof of Theorem 15.11, except that, since Φ is fat holomorphic, we can choose Ψ to be also, by Theorem 15.9. Hence we construct a fat holomorpic function g with the required properties. $\qquad\square$

The Implicit Function Theorem 339

15.4 The Range of an nc-Function

In Theorem 14.4, we proved that if φ is a free holomorphic function, then $\varphi(x)$ is always in $\mathrm{Alg}(x)$, the algebra generated by x. We shall see in Theorem 15.18 that this statement is not true for fine or fat holomorphic functions. There is nevertheless a range restriction that all nc-functions must satisfy: any value $\varphi(x)$ lies in the bicommutant of x, the commutant of the commutant of $\{x\}$, which is denoted by $\{x\}''$.

Lemma 15.15. *Let φ be an nc-function on a set Ω. For all $x \in \Omega$,*

$$\varphi(x) \in \{x\}''.$$

Proof. Let $S \in \{x\}'$. Then for t sufficiently small, $1 + tS$ is invertible, and

$$(1+tS)x(1+tS)^{-1} = x.$$

Since φ is nc, this means that $\varphi(x)$ also commutes with $1 + tS$, and hence with S. Therefore $\varphi(x) \in \{x\}''$. $\qquad\square$

When $d = 1$, the bicommutant is no larger than $\mathrm{Alg}(x)$, but for $d > 1$, it can be strictly bigger.

Lemma 15.16. *For any $x \in \mathbb{M}^1$ the double commutant of x equals $\mathrm{Alg}(x)$.*

Proof. See [96, theorem 13.22]. $\qquad\square$

Combining these two results, we get:

Proposition 15.17. *If φ is an nc-function on an nc-domain $\Omega \subseteq \mathbb{M}^1$, then $\varphi(x) \in \mathrm{Alg}(x)$.*

Here is an example that shows this fails when $d = 2$.

Theorem 15.18. *There exists a fat domain $\Omega \subseteq \mathbb{M}^2$ and a fat holomorphic function φ on Ω such that, for some $\lambda \in \Omega$, we have $\varphi(\lambda) \notin \mathrm{Alg}(\lambda)$.*

Proof. Let $\lambda \in \mathbb{M}_2^2$ be

$$\lambda = \left(\begin{bmatrix} 0 & 1 \\ 0 & 0 \end{bmatrix}, \begin{bmatrix} 1 & 0 \\ 0 & 0 \end{bmatrix} \right),$$

and let $\mu \in \mathbb{M}_2$ be

$$\mu = \begin{bmatrix} 0 & 0 \\ 1 & 0 \end{bmatrix}.$$

Define $p \in \mathcal{P}^3$ by

$$p(x,y,z) = (z)^2 + xz + zx + yz - \mathrm{id}. \tag{15.19}$$

340   *Non-Commutative Theory*

If $\lambda = (x,y)$ and $\mu = z$ are substituted in equation (15.19), we get $p(\lambda,\mu) = 0$. Let

$$a = \left(\begin{bmatrix} 0 & 1 \\ 0 & 0 \end{bmatrix}, \begin{bmatrix} 1 & 0 \\ 0 & 0 \end{bmatrix} \begin{bmatrix} 0 & 0 \\ 1 & 0 \end{bmatrix} \right). \tag{15.20}$$

It is easy to check that

$$\frac{\partial}{\partial z} p(a)[h] = \begin{bmatrix} h_{11} + h_{12} + h_{21} & h_{11} + h_{12} + h_{22} \\ h_{11} + h_{22} & h_{12} + h_{21} \end{bmatrix}. \tag{15.21}$$

From equation (15.21) we see that the derivative of p has full rank at a, so by Theorem 15.13 there is a fat neighborhood Ω of λ and a fat holomorphic function φ on Ω such that $\varphi(\lambda) = \mu$, which is not in $\mathrm{Alg}(\lambda)$. $\qquad\square$

We proved Theorem 15.18 by using the implicit function theorem; one can reverse this argument to show that you cannot have both an implicit function theorem and pointwise polynomial approximation for any admissible topology.

The following lemma was proved by Sylvester [192]; for a more recent discussion, see [49].

Lemma 15.22. *The matrix equation $AH - HB = 0$, for $A, B, H \in \mathbb{M}_n$, has a non-zero solution H if and only if $\sigma(A) \cap \sigma(B) \neq \emptyset$. The dimension of the set of solutions is $\#\{(\lambda,\mu): \lambda \in \sigma(A), \mu \in \sigma(B), \lambda = \mu\}$, where eigenvalues are counted with multiplicity.*

Let p be as in equation (15.19), and define

$$\Phi(x,y,z) = (x,y,p(x,y,z)).$$

As

$$\frac{\partial}{\partial z} p(x,y,z)[h] = (x+y+z)h + h(x+z),$$

it follows from Lemma 15.22 that the derivative of Φ is non-singular if and only if

$$\sigma(x+y+z) \cap \sigma(-x-z) = \emptyset. \tag{15.23}$$

Lemma 15.24. *Let a be as in equation (15.20). There is a free neighborhood of a on which equation (15.23) holds; moreover it is of the form B_δ where δ is a diagonal matrix of polynomials.*

Proof. The eigenvalues of $a^1 + a^2 + a^3$ are $(1 \pm \sqrt{5})/2$; call them c_1 and c_2. The eigenvalues of $a^1 + a^3$ are ± 1. Let $\varepsilon > 0$ be such that the closed disks of radius ε and centers $c_1, c_2, 1, -1$ are disjoint.

Let $\delta(x)$ be the 2×2 diagonal matrix with entries

$$M(x^1 + x^2 + x^3 - c_1)(x^1 + x^2 + x^3 - c_2) \text{ and } M(x^1 + x^3 - 1)(x^1 + x^3 + 1).$$

The Implicit Function Theorem 341

By choosing M large enough, one can ensure that if $x \in B_\delta$, then

$$\sigma(x^1 + x^2 + x^3) \subset \mathbb{D}(c_1, \varepsilon) \cup \mathbb{D}(c_2, \varepsilon) \text{ and } \sigma(x^1 + x^3) \subset \mathbb{D}(1, \varepsilon) \cup \mathbb{D}(-1, \varepsilon).$$

$\square$

If evaluation of the norms of polynomials is τ-continuous, then B_δ is τ-open, so we cannot have both an implicit function theorem and pointwise approximation.

Theorem 15.25. *Let τ be an nc topology defined on $\mathbb{M}^d$ for every d, and suppose that for each polynomial $q \in \mathcal{P}_d$, the map $x \mapsto \|q(x)\|$ is τ-continuous into $\mathbb{R}^+$. If every τ-holomorphic function is pointwise approximable by free polynomials, then there is no analog of Theorem 5.11 in which the fine topology is replaced by τ.*

15.5 Applications of the Implicit Function Theorem in Non-Commutative Algebraic Geometry

Example 15.26. Let us consider the free polynomial (11.7) from Section 11.5,

$$p(x, y) = x^3 + 2xy - 4yx + 5x - 6y + 7.$$

The partial derivative of p with respect to y is easy to calculate:

$$\frac{\partial p}{\partial y} p(x, y)[h] = 2xh - 4hx - 6h.$$

The partial derivative has full rank if the right-hand side is non-zero for all non-zero h. By Lemma 15.22, this occurs if and only if

$$\sigma(2x) \cap \sigma(4x + 6) = \emptyset, \tag{15.27}$$

that is, whenever λ is an eigenvalue of x, then $2\lambda + 3$ is not an eigenvalue. Generically, this happens. Let $(X, Y) \in \mathbb{M}^2$ be a pair of matrices satisfying $p(X, Y) = 0$, and assume X satisfies equation (15.27). We can apply Theorem 15.13 to conclude that there is a fat open set U in $\mathbb{M}^2$ containing (X, Y) and a fat holomorphic function g defined on W, the projection of U onto the first variable, such that $Z_p \cap U = \{(x, g(x)): x \in W\}$. By Proposition 15.17, this means that $Y \in \mathrm{Alg}(X)$, and therefore commutes Y with X.

Although, generically in X, the solutions of a (generic) polynomial equation in two variables commute, at some non-generic X's the solution set can be large.

342   *Non-Commutative Theory*

Example 15.28. Let $a, b, c \in \mathbb{C}$, and assume $b \neq -c$ and b/c is not a root of unity. Let

$$p(x, y) = ax^2 + bxy + cyx.$$

Let $X \in \mathbb{M}_n$. Let

$$\mathcal{Y} = \{Y \in \mathbb{M}_n : aX^2 + bXY + cYX = 0\}.$$

Let k be the number of common eigenvalues of bX and $-cX$, counting multiplicity.

(i) If $k = 0$, then $\mathcal{Y}$ has a unique element, which commutes with X.
(ii) If $k > 0$, then $\mathcal{Y}$ is a k-dimensional affine space in $\mathbb{M}_n$, and it contains a unique element that commutes with X.
(iii) For $n \geq 2$, the dimension of the set of non-commuting solutions in $\mathbb{M}_n^2$ of $p(x, y) = 0$ is exactly n^2, the same as the dimension of the set of commuting solutions.

We leave the proof as an exercise.

There are polynomials, like $(xy - yx)^2 - 1$, whose zero sets contain no commuting elements, but this is not generic in the coefficients. For convenience in the following theorem, we shall write points in $\mathbb{M}^d$ as $(x, y^1, \ldots, y^k)$, where $k = d - 1$.

Theorem 15.29. *Let $p_1, \ldots, p_k$ be free polynomials in $\mathcal{P}_d$ with the property that, when evaluated on d-tuples of complex numbers, none of them is constant in the last k variables. Let $p = (p_1, \ldots, p_k)^t$, and let*

$$\mathfrak{V} = \{(x, y^1, \ldots, y^k) : p(x, y^1, \ldots, y^k) = 0\}.$$

Let B be the finite (possibly empty) set

$$B = \bigcup_{j=1}^{k} \{x \in \mathbb{C} : p_j(x, y^1, \ldots, y^k) \neq 0 \text{ for all } y \in \mathbb{C}^k\}.$$

If X_0 in $\mathbb{M}_n$ has n linearly independent eigenvectors and $\sigma(X_0) \cap B = \emptyset$, then there exists Y_0 in $\mathbb{M}_n^k$ that satisfies $(X_0, Y_0) \in \mathfrak{V}$ and such that each element Y_0^j commutes with X_0.

Moreover, if (X_0, Y_0) is in $\mathfrak{V}$ and X_0 and Y_0 do not commute, then

$$(X_0, Y_0) \in V \cap \{(X, Y) : Dp(X, Y) \text{ does not have full rank on } 0 \times \mathbb{M}_n^k\}.$$

Proof. Exercise. $\square$

Question 15.30. With one non-commutative polynomial equation in two unknowns, we have shown that in some generic sense, the solutions commute.

The Implicit Function Theorem 343

With two equations in three unknowns, $p_1(x,y,z) = 0 = p_2(x,y,z)$, what restrictions apply? At non-degenerate points of the zero set where we can apply the implicit function theorem, we can conclude from Lemma 15.15 that z must lie in the double commutant of x and y. Sometimes the double commutant will be everything, but not always. What more can one say?

15.6 Additional Results

Helton, Klep, and McCullough proved that one does not even need an injectivity assumption to get an nc inverse function theorem; it is sufficient that the map be proper. The following theorem is proved in [104, theorem 3.1].

Theorem 15.31. *Let $\Omega \subset \mathbb{M}^d$ and $U \subset \mathbb{M}^\ell$ be nc-domains, and suppose $\Phi \colon \Omega \to U$ is an nc-map.*

(i) *If Φ is proper with respect to the finite topologies on Ω and U, then it is injective, and Φ^{-1} is nc.*

(ii) *If, for all $y \in U$, the set $\Phi^{-1}(y)$ has compact closure in Ω_n, where n is the level of y, then Φ is injective.*

(iii) *If $n = \ell$ and Φ is proper and finitely continuous, then so is Φ^{-1}.*

15.7 Historical Notes

The equivalence of (i) and (ii) in Theorem 15.1 is due to Pascoe [160]; the equivalence of (i) and (iii) is due to Helton, Klep, and McCullough [104]. Most of the results of Sections 15.2–15.5 are from [18].

Other approaches to inverse and implicit function theorems for nc-functions are given in [2, 149].

16

Non-Commutative Functional Calculus

Suppose that φ is a free holomorphic function in $H_1^\infty(B_\delta)$; then it has a realization formula as in equation (14.9):

$$\varphi(x) = \underset{1}{\overset{A}{\otimes}} + \underset{1}{\overset{B}{\otimes}} \underset{\delta(x)}{\overset{1}{\otimes}} \left[1 - \underset{1}{\overset{D}{\otimes}} \underset{\delta(x)}{\overset{1}{\otimes}} \right]^{-1} \underset{1}{\overset{C}{\otimes}} \qquad \text{for all } x \in B_\delta. \qquad (16.1)$$

The right-hand side of the formula makes perfect sense if one allows x to be a d-tuple of operators in $\mathcal{B}(\mathcal{H})$ for which $\|\delta(x)\| < 1$, even when $\mathcal{H}$ is infinite-dimensional. This allows us to define a functional calculus with the aid of free holomorphic functions.

The goal of this chapter is to make sense of the statement that equation (16.1) defines a free function of operators and to give an intrinsic characterization of the functions that arise from such a realization formula.

16.1 Nc Operator Functions

Let $\mathcal{H}$ be a fixed separable infinite-dimensional Hilbert space. The problem in trying to define an nc-function on $\mathcal{B}(\mathcal{H})^d$ is that there is no canonical way to identify a direct sum of two operators on $\mathcal{H}$ with another operator on $\mathcal{H}$. We shall navigate this obstacle by restricting ourselves to domains that are in some sense unitarily equivalent to their direct sums. We say that a subset $\Omega \subseteq \mathcal{B}(\mathcal{H})^d$ is *unitarily invariant* if, whenever $x \in \Omega$ and $u \in \mathcal{B}(\mathcal{H})$ is unitary, then $u^* x u \in \Omega$.

Recall that Δ° means the interior of Δ.

Definition 16.2. *A set* $\Omega \subseteq \mathcal{B}(\mathcal{H})^d$ *is called an* nc operator domain *if there exists a sequence* $\{\Delta_k\}_{k=1}^\infty$ *of subsets of* Ω *with with the following properties:*

(i) $\Delta_k \subseteq \Delta_{k+1}^\circ$ *for all* k;

(ii) $\Omega = \bigcup_{k=1}^\infty \Delta_k$;

Non-Commutative Functional Calculus

(iii) each Δ_k is bounded and unitarily invariant; and

(iv) each Δ_k is closed under countable direct sums: if x_j is a sequence in Δ_k of length $\ell \in \{1, 2, \ldots, \infty\}$, then there exists a unitary $U \colon \mathcal{H} \to \mathcal{H}^{(\ell)}$ such that

$$U^{-1} \begin{bmatrix} x_1 & 0 & \cdots \\ 0 & x_2 & \cdots \\ \cdots & \cdots & \ddots \end{bmatrix} U \in \Delta_k. \qquad (16.3)$$

Note that if equation (16.3) holds for some unitary $U \colon \mathcal{H} \to \mathcal{H}^{(\ell)}$, then it holds for every unitary $V \colon \mathcal{H} \to \mathcal{H}^{(\ell)}$, since Δ_k is unitarily invariant.

For δ a matrix of free polynomials, let

$$B_\delta^\sharp = \{x \in \mathcal{B}(\mathcal{H})^d \colon \|\delta(x)\| < 1\}.$$

These sets will be our prototypical examples of nc operator domains. Indeed, if

$$\Delta_k = \{x \in B_\delta^\sharp \colon \|\delta(x)\| \le 1 - 1/k, \text{ and } \|x\| \le k\},$$

then all of the properties of Definition 16.2 are satisfied.

Definition 16.4. Let $\Omega \subseteq \mathcal{B}(\mathcal{H})^d$ be an nc operator domain. A function $F \colon \Omega \to \mathcal{B}(\mathcal{H})$ is called an nc operator function if whenever $x, y \in \Omega$ and whenever $s \colon \mathcal{H} \to \mathcal{H}^{(2)}$ is a bounded invertible map with

$$s^{-1} \begin{bmatrix} x & 0 \\ 0 & y \end{bmatrix} s \in \Omega,$$

then

$$F\left(s^{-1} \begin{bmatrix} x & 0 \\ 0 & y \end{bmatrix} s \right) = s^{-1} \begin{bmatrix} F(x) & 0 \\ 0 & F(y) \end{bmatrix} s.$$

The proof of the next statement is similar to that of Lemma 12.5.

Lemma 16.5. Let $\Omega \subseteq \mathcal{B}(\mathcal{H})^d$ be an nc operator domain, and let $F \colon \Omega \to \mathcal{B}(\mathcal{H})$ be an nc operator function. Let $x, y \in \Omega$, let $L \in \mathcal{B}(\mathcal{H})$ and $s \colon \mathcal{H} \to \mathcal{H}^{(?)}$ be a bounded invertible map such that

$$s^{-1} \begin{bmatrix} x & Ly - xL \\ 0 & y \end{bmatrix} s \in \Omega.$$

Then

$$F\left(s^{-1} \begin{bmatrix} x & Ly - xL \\ 0 & y \end{bmatrix} s \right) = s^{-1} \begin{bmatrix} F(x) & LF(y) - F(x)L \\ 0 & F(y) \end{bmatrix} s.$$

In particular, nc operator functions are intertwining-preserving, and also preserve countable direct sums.

346   Non-Commutative Theory

Lemma 16.6. *Let* $\Omega \subseteq \mathcal{B}(\mathcal{H})^d$ *be an* nc *operator domain, and let* $F: \Omega \to \mathcal{B}(\mathcal{H})$ *be an* nc *operator domain. Let* $x_1, x_2, \ldots \in \Omega$, *and* $s: \mathcal{H} \to \mathcal{H}^{(\infty)}$ *be a bounded invertible map such that*

$$
y = s^{-1}
\begin{bmatrix}
x_1 & 0 & \cdots \\
0 & x_2 & \cdots \\
\cdots & \cdots & \ddots
\end{bmatrix}
s \in \Omega.
$$

Then

$$
F\left(s^{-1}
\begin{bmatrix}
x_1 & 0 & \cdots \\
0 & x_2 & \cdots \\
\cdots & \cdots & \ddots
\end{bmatrix}
s \right) = s^{-1}
\begin{bmatrix}
F(x_1) & 0 & \cdots \\
0 & F(x_2) & \cdots \\
\cdots & \cdots & \ddots
\end{bmatrix}
s.
$$

Proof. Let $\Gamma_n: \mathcal{H}^\infty \to \mathcal{H}$ be projection onto the nth component. Let $L_n = \Gamma_n s$. Then $L_n y = x_n L_n$, and the result follows. $\qquad\square$

In particular, since each Δ_k is closed under countable direct sums, it follows that an nc operator function must be bounded on Δ_k, since $\oplus F(x_n)$ is a bounded operator.

Let $\mathcal{D}$ be an open subset of a Banach space $\mathcal{X}$, and $f: \mathcal{D} \to Y$ a map into a Banach space Y. We say that f has a *Gâteaux derivative* at x if

$$
Df(x)[h] \stackrel{\text{def}}{=} \lim_{X \to 0} \frac{f(x + Xh) - f(x)}{X}
$$

exists for all $h \in \mathcal{X}$. If f has a Gâteaux derivative at every point of $\mathcal{D}$, it is Gâteaux holomorphic [73, lemma 3.3], that is, holomorphic on each 1-dimensional slice. If in addition f is locally bounded on $\mathcal{D}$, then it is Fréchet holomorphic [73, proposition 3.7], which means that, for each x, there is a neighborhood G of 0 such that the Taylor series

$$
f(x + h) = f(x) + \sum_{k=1}^{\infty} D^k f(x)[h, \ldots, h] \quad \text{for all } h \in G \tag{16.7}
$$

converges uniformly for all h in G. The kth derivative is a continuous k-linear map from $\mathcal{X}^k \to Y$, which is evaluated on the k-tuple $(h, h, \ldots, h)$.

The following lemma is the operator version of Theorem 12.17.

Lemma 16.8. *Let* Ω *be an* nc *operator domain in* $\mathcal{B}(\mathcal{H})^d$, *and let* $F: \Omega \to \mathcal{B}(\mathcal{H})$ *be an* nc *operator function. Then* F *is locally bounded on* Ω, *continuous, and Fréchet holomorphic.*

Proof. Local boundedness of F follows from the remark after Lemma 16.6. We can repeat the proof of Theorem 12.17 to conclude that F is Gâteaux differentiable at every point, and so, by local boundedness, it is Fréchet holomorphic. $\qquad\square$

Non-Commutative Functional Calculus   347

Lemma 16.9. *If F is an nc operator function on $\Omega \subseteq \mathcal{B}(\mathcal{H})^d$, and $P \in \mathcal{B}(\mathcal{H})$ is a projection, then, for all $c \in \Omega$ satisfying $c = cP$ (or $c = Pc$) and for all $a \in \mathcal{B}(\mathcal{H})^d$,*

$$a = PaP \;\Rightarrow\; F(a) = PF(a)P + P^{\perp}F(c)P^{\perp}. \qquad (16.10)$$

Proof. As $Pa = aP$, we get $PF(a) = F(a)P$. As $P^{\perp}a = 0 = cP^{\perp}$, we get $P^{\perp}F(a) = F(c)P^{\perp}$. Combining these statements, we deduce the implication (16.10). $\qquad\square$

We let m_α denote the Möbius map on $\mathbb{D}$ given by

$$m_\alpha(\zeta) = \frac{\zeta - \alpha}{1 - \bar{\alpha}\zeta}.$$

Lemma 16.11. *Let $\Omega \subseteq \mathcal{B}(\mathcal{H})^d$ be an nc operator domain containing 0, and assume F is an nc operator function from Ω to ball $\mathcal{B}(H)$. Then*

(i) *$F(0) = \alpha I_{\mathcal{H}}$, for some $\alpha \in \mathbb{D}$;*
(ii) *the map $H(x) \stackrel{\mathrm{def}}{=} m_\alpha \circ F(x)$ is an nc operator function on Ω that maps 0 to 0;*
(iii) *for any $a \in \Omega$ and any projection P,*

$$a = PaP \;\Rightarrow\; H(a) = PH(a)P; \; and$$

(iv) *$F = m_{-\alpha} \circ H$.*

Proof. (i) By Lemma 16.9 applied to $a = c = 0$, we find that $F(0)$ commutes with every projection P in $\mathcal{B}(\mathcal{H})$. Therefore it must be a scalar.

(ii) For all z in ball $\mathcal{B}(H)$,

$$m_\alpha(z) = -\alpha \; \mathrm{id}_{\mathcal{H}} + (1 - |\alpha|^2) \sum_{n=1}^{\infty} \bar{\alpha}^{n-1} z^n, \qquad (16.12)$$

where the series converges uniformly and absolutely on the ball of radius one. By (i), we have $H(0) = m_\alpha(\alpha \, \mathrm{id}_{\mathcal{H}}) = 0$. To calculate

$$m_\alpha \circ F\left(s^{-1}\begin{bmatrix} x & 0 \\ 0 & y \end{bmatrix}s\right) = m_\alpha\left(s^{-1}\begin{bmatrix} F(x) & 0 \\ 0 & F(y) \end{bmatrix}s\right),$$

we evaluate the terms in the series (16.12) and sum them to conclude that

$$H\left(s^{-1}\begin{bmatrix} x & 0 \\ 0 & y \end{bmatrix}s\right) = s^{-1}\begin{bmatrix} m_\alpha \circ F(x) & 0 \\ 0 & m_\alpha \circ F(y) \end{bmatrix}s$$

$$= s^{-1}\begin{bmatrix} H(x) & 0 \\ 0 & H(y) \end{bmatrix}s$$

and hence that H is an nc operator function.

348   *Non-Commutative Theory*

Now (iii) follows from Lemma 16.9 with $c = 0$. We deduce (iv) because $m_{-\alpha} \circ m_\alpha(z) = z$ for every $z \in \text{ball}\,\mathcal{B}(H)$. $\qquad\square$

We want to interface with the theory of nc-functions. The easiest way to do this is to make a blanket assumption that $0 \in \Omega$. Now we choose a fixed basis $\{e_n : n \geq 1\}$ for $\mathcal{H}$. Let P_n be the orthogonal projection from $\mathcal{H}$ onto $\vee_{k=1}^n \{e_k\}$. Let $\mathcal{M}_n = P_n \mathcal{B}(\mathcal{H}) P_n$; then we can identify $\mathbb{M}_n$ with $\mathcal{M}_n$ and $\mathbb{M}^d$ with

$$\mathcal{M}^d \overset{\text{def}}{=} \bigcup_{n=1}^\infty \mathcal{M}_n^d.$$

If Ω is an nc operator domain containing 0 and F is a bounded nc operator function on Ω, then, in the light of Lemma 16.11, it is a mild normalization to assume that $F(0) = 0$. With this assumption, F maps $\Omega \cap \mathcal{M}_n^d$ to $\mathcal{M}_n$. Thus, after the identification of $\mathcal{M}_n^d$ with $\mathbb{M}_n^d$, the restriction of F to $\Omega \cap \mathcal{M}_n^d$ is an nc-function $F^\flat$. To be explicit, if $x \in \mathbb{M}_n^d$, let $\iota_n(x)$ be the element of $\mathcal{B}(\mathcal{H})^d$ defined by

$$\langle (\iota_n(x))^r e_j, e_i \rangle = \begin{cases} x_{i,j}^r & \text{if } 1 \leq i, j \leq n \\ 0 & \text{if } i \text{ or } j > n. \end{cases}$$

(In other words, we fill out the infinite matrix with 0's.) Now define

$$\Omega^\flat = \bigcup_{n=1}^\infty \{x \in \mathbb{M}_n^d : \iota_n(x) \in \Omega\}.$$

Then we define $F^\flat$ on $\Omega^\flat$ by

$$F^\flat(x) = \iota_n^{-1}(F(\iota_n(x))) \qquad \text{for all } x \in \Omega_n^\flat, \tag{16.13}$$

where $\iota_n^{-1} \colon \mathcal{M}_n \to \mathbb{M}_n$.

Proposition 16.14. *Let Ω be an* nc *operator domain in $\mathcal{B}(\mathcal{H})^d$ containing 0, and let F be an* nc *operator function on Ω with $F(0) = 0$. Then $\Omega^\flat$ is an* nc-*domain in $\mathbb{M}^d$, and $F^\flat$ is an* nc-*function on $\Omega^\flat$.*

Proof. It is clear that $\Omega^\flat$ is finitely open. To see that it is closed with respect to direct sums, suppose x, y are in $\Omega_m^\flat$ and $\Omega_n^\flat$, respectively. Then $U(\iota_m(x) \oplus \iota_n(y))U^* = z$ for some unitary $U \colon \mathcal{H}^{(2)} \to \mathcal{H}$ and some $z \in \Omega$, since Ω is closed with respect to direct sums. Let $V \colon \mathcal{H} \to \mathcal{H}^{(2)}$ be a unitary satisfying $V^*(\iota_m(x) \oplus \iota_n(y))V = \iota_{m+n}(x \oplus y)$. Since Ω is unitarily invariant, $V^* U^* z U V = \iota_{m+n}(x \oplus y)$ is in Ω, so $x \oplus y$ is in $\Omega^\flat$.

($F^\flat$ is graded) For $x \in \Omega_n^\flat$, this follows from the definition (16.13).

(Intertwining preserving) Suppose $x \in \Omega_m^\flat$, $y \in \Omega_n^\flat$ and $L \colon \mathbb{C}^m \to \mathbb{C}^n$ intertwines x and y, and so $Lx^r = y^r L$ for each r from 1 to d. Then extending the

Non-Commutative Functional Calculus   349

matrix of L with zeroes, we get a finite-rank operator $\tilde{L}$ on $\mathcal{H}$ that maps $P_m \mathcal{H}$ to $P_n \mathcal{H}$, and intertwines $\iota_m(x)$ with $\iota_n(y)$. Since F is intertwining-preserving, we get $\tilde{L} F(\iota_m(x)) = F(\iota_n(y))\tilde{L}$, so that $L F^\flat(x) = F^\flat(y)L$, as required.   □

16.2 Polynomial Approximation and Power Series

From here on we shall dispense with the formalism of ι, and just think of $\Omega^\flat$ as the subset of Ω of d-tuples that are all 0 outside of some finite-dimensional block, and $F^\flat$ as the restriction of F to these elements. It is natural to ask which nc-functions on $\Omega^\flat$ arise this way.

To extend nc-functions to nc operator functions on open sets in $\mathcal{B}(\mathcal{H})^d$, we need some form of continuity that is weaker than norm continuity (since the norm closure of $\mathcal{M}^d$ is the set of compact d-tuples).

Definition 16.15. *Let* $F\colon \Omega \subseteq \mathcal{B}(\mathcal{H})^d \to \mathcal{B}(\mathcal{H})$. *We say* F *is* sequentially strong operator continuous *(SSOC) if, whenever* $x_n \to x$ *in the strong operator topology on* A, *then* $F(x_n)$ *tends to* $F(x)$ *in the strong operator topology on* $\mathcal{B}(\mathcal{H})$.

Multiplication is not continuous in the strong operator topology, but it is sequentially strong operator continuous, since it is continuous on bounded sets. Every free polynomial is SSOC and an nc operator function, and these properties will be inherited by limits.

We want to improve Lemma 16.8 so that the kth term in equation (16.7) is a free polynomial, homogeneous of degree k, in the entries of h. We need the following result, which was proved by Kaliuzhnyi-Verbovetskyi and Vinnikov [127, theorem 6.1].

Theorem 16.16. *Let*

$$\varphi\colon \mathbb{M}^d \to \mathbb{M}^1$$

be an nc-*function such that each matrix entry of* $\varphi(x)$ *is a polynomial of degree less than or equal to* N *in the entries of the matrices* $x^r, 1 \le r \le d$. *Then* φ *is a free polynomial of degree less than or equal to* N.

We extend this result to multilinear SSOC intertwining-preserving maps. Each h_j in the following statement is a d-tuple of operators, $(h_j^1, \ldots, h_j^d)$.

Lemma 16.17. *Let*

$$\Lambda\colon \mathcal{B}(\mathcal{H})^{dN} \to \mathcal{B}(\mathcal{H})$$

$$(h_1, \ldots, h_N) \mapsto \Lambda(h_1, \ldots, h_N)$$

350 Non-Commutative Theory

be a continuous N-linear map from $(\mathcal{B}(\mathcal{H})^d)^N$ to $\mathcal{B}(\mathcal{H})$ that is an nc *operator function and is SSOC. Then Λ is a homogeneous polynomial of degree N in the variables $h_1^1, \ldots, h_N^d$.*

Proof. By Proposition 16.14, if we restrict Λ to $\mathcal{M}^{dN}$, we get an nc-function. By Theorem 16.16, there is a free polynomial p of degree N that agrees with Λ on $\mathcal{M}^{dN}$. By homogeneity, p must be homogeneous of degree N. Define

$$\xi(h) = \Lambda(h) - p(h).$$

Then ξ vanishes on $(\mathcal{M}^d)^N$ and is SSOC. Since $(\mathcal{M}^d)^N$ is dense in $(\mathcal{B}(\mathcal{H})^d)^N$ for the strong operator topology, it follows that ξ is identically 0. $\quad\square$

In the next proposition, we show that the derivatives in the Taylor series (16.7) are free polynomials.

Proposition 16.18. *Let Ω be an* nc *operator domain that is a neighborhood of 0 in $\mathcal{B}(\mathcal{H})^d$, and let $F: \Omega \to \mathcal{B}(\mathcal{H})$ be an* nc *operator function that is sequentially strong operator continuous. Then there is an open set $\mathcal{D} \subseteq \Omega$ containing 0 and homogeneous free polynomials P_k of degree k such that*

$$F(x) = F(0) + \sum_{k=1}^{\infty} P_k(x) \quad \text{for all } x \in \mathcal{D}, \tag{16.19}$$

where the convergence is uniform for $x \in \mathcal{D}$.

Proof. By Lemma 16.8, F is Fréchet holomorphic. Therefore there is some open ball $B(0, R) \subseteq \Omega$ such that

$$F(h) = F(0) + \sum_{k=1}^{\infty} D^k F(0)[h, \ldots, h] \quad \text{for all } h \in B(0, R),$$

and the convergence is uniform for $h \in B(0, R)$. If we let $\mathcal{D}$ be $B(0, R/2)$, we have, for each k, a uniform bound on $D^k F$ on all of $\mathcal{D}$.

We must show that each $D^k F(0)[h, \ldots, h]$ is a free polynomial in h.

Claim 1. For each $k \in \mathbb{N}$, the function

$$G^k: (h^0, \ldots, h^k) \mapsto D^k F(h^0)[h^1, \ldots, h^k] \tag{16.20}$$

is an nc operator function on $\mathcal{D} \times (\mathcal{B}(\mathcal{H})^d)^k \subseteq (\mathcal{B}(\mathcal{H})^d)^{k+1}$.

When $k = 1$, we have

$$DF(h^0)[h^1] = \lim_{t \to 0} \frac{1}{t}[F(h^0 + th^1) - F(h^0)]. \tag{16.21}$$

As F is an nc operator function, so is the right-hand side of (16.21). Indeed, we need to check the condition of Definition 16.4.

We have

$$\lim_{t \to 0} \frac{1}{t}\left[F\left(s^{-1}\begin{bmatrix} h_1^0 + th_1^1 & 0 \\ 0 & h_2^0 + th_2^1 \end{bmatrix} s\right) - F\left(s^{-1}\begin{bmatrix} h_1^0 & 0 \\ 0 & h_2^0 \end{bmatrix} s\right)\right]$$

$$= \lim_{t \to 0} \frac{1}{t}\left[s^{-1}\left(\begin{bmatrix} F(h_1^0 + th_1^1) & 0 \\ 0 & F(h_2^0 + th_2^1) \end{bmatrix} - \begin{bmatrix} F(h_1^0) & 0 \\ 0 & F(h_2^0) \end{bmatrix}\right) s\right]$$

$$= s^{-1}\left(\begin{bmatrix} \lim_{t \to 0} \frac{1}{t}[F(h_1^0 + th_1^1) - F(h_1^0)] & 0 \\ 0 & \lim_{t \to 0} \frac{1}{t}[F(h_2^0 + th_2^1) - F(h_2^0)] \end{bmatrix}\right) s.$$

So G^1 is an nc operator function.

For $k > 1$,

$$D^k F(h^0)[h^1, \dots, h^k] = \lim_{t \to 0} \frac{1}{t}\big[D^{k-1} F(h^0 + th^k)[h^1, \dots, h^{k-1}]$$

$$- D^{k-1} F(h^0)[h^1, \dots, h^{k-1}]\big].$$

By induction, these are all nc operator functions.

Claim 2. For each $k \in \mathbb{N}$, the function G^k from the display (16.20) is SSOC on $\mathcal{D} \times (\mathcal{B}(\mathcal{H})^d)^k$.

Again we show this by induction on k. Let $G^0 \overset{\text{def}}{=} F$, which is SSOC on $2\mathcal{D} \subseteq \Omega$ by hypothesis. We have G^{k-1} is an nc operator function on the set $\mathcal{D}^k$ and is bounded on this set (since $\mathcal{D}$ is bounded and all of F's derivatives are bounded on $\mathcal{D}$). By Lemma 16.8 G^{k-1} is Fréchet differentiable. Suppose

$$\text{SOT} \lim_{n \to \infty} h_n^j = h^j \qquad \text{for } 0 \le j \le k,$$

where each h_n^j and h^j is in $\mathcal{D}$. Let h denote the $(k+1)$-tuple $(h^0, \dots, h^k)$ in $\mathcal{D}^{k+1}$, and let $\tilde{h}$ denote the k-tuple $(h^0, \dots, h^{k-1})$; similarly, let h_n denote $(h_n^0, \dots, h_n^k)$ and $\tilde{h}_n$ denote $(h_n^0, \dots, h_n^{k-1})$. There exists some unitary u so that $y = u^*(\tilde{h} \oplus \tilde{h}_1 \oplus \tilde{h}_2 \oplus \cdots)u$ is in $\overline{\mathcal{D}^k}$. The argument in Claim 1 actually showed that G^{k-1} is an nc operator function on $(2-\varepsilon)\mathcal{D} \times (\mathcal{B}(\mathcal{H})^d)^{k-1}$ for every $\varepsilon > 0$, so in particular G^{k-1} is differentiable at y. Therefore the diagonal operator with entries

$$\frac{1}{t}[G^{k-1}(h^0 + th^k, h^1, \dots, h^{k-1}) - G^{k-1}(h^0, h^1, \dots, h^{k-1})],$$

$$\frac{1}{t}[G^{k-1}(h_1^0 + th_1^k, h_1^1, \dots, h_1^{k-1}) - G^{k-1}(h_1^0, h_1^1, \dots, h_1^{k-1})],$$

$$\dots \tag{16.22}$$

has a limit in the norm topology as $t \to 0$.

Let $\varepsilon > 0$, and let $v \in \mathcal{H}$ have $\|v\| \le 1$. Choose t sufficiently close to 0 that each of the difference quotients (16.22) is within $\varepsilon/3$ of its limit (which is G^k evaluated at the appropriate h or h_n). Let n be large enough that

352 Non-Commutative Theory

$$\left\| \left[G^{k-1}(h^0+th^k,h^1,\ldots,h^{k-1}) - G^{k-1}(h_n^0+th_n^k,h_n^1,\ldots,h_n^{k-1}) \right]v \right\|$$

$$+ \left\| \left[G^{k-1}(h^0,h^1,\ldots,h^{k-1}) - G^{k-1}(h_n^0,h_n^1,\ldots,h_n^{k-1}) \right]v \right\|$$

$$\le \frac{\varepsilon t}{3}.$$

Then

$$\left\| \left[G^k(h^0,\ldots,h^k) - G^k(h_n^0,\ldots,h_n^k) \right]v \right\| \le \varepsilon.$$

So each G^k is SSOC on $\mathcal{D}^{k+1}$. As G^k is linear in the last k variables, it is SSOC on $\mathcal{D} \times (\mathcal{B}(\mathcal{H})^d)^k$ as claimed.

Therefore for each k, the map

$$(h^1,\ldots,h^k) \mapsto D^k F(0)[h^1,\ldots,h^k]$$

is a linear nc operator function that is SSOC in a neighborhood of 0, and so, by Lemma 16.17, is a free polynomial. $\square$

We can extend the convergence in Proposition 16.18 from a neighborhood of 0 to the whole set, provided it is balanced. A set Ω in a vector space is called *balanced* if whenever $x \in \Omega$, then $\alpha x \in \Omega$ for all α in the closed unit disc $\overline{\mathbb{D}}$. The equivalence of (i) and (iii) should be compared with Theorem 14.39.

Theorem 16.23. *Let Ω be a balanced* nc *operator domain in $\mathcal{B}(\mathcal{H})^d$, and let $F\colon \Omega \to \mathcal{B}(\mathcal{H})$. The following statements are equivalent.*

(i) *The function F is an nc operator function and is sequentially strong operator continuous;*

(ii) *there is a power series expansion*

$$\sum_{k=0}^{\infty} P_k(x) \tag{16.24}$$

that converges absolutely at each point $x \in \Omega$ to $F(x)$, where each P_k is a homogeneous free polynomial of degree k; and

(iii) *the function F is uniformly approximable on triples of elements of Ω by free polynomials.*

Proof. (i) $\Rightarrow$ (ii). By Proposition 16.18, F has a power series at 0 of the form (16.24). We must show the series converges on all of Ω.

Fix $x \in \Omega$. Since Ω is open and balanced, there exists $r > 1$ such that $Xx \in \Omega$ for every $X \in \mathbb{D}(0,r)$. As each P_k is homogeneous, for X in a neighborhood of 0,

$$F(Xx) = \sum_{k=0}^{\infty} P_k(Xx) = \sum_{k=0}^{\infty} X^k P_k(x). \tag{16.25}$$

Non-Commutative Functional Calculus

Therefore the function $\psi\colon X \mapsto F(Xx)$ is analytic on $\mathbb{D}(0,r)$, with values in $\mathcal{B}(\mathcal{H})$, and its power series expansion at 0 is given by equation (16.25). Let $M = \sup\{\|F(Xx)\|\colon |X| \le \frac{1+r}{2}\}$. Since F is SSOC, we have $M < \infty$. Let $s = \frac{1+r}{2}$. By the Cauchy integral formula, since $\|F\|$ is bounded by M,

$$\left\|\frac{d^k\psi}{dX^k}(0)\right\| \le M\frac{k!}{s^k}. \tag{16.26}$$

Comparing the formulae (16.25) and (16.26), we conclude that

$$\|P_k(x)\| \le \frac{M}{s^k},$$

and so the power series (16.25) converges uniformly and absolutely on the closed unit disc.

(ii) $\Rightarrow$ (iii). Obvious.

(iii) $\Rightarrow$ (i)(nc operator function). Suppose x, y, z are in Ω, where

$$z = s^{-1}\begin{bmatrix} x & 0 \\ 0 & y \end{bmatrix}s.$$

Let p_j be a sequence of free polynomials that approximates F on $\{x, y, z\}$. Then

$$F\left(s^{-1}\begin{bmatrix} x & 0 \\ 0 & y \end{bmatrix}s\right) = \lim_j p_j\left(s^{-1}\begin{bmatrix} x & 0 \\ 0 & y \end{bmatrix}s\right)$$

$$= s^{-1}\begin{bmatrix} \lim_j p_j(x) & 0 \\ 0 & \lim_j p_j(y) \end{bmatrix}s$$

$$= s^{-1}\begin{bmatrix} F(x) & 0 \\ 0 & F(y) \end{bmatrix}s.$$

(SSOC). Suppose x_n in Ω converges to x in Ω in the SOT. Let v be any unit vector, and $\varepsilon > 0$. Let $z = U(\oplus[x_n])U^*$, and choose $p \in \mathcal{P}_d$ so that $\|p(z) - F(z)\|$ and $\|p(x) - F(x)\|$ are both less than $\varepsilon/3$. Then by Lemma 16.6, we have $\|p(x_n) - F(x_n)\| < \varepsilon/3$ for every n. Now choose N so that for $n \ge N$, we have $\|[p(x) - p(x_n)]v\| < \varepsilon/3$. Then

$$\|(F(x) - F(x_n))v\| \le \|(F(x) - p(x))v\| + \|(p(x) - p(x_n))v\|$$
$$+ \|(p(x_n) - F(x_n))v\|,$$

and all three terms on the right are less than $\varepsilon/3$ for $n \ge N$. $\qquad\square$

One consequence is that, in this context, the image of a SSOC nc operator function is always in the norm-closed algebra generated by the argument (cf. Theorem 14.4).

354   *Non-Commutative Theory*

Corollary 16.27. *Assume that Ω is a balanced* nc *operator domain, and that* $F \colon \Omega \to \mathcal{B}(\mathcal{H})$ *is a SSOC* nc *operator function. Then, for each* $x \in \Omega$, *the operator* $F(x)$ *is in the norm-closed unital algebra generated by* $x^1, \ldots, x^d$.

Remark 16.28. Observe that the proof of (iii)$\Rightarrow$ (i) in Theorem 16.23 does not require that Ω be balanced.

16.3 Extending Free Functions to Operators

If φ is in $H_1^\infty(B_\delta)$ and has a free realization as in equation (16.1), then one can define a function $\varphi^\sharp$ on $B_\delta^\sharp$ by

$$\varphi^\sharp(X) = \overset{A}{\underset{1}{\otimes}} + \overset{B}{\underset{1}{\otimes}} \underset{\delta(X)}{\overset{1}{\otimes}} \left[1 - \overset{D}{\underset{1}{\otimes}} \underset{\delta(X)}{\overset{1}{\otimes}} \right]^{-1} \overset{C}{\underset{1}{\otimes}}.$$

By exactly the same calculation as in Theorem 14.10, we can show that $1 - \varphi^\sharp(X)^* \varphi^\sharp(X)$ is positive, so that $\varphi^\sharp$ maps $B_\delta^\sharp$ into the unit ball of $\mathcal{B}(\mathcal{H})$.

Theorem 16.29. *Let δ be an $I \times J$ matrix of free polynomials, and assume that $B_\delta^\sharp$ is connected and contains 0. Let $F \colon B_\delta^\sharp \to \text{ball}\,\mathcal{B}(H)$. The following statements are equivalent.*

(i) *The function F is a sequentially strong operator continuous* nc *operator function;*

(ii) *there exists an auxiliary Hilbert space $\mathcal{M}$ and an isometry*

$$\begin{bmatrix} A & B \\ C & D \end{bmatrix} \colon \mathbb{C} \oplus \mathcal{M}^I \to \mathbb{C} \oplus \mathcal{M}^J \tag{16.30}$$

such that for all X in $B_\delta^\sharp$

$$F(X) = \overset{A}{\underset{1}{\otimes}} + \overset{B}{\underset{1}{\otimes}} \underset{\delta(X)}{\overset{1}{\otimes}} \left[1 - \overset{D}{\underset{1}{\otimes}} \underset{\delta(X)}{\overset{1}{\otimes}} \right]^{-1} \overset{C}{\underset{1}{\otimes}}; \text{ and} \tag{16.31}$$

(iii) *for each $t > 1$, the function F is uniformly approximable by free polynomials on $B_{t\delta}^\sharp$.*

Proof. Assume F is non-constant, since otherwise the assertion is trivial.

(i) $\Rightarrow$ (ii). By Lemma 16.11 and Proposition 16.14, there exists $A \in \mathbb{D}$ such that if $\Phi \overset{\text{def}}{=} m_A \circ F$, then $\Phi^\flat$ is a free holomorphic function on B_δ that is bounded by 1 in norm, and that maps 0 to 0.

Let $\varphi = m_{-A} \circ \Phi^\flat$. This is a function in $H_1^\infty(B_\delta)$ that maps 0 to A and hence has a realization formula in terms of some isometry of the form (16.30). Using this isometry, define a function $H(X)$ by the right-hand side of equation (16.31). Then $H \colon B_\delta^\sharp \to \text{ball}\,\mathcal{B}(H)$. Expanding the inverse in equation

Non-Commutative Functional Calculus 355

(16.31) in a Neumann series, we see that this series converges uniformly on $B_{t\delta}^\sharp$ for every $t > 1$. In particular, for any pair of points, the partial sums of this series give a sequence of polynomials that uniformly approximate H, so by Theorem 16.23 we deduce that H is a SSOC nc operator function.

To see that F and H coincide, consider the difference

$$\xi(X) \stackrel{\text{def}}{=} m_A \circ F(X) - m_A \circ H(X).$$

Then $\xi^\flat = \Phi^\flat - \Phi^\flat$ vanishes on B_δ. Moreover, the function ξ is a SSOC nc operator function since both F and H are. Therefore ξ vanishes on the SOT sequential closure of $B_\delta^\sharp \cap \mathcal{M}^d$. The set $B_\delta^\sharp$ contains an open non-empty ball centered at 0, since $0 \in B_\delta^\sharp$ and δ is continuous, so ξ vanishes on this open ball. By Lemma 16.8, ξ is holomorphic; therefore it vanishes identically on all of $B_\delta^\sharp$.

(ii) $\Rightarrow$ (iii). The Neumann series for $\left[1 - \overset{D}{\underset{1}{\otimes}} \ \overset{1}{\underset{\delta(X)}{\otimes}} \right]^{-1}$ converges uniformly and absolutely on $B_{t\delta}^\sharp$, and each partial sum in the expansion of equation (16.31) gives a free polynomial.

(iii) $\Rightarrow$ (i). This follows from the proof of (iii) $\Rightarrow$ (i) in Theorem 16.23 and Remark 16.28. $\qquad\square$

One consequence is that φ and $\varphi^\sharp$ uniquely determine each other, at least under the assumptions of the theorem.

Corollary 16.32. *Assume $B_\delta^\sharp$ is connected and contains 0. Then every bounded free holomorphic function φ on B_δ has a unique extension to a function $\varphi^\sharp$ on $B_\delta^\sharp$ that is a SSOC nc operator function.*

Proof. Without loss of generality, assume $\varphi \in H_1^\infty(B_\delta)$ and is non-constant; by Theorem 14.33 it has a realization formula of the form (16.1). One extension F is then given by equation (16.31). If there were another, H say, proceed as in the proof of (i) $\Rightarrow$ (ii) to consider $\xi(X) = m_A \circ F(X) - m_A \circ H(X)$, where $A = \varphi(0)$. Then ξ is a SSOC nc operator function that vanishes on $B_\delta^\sharp \cap \mathcal{M}^d$, so identically is zero as before. $\qquad\square$

Note that although we use the realization of φ to extend to $\varphi^\sharp$, Corollary 16.32 states that the extension does not depend on the choice of model for φ.

Some assumptions are needed on δ for $H^\infty(B_\delta)$ to determine $H^\infty(B_\delta^\sharp)$, as the following example shows.

Example 16.33. Let

$$\delta(x^1, x^2) = 1 - (x^1 x^2 - x^2 x^1).$$

356 *Non-Commutative Theory*

Then B_δ is empty, since for any $x \in \mathbb{M}^2$, the normalized trace of $\delta(x)$ is 1, so $\delta(x)$ cannot be a strict contraction.

However, $B_\delta^\sharp$ is not empty. Let P be a projection onto a subspace of infinite dimension and codimension. Every operator that is not a scalar plus a compact is a commutator in $\mathcal{B}(\mathcal{H})$ [55], so one can choose x so that $[x^1, x^2] = I - \frac{1}{2}P$. For this x, $\|\delta(x)\| = \frac{1}{2} < 1$.

16.4 Nc Functional Calculus in Banach Algebras

In Section 16.3 we used model formulas to extend free functions to d-tuples in $\mathcal{B}(\mathcal{H})$. What about d-tuples in $\mathcal{B}(\mathcal{X})$ for a Banach space $\mathcal{X}$? This can be done too, though, not surprisingly, it is more complicated. One approach to this non-commutative functional calculus is given in [20].

In brief, the idea is as follows. First we give some definitions from the theory of tensor products on Banach spaces [177, 71]. A *reasonable cross norm* on the algebraic tensor product $\mathcal{X} \otimes \mathcal{Y}$ of two Banach spaces is a norm τ satisfying

(i) For every $x \in \mathcal{X}$, $y \in \mathcal{Y}$, we have $\tau(x \otimes y) = \|x\| \|y\|$.
(ii) For every $x^* \in \mathcal{X}^*$, $y^* \in \mathcal{Y}^*$, we have $\|x^* \otimes y^*\|_{(\mathcal{X} \otimes Y, \tau)^*} = \|x^*\| \|y^*\|$.

A *uniform cross norm* is an assignment to each pair of Banach spaces $\mathcal{X}, \mathcal{Y}$ a reasonable cross-norm on $\mathcal{X} \otimes \mathcal{Y}$ such that if $R: \mathcal{X}_1 \to \mathcal{X}_2$ and $S: \mathcal{Y}_1 \to \mathcal{Y}_2$ are bounded linear operators, then

$$\|R \otimes S\|_{\mathcal{X}_1 \otimes \mathcal{Y}_1 \to \mathcal{X}_2 \otimes \mathcal{Y}_2} \leq \|R\| \|S\|.$$

Definition 16.34. *Let $\mathcal{X}$ be a Banach space. A Hilbert tensor norm on $\mathcal{X}$ is an assignment of a reasonable cross norm h to $\mathcal{X} \otimes \mathcal{K}$ for every Hilbert space $\mathcal{K}$ with the property: if $\mathcal{K}_1$ and $\mathcal{K}_2$ are Hilbert spaces and $R: \mathcal{X} \to \mathcal{X}$ and $S: \mathcal{K}_1 \to \mathcal{K}_2$ are bounded linear operators, then*

$$\|R \otimes S\|_{\mathcal{L}(\mathcal{X} \otimes_h \mathcal{K}_1, \, \mathcal{X} \otimes_h \mathcal{K}_2)} \leq \|R\|_{\mathcal{L}(\mathcal{X})} \|S\|_{\mathcal{L}(\mathcal{K}_1, \mathcal{K}_2)}. \tag{16.35}$$

Any uniform cross norm, like the injective or projective tensor product norms, is a Hilbert tensor norm, but there are others. Most importantly, if $\mathcal{X}$ is itself a Hilbert space, then the Hilbert space tensor product (discussed in Section 3.4) is a Hilbert tensor norm.

For the rest of this section, we shall use $\otimes$ without a subscript to denote the Hilbert space tensor product of two Hilbert spaces, and $\otimes_h$ to denote the tensor product with a Hilbert tensor norm.

Let δ be an $I \times J$ matrix of free polynomials, and let $E_{ij}: \mathbb{C}^J \to \mathbb{C}^I$ be the $I \times J$ matrix with 1 in the (i, j) entry, and 0 elsewhere. Let φ be in $H_1^\infty(B_\delta)$, with a representation as in equation (16.1).

Non-Commutative Functional Calculus
357

Suppose T is a d-tuple of operators on a Banach space $\mathcal{X}$. To make sense of $\varphi(T)$ by the substitution of T into equation (16.1), we need to know that

$$\|\delta(T)\|_\bullet \overset{\text{def}}{=} \inf_h \sup_{\mathcal{K}} \left\| \sum_{i,j} \delta_{ij}(T) \otimes_h (E_{ij} \otimes \mathrm{id}_{\mathcal{K}}) \right\| < 1, \qquad (16.36)$$

where h ranges over Hilbert tensor norms and $\mathcal{K}$ ranges over Hilbert spaces. (Notice that we evaluate the tensor products from right to left, so we do all the Hilbert space tensor products first.) If this can be done, then one can use the formula (16.1) to define $\varphi(T)$, and this will extend the functional calculus on T to $H^\infty(B_\delta)$. To do this, choose a Hilbert tensor norm h so that the inequality (16.36) holds. Assume φ in $H^\infty(B_\delta)$ has a model of the form (16.1), where

$$\begin{bmatrix} A & B \\ C & D \end{bmatrix} : \mathbb{C} \oplus (\mathbb{C}^I \otimes \mathcal{M}) \to \mathbb{C} \oplus (\mathbb{C}^J \otimes \mathcal{M}).$$

Let

$$R = \sum_{i,j} \delta_{ij}(T) \otimes_h (E_{ij} \otimes \mathrm{id}_{\mathcal{M}}).$$

Then we define

$$\varphi(T) = A\,\mathrm{id}_{\mathcal{X}} + (\mathrm{id}_{\mathcal{X}} \otimes_h B)R\big[\mathrm{id}_x \otimes_h (\mathrm{id}_{\mathbb{C}^J \otimes \mathcal{M}}) - (\mathrm{id}_x \otimes_h D)R\big]^{-1} \mathrm{id}_{\mathcal{X}} \otimes_h C.$$

One can no longer use the Hilbert space argument to estimate the norm of $\varphi(T)$, but the Neumann series will still converge provided $\|R\| < 1$, so that we obtain the weaker bound

$$\|\varphi(T)\| \le \frac{1}{1 - \|\delta(T)\|_\bullet}.$$

16.5 Historical Notes

Definitions 16.2 and 16.4 and Lemmas 16.5 and 16.6 are due to Mark Mancuso in [149]. In that paper, Mancuso also proves the inverse function and implicit function theorems for nc operator functions.

Theorem 16.16 is originally due to Kaliuzhnyi-Verbovetskyi and Vinnikov [127, theorem 6.1]; an alternative proof was given by by I. Klep and S. Spenko [130, proposition 3.1].

Proposition 16.18 is an nc operator function version of [127, theorem 7.4]. Theorems 16.23 and 16.29 are from [15]. Section 16.4 is fleshed out in [20].

Notation

$\mathbb{C}$	The complex numbers
$\mathbb{D}$	The open unit disc in $\mathbb{C}$ centered at the origin
$\mathbb{N}$	$\{1,2,3,\ldots\}$
$\mathbb{R}$	The real numbers
$\mathbb{T}$	The unit circle
$\mathcal{X}, \mathcal{Y}$	Normed spaces (may be Hilbert)
$\mathcal{L}, \mathcal{M}, \mathcal{N}, \mathcal{H}, \mathcal{K}$	Hilbert spaces
ball $\mathcal{X}$	$\{x \in \mathcal{X} : \|x\| \leq 1\}$
T	operator or operator tuple
$\sigma(T)$	spectrum of T
ran(T)	range of T
z_T	$z^1 T_1 + z^2 T_2$ (p. 162)
$[S]$	If $S \subseteq \mathcal{X}$, then $[S]$ is the closed linear span of S
$[u_\lambda : \lambda \in \Omega]$	The closed linear span of $\{u_\lambda : \lambda \in \Omega\}$
s_λ	Szegő kernel (1.30)
$\varphi^\vee(\lambda)$	$\overline{\varphi(\bar{\lambda})}$
U	Open set
K	Compact set
Hol(U)	Holomorphic functions defined on U
Hol(K)	Holomorphic functions defined on a neighborhood of K
Rat(K)	Uniform closure of the rational functions with poles off K
$\|\varphi\|_U$	$\sup_{\lambda \in U} |\varphi(\lambda)|$
$H^\infty(U)$	Bounded holomorphic functions on U equipped with the sup norm
$\mathscr{S}(U)$	The Schur class of U, that is, ball $H^\infty(U)$

Notation

$\mathcal{L}_n(E)$	The Löwner class of E (p. 254)
$\mathcal{P}(E)$	The Pick class of E (p. 148)
$\mathcal{L}_n^d(E)$	The Löwner class of E in d variables (p. 274)
J_φ	Julia quotient of φ
$G = \{(\lambda^1 + \lambda^2,$ $\lambda^1\lambda^2)\colon \lambda \in \mathbb{D}^2\},$	the symmetrized bidisc
$\int_\gamma^{(4)} \ldots d\lambda \, d\overline{\mu}$	$\int_\gamma \int_\gamma \int_\gamma \int_\gamma \ldots d\lambda^1 \, d\lambda^2 \, d\overline{\mu^1} \, d\overline{\mu^2}$ (p. 179)
$d(\lambda_1, \lambda_2)$	Pseudo-hyperbolic metric (8.7)
d_U	Carathéodory distance function (8.9)
K_U	Kobayashi distance (8.17)
δ_U	Lempert function (8.16)
$\mathcal{P}_d$	Non-commutative polynomials in d variables
$\mathbb{M}_n$	Square matrices of dimension n (p. 286)
SAM_n	Self-adjoint matrices of dimension n
$CSAM_n^d$	Commuting self-adjoint d-tuples of $n \times n$ matrices
$\mathbb{M}^d$	$\cup_{n=1}^\infty \mathbb{M}_n^d$ (p. 286)
SAM^d	Self-adjoint d-tuples in $\mathbb{M}^d$
CM^d	Commuting d-tuples in $\mathbb{M}^d$
$\underset{s}{\overset{T}{\otimes}}$	$T \otimes s$ (p. 295)
$D\varphi(a)[h]$	Derivative of φ at a in the direction h
$x^{(k)}$	$\oplus_1^k x$, the ampliation of x
$F(x, r)$	$\cup_{k=1}^\infty \{y \in \mathbb{M}_{nk}^d \colon \|y - x^{(k)}\| < r\}$
B_δ	$\{x \in \mathbb{M}^d \colon \|\delta(x)\| < 1\}$
$\mathrm{Hol}(\Omega, \mathcal{H})$	Holomorphic functions from Ω to $\mathcal{H}$ (p. 302)
$\mathrm{Holg}_\mathcal{H}(\Omega)$	Holomorphic graded $\mathcal{H}$-valued functions on Ω (p. 306)
$\mathrm{Holg}_\mathcal{H}^{nc}(\Omega)$	Holomorphic graded $\mathcal{H}$-valued nc functions on Ω (p. 306)
$\mathrm{Holg}_\mathcal{H}^\tau(\Omega)$	τ-holomorphic $\mathcal{H}$-valued functions on Ω (p. 306)
$\mathrm{Herg}(\Omega)$	Hereditary graded holomorphic functions on Ω (p. 311)
$H^\infty(B_\delta)$	Bounded nc-functions on B_δ
$H_1^\infty(B_\delta)$	Closed unit ball of $H^\infty(B_\delta)$
$\mathrm{Alg}(x)$	The algebra generated by x
I_λ	All polynomials that vanish on λ
V_λ	Vanishing locus of I_λ

$H\varphi(a)[h,k]$	Hessian of φ
Z_f	Zero set of f
P_n	Projection onto first n basis vectors
$\mathcal{M}_n$	$P_n \mathcal{B}(\mathcal{H}) P_n$
$\mathcal{M}^d$	$\cup_{n=1}^\infty \mathcal{M}_n^d$
$B_\delta^\sharp$	$\{x \in \mathcal{B}(\mathcal{H})^d : \|\delta(x)\| < 1\}$

Bibliography

[1] Marco Abate. The Julia-Wolff-Carathéodory theorem in polydisks. *J. Anal. Math.*, 74:275–306, 1998.

[2] Gulnara Abduvalieva and Dmitry S. Kaliuzhnyi-Verbovetskyi. Implicit/inverse function theorems for free noncommutative functions. *J. Funct. Anal.*, 269(9):2813–2844, 2015.

[3] Vadym M. Adamian, Damir Z. Arov, and Mark G. Krein. On infinite Hankel matrices and generalized problems of Carathéodory-Fejér and F. Riesz. *Funkcional. Anal. Prilozen.*, 2(1):1–19, 1968.

[4] Vadym M. Adamian, Damir Z. Arov, and Mark G. Krein. Analytic properties of Schmidt pairs for a Hankel operator and the generalized Schur-Takagi problem. *Math. USSR. Sb.*, 15:31–73, 1971.

[5] Jim Agler. Some interpolation theorems of Nevanlinna-Pick type. Preprint, 1988.

[6] Jim Agler. The Arveson extension theorem and coanalytic models. *Integral Eq. Oper. Theory*, 5:608–631, 1982.

[7] Jim Agler. On the representation of certain holomorphic functions defined on a polydisc. In *Operator Theory: Advances and Applications*, vol. 48, pages 47–66. Birkhäuser, Basel, 1990.

[8] Jim Agler. Operator theory and the Carathéodory metric. *Inventiones Math.*, 101:483–500, 1991.

[9] Jim Agler, John Harland, and Benjamin J. Raphael. Classical function theory, operator dilation theory, and machine computation on multiply-connected domains. *Mem. Amer. Math. Soc.*, 191(892):viii, 159, 2008.

[10] Jim Agler and John E. McCarthy. Nevanlinna-Pick interpolation on the bidisk. *J. Reine Angew. Math.*, 506:191–204, 1999.

[11] Jim Agler and John E. McCarthy. *Pick Interpolation and Hilbert Function Spaces. Graduate Studies in Mathematics*, vol. 44. American Mathematical Society, Providence, RI, 2002.

[12] Jim Agler and John E. McCarthy. Norm preserving extensions of holomorphic functions from subvarieties of the bidisk. *Ann. Math. (2)*, 157(1):289–312, 2003.

[13] Jim Agler and John E. McCarthy. Distinguished varieties. *Acta Math.*, 194:133–153, 2005.

[14] Jim Agler and John E. McCarthy. Global holomorphic functions in several non-commuting variables. *Canad. J. Math.*, 67(2):241–285, 2015.

Bibliography

[15] Jim Agler and John E. McCarthy. Non-commutative holomorphic functions on operator domains. *European J. Math*, 1(4):731–745, 2015.

[16] Jim Agler and John E. McCarthy. Operator theory and the Oka extension theorem. *Hiroshima Math. J.*, 45(1):9–34, 2015.

[17] Jim Agler and John E. McCarthy. Pick interpolation for free holomorphic functions. *Amer. J. Math.*, 137(6):1685–1701, 2015.

[18] Jim Agler and John E. McCarthy. The implicit function theorem and free algebraic sets. *Trans. Amer. Math. Soc.*, 368(5):3157–3175, 2016.

[19] Jim Agler and John E. McCarthy. Wandering Montel theorems for Hilbert space valued holomorphic functions. *Proc. Amer. Math. Soc.*, 146(10):4353–4367, 2018.

[20] Jim Agler and John E. McCarthy. Non-commutative functional calculus. *J. Anal.*, 137(1):211–229, 2019.

[21] Jim Agler, John E. McCarthy, and Nicholas Young. On the representation of holomorphic functions on polyhedra. *Michigan Math. J.*, 62(4):675–689, 2013.

[22] Jim Agler, John E. McCarthy, and Nicholas Young. Facial behavior of analytic functions on the bidisk. *Bull. Lond. Math. Soc.*, 43:478–494, 2011.

[23] Jim Agler, John E. McCarthy, and Nicholas Young. A Carathéodory theorem for the bidisk via Hilbert space methods. *Math. Ann.*, 352(3):581–624, 2012.

[24] Jim Agler, John E. McCarthy, and Nicholas Young. Operator monotone functions and Löwner functions of several variables. *Ann. Math.*, 176(3):1783–1826, 2012.

[25] Jim Agler, Ryan Tully-Doyle, and Nicholas Young. Boundary behavior of analytic functions of two variables via generalized models. *Indag. Math. (n.s.)*, 23(4):995–1027, 2012.

[26] Jim Agler, Ryan Tully-Doyle, and Nicholas Young. Nevanlinna representations in several variables. *J. Funct. Anal.*, 270(8):3000–3046, 2016.

[27] Jim Agler and Nicholas Young. A commutant lifting theorem for a domain in $\mathbf{C}^2$ and spectral interpolation. *J. Funct. Anal.*, 161(2):452–477, 1999.

[28] Jim Agler and Nicholas Young. The hyperbolic geometry of the symmetrized bidisc. *J. Geom. Anal.*, 14(3):375–403, 2004.

[29] Jim Agler and Nicholas Young. Symmetric functions of two noncommuting variables. *J. Funct. Anal.*, 266(9):5709–5732, 2014.

[30] Jim Agler and Nicholas Young. Realization of functions on the symmetrized bidisc. *J. Math. Anal. Applic.*, 453(9):227–240, 2017.

[31] Daniel Alpay and Dmitry S. Kalyuzhnyi-Verbovetzkii. Matrix-J-unitary non-commutative rational formal power series. In *The State Space Method Generalizations and Applications. Operator Theory: Advances and Applications*, vol. 161, pages 49–113. Birkhäuser, Basel, 2006.

[32] Calin Ambrozie, Miroslav Engliš, and Vladimír Müller. Operator tuples and analytic models over general domains in $\mathbf{C}^n$. *J. Oper. Theory*, 47:287–302, 2002.

[33] Tsuyoshi Andô. On a pair of commutative contractions. *Acta Sci. Math. (Szeged)*, 24:88–90, 1963.

[34] Wolfgang Arendt and Nikolaĭ Nikol'skiĭ. Vector-valued holomorphic functions revisited. *Math. Z.*, 234(4):777–805, 2000.

[35] Nachman Aronszajn. Theory of reproducing kernels. *Trans. Amer. Math. Soc.*, 68:337–404, 1950.

[36] William Arveson. Subalgebras of C*-algebras. *Acta Math.*, 123:141–224, 1969.

Bibliography 363

[37] William Arveson. Subalgebras of C^*-algebras. II. *Acta Math.*, 128(3–4):271–308, 1972.

[38] Meric Augat, Sriram Balasubramanian, and Scott McCullough. Compact sets in the free topology. *Linear Algebra Appl.*, 506:6–9, 2016.

[39] Catalin Badea, Bernhard Beckermann, and Michel Crouzeix. Intersections of several disks of the Riemann sphere as K-spectral sets. *Commun. Pure Appl. Anal.*, 8(1):37–54, 2009.

[40] Sriram Balasubramanian. Toeplitz corona and the Douglas property for free functions. *J. Math. Anal. Appl.*, 428(1):1–11, 2015.

[41] Joseph A. Ball, Israel Gohberg, and Leiba Rodman. *Interpolation of Rational Matrix Functions*. Birkhäuser, Basel, 1990.

[42] Joseph A. Ball, Gilbert Groenewald, and Tanit Malakorn. Conservative structured noncommutative multidimensional linear systems. In *The State Space Method Generalizations and Applications. Operator Theory: Advances and Applications*, vol. 161, pages 179–223. Birkhäuser, Basel, 2006.

[43] Joseph A. Ball, Gregory Marx, and Victor Vinnikov. Interpolation and transfer-function realization for the noncommutative Schur-Agler class. In *Operator Theory in Different Settings and Related Applications. Operator Theory: Advances and Applications*, vol. 262, pages 23–116. Birkhäuser/Springer, Cham, Switzerland, 2018.

[44] Joseph A. Ball, Cora Sadosky, and Victor Vinnikov. Scattering systems with several evolutions and multidimensional input/state/output systems. *Integral Eq. Oper. Theory*, 52:323–393, 2005.

[45] Joseph A. Ball and Tavan T. Trent. Unitary colligations, reproducing kernel Hilbert spaces, and Nevanlinna-Pick interpolation in several variables. *J. Funct. Anal.*, 197:1–61, 1998.

[46] Stephen Barnett. *Matrices in Control Theory: With Applications to Linear Programming*. Van Nostrand Reinhold, London, New York, and Toronto, 1971.

[47] H. Bercovici, C. Foias, and A. Tannenbaum. Spectral variants of the Nevanlinna-Pick interpolation problem. In *Signal Processing, Scattering and Operator Theory, and Numerical Methods (Amsterdam, 1989). Program Systems Control Theory*, vol. 5, pages 23–45. Birkhäuser, Boston, 1990.

[48] Arne Beurling. On two problems concerning linear transformations in Hilbert space. *Acta Math.*, 81:239–255, 1949.

[49] Rajendra Bhatia and Peter Rosenthal. How and why to solve the operator equation $AX - XB = Y$. *Bull. London Math. Soc.*, 29(1):1–21, 1997.

[50] Tirthankar Bhattacharyya, Sourav Pal, and Subrata Shyam Roy. Dilations of Γ-contractions by solving operator equations. *Adv. Math.*, 230(2):577–606, 2012.

[51] Kelly Bickel. Differentiating matrix functions. *Oper. Matrices*, 7(1):71–90, 2013.

[52] Kelly Bickel and Greg Knese. Inner functions on the bidisk and associated Hilbert spaces. *J. Funct. Anal.*, 265(11):2753–2790, 2013.

[53] Frank Bonsall and John Duncan. *Complete Normed Algebras*. Springer, New York and Berlin, 1973.

[54] Stephen Boyd, Laurent El Ghaoui, Eric Feron, and Venkataramanan Balakrishnan. *Linear Matrix Inequalities in System and Control Theory. SIAM Studies in Applied Mathematics*, vol. 15. Society for Industrial and Applied Mathematics (SIAM), Philadelphia, 1994.

364 *Bibliography*

[55] Arlen Brown and Carl Pearcy. Structure of commutators of operators. *Ann. Math. (2)*, 82:112–127, 1965.

[56] Constantin Carathéodory. Über die Winkelderivierten von beschränkten analytischen Funktionen. *Sitzungsber Preuss. Akad. Wissens.* pages 39–52, 1929.

[57] Lennart Carleson. Interpolations by bounded analytic functions and the corona problem. *Ann. Math.*, 76:547–559, 1962.

[58] Henri Cartan. *Séminaire Henri Cartan 1951/2*. W. A. Benjamin, New York, 1967.

[59] Jakob Cimpric, J. William Helton, Scott McCullough, and Christopher Nelson. A noncommutative real nullstellensatz corresponds to a noncommutative real ideal: Algorithms. *Proc. Lond. Math. Soc. (3)*, 106(5):1060–1086, 2013.

[60] John B. Conway. *A Course in Functional Analysis*. Springer-Verlag, New York, 1985.

[61] Constantin Costara. The symmetrized bidisc and Lempert's theorem. *Bull. London Math. Soc.*, 36(5):656–662, 2004.

[62] Michael J. Crabb and Alexander M. Davie. Von Neumann's inequality for Hilbert space operators. *Bull. London Math. Soc.*, 7:49–50, 1975.

[63] Michel Crouzeix. Bounds for analytical functions of matrices. *Integral Eq. Oper. Theory*, 48(4):461–477, 2004.

[64] Michel Crouzeix. A functional calculus based on the numerical range: applications. *Linear Multilinear Algebra*, 56(1-2):81–103, 2008.

[65] Michel Crouzeix. The annulus as a K-spectral set. *Commun. Pure Appl. Anal.*, 11(6):2291–2303, 2012.

[66] Michel Crouzeix and Cesar Palencia. The numerical range is a $(1 + \sqrt{2})$-spectral set. *SIAM J. Matrix Anal. Appl.*, 38(2):649–655, 2017.

[67] Raúl E. Curto. Applications of several complex variables to multiparameter spectral theory. In J. B. Conway and B. B. Morrel, eds., *Surveys of Some Recent Results in Operator Theory*, pages 25–90. Longman Scientific, Harlow, UK 1988.

[68] Louis de Branges. A proof of the Bieberbach conjecture. *Acta Math.*, 154:137–152, 1985.

[69] Louis de Branges and James Rovnyak. *Square Summable Power Series*. Holt, Rinehart, and Winston, New York, 1966.

[70] Bernard Delyon and Francois Delyon. Generalizations of von Neumann's spectral sets and integral representations of operators. *Bull. Soc. Math. France*, 127:25–41, 1999.

[71] Joe Diestel, Jan H. Fourie, and Johan Swart. *The Metric Theory of Tensor Products*. American Mathematical Society, Providence, RI, 2008.

[72] Jean Dieudonné. *History of Functional Analysis*. In L. Nachbin, ed., Notas de Matematica (77). North-Holland Mathematics Studies, book 49. North-Holland, Amsterdam, New York, and Oxford, 1981.

[73] Seán Dineen. *Complex Analysis on Infinite-Dimensional Spaces*. Springer Monographs in Mathematics. Springer-Verlag, London, 1999.

[74] Jacques Dixmier. *C^*-algebras*. North-Holland Mathematical Library, vol. 15. North-Holland, Amsterdam, New York, and Oxford, 1977. (Translated from the French by Francis Jellett.)

Bibliography   365

[75] Jacques Dixmier. *Les algèbres d'opérateurs dans l'espace hilbertien (algèbres de von Neumann)*. Les Grands Classiques Gauthier-Villars. [Gauthier-Villars Great Classics]. Éditions Jacques Gabay, Paris, 1996. (Reprint of the 2nd ed., 1969.)

[76] William F. Donoghue. *Monotone Matrix Functions and Analytic Continuation*. Springer, Berlin, 1974.

[77] Ronald G. Douglas and Vern I. Paulsen. Completely bounded maps and hypo-Dirichlet algebras. *Acta Sci. Math. (Szeged)*, 50(1–2):143–157, 1986.

[78] John Doyle. Analysis of feedback systems with structured uncertainties. *IEE Proc.*, 129:242–250, 1982.

[79] John Doyle and Gunter Stein. Multivariable feedback design: Concepts for a classical/modern synthesis. *IEEE Trans. Automatic Control*, 26:4–13, 1981.

[80] Michael A. Dritschel and Scott McCullough. The failure of rational dilation on a triply connected domain. *J. Amer. Math. Soc.*, 18(4):873–918, 2005.

[81] Scott W. Drury. Remarks on von Neumann's inequality. In R. C. Blei and S. J. Sidney, eds. *Banach Spaces, Harmonic Analysis, and Probability Theory*, pages 14–32. *Lecture Notes in Mathematics*, vol. 995. Springer Verlag, Berlin, 1983.

[82] Richard J. Duffin. Infinite programs. In *Linear Inequalities and Related Systems. Annals of Mathematics Studies*, no. 38, pages 157–170. Princeton University Press, Princeton, NJ, 1956.

[83] Geir Dullerud and Fernando Paganini. *A Course in Robust Control Theory*. Springer Verlag, New York, 2000.

[84] Peter L. Duren. *Theory of H^p Spaces*. Academic Press, New York, 1970.

[85] Harry Dym. *J Contractive Matrix Functions, Reproducing Kernel Hilbert Spaces and Interpolation. CBMS Regional Conference Series in Mathematics*, vol. 71. Published for the Conference Board of the Mathematical Sciences, Washington, DC by the American Mathematical Society, Providence, RI, 1989.

[86] Armen Edigarian and Włodzimierz Zwonek. Geometry of the symmetrized polydisc. *Arch. Math. (Basel)*, 84(4):364–374, 2005.

[87] Edward G. Effros and Zhong-Jin Ruan. Operator space tensor products and Hopf convolution algebras. *J. Operator Theory*, 50(1):131–156, 2003.

[88] David Eisenbud and Melvin Hochster. A Nullstellensatz with nilpotents and Zariski's main lemma on holomorphic functions. *J. Algebra*, 58(1):157–161, 1979.

[89] Jörg Eschmeier and Mihai Putinar. *Spectral Decompositions and Analytic Sheaves*. Oxford University Press, Oxford, 1996.

[90] Ciprian Foias and Arthur Frazho. *The Commutant Lifting Approach to Interpolation Problems*. Birkhäuser, Basel, 1990.

[91] Bent Fuglede. A commutativity theorem for normal operators. *Proc. Nat. Acad. Sci.*, 36:35–40, 1950.

[92] Theodore W. Gamelin. *Uniform Algebras*. Chelsea, New York, 1984.

[93] John B. Garnett. *Bounded Analytic Functions*. Academic Press, New York, 1981.

[94] Bernd Gärtner and Jiří Matoušek. *Approximation Algorithms and Semidefinite Programming*. Springer, Heidelberg, 2012.

[95] Anne Greenbaum and Michael L. Overton. Numerical investigation of Crouzeix's conjecture. *Linear Algebra Appl.*, 542:225–245, 2018.

[96] Werner Greub. *Linear Algebra. Graduate Texts in Mathematics*, no. 23. Springer-Verlag, New York and Berlin, 4th ed., 1975.

366 Bibliography

[97] Werner Greub. *Multilinear Algebra*. Springer-Verlag, New York and Heidelberg, 2nd ed., 1978. (Universitext).

[98] Paul R. Halmos. Shifts on Hilbert spaces. *J. Reine Angew. Math.*, 208:102–112, 1961.

[99] Paul R. Halmos. Ten problems in Hilbert space. *Bull. Amer. Math. Soc.*, 76:887–933, 1970.

[100] Alexander Ya. Helemskii. *The Homology of Banach and Topological Algebras. Mathematics and Its Applications (Soviet Series)*, vol. 41. Kluwer Academic, Dordrecht, The Netherlands, 1989. (Translated from the Russian by Alan West.)

[101] J. William Helton. Discrete-time systems, operator models and scattering theory. *J. Funct. Anal.*, 16:15–38, 1974.

[102] J. William Helton. "Positive" noncommutative polynomials are sums of squares. *Ann. Math. (2)*, 156(2):675–694, 2002.

[103] J. William Helton, Igor Klep, and Scott McCullough. Analytic mappings between noncommutative pencil balls. *J. Math. Anal. Appl.*, 376(2):407–428, 2011.

[104] J. William Helton, Igor Klep, and Scott McCullough. Proper analytic free maps. *J. Funct. Anal.*, 260(5):1476–1490, 2011.

[105] J. William Helton, Igor Klep, and Scott McCullough. Convexity and semidefinite programming in dimension-free matrix unknowns. In *Handbook on Semidefinite, Conic and Polynomial Optimization. International Series in Operations Research and Management Science*, vol. 166, pages 377–405. Springer, New York, 2012.

[106] J. William Helton, Igor Klep, and Scott McCullough. Free analysis, convexity and LMI domains. In *Operator Theory: Advances and Applications*, vol. 222, pages 195–219. Springer, Basel, 2012.

[107] J. William Helton and Scott McCullough. Every convex free basic semialgebraic set has an LMI representation. *Ann. Math. (2)*, 176(2):979–1013, 2012.

[108] J. William Helton and Scott McCullough. Free convex sets defined by rational expressions have LMI representations. *J. Convex Anal.*, 21(2):425–448, 2014.

[109] J. William Helton, Scott McCullough, Mihai Putinar, and Victor Vinnikov. Convex matrix inequalities versus linear matrix inequalities. *IEEE Trans. Automat. Control*, 54(5):952–964, 2009.

[110] J. William Helton and Scott A. McCullough. A Positivstellensatz for noncommutative polynomials. *Trans. Amer. Math. Soc.*, 356(9):3721–3737 (electronic), 2004.

[111] J. William Helton and Mihai Putinar. Positive polynomials in scalar and matrix variables, the spectral theorem, and optimization. In *Operator Theory, Structured Matrices, and Dilations. Theta Series in Advanced Mathematics*, vol. 7, pages 229–306. Theta, Bucharest, 2007.

[112] J. William Helton and Victor Vinnikov. Linear matrix inequality representation of sets. *Comm. Pure Appl. Math.*, 60(5):654–674, 2007.

[113] Guennadi M. Henkin and Pierre L. Polyakov. Prolongement des fonctions holomorphes bornées d'une sous-variétée du polydisque. *Comptes Rendus Acad. Sci. Paris Sér. I Math.*, 298(10):221–224, 1984.

[114] Gustav Herglotz. Über Potenzreihen mit positivem, reellem Teil im Einheitskreis. *Leipz. Bericht*, 63:501–511, 1911.

Bibliography 367

[115] David Hilbert. *Grundzüge einer allgemeinen Theorie der linearen Integralgleichungen*. B. G. Teubner, Leipzig, Germany 1912.

[116] Kenneth Hoffman. *Banach Spaces of Analytic Functions*. Prentice-Hall, Englewood Cliffs, NJ, 1962.

[117] John A. Holbrook. Polynomials in a matrix and its commutant. *Linear Algebra Appl.*, 48:293–301, 1982.

[118] Lars Hörmander. *An Introduction to Complex Analysis in Several Variables*. North Holland, Amsterdam, 1973.

[119] Vlad Ionescu, Cristian Oară, and Martin Weiss. *Generalized Riccati Theory and Robust Control*. John Wiley, Chichester, UK, 1999. (A Popov function approach.)

[120] Farhad Jafari. Angular derivatives in polydisks. *Indian J. Math.*, 35:197–212, 1993.

[121] Hans Jarchow. *Locally Convex Spaces*. Teubner, Stuttgart, 1981.

[122] Marek Jarnicki and Peter Pflug. *Invariant Distances and Metrics in Complex Analysis. De Gruyter Expositions in Mathematics*, vol. 9. Walter de Gruyter, Berlin, extended ed., 2013.

[123] Gaston Julia. Extension nouvelle d'un lemme de Schwarz. *Acta Math.*, 42:349–355, 1920.

[124] Marinus A. Kaashoek and James Rovnyak. On the preceding paper by R. B. Leech. *Integral Eq. Oper. Theory*, 78(1):75–77, 2014.

[125] Thomas Kailath. *Linear systems*. Prentice-Hall Information and System Sciences Series. Prentice-Hall, Englewood Cliffs, NJ, 1980.

[126] Dmitry S. Kaliuzhnyi-Verbovetskyi and Victor Vinnikov. Singularities of rational functions and minimal factorizations: The noncommutative and the commutative setting. *Linear Algebra Appl.*, 430(4):869–889, 2009.

[127] Dmitry S. Kaliuzhnyi-Verbovetskyi and Victor Vinnikov. *Foundations of Free Non-Commutative Function Theory*. American Mathematical Society, Providence, RI, 2014.

[128] Tosio Kato. *Perturbation Theory for Linear Operators*. Springer, Berlin, 1966.

[129] John L. Kelley. *General topology*. Springer-Verlag, New York-Berlin, 1975. Reprint of the 1955 edition [Van Nostrand, Toronto, Ont.], Graduate Texts in Mathematics, No. 27.

[130] Igor Klep and Spela Spenko. Free function theory through matrix invariants. *Canad. J. Math.*, 69(2):408–433, 2017.

[131] Greg Knese. Polynomials defining distinguished varieties. *Trans. Amer. Math. Soc.*, 362(11):5635–5655, 2010.

[132] Greg Knese. Rational inner functions in the Schur-Agler class of the polydisk. *Publ. Mat.*, 55(2):343–357, 2011.

[133] Shoshichi Kobayashi. *Hyperbolic Complex Spaces. Grundlehren der Mathematischen Wissenschaften [Fundamental Principles of Mathematical Sciences]*, vol. 318. Springer-Verlag, Berlin, 1998.

[134] Paul Koosis. *An Introduction to H^p Spaces. London Mathematical Society Lecture Notes*, vol. 40. Cambridge University Press, Cambridge, 1980.

[135] Łukasz Kosiński and Wlodzimierz Zwonek. Nevanlinna-Pick problem and uniqueness of left inverses in convex domains, symmetrized bidisc and tetrablock. *J. Geom. Anal.*, 26(3):1863–1890, 2016.

368 Bibliography

[136] Steven G. Krantz. *Function Theory of Several Complex Variables*. Wiley, New York, 1982.

[137] Steven G. Krantz. *Function Theory of Several Complex Variables*. 2nd ed. Wadsworth-Cole, Belmont, CA, 1992.

[138] Anton Kummert. Synthesis of two-dimensional lossless m-ports with prescribed scattering matrix. *Circ. Syst. Signal Process.*, 8(1):97–119, 1989.

[139] Peter Lax. Translation invariant spaces. *Acta Math.*, 101:163–178, 1959.

[140] Peter Lax. *Functional Analysis*. Wiley, New York, 2002.

[141] Peter Lax and Ralph S. Phillips. Scattering theory. *Rocky Mountain J. Math.*, 1:173–224, 1971.

[142] Robert B. Leech. Factorization of analytic functions and operator inequalities. *Integral Eq. Oper. Theory*, 78(1):71–73, 2014.

[143] Laszlo Lempert. La métrique de Kobayashi et la représentation des domaines sur la boule. *Bull. Soc. Math. France*, 109:427–484, 1981.

[144] Lynn H. Loomis. *An Introduction to Abstract Harmonic Analysis*. D. Van Nostrand, Toronto, New York, and London, 1953.

[145] Karl Löwner. Über monotone Matrixfunktionen. *Math. Z.*, 38:177–216, 1934.

[146] Annemarie Luger and Mitja Nedic. A characterization of Herglotz-Nevanlinna functions in two variables via integral representations. *Ark. Mat.*, 55(1):199–216, 2017.

[147] John E. McCarthy and James E. Pascoe. The Julia-Carathéodory theorem on the bidisk revisited. *Acta Sci. Math. (Szeged)*, 83(1–2):165–175, 2017.

[148] John E. McCarthy and Richard Timoney. Non-commutative automorphisms of bounded non-commutative domains. *Proc. Roy. Soc. Edinburgh Sect. A*, 146(5):1037–1045, 2016.

[149] Mark Mancuso. Inverse and implicit function theorems for noncommutative functions on operator domains. To appear in *J. Oper. Theory*.

[150] Scott McCullough and Mihai Putinar. Noncommutative sums of squares. *Pacific J. Math.*, 218(1):167–171, 2005.

[151] James Mercer. Functions of positive and negative type and their connection with the theory of integral equations. *Philos. Trans. Roy. Soc. London Ser. A*, 209:415–446, 1909.

[152] Eliakim H. Moore. *General Analysis*. The American Philosophy Society, Philadelphia, PA, part I, 1935; part II, 1939.

[153] Chaim Herman Müntz. Über den Approximationssatz von Weierstrass. In *H. A. Schwartz Festschrift*, pages 303–312. Berlin, 1914.

[154] Mark A. Naĭmark. Positive definite operator functions on a commutative group. *Izv. Akad. Nauk SSSR*, 7:237–244, 1943.

[155] Rolf Nevanlinna. Asymptotische Entwicklungen beschränkter Funktionen und das Stieltjessche Momentproblem. *Ann. Acad. Sci. Fenn. Ser. A*, 18:1–53, 1922.

[156] Rolf Nevanlinna. Über beschränkte Funktionen. *Ann. Acad. Sci. Fenn. Ser. A*, 32(7):7–75, 1929.

[157] Nikolaĭ Nikol'skiĭ and Vasily Vasyunin. Elements of spectral theory in terms of the free function model Part I: Basic constructions. In S. Axler, J. E. McCarthy, and D. Sarason, eds., *Holomorphic Spaces*, vol. 33, pages 211–302. Mathematical Sciences Research Institute Publications, Cambridge University Press, Cambridge, 1998.

Bibliography 369

[158] Nikolaĭ K. Nikol'skiĭ. *Treatise on the Shift Operator: Spectral Function Theory.* *Grundlehren der mathematischen Wissenschaften*, vol. 273. Springer-Verlag, Berlin, 1985.

[159] James E. Pascoe. Note on Löwner's theorem on matrix monotone functions in several commuting variables of Agler, McCarthy and Young. arXiv:1409.2605.

[160] James E. Pascoe. The inverse function theorem and the Jacobian conjecture for free analysis. *Math. Z.*, 278(3-4):987–994, 2014.

[161] Vern I. Paulsen. Every completely polynomially bounded operator is similar to a contraction. *J. Funct. Anal.*, 55:1–17, 1984.

[162] Vern I. Paulsen. *Completely Bounded Maps and Operator Algebras.* Cambridge University Press, Cambridge, 2002.

[163] Georg Pick. Über die Beschränkungen analytischer Funktionen, welche durch vorgegebene Funktionswerte bewirkt werden. *Math. Ann.*, 77:7–23, 1916.

[164] Gilles Pisier. A polynomially bounded operator on Hilbert space which is not similar to a contraction. *J. Amer. Math. Soc.*, 10:351–369, 1997.

[165] Gilles Pisier. *Introduction to Operator Space Theory. London Mathematical Society Lecture Note Series*, vol. 294. Cambridge University Press, Cambridge, 2003.

[166] Gelu Popescu. Free holomorphic functions on the unit ball of $B(\mathcal{H})^n$. *J. Funct. Anal.*, 241(1):268–333, 2006.

[167] Gelu Popescu. Free holomorphic functions and interpolation. *Math. Ann.*, 342(1):1–30, 2008.

[168] Gelu Popescu. Free holomorphic automorphisms of the unit ball of $B(\mathcal{H})^n$. *J. Reine Angew. Math.*, 638:119–168, 2010.

[169] Gelu Popescu. Free biholomorphic classification of noncommutative domains. *Int. Math. Res. Not. IMRN*, (4):784–850, 2011.

[170] Claudio Procesi. A noncommutative Hilbert Nullstellensatz. *Rend. Mat. Appl. (5)*, 25(1–2):17–21, 1966.

[171] Mihai Putinar. Uniqueness of Taylor's functional calculus. *Proc. Amer. Math. Soc.*, 89:647–650, 1983.

[172] Frigyes Riesz and Béla Sz.-Nagy. *Functional Analysis.* Dover, New York, 1990.

[173] Marvin Rosenblum and James Rovnyak. *Hardy Classes and Operator Theory.* Oxford Mathematical Monographs, Oxford Science Publications. Clarendon Press, Oxford University Press, New York, 1985.

[174] Walter Rudin. *Function Theory in Polydiscs.* Benjamin, New York, 1969.

[175] Walter Rudin. *Real and Complex Analysis.* McGraw-Hill, New York, 1986.

[176] Walter Rudin. *Functional Analysis.* McGraw-Hill, New York, 1991.

[177] Raymond A. Ryan. *Introduction to Tensor Products of Banach Spaces.* Springer Monographs in Mathematics. Springer-Verlag London, London, 2002.

[178] Guy Salomon, Orr M. Shalit, and Eli Shamovich. Algebras of bounded noncommutative analytic functions on subvarieties of the noncommutative unit ball. *Trans. Amer. Math. Soc.*, 370(12):8639–8690, 2018.

[179] Donald Sarason. On spectral sets having connected complement. *Acta Sci. Math.*, 26:289–299, 1965.

[180] Donald Sarason. Generalized interpolation in H^∞. *Trans. Amer. Math. Soc.*, 127:179–203, 1967.

[181] Donald Sarason. *Sub-Hardy Hilbert Spaces in the Unit Disk.* University of Arkansas Lecture Notes, Wiley, New York, 1994.

370 Bibliography

[182] Donald Sarason. Nevanlinna-Pick interpolation with boundary data. *Integral Eq. Oper. Theory*, 30:231–250, 1998.

[183] Jaydeb Sarkar. Operator theory on symmetrized bidisc. *Indiana Univ. Math. J.*, 64(3):847–873, 2015.

[184] Issai Schur. Über Potenzreihen, die im Innern des Einheitskreises beschränkt sind I. *J. Reine Angew. Math.*, 147:205–232, 1917.

[185] Issai Schur. Über Potenzreihen, die im Innern des Einheitskreises beschränkt sind II. *J. Reine Angew. Math.*, 148:122–145, 1918.

[186] Allen L. Shields. Weighted shift operators and analytic function theory. In *Topics in Operator Theory. Mathematical Surveys and Monographs*, vol. 13, pages 49–128. American Mathematical Society, Providence, RI, 1974.

[187] Georgi E. Shilov. On the decomposition of a commutative normed ring into a direct sum of ideals. *Amer. Math. Soc. Transl. (2)*, 1:37–48, 1955.

[188] Vladimir I. Smirnov. Sur les valeurs limites des fonctions, régulières à l'intérieur d'un cercle. *J. Soc. Phys.-Math. Léningrade*, 2(2):22–37, 1929.

[189] Vladimir I. Smirnov. Sur les formules de Cauchy et de Green et quelques problèmes qui s'y rattachent. *J. Soc. Phys.-Math. Léningrade*, 3:338–372, 1932.

[190] Roger R. Smith. Completely bounded maps between C*-algebras. *J. Lond. Math. Soc.*, 27:157–166, 1983.

[191] Marshall H. Stone. *Linear Transformations in Hilbert Space*. New York, 1932.

[192] James J. Sylvester. Sur l'équations en matrices px = xq. *C. R. Acad. Sci. Paris*, 99:67–71, 1884.

[193] Béla Szőkefalvi-Nagy. Sur les contractions de l'éspace de Hilbert. *Acta Sci. Math.*, 15:87–92, 1953.

[194] Béla Szőkefalvi-Nagy and Ciprian Foias. *Harmonic Analysis of Operators on Hilbert Space*. North Holland, Amsterdam, 1970.

[195] Béla Szőkefalvi-Nagy and Ciprian Foias. *Harmonic Analysis of Operators on Hilbert Space, 2nd ed.* Springer, New York, 2010.

[196] Otto Szász. Über die Approximation stetiger Funktionen durch lineare Aggregate von Potenzen. *Math. Ann.*, 77(4):482–496, 1916.

[197] Joseph L. Taylor. The analytic-functional calculus for several commuting operators. *Acta Math.*, 125:1–38, 1970.

[198] Joseph L. Taylor. A joint spectrum for several commuting operators. *J. Functional Analysis*, 6:172–191, 1970.

[199] Joseph L. Taylor. A general framework for a multi-operator functional calculus. *Adv. Math.*, 9:183–252, 1972.

[200] Joseph L. Taylor. Functions of several noncommuting variables. *Bull. Amer. Math. Soc.*, 79:1–34, 1973.

[201] Joseph L. Taylor. *Several Complex Variables with Connections to Algebraic Geometry and Lie Groups. Graduate Studies in Mathematics*, vol. 46. American Mathematical Society, Providence, RI, 2002.

[202] Vadim A. Tolokonnikov. Estimates in the Carleson corona theorem, ideals of the algebra H^∞, a problem of Sz.-Nagy. *Zap. Nauchn. Sem. Leningrad. Otdel. Mat. Inst. Steklov. (LOMI)*, 113:178–198, 267, 1981. (Investigations on linear operators and the theory of functions, XI.)

[203] Sergei Treil. Estimates in the corona theorem and ideals of H^∞: A problem of T. Wolff. *J. Anal. Math.*, 87:481–495, 2002. (Dedicated to the memory of Thomas H. Wolff.)

Bibliography

[204] Maria Trybula. Invariant metrics on the symmetrized bidisc. *Complex Var. Elliptic Equ.*, 60(4):559–565, 2015.

[205] Ryan Tully-Doyle. Analytic functions on the bidisk at boundary singularities via Hilbert space methods. *Oper. Matrices*, 11(1):55–70, 2017.

[206] N. T. Varopoulos. On an inequality of von Neumann and an application of the metric theory of tensor products to operators theory. *J. Funct. Anal.*, 16:83–100, 1974.

[207] Dan Voiculescu. Free analysis questions. I. Duality transform for the coalgebra of $\partial_{X:B}$. *Int. Math. Res. Not.*, (16):793–822, 2004.

[208] Dan Voiculescu. Free analysis questions II: The Grassmannian completion and the series expansions at the origin. *J. Reine Angew. Math.*, 645:155–236, 2010.

[209] John von Neumann. Allgemeine Eigenwerttheorie Hermitescher Funktionaloperatoren. *Math. Ann.*, 102:49–131, 1929.

[210] John von Neumann. Eine Spektraltheorie für allgemeine Operatoren eines unitären Raumes. *Math. Nachr.*, 4:258–281, 1951.

[211] Edward Waring. *Meditationes Algebraicæ*. American Mathematical Society, Providence, RI, 1991. (Translated from the Latin, edited, and with a foreword by Dennis Weeks, with an appendix by Franz X. Mayer; translated from the German by Weeks.)

[212] John Wermer. *Banach Algebras and Several Complex Variables*. Markham, Chicago, 1971.

[213] Kazimierz Wlodarczyk. Julia's lemma and Wolff's theorem for J^*-algebras. *Proc. Amer. Math. Soc.*, 99:472–476, 1987.

[214] Nicholas Young. *An Introduction to Hilbert Space*. Cambridge Mathematical Textbooks. Cambridge University Press, Cambridge, 1988.

[215] Nicholas Young. Some analysable instances of μ-synthesis. In *Mathematical Methods in Systems, Optimization, and Control. Operator Theory: Advances and Applications*, vol. 222, pages 351–368. Birkhäuser/Springer Basel, 2012.

Index

$x \oplus y$, 286
$\ominus$, 27
$\otimes$, 11, 78, 79, 91, 92
$\hat{\otimes}$, 97
$s_\lambda \otimes x$, 73
$E \boxtimes F$, 291
$\| \ \|$, 3
$\varphi^\vee$, 60
h^*, 63
∂X, 22
$\Omega^\flat$, 348
$E^\perp$, 27
$E^{[2]}$, 291
$F^\flat$, 348
$[S]$, 39
$S \xrightarrow{\text{nt}} \tau$, 131

$\sigma(T)$, 8
$\sigma_{\text{alg}}(T)$, 8
Γ, 177

b_α, 50
ball $\mathcal{X}$, 3
ℓ^2, 13
q_α, 50
$r \cdot \Gamma$, 177
s_U, 173
z_T, 162
$A(r)$, 226
$\text{Alg}(x)$, 313
$\mathcal{B}(\mathcal{X})$, 4
$\mathcal{B}(\mathcal{X}, \mathcal{Y})$, 4
B_δ, 299
$B_\delta^\sharp$, 345
$\text{Ball}(a, r)$, 336
$C(K)$, 12
$C^*(N)$, 12

CM^d, 286
$CSAM_n^d$, 269
$D\varphi(x)[h]$, 298
$\mathcal{D}_T$, 23
$F(a, r)$, 299
$\mathcal{F}(d)$, 243
$\mathcal{F}_{\text{dp}}$, 226
$\mathscr{H}(U)$, 148
H^2, 14
$\text{H}^2_{\mathcal{H}}$, 72
$\mathbb{H}$, 148
$H^\infty(B_\delta)$, 313
$H_1^\infty(B_\delta)$, 313
$\text{H}^\infty(U)$, 19
$\text{H}^\infty_{\mathcal{B}(\mathcal{H})}(\mathbb{D})$, 24
$\text{H}^\infty_{\mathcal{B}(\mathcal{H}, \mathcal{K})}(\mathbb{D})$, 24
$\text{Her}(\mathbb{D})$, 62
$\text{Her}(\mathbb{D}^2)$, 93
$\text{Her}_{\mathcal{B}(\mathcal{X})}(\mathbb{D}^2)$, 114
$\text{Herg}(\Omega)$, 311
$\text{Hol}(K)$, 18
$\text{Hol}(U)$, 5
$\text{Hol}(\Omega, \mathcal{H})$, 302
$\text{Holg}(\Omega)$, 306
$\text{Holg}_{\mathcal{H}}(\Omega)$, 306
$\text{Holg}_{\mathcal{H}}^{\tau}(\Omega)$, 306
$\text{Holg}_{\mathcal{H}}^{\text{nc}}(\Omega)$, 306
I_λ, 319
K_δ, 325
$\mathcal{L}^d$, 278
$\mathcal{L}(E)$, 278
$\mathcal{L}_\partial(E)$, 278
$\mathcal{L}_n(E)$, 254
$\mathcal{M}_n$, 348
$\mathcal{M}^d$, 346
$\mathbb{M}_n$, 286

372

Index

373

$\mathbb{M}^d$, 286
M_z, 15
$\mathcal{P}$, 95
P_+, 90
$\mathcal{P}_d$, 286
P_M, 11
P_n, 348
$\mathcal{P}(E)$, 255
$\mathscr{P}(U)$, 148
Pos_n, 266
$\text{Rat}(K)$, 5
$\text{Re}\,T$, 150
$\mathscr{S}$, 20
$\mathscr{S}(U)$, 19
$\mathscr{S}_{\mathcal{B}(\mathcal{H},\mathcal{K})}(\mathbb{D})$, 24
SAM^d, 286
V_λ, 319

admissible kernel, 113
ampliation, 299
angle between subspaces, 190
annulus, 226
Arveson, 21
automorphism, 192

B-point, 134, 145, 262
    theorem on $\mathbb{D}$, 134
    theorem on $\mathbb{D}^2$, 146
balanced, 352
Bickel's theorems, 282
bicommutant, 339
Blaschke factor, 50
    product, 49

C-point, 134, 145
    theorem on $\mathbb{D}$, 135
    theorem on $\mathbb{D}^2$, 146
calcular algebra, 220
carapoint, 132, 144
Carathéodory distance function, 192
Carathéodory–Julia theorem, 131, 132
    bidisc, 147
Carleson's theorem, 86
character, 194
characteristic operator function, 23
completely non-singular, 334
cone, 95
contraction, 10
    model of, 23
corona model, 87
corona problem, 86
Corona theorem, 86

de Branges–Rovnyak space, 133
derivative, 18
differentiable, 18
dilation
    boundary, 22, 226
    power, 21, 23
disjoint union topology, 292
distinguished variety, 209
dp-model, 229
dual problem, 263
Duffin's strong duality theorem, 262
dyad, 64

extension, 209
extension step, 43

Farkas's lemma, 263, 267
fat topology, 299
feasible, 262
fine topology, 299
finite topology, 292
finitely locally bounded, 295, 297
finitely open topology, 292
Fréchet derivative, 18
Fréchet holomorphic, 346
free holomorphic function, 313
    model, 315
    polynomial, 285
    realization, 315
    topology, 299
functional calculus, 4
    Taylor, 8

Gâteaux derivative, 346
generic d-tuple, 270
graded, 291
graded $\mathcal{B}(\mathcal{K}, \mathcal{L})$-valued function, 295
graded vector-valued function, 295
gramian, 35

Hahn–Banach theorem, 104, 124
Hardy space, 14
    vector-valued, 72
hereditary function, 62
    functional calculus, 62, 177
    operator-valued, 81
Herglotz class, 148
Hessian, 332
Hilbert
    space, 9, 13
        reproducing kernel, 37, 73
    tensor norm, 356
holomorphic, 18

374   Index

implicit function theorem, 337, 338, 357
infeasible, 262
inner function, 16
  matrix-valued, 82, 82, 205
  operator-valued, 82
invariant distance, 192
invariant subspace, 15
  problem, 16
inverse function theorem, 330, 336, 357
invertible, 4
involution, 63
isometry, 10

joint
  eigenvalue, 7
  eigenvector, 7
Julia quotient, 131
  on $\mathbb{D}^2$, 144
Julia's inequality, 147

kernel, 33
  positive semi-definite, 34
  reproducing, 38
Kobayashi distance, 196, 196

Löwner class, 254, 278
  in d variables, 275
  of E, 275, 278
Leech's theorem, 90, 92, 328
Lempert function, 196
Lempert's theorem, 195
level, 286, 291
limit feasible, 265
limit sequence, 265
limit value, 265
LMI, 288
lurking isometry, 39

Müntz–Szász interpolation problem, 56
Müntz–Szász theorem, 56
matrix monotone, 254
McMillan degree, 82, 205
model, 37
  corona, 87, 112
  formula, 33, 37, 71, 74
  of a hereditary function on $\mathbb{D}$, 87
  of a hereditary function on $\mathbb{D}^2$, 112
  on $\mathbb{D}^2$, 99
  space, 48
  wedge, 102, 121
Moore's theorem, 35, 73

n-matrix monotone, 254
nc operator domain, 345

nc operator function, 345
nc topology, 298
nc-domain, 292
nc-function, 293, 294
nc-model, 315
nc-set, 292
network realization formula, 24, 44
Nevanlinna–Pick problem, 52
  spectral, 186
non-commutative polynomial, 285
non-tangentially, 131
numerical range, 233
  joint, 199

Oka–Weil theorem, 9
operator, 4
  contractive, 10
  creation, 214
  isometric, 10
  Laurent, 209
  monotone, 254
  multiplication, 25
  normal, 12
  self-adjoint, 12
  Toeplitz, 90
  unitary, 12

Pick class, 148
  of an interval, 255
Pick interpolation theorem, 29, 52, 109
Pick's lemma, 34, 72
polynomial dilation hull, 327
polynomial hull, 19, 327
polynomial polyhedron, 299
polynomially convex, 19, 177
positive cone, 95
positive semi-definite, 9
primal problem, 263
projection, 11
projective tensor product, 97
pseudo-hyperbolic metric, 192

realization, 45
  free, 315
  contractive, 48
  formula, 44
  isometric, 48
  problem, 44
reasonable cross norm, 356
Rellich's theorem, 282
reproducing kernel
  property, 15

Index 375

resolvent
    identity, 42
Riesz–Dunford functional calculus, 6
    enhanced, 81

Schur class, 19
Schur product, 258
sequentially strong operator
    continuous, 349
set theory, 221
    Morse–Kelley, 222
slack variables, 263
spectral
    complete spectral set, 22
    domain, 20
        complete, 193
    measure, 12
    set, 17
    theorem, 12
spectrum, 4
    of normal tuple, 194
    algebraic, 8
    Taylor, 8, 214
SSOC, 349
stable, 45
state space, 48
strong duality, 266
    theorem, 264
system, 44
Szegő kernel, 15, 207
    normalized, 50

taut, 37, 46
Taylor–Taylor formula, 300
tensor product, 356
    Hilbert, 79
    projective, 97
Toeplitz corona theorem, 90–92
    operator, 90
topology
    disjoint union, 292
    fat, 299
    fine, 299
    finite, 292
    free, 299
    nc, 298
    uniformly open, 299
transfer function, 44, 45
trivial extension, 37

uniform cross norm, 356
uniformly open topology, 299
unilateral shift, 14
unitarily invariant domain, 344
unitary envelope, 294

value, 263
von Neumann–Wold decomposition, 195

wandering isometries, 304, 307
weak duality theorem, 263
weakly holomorphic, 65
wedge, 102